ANILINE
AND ITS
ANALOGS

INDIA · SINGAPORE · MALAYSIA

ANILINE
— AND ITS —
ANALOGS

JAGANNATH B. LAMTURE, PH.D.

INDIA • SINGAPORE • MALAYSIA

Notion Press

Old No. 38, New No. 6
McNichols Road, Chetpet
Chennai - 600 031

First Published by Notion Press 2018
Copyright © Jagannath B. Lamture 2018
All Rights Reserved.

ISBN 978-1-64249-257-6

Dedication

This book is dedicated to my Family, Parents, Teachers and all the Scientists in the World who have devoted their lives in the interest of improving life for mankind through the development of New Methods and Technologies in Synthetic Organic Chemistry and Medicinal Chemistry.

CONTENTS

Chapter-3 Haloanilines (HANs)

Chapter-4 Substituted Aniline Analogs-I

Chapter-5 Substituted Aniline Analogs-II

Chapter-6 CTH of Nitroarenes Using Different Techniques

REVIEWS OF BOOK ON "ANILINE AND ITS ANALOGS"

1). Reviewer: Dr. Navnath Shinde

I am very happy and excited to write this mail to you regarding your book, "Aniline and its Analogs". After having reviewed the book, the first thing came to my mind was a quote which I had read sometimes ago and it goes as, "The discovery consists of seeing what everybody has seen and thinking of what nobody has thought". This book is very innovative, informative and voluminous covering details about the old and the new developments about aniline and its analogs during last 200 years. It covers a vast data till date about Aniline which is a common and basic chemical, but along with its derivatives they are highly important in pharmaceutical, agrochemical, polymers, paints and dyes and fine chemical industries. This book is an attempt to bring together various chemical strategies used in many chemical reactions, processes and technological options.

All chapters are filled with fresh and most updated information along with relevant and authenticated scientific references about aniline and its analogs, which is very helpful for current industrial process chemistry, academics and research students in universities and industries. The vast information covered in is useful in manufacturing of APIs, API Intermediates, and other products which find applications in our daily life. This book is very well written in simple language with a very good sequence of reactions, which are properly arranged and easy to understand for everyone who wants to enrich and update their knowledge about synthetic organic chemistry, especially about the reduction options for transforming nitroarenes into aniline and its analogs.

Sir, I would like to express my sincere thanks to you for giving me an opportunity to review your book as a rich source of most advanced chemical technologies and enjoy and utilize such uniquely compiled treasure of knowledge.

Thanks and Regards,
Dr. Navnath Shinde, M. Sc. Ph. D.
Senior Scientist, Research and Development
Piramal Pharmaceuticals, Mumbai
Part of Piramal Enterprises Ltd: Mumbai, India.
Email: navnath1983@gmail.com,   navnath.shinde@piramal.com.

2). Reviewer: Dr. Valmik Dhakane

Chemistry is the most important part of human life. Chemicals used in making life saving drugs, building infrastructure like roads, canals, in electronics like mobile phones, televisions and in power generation to nuclear weapons. Due to this life becomes more and more comfortable. Today the average life span of human being is increased due to availability of medicines. Food production is increased tremendously by the use of agrochemicals. Hence, aniline and its derivatives play a major role in synthesis of medicines, agrochemicals, dyes, pigments, rubber products, plastics. Some of the examples of agrochemicals are Metalaxyl, Sulfosulfuron., drugs like Paracetamol, Clonazepam and Xanax. There are many dyes and pigments like PY-24, ANTS, CF Dye and various hair dyes made from aniline and its analogs. Due to versatile applications of aniline and its derivatives "ANILINE AND ITS ANALOGS" book will be very useful for data sharing. This book covers most of the developments regarding aniline and will be the useful source of information for college, university students, teachers and scientists in academics. This book is very useful for industrial scientists as it covers preparation of most of the building blocks of different chemistries with excellent yields.

Dear Sir, Thank you very much for giving me an opportunity to write about your book. The book is really useful for all the horizons of chemistry.

Thanks and Regards,
Dr. Valmik D. Dhakane, M. Sc. Ph. D.
General Manager, Research and Development,
Godrej Agrovet-Astec Life Sciences Ltd, Mumbai, India
Email: valmik.dhakane@godrejastec.com,   valmikdhakane@gmail.com.

3). Reviewer: Dr. Prakash P. Wadgaonkar

It was a great delight to read the book entitled "Aniline and its Analogs" written by a veteran in Organic Chemistry who has a rich experience working both in Academia and Industry.

Aromatic amines are highly useful chemicals which find numerous applications in a wide spectrum of industries. The major industries that make use of anilines and their derivatives include, dye stuffs, pigments, colorants, agrochemicals, pharmaceuticals, polymers, rubber industry and so on.

The book begins with a historical perspective and goes on to analyze the important developments that have taken place in the Academic and Industrial worlds in terms of their methods of synthesis and practices. Every chapter is rich in its contents and the information given is very valuable and comes handy for a practitioner of aromatic amine chemistry.

In summary, this book is a unique compendium of synthetic and process chemistry of aromatic amines viewed through the eyes of an experienced chemist. There is simply no other publication quite like this and it is a credit to the Author's dedication who has taken tremendous efforts to bring out such a valuable treatise. There is so much valuable information in this book that such a short book review can not truly do justice to it.

The book is highly recommended for students, researchers and industrial R and D personnel and a must for every individual and library.

Thanks and Regards,
Dr. Prakash P. Wadgaonkar
Scientist,
Polymer Science and Engineering Division.
CSIR-National Chemical Laboratory,
Dr. Homi Bhabha Road,
Pune 411008, Maharashtra, India.
pp.wadgaonkar@ncl.res.in

INTRODUCTION

In general, Chemistry plays very important roles in human life as it is one of the most important disciplines of science involving all living beings. Mankind has made significant improvements in many fields as electronics, agriculture, pharmaceuticals, aerospace, infrastructure, etc. Some of the basic chemicals have been used in all these fields. "ANILINE AND ITS ANALOGS" are among them and are commonly used in the commercial productions of many useful products as agro-products as herbicides and pesticides, dyes and paints, polymers as polyanilines and polyurethanes, rubber additives, rocket fuels, pharmaceuticals as Paracetamol, Anticancer drugs as Erlotinib, Bicalutamide, Anti-anxiety drugs as Clonezapam, Xanax, Fluoroquinone based Antibiotics as Gatifloxacin, Ciprofloxacin, Norfloxacin, Linezolid, etc. Average human life is increased by the availability of effective medicines, thanks to untiring efforts by the scientists all over the world. With such immense applications, Aniline and its Analogs have unambiguously occupied an important place in human life and hence they are required to be produced at commercial scales. Although discovered and identified during 1826–1843, Aniline was first synthesized in laboratories and then commercially produced using Bechamp process since 1854. Since then a lot of progress has been made in developing new methodologies for Aniline (AN) synthesis and manufacturing both at laboratory and industrial scales. However, some of the old methods have certain limitations as exothermic reactions, use of huge amounts of acids and metals, large quantities of bases for neutralization of acids and generation of high volumes of waste leading to waste disposal and environmental concerns. All these steps involve high energy consumptions ultimately leading to high cost of AN productions at commercial scales. Therefore, constant efforts are being made by the scientists all over the world to develop commercial processes, especially addressing the above listed concerns, so that the most cost-effective, safe, non-hazardous and environmental friendly technologies are available for manufacturing aniline and its analogs. We have tried to cover the old, historical and the most recent developments about aniline and its analogs through June, 2017.

The readers must note that i) some of the information is likely to be seen under two different sections as it would be relevant at both the places and ii) the scientific discovery from different groups in the world focuses on some of the general aspects as the factors responsible for the particle size of the catalysts and their activity, role of solid supports, the % substrate conversions, and the yield and selectivity of the end products. Therefore, we have covered most of the important information published in different sources as journals, books, internet especially Google, keeping in mind the common objectives of new process developments based on above parameters as yield, selectivity etc. Throughout the text, we have discussed the results published by individual groups. However, it is not possible to cover and discuss in detail all the vast amount of data available in the literature. Therefore, in order to justify the author's efforts and also giving the updated details for the readers' benefits, we have highlighted some of the most recent work only in the form of publication titles and relevant references. The readers can visit the references of their interests for further details. The information covered in this book will serve as a rich source of knowledge for college, university students and professors of organic chemistry at all levels and researchers in academics and chemical/pharmaceutical industries.

CHAPTER-1

ANILINE

1.00.00: Aniline: Global Scenario:

"ANILINE" is a simple and attractive name. However, by no means Aniline (AN) is a "Lean" molecule as its annual demand is in multimillion tons and billions of dollars in sales per year. According to a report, "Aniline: 2014 World Market Outlook and Forecast up to 2018" by Merchant Research & Consulting and recently published by Market Publishers Ltd, the following data is available. As of 2011, the world production capacity of aniline exceeded 6.47 million tons, with Asia calling for more than 45% of the world's total capacity. Besides, in the same year the world's aniline production surpassed the 4.7 million tons mark registering a 5.4% yearly increase. The world aniline production is poised for robust growth to go beyond 5.62 million tons in 2016 (1a (i)). In 2011, the top five aniline manufacturing countries (China, USA, Belgium, Japan and Germany) comprised nearly 78.3% of the global AN production volume. The dominant players in the worldwide AN market are Huntsman, DuPont, BASF, Bayer, Yantai Wanhua Polyurethane, Dow, Shandong Jinling Chemicals and Tosoh Corporation. According to Vibrant Gujrat Report (Jan 10–13, 2017) the aniline annual production is estimated to be more than 8 million tons from about 6 million tons in 2014. According to Transperancy Market Research Report on May 24, 2015, aniline market is going to reach about US$ 16bn by 2020 from $10bn in 2013–2014 (1a(ii and iii)).

1.01.00: Importance of Aniline (AN):

Aniline and its Derivatives: Uses and Applications: The importance of anything (plants, animals, humans and materials) in the world depends upon their uses in general and especially in one or more segments or facets of human life. Aniline being the first member of the family of aromatic amines, the importance of aniline stems from the fact that it has tremendous applications in human life, as it is the primary and basic raw material for many products required on daily basis such as agro products *as* crop protection agents, polymers, dyes and paints, explosives, fabrics, rubber industry, automotive industry, packaging and pharmaceutical products which work as life savings drugs for mankind. Major part of aniline (about 80%) goes into the production of methylene diphenyl diisocyanate (MDI), which is used to manufacture polyurethane foams. Its non-foam applications include paints, coatings, adhesives, sealants and elastomers. Polyurethanes are most commonly used as a foaming material in furniture, and in the rubber industry for the production of vulcanization accelerators and anti-degradants. Aniline is used in insulation industry at large scale as it enhances the electrical strength of the insulation materials and as a raw material in making intermediates and finished drugs. Aniline usess in drugs is attributed to a lone pair of electrons on nitrogen which inhibits the activities of some enzymes thereby enhancing the efficacy of the drug molecules. One of the drugs that are produced in tons of quantities at commercial scale is "Paracetamol or Acetaminophen", which is an antipyretic/anti-inflammatory drug with its annual volume at about 150 kilo tons in 2014. According to the Zion Research Report (Feb 03, 2016 Globe Newswire, Florida) global acetaminophen market was valued at around US$ 801.3 million in 2014 and is expected to reach US$ 999.4 million in 2020 (1b).

Therefore, before we delve into multiple aspects of aniline such as the history of aniline discovery, physical and chemical properties, chemical reactions, reaction products and the developments in terms of different new and novel processes of synthesizing and manufacturing of such an important and versatile chemical entity known as "ANILINE", let us briefly and quickly look into its uses and applications in different industries (1c, 1d).

1.02.00: Aniline: Industrial Applications:

I): Polymers:

The largest application of aniline is for the preparation of methylene di-aniline and related compounds through its condensation with formaldehyde. Further, the diamines are condensed with phosgene to give methylene diphenyl diisocyanate (MDI), a precursor to polyurethanes. A major portion of aniline volume is used in making MDI, and to a smaller scale in the production of an intrinsically conducting polymer "Polyaniline".

II): Dyes:

In 1856, during the synthesis of quinine, August W. von-Hofmann and his student William Henry Perkin serendipitously discovered mauveine known as AN purple or Perkin's mauve, which was the first synthetic organic chemical dye. Later, Perkin used the process/technology in the production of this dye at commercial scale. In the fifties of ninetieth century aniline was expensive. Therefore, in 1854 it was prepared commercially at ton scale using a nitrobenzene (NB) reduction method reported by Antoine Bechamp (2), which enabled the evolution of a massive dye industry in Germany. Today, the name of BASF (English: Baden Aniline and Soda Factory, adopted in 1973; originally *Badische Anilin- und Soda-Fabrik,* established in 1865) is among the largest chemical suppliers in the world. BASF echoes the legacy of the synthetic dye industry, built via AN dyes and extended via its related azo dyes. The first azo dye was aniline yellow. Other aniline dyes are indigo (blue jean dye), fuchsine, safranine, aniline purple and induline, which are used as wood stains, varnishes, wall paints and for vehicles and appliances.

III): Rubber & Other Industries:

The applications of aniline include rubber processing chemicals (9%), herbicides (2%), MDI (~80%) and dyes and pigments (2%). As additives to rubber, aniline derivatives such as phenylenediamines and diphenylamine serve as antioxidants, activators, and accelerators. Diphenylamine is also used as a raw material for the rubber, petroleum, plastics, agricultural, explosives, and chemical industries. A derivative of aniline, acetoacetanilide (AAA), is used for a variety of applications such as i) an inhibitor of H_2O_2, ii) as a stabilizer in cellulose ester varnishes, iii) as an intermediate in dyes and camphor synthesis iv) about 10,000 tons are used in the production of pigments. Aniline is also used in the manufacturing of isocyanates which are the raw materials for polyurethanes, which in turn are used in the manufacturing of spandex fibres for athlets and foam insulators. The latter are used as i) building materials and ii) in the constructions of refrigerators (3).

IV): Aniline Derivatives in the Development of Medicines:

Structurally, aniline is a simple aromatic amine with amino group attached to one of the carbons in a six membered benzene ring. It undergoes multiple reactions, as additions and substitutions, involving both the amine group and the benzene ring. The end products find applications in various industries as mentioned earlier. Besides, aniline and its derivatives find applications in the manufacturing of wide range of pharmaceutical products. For example Prontosil, a AN based azo red dye, is an antibacterial drug discovered in 1932 by a research team at the Bayer Laboratories in Germany and introduced in 1935. M. M. Khan Shahid, University of Punjab, Lahore, has given a list of aniline reaction products in the pharmaceutical industry, which is published in the Journal Science on September 15, 2015 (4).

Aniline has wide range of pharmaceutical products as given below:

1). Acetanilide: A key intermediate for the manufacture of sulfa drugs and a precursor in the synthesis of Penicillin.

2). Paracetamol (PA): It is also known as acetaminophen, *p*-acetaminophenol and the brand name is Tylenol. It is an illustrative example of the drugs prepared from AN. Today PA in terms of its volume with 150000 tons yearly production and with total revenue of $801 million in 2014 is probably the world's largest drug produced worldwide and it is estimated to grow to about 1 billion dollars in sales by 2020. Key application markets for PA include pharmaceuticals, dye industry and chemical industry. Pharmaceutical segment was the largest application market for PA in 2014, which accounted for more than 86%

share of global consumption. Rising need of medicines for a pain relief and increasing health awareness is a major driving factor for pharmaceuticals market (4c).

3). 3-Trifluromethylaniline: It is an intermediate for herbicides and fungicides. There are dozens of herbicides based on aniline and its derivatives as Sulfonanilide Metosulam, Cloransulam, Metolachlor, Pentanochlor, etc.

4). Sulfonamides: Aniline derivatives of sulfonamide, as Sulfafurazol, Sulfadiazine, Sulfacetamide, Sulfisomidine, etc. are anti-microbial drugs with short action period.

5). Sulfapyridine: For treating linear IgA dermatosis and as a good antibacterial.

6). Sulfacetamide: Used in the treatment of pityriasis versicolor (a fungal infection of the skin) and anti-inflammatory in conjunctivitis.

7). Sulfanilamide: It is sulfonamide antibiotic/antifungal and is used for treating i) certain vaginal yeast infections and ii) in the treatment and prophylaxis of certain infections as it blocks the growth activity of certain bacteria (4b).

8). Succinyl Sulfathiazole: Ultra long acting drug for antibacterial activity in GIT.

9). Anti-Cancer Drugs: Bicalutamide, Nilutamide and many more in this category are being developed for prostate cancer treatment in men and Erlotinib is used in the treatment of non-small cell lung cancer (NSCLC), pancreatic cancer and several other types of cancers. Some of the aniline based Asiatic acid derivatives have found applications as potential anti-cancer agents (5).

10). Agro-Products: Aniline is used in the manufacturing of agro-products (6) as fungicides and herbicides. KOH-Aniline blue fluorescence is used in studying plant-fungal interactions.

11). Amongst other inorganic and organic acids, aniline acid derivatives as 4-aminophenylacetic acid have found applications as antimicrobial agents (7).

12). Ackermann et al. have reported dozens of aniline derivatives, their manufacturing processes and uses as pharmaceutical agents. The authors also have cited earlier work which covers drugs as anti-diabetics, anti-infective agents, TNF-inhibitors, peroxime proliferator activated receptor modulators, aniline acid derivatives and their use in the preparation of medicines (8).

13). Alexander Fleming discovered penicillin in 1928, the world's first antibiotics. In 1932 azo-dyes were found useful as antibiotics, followed by the discovery of sulfa drugs. By the 1940s, over 500 related sulfa drugs were produced. Medications were in high demand during World War II (1939–1945). Some of these first miracle drugs as chemotherapeutic agents with wide effectiveness propelled the American pharmaceutical industry. Paul Ehrlich had coined the term chemotherapy for his magic bullet approach to medicine. After World War II, Cornelius P. Rhoads introduced the chemotherapeutic approach to cancer treatment (9a). Recently, it is found that a dye containing N, N-dihydroxyethylaniline as a coupler had the highest activity against the bacteria and fungi (9b).

V): Rocket Fuel:

In the 1940s and early 1950s, aniline was used with nitric acid or dinitrogen tetroxide as rocket fuel for small missiles and Jet Assisted Take-Off (JATO). The two fuel components are hypergolic in nature, producing a violent reaction on contact. Aniline was later replaced by hydrazine (10).

References:

1a).i). http://www.greenatom.earth/news/world-aniline-production-to-go-beyond-562-million-tonnes-in. ii). https://vibrantgujarat.com/writereaddata/images/pdf/projectprofiles/Manufacturing-of-Aniline.pdf.

iiia). http://www.transparencymarketresearch.com/pressrelease/aniline-market.htm.

iiib). https://globenewswire.com/newsrelease/2015/05/12/734906/10133971/en/Aniline-Market-will-reach-a-total-market-value-of-US-15-66-billion-by-2020Transparency-Market-Research.html. iv).http://www.military-technologies.net/2017/05/08/aniline-market-size-share-growth-trends-demand-and-forecast-to-2021.

1b). http://www.marketresearchstore.com/report/acetaminophen-market-z45985

1c). Applications: A. S. Travis, *Manufacture and Uses of the Anilines*: A Vast Array of Processes and Products, Anilines (2007). John Wiley and Sons Ltd: (2009).

1d). H. Kaminaka, *J. Synthetic Organic Chemistry,* Japan, 34(10):758-764 91976).

2). A. Bechamp, *Annales de Chemie et de Physique*, 3rd Series, 42: 186-196 (1854).

3) (i). http://study.com/academy/lesson/aniline-structure-formula-uses.html.

 (ii). https://www.chemours.com/Aniline/en_US/uses_apps/index.html.

 (iii). http://product-finder.basf.com/group/corporate/product-finder/en/brand/ANILINE.

4a). http://www.slideshare.net/MMK_Shahid/reactions-and-pharmaceutical-applications-of-aniline. 4b). https://www.drugs.com/cdi/sulfanilamide.html

4c).i). http://www.marketresearchstore.com/report/acetaminophen-market-z45985.

 ii). https://globenewswire.com/newsrelease/2016/02/03/807269/0/en/Acetaminophen-Paracetamol-Market-Set-for-Explosive-Growth-To-Reach-Around-USD-999-4-Million-Globally-by-2020-Growing-at-3-8-CAGR-MarketResearchStore-Com.html.

5). i). J. F. Li et al. *Eur. J. Medicinal Chemistry*, 30(86):175-188 (2014), ii). Y. Q. Meng et al. *J. Asian Natural Products Research,* 14(09):844-855(2012), iii). H. C. Huang et al. *Eur. J. Medicinal Chemistry*, 69:508-520 (2013).

6). M. E. Hood and H.D. Shew, *Phytopathalogy*, 86:704-708 (1996).

7). A. H. Bedair, et al. *Acta. Pharmaceutica,* 56(03):273-84 (2006).

8). (i). J. Akerman et al. United States Patent 7345067, (2008).

  (ii). http://www.freepatentsonline.com/7345067.html.

  (iii). http://encyclopedia2.thefreedictionary.com/Aniline+compounds.

9a).i) https://en.wikipedia.org/wiki/Cornelius_P._Rhoads.

9b). H. Shaki, et al. *Biotechnology Progress*, 31(04):1086-95 (2015).

10a). W. P. Berggren, "The Acid-Aniline Rocket Engine", *Journal of the American Rocket Society,* 00(73): 17-30 (1948). 10b).http://dx.doi.org/10.2514/8.4200 10c). https://en.wikipedia.org/wiki/Aniline

1.03.00: Aniline Discovery and Identifications:

Aniline (AN) is the first member of the aromatic amines and is also called Benzenamine, Phenylamine or Aminobenzene. It has a molecular formula as $C_6H_5NH_2$ and mass of 93.13g/mole, with CAS No. (62-53-3). Physically, it is a colorless to pale yellow liquid with rotten-fish odor and other physical characteristics as i) density of 1.02g/mL, ii) b.p.184C, iii) basicity PKb = 9.3, indicating AN as a moderate base. Based on the data about the uses and applications of AN, it is unambiguously convincing that aniline is one of the most important products in human life. Therefore, we will look into its history of discovery, properties, and some of the novel methods for its synthesis in laboratories and at industrial scales as well. Aniline was first prepared by Unverdorben (1826) by dry distillation of indigo (1). It was also obtained by coal tar distillation in (1834) by F. Runge (2). In 1840, Carl Julius Fritzsche (3) also obtained aniline from indigo by heating it with potash and he named it "Aniline", after an indigo yielding plant "Anil". In 1842, Zinin (4a) and Fritzsche (4b) reduced NB and obtained a base that they named it as "benzida". Further, Anon in 1842 reported the use of aromatic nitrates, such as NB and 1-nitronaphthaline, as the starting materials for aniline and napthylamine using hydrogen sulfide (5) as the reducing agent. Hofmann, in 1843, obtained aniline by the reduction of NB and confirmed that all these products obtained earlier by different groups are the same and he named them phenylamine or aniline (6). In the laboratory, it can be prepared by the reduction of NB with Sn/HCl or Fe/HCl. A. Béchamp (1854) studied the actions of iron proto salts (FeO and not Fe_2O_3) on nitronaphthalene and NB and reported a new method of forming synthetic organic bases naphthylamine and aniline. NB is a classical feedstock for aniline manufacturing (5, 6, 7).

Recently, chlorobenzene (CB) and phenol are being used to a lesser extent in AN manufacturing processes in several countries. For example, a high catalytic activity for the synthesis of aniline from phenol and ammonia has been found on gallium-containing zeolite materials (8). Here phenol serves as the starting material for AN. There are many ways of producing phenol, but one of the most recent processes is as a one-step conversion of benzene to phenol with a palladium membrane (9). The existing phenol production processes tend to be energy-consuming and produce unwanted by-products. However, the palladium membrane method is an efficient process using a shell-and-tube reactor, in which a gaseous mixture of benzene and oxygen is fed into a porous alumina tube coated with a palladium thin layer and hydrogen is fed into the shell. Hydrogen dissociated on the palladium layer surface permeates onto the back of the tube and reacts with oxygen to give active oxygen species which attack benzene to produce phenol. This one-step process attained phenol formation selectivities of 80 to 97% at benzene conversions of 2 to 16%, below 250°C. The phenol yield was 1.5 kilograms per kilogram of the catalyst/hour at 150°C. Some other companies manufacture aniline by direct amination of benzene, which is covered briefly in the following sections.

The reduction of NB with iron turnings and water in the presence of small amounts of hydrochloric acid is the oldest form of industrial aniline manufacturing process. It would have been replaced much earlier by more economical reduction methods, if it had not been possible to obtain valuable iron oxide pigments from the resulting iron oxide sludge in this process.

References:

1). Otto Unverdorben. *Annalen der Physik und Chemie*, 08: 397-410 (1826).

2). F. F. Runge, *Annalen der Physik & Chemie*, 31(65-77):513-524; P308-332(1834).

3). J. Fritzsche, *Bulletin Scientifique* [publié par l'Académie Impériale des Sciences de Saint-Petersbourg], 07(12): 161-165 (1840). Reprinted in: J. Fritzsche, *Justus Liebigs Annalen der Chemie*, 36(01):84-90(1840); J. Fritzsche, *Journal für praktische Chemie*, 20: 453-457 (1840). In a postscript to this article, Erdmann (one of the journal's editors) argues that aniline and the "Cristallin", which was found by Unverdorben in 1826, are the same substance; see pages 457-459.

4a). N. Zinin, *Académie Impériale des Sciences de Saint-Petersbourg*], 10(18):272-285(1842). Reprinted in: N. Zinin, *Journal für Praktische Chemie*, 27(01):140–153 (1842). 4b). Fritzsche, Zinin's colleague, soon recognized that "benzidam" was actually aniline. See: *Fritzsche Bulletin Scientifique*, 10: 352(1842).

5). Anon, *Annalen der Chemie und Pharmacie*, 44: 283-287 (1842).

6). A. W. Hofmann, "Chemische Untersuchung der Organischen Basen im Steinkohlen-Theeröl" (Chemical Investigation of Organic Bases in Coal Tar Oil), *Annalen der Chemie und Pharmacie*, 47: 37-87(1843). On Page 48, Hofmann argues that krystallin, kyanol, benzidam, and aniline are identical or the same.

7). A. Bechamp, *Annal de Chemie et de Physique*, 3rd Series, 42:186 –196 (1854).

8). N. Katada et al. *Applied Catalysis A: General*, 180 (1999).

http://chempedia.info/info/151260/.

9). S. Niwa, M. Eswaramoorthy, et al. *Science*, 295(5552):105-107 (2002).

http://www.scielo.org.za/scielo.php?script=sci_arttext&pid=S181679502011000100005.

1.04.00: Laboratory Practical Synthesis of Nitrobenzene (NB):

From the literature, it is evident that aniline synthesis involves a two stage process. First, benzene is nitrated to mono-nitrobenzene (MNB) using the nitrating mixture, and secondly the MNB is further reduced to aniline using different reducing agents, especially metals and aqueous HCl. In the early days, the reduction processes involved the use of metals (Sn, Fe, Zn, etc.), metal salts and mineral acids. In the laboratory, the nitrating mixture is prepared by the slow addition of concentrated H_2SO_4 to concentrated HNO_3 at 5–10°C. Benzene is added slowly to the nitrating mixture, which is carefully maintained below 55–60°C. The reaction mixture is further stirred at 60°C for an hour and then cooled to room temperature. NB is isolated in good yields (~85%) as a pale yellow liquid by regular aqueous/organic bilayer separation, cold water wash of organic layer, drying on anhydrous sodium sulfate or dry calcium chloride, solvent removal, and the distillation of the residue at 200–214°C (Scheme-1) (1).

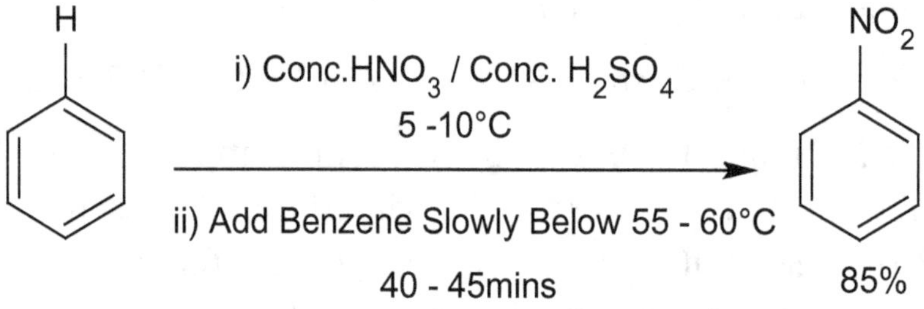

i) Conc.HNO_3 / Conc. H_2SO_4
5 -10°C

ii) Add Benzene Slowly Below 55 - 60°C

40 - 45mins

85% --Scheme-1

There has been a huge progress in the nitration processes of benzene using reagents other than nitrating mixtures (HNO_3+H_2SO_4), such as solid catalysts and silica as a solid phase. Other alternatives as continuous flow micro-reactors (2a-d) have emerged as a substitute to an exothermic nitrating mixture and their applications in drug discovery (2e). Also, vapor phase nitration of benzene has been carried out using dilute nitric acid as the nitrating agent over a mixed oxide (Fe/Mo/SiO_2) solid acid catalyst with more than 80% benzene conversion and 99% selectivity to MNB (3). High activity and *para*-selectivity in the nitration of aromatic compounds is achieved by the high density of acidic sites and ready formation of the *p*-isomer in the pores of zeolite beta with low Si/Al ratio, as revealed by molecular modeling studies (4). Nitroaromatic compounds (NACs) play an important role in the chemical industry as so many of these products are required as intermediates for many end products with commercial applications. D. R. Hatter has summarized the applications and importance of NACs in the chemical industry (5).

References:

1a). P. R. Buckland & R. N. Gourley, US 4,994,576: (1991). 1b). Morrison and Boyd, *Organic Chemistry*: 5th Edition, 939-943 (1987). 1c).Vogel, *Practical Organic Chemistry*: 5th Edition, 854 (1989).

2a). J. R. Burns & C. Ramshaw, *J. Chemical Engg & Communications*: 89(12):1611-1628 (2002). 2b). K. Boodhoo at: http://pig.ncl.ac.uk/micro-reactors.htm and references cited therein. 2c)(i) A. A. Kulkarni, *Beilstein J. Organic Chem*istry, 10: 405–424(2014). ii). A. A. Kulkarni et al. WO 2015011729 A1(2015). 2d). M. Brivio et al. *J. Lab Chip* (RSC), 06:329–344 (2006). 2e). P. Watts & S. J. Haswell, *Drug Discovery Today*, 08(13), 586–593 (2003).

3). V. Mane et al. *J. of Applied Chemistry*, 07(07):50-57 (2014).

4). B. M. Choudary, et al. *Chem. Communs*, 25-26 (2000). www.iosrjournals.org.

5). D. R. Hatter, Chapter-1 in a Book, *"Toxicity of Nitro-aromatic Compounds"*, by Douglas E. Rickert, CRC Press, - Medical - 295 pages, 01-Jan-1985.

1.05.00: Laboratory Preparation of Aniline from NB Using Sn/HCl:

NB reduction was originally elucidated by Haber, which follows the sequential reduction of NB (A), first to nitrosobenzene(B), next to phenylhydroxylamine (C), and finally to aniline (D) (3c(i)) (Scheme-2).

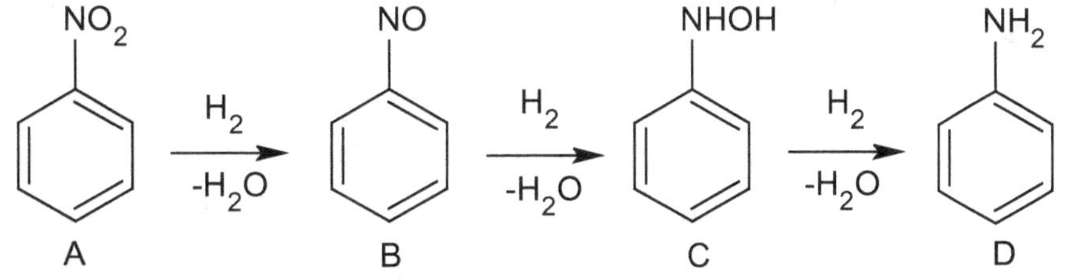

Scheme - 2: Intermediate Steps in the Reduction of Nitrobenzene to Aniline --Scheme-2

The stoichiometric composition of the complete reduction product of NB ranges from 27.9% water: 72.1% aniline to 14.3% water: 85.7% AN, depending upon the reaction conditions as temperatures and hydrogen pressures (1). The catalytic reduction of NB to AN can be done either in gas-phase or in liquid-phase hydrogenations. Recently, the liquid-phase hydrogenation of NB to AN with the use of Pd- and Pt-based catalysts or their combinations has been reported (2a). Exactly the same mechanistic pathway is reported by a Chilean group for the reduction of NB to AN, where they have used zirconia supported gold (1% Au/ZrO_2) as a catalyst (3a-c). Different procedures to obtain highly dispersed species of gold with nanoparticle (NP) sizes also have been reported (2b). In another case, the Au-NPs were prepared from $AuCl_4$ and impregnated over urea or cetyltrimethylammonium bromide (CTMABr) as a surfactant and Ir/ZrO_2 as a catalyst (3a-c). In another case, nanoporous silver was used as the catalyst for the selective reduction of aromatic nitro compounds (ANCs) or nitro aromatic compound (NACs), even in the presence of some sensitive functional groups under mild conditions with excellent yields (3d). Copper salt-amine complex catalyst has been used for a long time for the catalytic reduction of nitroarenes (NAs) to the corresponding amines (4). Synthesis and the reduction of NB using Sn/HCl are reported (Scheme-3a, b, c) (5).

$$Sn/HCl + C_6H_5NO_2 + 6H \longrightarrow C_6H_5NH_2 + 2H_2O \text{ —(Scheme-3a)}$$

Aniline produced combines with H_2SnCl_6 ($SnCl_4$ + 2HCl) to form a double salt.

$$2C_6H_5NH_2 + SnCl_4 + 2HCl \longrightarrow (C_6H_5NH_3)_2 SnCl_6 \text{ (Double Salt)—(3b)}$$

Aniline is obtained by treating this double salt with conc. NaOH solution.

$$(C_6H_5NH_3)_2SnCl_6 + 8NaOH \longrightarrow 2C_6H_5NH_2 + 6NaCl + Na_2SnO_3 + 5H_2O \text{—(3c)}$$

During the reduction of NB using Sn/Hcl mixture the tin metal to substrate ratio is 1.5. To this mixture, HCl is added slowly under controlled temperature and the stirring of the reaction mixture is continued for about an hour. The mixture is cooled to room temperature and aq. NaOH solution is added slowly (Scheme-4).

--Scheme-4

Aniline is extracted in an organic solvent from the aqueous reaction mass. The combined organic extract is washed with water, dried over anhydrous potassium carbonate, solvent removed and the residue is distilled (bp 184-186°C) to obtain pure AN (97%) (1a). Aniline is also prepared by a modified Jones reduction method (6).

References:

1). a). A. M. Strätz, *Catalysis of Organic Reactions*, Vol. 18, p. 337, Dekker (1984).

b). http://www.mt.com/dam/mt_ext_files/Editorial/Generic/3/Hydrogenation-Nitrobenzene-Aniline-Mass-Transfer_Editorial-Generic_1204310241270.

c). http://www.mt.com/us/en/home/library/applications/automated-reactors/ Hydrogenation-Nitrobenzene-Aniline-Mass-Transfer.html

d). http://www.chemguide.co.uk/organicprops/aniline/preparation.html

2a). *J. Pasek & M. Petrisko, EP 2471768 A1, July 04, (2012)*, 2b).H. K. Kadam & S. G. Tilve, *RSC Advances*: 05:83391-83407(2015).

3a). S. Gomez, et al. *J. Chilean Chemical Society*: 57(02):1194-1198 (2012). 3b). C. H. Campos et al. *Hindawi Journal of Chemistry*, Vol. 2017, 9 pages (2017). 3c). C. H. Campos et al. *Catalysis Today*: 213, 93–100 (2013), 3d). Z. Li et al. *RSC Advances*, 5, 30062-30066 (2015). 4). H. R. Appell, US 3290377 A, (1965).

5). i) A. Vogel, *Practical Organic Chemistry*, 5th Ed. John Wiley & Sons, New York, P854 and 892(1989).

6) U. S. Stubbs Jr. & C. F. Atkins, *J. Chemical Education*, 36 (12), p 611 (1959).

1.06.00: Nitrobenzene Reduction Using Fe (II)/HCl:

The reduction of NACs is a well-known reaction for a long time now (1a, 1b). The reduction of NB to AN was first performed by Nikolay Zinin in 1842 using inorganic sulfide as a reductant and later Bechamp in 1854 used iron/HCl. However, the increasing demand for aniline has far surpassed its availability in the market for the pigments, herbicides etc., so that not only the catalytic hydrogenation processes (both liquid- and gas-phase) but also other feed stocks have been used for AN production at commercial scales. As is evident from the fact that even in the current times iron and iron-organo complexes find interest in the reduction processes of NAs. Reduction of NACs to the aromatic amines by iron catalyst in aqueous acid is generally referred to as Bechamp reduction. On a commercial scale, aniline is obtained by reducing NB with iron filings/iron oxide and aq. HCl at ~100°C (Scheme-5). The crude product was first neutralized with an aqueous inorganic base and then distilled to obtain AN in about 95%–98%. Iron quality directly affects the quality and the yield of the resulting AN. This method is routinely in use due to the existence of equipment for commercial productions.

--Scheme-5

Old Commercial Manufacturing of AN by Reduction of NB

However, it has serious limitations as, i) it is difficult to control the reaction heat, ii) consumes large quantity of iron and its recovery is difficult, iii) it causes serious environmental pollution, iv) equipment corrosion is a serious problem, v) the high cost of operation and maintenance is difficult to a continuous production, vi) the reaction was slow, and it was difficult to separate the product. Fe/HCl is the oldest industrial AN manufacturing process and still being used by some companies for its valuable iron oxide pigments resulting from iron oxide sludge from this method (2).

1.07.00: Efficient NB Reduction Methods with Iron Metal or Its Salts:

Reduction of NACs in $Fe(0)$-CO_2-H_2O system with implications for groundwater remediations with Iron metal is reported (3). Also, in-situ generated Iron Oxide nanocrystals as efficient and selective catalyst, under a Continuous Flow Method, has been applied to the selective reduction of NAs to ANs (4). In a solution containing the NA, Fe_3O_4 nanocrystals were generated in situ from an inexpensive Fe precursor using hydrazine hydrate as the reducing agent at elevated temperatures. The reaction is performed at 150°C (MWs) for 2–6 minutes. The isolated yields of the different AN derivatives are excellent (95–99%)(Scheme-6).

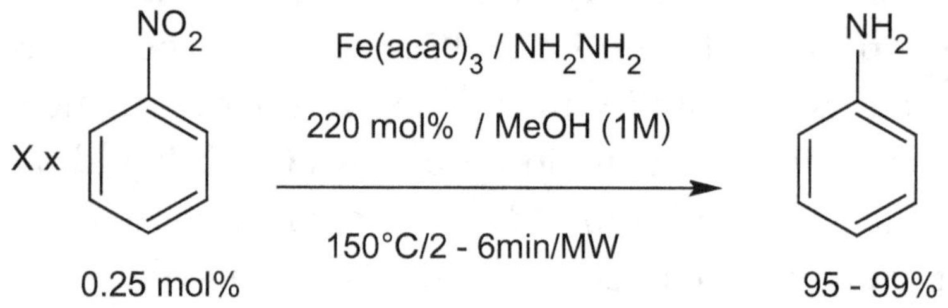

--Scheme-6

Some other efficient methods have been reported for the reduction of NAs to the corresponding ANs. For example:

i). Desai et al. have developed an efficient and economical process for the reduction of NAs to ANs in good yields using FeS-NH_4Cl-$MeOH$-H_2O system (5).

ii). An efficient $Fe/CaCl_2$ also system enables the reduction of NAs and reductive cleavage of azo compounds by catalytic transfer hydrogenation in the presence of sensitive functional groups including halides, carbonyl, aldehyde, acetyl, nitrile, and ester etc. with excellent yields. The simple experimental procedure and easy purification process makes this protocol advantageous (6).

iii). Reduction of o-nitroarylcarbaldehydes to o-aminoarylcarbaldehydes with Fe/HCl followed by in situ condensation of the resulting amines with ketones or aldehydes (Friedlaender quinoline synthesis) gives mono- or disubstituted quinolines (7).

iv). A tandem condensation of a cyanoimidate with an amine followed by reductive cyclization in an Iron-HCl system enables an efficient route to N^4-substituted

2, 4-diaminoquinazolines. Additional *N*-alkylation can produce two fused heterocycles (8).

v). The selective reduction of NAs to the corresponding ANs with Iron Powder or SnCl$_2$ in the presence of sensitive functionalities under ultrasonic irradiation at 35 kHz is reported in yields of 39–98%. Iron powder proved superior to SnCl$_2$ with high tolerance of sensitive functional groups and high yields of the desired aryl amines in relatively short reaction times. Simple experimental procedure and purification make the iron reduction of NACs advantageous over other methods (1b).

1.08.00: Role of a Reagent and the Solvent System:

NAs are reduced to corresponding ANs with FeS-NH$_4$Cl-CH$_3$OH-H$_2$O system in good yields (5). But the replacement of FeS with just iron powder under similar experimental conditions did not yield good results. For example, the reduction of NAs with Fe–NH$_4$Cl–EtOH–H$_2$O hardly furnished the corresponding products. This is attributed to low solubility of NAs in EtOH-H$_2$O system. However, the reduction of NAs with Fe–NH$_4$Cl–Acetone–H$_2$O system smoothly yields the corresponding ANs, without affecting the reducible or hydrolysable groups (9).

References:

1a). AN synthesis with iron powder and HCl I) A. Béchamp, *Annales de chimie et de physique* 42: 186–196, (1854). ii) *Organic Reactions*, 02, 428 (1944). iii). *Annal. d. Chemie u. Pharm.*, vol. XCII, Issue 3, pp. 401-403. 1b). A. B. Gamble, et al. *Synthetic Communs*: 37:2777-2786 (2007). 1c). Kent and *Riegel's Handbook of Industrial Chemistry and Biotechnology*: James A. Kent, Springer Science & Business Media, 27-May-2010 - Science - 1875 pages, pp 1073.

2). i). J. Laux, US 1793942 A, 1931. ii). J. Laux, US 1,849,428, (1932).

3). Ph. D. thesis by Abinash Agrawal, University of North Carolina, Chappel Hill, 1990. http://www.dtic.mil/dtic/tr/fulltext/u2/a350485.pdf.

4). D. Cantillo, et al. *Angewandte Chemie Intl. Edition*, 51, 10190 –10193 (2012).

5). D. G. Desai, et al. *Synthetic Communications*, 31(8), 1249-1251 (2001).

6). S. Chandrappa, et al. *Synlett*, 3019-3022 (2010).

7a). A. H. Li, et al. *Synthesis*: 1629-1632(2010).

8). i). P. Yin, et al. *J. Organic Chemistry*, 77:2649-2658(2012).

9). P. Xiao et al. *Synthetic Communications*, 40(05):661-665 (2010).

1.09.00: Other Nitrobenzene Reducing Agents:

1.09.10: Known Reduction Processes:

The metal-acid catalyzed NB reduction methods have been replaced by more efficient and new processes.

1.09.11: Aniline by Reduction of NB with Metals and Acids:

As stated above, commercial AN synthesis involves two steps: i) nitration of benzene and ii) reduction of resulting NB with metal and an aqueous HCl. Mechanistically the needed electrons are derived from the metals used in the reduction process and the ammonium ion is formed in the presence of an acid (Scheme-7).

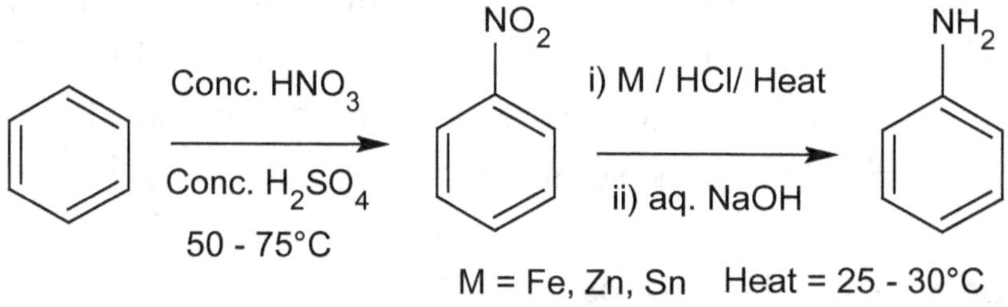

--Scheme-7

At the end, it is neutralized by an aqueous inorganic base to obtain the free AN base. There are other metal based catalysts used for the vapor phase reduction of NB to AN. See the following references:

1). Di-Nickel Phosphide Catalyst: Brown et al. have reported catalytic preparation of aniline using Di-Nickel Phosphide as a heterogeneous catalyst. The vapor phase reduction of NB with di-nickel-phosphide as a catalyst in the presence of hydrogen and water smoothly yields AN (1).

2a). Zn/HCl: Metal-Mineral acid combination is used for the reduction of NB to Aniline for a long time now (2a).

2b). NB Reduction Using Fe/Zn: Rapid and inexpensive method for the chemoselective reduction of NAs to ANs in good yields using Ferric Chloride - Zinc - Dimethyl Formamide - Water System (2b) and also chemoselective reduction of NBs with metallic Zn in near critical water are achieved (2d).

Some of these metals (as Iron, zinc, tin, etc. and inorganic acids as HCl) reduce the NAs to azoarenes, or azoxyarenes, hydroxylamines, hydrazines, at intermediate stages in the preparation of ANs are reported. The main shortcomings of these reactions include the slow reaction rate and costly steam distillation. Some neutral organic solvents such as acetonitrile and propylene carbonate have been added and found to give better results (3a-c). Environmentally benign and selective reduction of NAs with Fe in pressurized CO_2–H_2O medium is achieved (3d).

3a). Sodium Borohydride (SBH, NaBH$_4$) as a Reducing Agent: SBH in the presence of charcoal (0.4–0.8 g) reduces NAs to their corresponding amines in a mixture of H$_2$O-THF (1:0.5 mL) at 50–60°C with high to excellent yields (4).

SBH in the presence of catalytic amounts of Ni(OAc)$_2$. 4H$_2$O reduces varieties of NAs to their corresponding amines in a mixture of CH$_3$CN and H$_2$O (3.0:0.3 ml) at room temperature with high to excellent yields (up to 90-99%) of products in 20 -60 minutes (5). Upon treatment with KBH$_4$-CuCl, ANCs afforded cleanly primary amines in high yields. Similarly nitroso-, azoxy- and azobenzenes were reduced by the same reagent system to give ANs (6). NAs can be conveniently reduced to primary amines in good to excellent yields by SBH in the presence of bismuth chloride or antimony chloride (NaBH$_4$-SbCl$_3$ OR NaBH$_4$-BiCl$_3$) as a novel reduction system (7). Selective reduction of NACs with SBH using Titanium (II) reagents as a catalyst can be achieved in good yields too (8). A new, efficient, chemoselective, green and practical method for the room-temperature reduction of NAs employing FeSO$_4$- 7H$_2$O, NaBH$_4$, H$_3$PW$_{12}$O$_{40}$ system in H$_2$O under mild conditions is reported. The method is simple, inexpensive, easily scalable and applicable for the large scale preparation of different substituted ANs (9).

3b). Metal-Nanoparticles-NaBH$_4$: NAs reductions using metal-nanoparticles (M-NPs) in aqueous SBH has been reported (10a, 10b, 10d). In another case, novel Cu-NPs were synthesized from cupric sulfate using hydrazine as reducing reagent. A series of ANCs were reacted with SBH in the presence of these Cu-NPs as catalyst to yield the ANs in high yields (10c). Boron exchange resin (BER) in the presence of metal salts as CuBr, CuSO$_4$, PdCl$_2$, gave ANs in high yields and 100% purity (10d).

4). Formic Acid and Formates as Hydrogen Source: Metal catalyzed selective reduction of NAs with formic acid (i), ammonium formate (ii) and hydrazonium monoformate (iii) is reported in detail (11). Catalytic transfer hydrogenation of ANCs using recyclable polymer-supported formates as hydrogen donor and Pd-C as a catalyst produces corresponding aromatic amines in excellent yields (90–98%) (12).

4a). Nano Particles as Metal Catalysts: 4a (i). A highly chemo- and regioselective reduction of a wide variety of ANCs to the corresponding amines has been achieved. A combination of Au-NPs supported on titania and ammonium formate (HCOONH$_4$) in ethanol at room temperature are optimal conditions (13).

4a (ii): Pd-NPs: Pd-NPs on amino-functionalized siliceous mesocellular foam as an efficient heterogeneous catalyst for the transfer hydrogenation of NAs selectively gives ANs in excellent isolated yields (98%). The catalyst displayed excellent recyclability over five cycles and negligible leaching of metal into solution, which makes it an eco-friendly and economic catalyst even at commercial scale (14).

4a (iii): Combined Microwaves/Ultrasound (CMUI): This hybrid technology of combined MW-US has found applications in preparing nano catalysts and their applications in green chemistry. Based on such new, innovative ideas, nanocatalysts are efficiently prepared. For example, an efficient and selective reduction of NAs to ANs in excellent yields (89-96%) is accomplished using hydrazine hydrate as a source of hydrogen in the preparation of Cu-NPs (Scheme-8).

--Scheme-8

These results reveal the synergetic effect of microwave and ultrasound on the synthesis of Cu-NPs (15, 16).

References:

1a). N. P. Sweeny, et al. *J. American Chemical Society*, 80 (04):799-800 (1958). 11b). O. W. Brown, et al. *J. Physical Chemistry*, 26(02):161-190 (1922).

2a). (i) M. A. Aramendia, et al. *Reaction Kinetics & Catalysis Letters*, 14(4) 489-493 (1980); ii) T. Tsukinoki & H. Tsuzuki, *Green Chemistry*, 03:37-38(2001), iii) Reduction of 2, 4-Dinitrotoluene, G. Neri et al. *Applied Catalysis A: General*, 208, 307–316 (2001). 2b). D. G. Desai, et al. *Synthetic Communications*, 29(06):1033-1036(1999). 2c). http://www.qa-show.com/wiki/16559. 2d). C. Boix, et al. *J. Chemical Society, Perkin Trans*: 01:1487-1490 (1999).

3a). Comprehensive Organic Name Reactions and Reagents, Published Online: 15 SEP 2010. DOI: 10.1002/9780470638859.conrr063. 3b). N. M. Yoon et al. *Bulletin Korean Chem Soc*: 149: 281-283(1993), 3c).http://en.wikipedia.org/wiki/Reduction. 3d). G. Gao, et al. *Green Chemistry*, 10:439-441(2008).

4). B. Zeynizadeh, et al. *Synthetic Communications*, 36(18):2699-2704 (2006).

5a). D. Setamdideh et al. *Oriental J. Chemistry*: 27(3):991-996 (2011).

6). Y. He et al. *Synthetic Communications*, 19(17):3047-3050 (1989).

7). P. D. Ren et al. *Synthetic Communications*: 25(23), 3799-3803 (1995).

8). J. George, et al. *Synthetic Communications*: 13(06):495-499(1983).

9). R. Fazaeli et al. *J. Nanostructures*, 01(01):21-26 (2011).

10). a). T. Aditya, et al. *Chemical Communications*, 51:9410-9431(2015), b). R. Kaur et al. *Chemistry An Asian Journal*, 09(01):189-198(2014), c(i)). Z. Duan et al. *Bull. Korean Chemical Society*, 33(12):4003-4006 (2012). c(ii). S. Megarajan et al. *RSC Advances*, 06:103065-103071(2016). d). S. Wunder et al. *J. Physical Chemistry C*,

114 (19): 8814–8820(2010). e). B. Zeynizadeh, *J. Colloid & Interface Sciences*, in Press, Available Online March 07, 2017. f). Z. H. Farooqui et al. *Materials Science-Poland*, 33(03):627-634(2015). g). R. Kaur et al. *Chemistry An Asian J.* 09:189-198(2014). h). NA reductions using $NaBH_4$ and $CuBr_2$. http://shodhganga.inflibnet.ac.in/bitstream/10603/134675/14/14_chapter%204.pdf i). http://reag.paperplane.io/00003086.htm

11a). D. C. Gowda, et al. *Synthetic Communications*, 30(16):2889-2895(2000). 11b). D. C. Gowda & B. Mahesh, *Synthetic Communications*, 30(20):3639-3644 (2000). 11c). S. Gowda et al. *Synthetic Communications*, 33(02):281-289 (2003). 11d) K. Abiraj et al. *Synthesis & Reactivity in Inorganic and Metal Organic Chemistry*, 32(08):1409-1417 (2002). 12). K. Abiraj et al. *Synthetic Communications*, 35(02):223-230 (2005).

13a). X B. Lou et al. *Adv. Synthesis & Catalysis*, 353(02-03):281–286(2011). 13b). S. Fountoulaki et al. *ACS Catalysis*, 04(10):3504-3511(2014). 13c). A. Liu et al. *ACS Catalysis*, 06 (05):3084–3091(2016). 13d). Nano-Au/MgO-H_2O/RT: K. Layek et al. *Green Chemistry*, 14, 3164-3174 (2012).

14a). O. Verho et al. *ChemCatChem*, 06(01):205–211(2014).

14b). O. Verho et al. *ChemCatChem*, 06(11):3153-3159 (2014).

14c). J. Safari et al. *J. Molecular Structures*: 1125:772-776 (2016).

15). H. Feng, Y. Li, et al. *Sustainable Chemical Processes*, 02(01):01-14(2014).

16). http://www.organic-chemistry.org/synthesis/C1N/amines/arylamines.shtm

1.10.00: Commercial Manufacturing of Aniline:

Zinin (1842) was the first one to prepare aniline by the reduction of NB to AN using negative divalent sulfur as sulfide, sulfhydrate and polysulfide.

$$ArNo_2 + 3H_2S \text{ (or } (NH_4)_2S) \rightarrow ArNH_2 + 3S + 2H_2O$$

$$4C_6H_5NO_2 + 6S^{2-} + 7H_2O \rightarrow 4C_6H_5NH_2 + 3S_2O_3^{2-} + 6OH^-$$

This process is used both for laboratory preparation and for industrial manufacturing of AN from NB, although it is less economical at commercial scale compared to iron as a catalyst (1). After Perkin and Bechamp in 1850s, as Kaminaka has pointed out in his article about the historical developments of different processes of AN manufacturing as listed above, i) first AN was produced in 1930 by amination of CB, ii) then by hydrogenations of NB in 1950s, which are based on innovative technologies such as vapor phase, liquid phase, fixed catalyst-bed, fluidized catalyst-bed reactors etc. Amination of phenol was invented in 1960s and the first plant was started in Japan. Generally speaking, we may well say that what sort of manufacturing processes are employed at a certain era depends on various factors such as market size of the product which influences the scale of the production facilities, availability of raw materials, equipment and the control system at such time. Kaminaka also gives the list of different companies in USA, Japan and Europe, with their annual production capacities for aniline at commercial scale (2).

1.10:10: Different Methods for Aniline Manufacturing:

With developments in methods and technologies, aniline is manufactured on a commercial scale using the following processes:

1). Catalytic hydrogenation of NB over metal catalysts or metal/solid supported catalysts both under liquid and gaseous phases. The main process of aniline manufacturing is the catalytic hydrogenation of NB at 270°C and above atmospheric pressure over Cu/Silica catalyst, which produces AN in 95%.

2). Amination of CB with aq. NH_3 over Cu catalyst at 210°C/65atm gives AN in 96%.

3). Amination of phenol with ammonia and suitable catalysts, and

4). Direct amination of benzene (DAB) catalyzed by different catalysts.

Different processes of aniline manufacturing by different companies depend upon the availability of raw materials in local market, production facilities and the cost of productions and have been reported in detail by different authors (3a-3d). Generally, AN is manufactured at commercial scale in two steps: i) by the direct nitration of benzene to NB and ii) by the reduction of NB with metals and aq. HCl. Especially, the reduction of NB with Fe/Aq.HCl is the oldest AN manufacturing process (See Scheme-7). Mechanistically the needed electrons are derived from the metals used in the reduction process and the ammonium ion is formed in the presence of an acid, which is neutralized by an aqueous inorganic base to AN.

$$C_6H5NO_2 + 3M + 6 H^+ \rightarrow C6H5NH_2 + 3 M^{2+} + 2 H_2O$$

There are other metal based catalysts as well, which are used for the vapor phase reduction of NB. Aniline with 99.5% yield and almost 100% purity is produced by liquid phase hydrogenation of NB over metal oxide catalysts (4a). Similar results are obtained by using 7% of nickel instead of copper and 10% of molybdenum instead of chromium, barium and zinc. Aniline is obtained in 98% yield with high purity by reducing NB with Fe or Zn and trifluoroacetic acid at 70°C (4b). Heterogenous catalytic hydrogenation of NB over metal (Pd, V, Pb and Al_2O_3) coated over monolith honey combs yielded aniline with selectivity more than 99.9% (5).

Besides what we have seen earlier about NB reduction to AN, either by metal/acid combination or catalytic hydrogenation processes, AN is commercially obtained by

i) the reaction between aryl halides and NH_3 in the presence of a metal catalyst and ii) amination of phenols catalyzed by metal or metal salts/oxides (Scheme-9) (6-16).

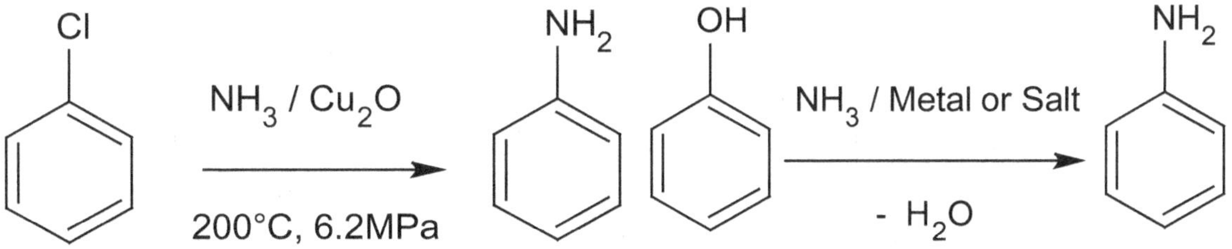

--Scheme-9

The above schemes are reported in most of the organic chemistry text books, journals and also available on internet: Wikipedia, Jan 2012. During early days of process discoveries, Mitsui Petrochemicals Unit (17) and Worrel (18) reported preparation of aniline from phenol and ammonia. Especially, Worrel prepared aromatic amines by reacting phenol with aluminium nitride and ammonia at temperatures from 200-600°C. For example, 2, 6-dimethyl-phenol reacts with aluminium nitride and ammonia to yield 2, 6-dimethylaniline. The reduction of nitroarenes (NAs) is conducted on an industrial scale using different catalysts such as Catalytic Hydrogenation over Metal or Metal salts, Metal/HCl or TFA, Sodiumhydrosulfite, Silanes, Pd/C/H_2, Raney/Ni/H_2, Pt/C/ H_2 etc. (19). The reduction *of* NACs to the corresponding ANs by CO/H_2O is a topic of high interest for potential industrial applications. Hence, CO + H_2O system along with metal carbonyl complexes as $Ru_3(CO)_{12}$ and diamine chelates is used for the reduction of NAs (20). Some other catalyst combinations with CO+H_2O are as follows: CO+H_2O/ $Ru(CO)_{12}$, CO+H_2O/Se, CO+H_2O/cis-[$Rh(CO)_2(amine)_2$]PF_6, CO+H_2O/$Rh_4(CO)_{12}$-9,10-diaminophenanthrene , CO+H_2O/Gold, etc.

1.10.11: A Comprehensive Look at AN Manufacturing Processes:

Let us look in detail at all these different processes/technologies for manufacturing of AN using different raw materials as i) nitrobenzene (NB), ii) chlorobenzene (CB), iii) phenol and iv) benzene.

1.10.12: Catalytic Hydrogenations of Nitrobenzene (NB):

We have seen earlier that aniline worldwide demand is steadily growing and likely to reach to about US$16bn in sales by 2020. The following sections highlight the old and new developments in aniline manufacturing processes.

Since Zinin (1842) and Bechamp (1854) methods, over a period of time many processes have been developed as listed above, for reducing NB to AN. In the olden days, based on Bechamp's process, commercially aniline was manufactured by reducing NB with iron filings and aq. HCl (Fe/Aq.HCl) (Scheme-7). AN is isolated by neutralizing the reaction mass with aq. NaOH, solvent extraction and distillation of the residue. This methodology was followed for quite some time. Later the NB reduction with iron was substituted by a catalytic reduction in 1871-1872, and it is still the most utilized procedure for the preparation/manufacturing of aniline and some of its analogs as well (21). Some time ago, copper catalysts (CuO, $CuCrO_2$ or $CuCrO_2$/Solid Supports/Promoters) were used in the manufacturing of AN by hydrogenation of NB. The catalysts were found stable up to 190°C and gave theoretical yields of AN (22). The industrial applications were enhanced by: i). the availability of the raw materials from the petrochemical operations, ii). the control over the catalyst poisoning in the feed. iii). the ease of maintenance of the temperature ranges in the fluidized bed reactors.

The catalytic hydrogenation of NB can be carried out either in the gas/vapor phase (G-S) or in the liquid phase (L-S). Copper catalysts are favored for the G-S and noble metals (mainly Pd and Pt) are preferred for the L-S system. Besides high yields of AN (more than 99%), the inevitable requirement of this process is the utilization of the reaction heat, which is more than 500 kJ/mol. There are several metal based catalysts which are used for the vapor phase or liquid phase hydrogenation of NB to AN. However, the deactivation of catalysts and relationships with their physicochemical features, as well as reactor arrangement and its influence on mass and heat transport are key factors of common concerns which need to be addressed. Necessity to combine theoretical (calculated) results with those from a model hydrogenation unit is also a key factor for design of efficient technology (23).

The rhodium cluster complex, $Rh_6(CO)_{16}$, has been found to catalyze the homogeneous reduction of NB to AN at temperatures above 80°C in the presence of N,N dimethylbenzylamine (DMBA) using any one of the reducing gases as 1) H_2/CO, 2) H_2, 3) CO/H_2O (Scheme-10).

was followed by a catalytic reduction in 1871-1872.

The reductions are highly selective and aniline was the only product detected and an evidence for water gas shift reaction (WGSR) was observed (24). $Rh_3(CO)_{12}$ in the presence of phenathrolines and CO+H_2O or [CO+H_2]+H_2O as a source of hydrogen gives AN in 97-98% yields (25). Activation of $Rh_6(CO)_{16}$ with 1,10-phenanthroline and substituted derivatives in the catalytic reduction of NB to AN with CO+H_2O has led to complete conversion of NB and isolation of AN in quantitative yields. The most active system is obtained using the ligand 3, 4, 5, 6, 7, 8-Me_6phen in a Chelate: Rh ratio of 3. Yields of 100% (substrate/rhodium = 1000) are easily achieved at 165°C without any basic co-catalyst (26). This process does have, if already not, a potential for adopting it to the industrial production of AN.

References:

1a). N. Zinin, *J. Practical Chemistry*, 27(01):149(1842).

1b). http://encyclopedia2.thefreedictionary.com/Zinin+Reaction

1c). http://infinity.usanethosting.com/Tuition/ZininReduction.pdf

2). H. Kaminaka, *J. Synthetic Organic Chemistry*, 34(10):758-764 (1976).

3a). Industrial Organic Chemistry, K. Weissermel & H. Arpe, John Wiley & Sons, - Science - 491 pages (2003), PP 376-379. 3b). Manufacture and Uses of the Anilines: A Vast Array of Processes and Products: ANILINES: Anthony S. Travis, John Wiley and Sons, (2007). http://www.civilengineeringhandbook.tk/petrochemical-processes/aniline-c6h5nh2.html. DOI: 10.1002/9780470682531.pat0395.

3c). AN Manufacturing Plants: http://www.chpn.cz/sites/chpn.cz/files/chpn_aniline.pdf 3d).AN-Manufacturing by catalytic hydrogenation, amination of CB or phenol: http://chempedia.info/page/09518102015508624220910108105916912709723800 26090/.

4a). Buckland et al.US-4994576 (1991): 4b). Wegerich et al. US 3136818 A(1964).

5a). Lehner et al. US0080293A1 (2005): 5b).M. Dugal et al. US7692042B2 (2010).

6). C. Grundmann, *Angewandte Chemie,* 62 (23-24): 558–560 (1950).

7). P. M. G. Bavin, *Organic Synthesis, Coll. Vol.* 05: 30(1973).

8). C. F. H. Allen, J. VanAllan, *Organic Synthesis, Coll. Vol.* 03:63(1955).

9). B. A. Fox, T. L. Threlfall, *Organic Synthesis, Coll. Vol.* 05:346(1973).

10). NaHSO$_3$: T. C. Redemann, et al. *Organic Synthesis: Coll. Vol.* 03: 69(1965).

11). For Ti(II)Cl$_2$: M. M. Faul, O. R. Thiel, *"Encyclopedia of Reagents for Organic Synthesis"*: (2005). doi:10.1002/047084289X.rt112.pub2.ISBN 9780470842898

12). J. Wisniak , M. Klein, *Ind. Eng. Chem. Prod. Res. Dev,* 23(01):44–50 (1984).

13). J. F. Knifton, *J. Organic Chemistry,* 40:519-520(1975).

14). A. S. Travis, "Manufacture & uses of the anilines: "A Vast Array of Processes & Products", in Z. Rappoport, *The chemistry of Anilines Part-1:* John Wiley (2007).

15). S. D. Naik, et al. *PTC in Chemistry. AIChE Journal,* 44(03):612-643 (1998).

16). K. Swaminathan, et al. *J Applied Electrochemistry* 2: 169(1972).

17). Mitsui Petrochemicals Ltd. *Chemical & Engineering News,* 47 (14), 44 (1969).

18). C. J. Worrel, US 3965182 (1976).

19a). http://research.omicsgroup.org/index.php/Reduction_of_nitro_compounds.

19b). G. Booth, "Nitro Compounds, Aromatic", In: *Ullmann's Encyclopedia of Industrial Chemistry*: John Wiley & Sons: New York, (2007).

20a). F. Ragaini, et al. *J. Molecular Catalysis A: Chemical*, 174(01-02):151-157

(2001): 20b). E. Allessio et al. *J. Molecular Catalysis*, 29(01):77-98 (1985), 20c).

L. He, et al. *Angewandte Chemie Intl. Edition*, 48:9538-9541 (2009).

21a). H. Kolbe, *J. Praktical Chemistry:* 04:418-419(1871). 21b). M. Saytzeff, *J. Praktical Chemistry:* 06:128-136(1872).

22). K. H. Gharda, et al. *Industrial Engineering & Chemistry*, 52 (05):417-420 (1960).

23a). M. Kralik et al et al. Aniline - catalysis and chemical engineering, Editor: Markoš, J., In Proceedings of the 41st International Conference of Slovak Society of Chemical Engineering, Tatranské Matliare, Slovakia, 723–733, (2014). 23b). M. Kralik et al. 41st Intl Conf. of SSCHE, May 26-30, 2014: Slovakia.

24). R. C. Ryan et al. *J. Molecular Catalysis:* 5 319 – 330(1979).

25). G. Mestroni et al. EP0097592A2 (1984). http://www.chpn.cz/en/produkty/anilin

26). E. Alessio, et al. *J. Molecular Catalysis* 22(3): 327-339 (1984).

1.10.13: MNCs in AN Manufacturing with Selectivity and Yields:

We have seen earlier that there are some methods based on the use of a particular metal catalyst, where aniline is obtained in 99.5% yield and 100% selectivity. Cu, Ni, Pt, and Pd metals on activated support (carbon or silica or alumina) in combination with other metals as modifiers/promoters are commonly used as catalysts for NB hydrogenations yielding high catalyst activity and aniline selectivity. Copper chromate and Nickel supported on Kieselguhr are comprehensively used for high pressure hydrogenations reactions. Copper chromium oxide is an exceptionally effective catalyst for the reduction of nitro group. The initial reduction of NB with batch process or through gas phase hydrogenation gave AN in about 99% yields. Both BASF and American Cyanamide operate the same process with fluidized bed reactors and Cu/SiO_2 catalyst achieving about 99% aniline yields. The modern catalytic gas-phase hydrogenation processes for NB can be carried out using a fixed-bed or a fluidized-bed reactor: Bayer and Allied Work has achieved this with NiS catalysts at 300–475°C in a fixed bed reactor (27).

The activation of hydrogenation catalysts with Cu or Cr and the use of different supports and catalyst sulfidization methods with sulfate, H_2S or CS_2, etc. belong to the expertise of the corresponding firms. The selectivity to aniline is >99.9% and the catalyst can be regenerated and reused. Similar processes are operated by Lonza with Cu on pumice, by ICI with Cu, Mn, or Fe catalysts with various modifications involving other metals, and by Sumitomo with a Cu-Cr system. BASF, Cyanamid and Lonza use the gas phase hydrogenation of NB. The BASF catalyst consists of Cu, Cr, Ba, and Zn oxides on a

SiO$_2$ support; the Cyanamid catalyst consists of Cu/SiO$_2$. The hydrogenation is conducted at 270-290°C and 1-5 bar in the presence of a large excess of hydrogen (H$_2$: NB~ 9:1), and the selectivity to aniline is 99.5% at the complete or quantitative conversion of NB (28). Based on their in-house "Process Technologies", Johnson Matthey has used **PRICAT™** NI 55/5-P for liquid phase slurry hydrogenation of NB to AN as their preferable choice of catalyst. It's a unique manufacturing process and the properties give this catalyst excellent activity, selectivity and a long lifetime (29).

Recently, a convenient and stable iron oxide (Fe$_2$O$_3$)-based catalyst has been used as a more earth-abundant alternative for this transformation. Pyrolysis of iron-phenanthroline complexes on carbon furnishes a unique structure in which the active Fe$_2$O$_3$ particles are surrounded by a nitrogen-doped carbon layer. Highly selective hydrogenation of numerous structurally diverse NAs (>80 examples) proceeded in good to excellent yields (90-99%) under industrially viable conditions (30). Fe-NPs in water at room temperature are also found highly selective in the reduction of NAs. A wide spectrum of reducible functionalities remained inert under the reaction conditions (31). Non-precious metal NPs (Co@CNPs or CoNi@CNPs) coated on carbon serve as good catalyst for chemoslective NA reductions to ANs (32). Dehydrogention/hydrogenation strategy is used for chemoselective reduction of NAs to ANs in yields up to 99% over Pd-NPs in 10-20 minutes at room temperature in aqueous methanol (33). At industrial scale, copper catalyst is used as a rapid method for estimating the working time for aniline production (34). Similarly, Au-NPs dispersed over Co-loaded SBA-15 are used as an efficient catalyst for making aniline by hydrogenation of NB. The activity of nanocatalysts is due to the large number of small size metal atoms at the surface providing maximum possible surface area which accelerates the rate of reactions compared to the bulk catalysts (35).

The technology to produce aniline by means of catalytic hydrogenation of NB has been jointly developed by Borsod Chem MCHZ Ostrava (S.R.O., formerly Moravské chemické závody) and the Institute of Chemical Technology Prague. Most recently, the aniline plant has been installed at Slovakia in 2013. This process has distinctive technical and economic features comprising three basic technological units: i) the hydrogenation train(s), ii) the aniline rectification and iii) the AN-Water treatment plant (26). Kralik et al. have covered different catalyst systems for industrial production of AN. It is clear from the data, as shown in the Table-1 below, that the use of Cu/Solid Support (BASF, BC-MCHZ etc.) and Pd-Pt-Fe (DuPont) give excellent results (Scheme-11).

Nitrobenzene Conversion = 100%   --Scheme-11

Especially this process provides high aniline selectivity and 100% NB conversions per pass under low pressure of 17bar in a Plug-Flow Reactor (26). Some other catalyst combinations, as listed below, have produced AN in quantitative yield: 1) Chinese Patent: CN 1657162 A (August 24, 2005), 2) United Chemical Company's Patent US2822397 (AN-Yield 99.5%), 3) Patent US4265834 (AN-Yield 99.85%), 4) Patent US3504035 (AN >99%). The recent surge in nanochemistry has led to the development of some interesting applications in nitro reduction processes, which are classified based on the source of hydrogen and the mechanism of reduction (36).

1.10.14: Aniline Manufacturers Worldwide:

Table-1: Some of the Major Chemical Companies Manufacturing Aniline:

BASF	Conner Chemicals	Shandong Jinling Chemical	Tianji
BAYER	Lanzhou Chemicals	Sinopec Nanjing Chemicals	Tosoh
Chemours*	MCHZ	SP Chemicals	Volzhsky Orgsynthese
Huntsman	Shandong Haihua	Sumitomo Chemicals	Wanhua Chemicals

*Chemours: A US based AN manufacturing multinational company, which is bought by Dow Chemical Co. in 2016.

References:

27). H. J. Arpe, *Industrial Organic Chemistry* (3rd Ed.), Weinheim: VCH, 375 (1997).

28a). C. A. Heaton, Y. 1975, Michigan, Ann Arbor Science of Modern Industrial Chemist, Blackie: Glasgow, 392-393(1991). 28b). http://shodhganga.inflibnet.ac.in/bitstream/10603/27786/6/06_chapter_1.pdf

28c). http://shodhganga.inflibnet.ac.in/bitstream/10603/13074/16/16-chapter%206.pdf 28d).http://www.lookchem.com/Chempedia/Chemical-Technology/Organic-Chemical-Technology/7791.html.

28e). Catalytic Hydrogenation of Aromatic Compounds in the Liquid Phase M. Králik et al. *J. Chemistry& Chemical Eng.* 6, 1074-1082 (2012) and references 17-20.

28f). Industrial scale production of aniline using different catalysts, S. Diao et al. *Applied Catalysis A General,* 286(1):30-35 (2005).

29). http://www.jmprotech.com/catalysts-for-aniline-production-johnson-matthey.

30a). R. V. Jagadeesh et al. *Science,* 342(6162):1073-1076 (2013). 30b). *Nature Protocols* 10(04):548-557 (2015). 30c). *Nature Protocols,* 10(6):916-926 (2015). 30d).R. V. Jagadeesh, et al. *ChemCommun* (Camb): 47(39):10972-10974 (2011). 30e). S. Fountoulaki et al. *Chemistry: A European J.,* 22(13):4600–4607(2016).

31). R. Dey, et al. *Chemical Communications,* (Camb), 48(64):7982-7984 (2012).

32). L. Liu, F. Gao, et al. *J. Catalysis,* 350:218–225 (2017).

33a). N. M. Patil et al. *RSC Advances*: 05:86529-86535(2015).33b).

http://www.radiantinsights.com/research/global-aniline-sales-market-report-2016

34). L. Petrov et al. *Applied Catalysis,* 59(1), 31-43 (1990).

35a). S. Viswanathan, et al. *J. Porous Materials:* 21(03)251-262 (2014).

35b). Production of Aniline: A. Wegerich, et al. US-3136818A (1964)

35c). https://www.icis.com/resources/news/2004/02/27/560820/aniline-the-builder/

36). Review on NB Reduction Methodologies: H. K. Kadam & S. G. Tilve, *RSC Advances,* 05:83391-83407 (2015).

For full market reports on Aniline Industry Trends visit the following sites, with forecast from 2016-2021 (37).

37a). http://www.advfn.com/news_Aniline-Industry-Trends-2016-2021-Forecast-Analysi_71020495.html. 37b). http://chempedia.info/info/151260/. 37c). http://www.mynewsdesk.com/us/pressreleases/global-aniline-market-2016-analysis-research-trends-growth-and-forecast-2021-1492637. 37d). http://www.chemengonline.com/aniline-production-nitrobenzene-liquid-phase-intratec-solutions/?printmode=1. *E-mail: chaas@nexant.com or* jplotkin@nexant.com. 37e). www.intratec.us/products/aniline-production-processes 37f). https://www.researchgate.net/figure/254216716_fig2_FIGURE-3-Aniline-production-process-hydrogenation-of-nitrobenzene-Stream-1.

1.11.00: New Developments in Aniline Synthesis/Manufacturing:

The existing methods of manufacturing aniline by vapor or liquid phase hydrogenation of NB are fairly good, with some of them being really good. However, these methods have certain limitations as: i) nitration of benzene using nitrating mixture, which is exothermic in nature, highly acidic and corrosive, cause environmental pollution, involve extensive acid base neutralization and work up and hence not so cost-effective. Therefore, always there is constant need to either improve the existing manufacturing processes or develop new ones which will meet the above listed concerns. The emphasis of new methods development would remain as simple process, high yields and about 100% aniline selectivity with almost quantitative NB conversions. Keeping in mind these aspects, the chemists/scientists all over the world strive to do their best in achieving these objectives.

Some of the most recent developments in this context have been briefly summarized below by using catalytic hydrogenations over different catalysts and a particular set of reaction conditions. For the sake of convenience & in order to give the year wise progressive account of these new developments, for aniline preparations at laboratory scale and commercial productions, the references are placed next to the descriptive part of the published data.

In the early days of aniline history, NB is reduced using a mixture of hydrogen and ammonia gases. NB is pre-hated to its vaporization temperatures and passed over the nickel chromite catalyst along with ammonia and hydrogen gases, which gave 99.1% aniline. M. L Huber, et al. US 2739985 A, (1956). Later on, the main process for aniline manufacturing is the catalytic hydrogenation over Cu/Silica catalyst at 270°C. Aniline was isolated in 95% yield. Kohler et al. have described the process for Cu/SiO$_2$ catalyst preparation using ion-exchange resin. M. A. Kohler, J.C. Lee et al. *Applied Catalysis*, 31(02) 309-321 (1987).

The Pd-B/SiO$_2$ amorphous catalyst exhibited higher activity than the corresponding Pd/SiO$_2$ and crystallized Pd-B/SiO$_2$ catalysts during the liquid phase hydrogenation of NB to AN, indicating the promoting effect of the Pd-B alloy. The fact that the activity per g Pd of the Pd-B/SiO$_2$ amorphous catalyst was higher than that of the unsupported Pd-B amorphous catalyst also demonstrated the promoting effect of the silica support. X. Yu, et al. *Applied Catalysis A: General*, 202(01):17-22(2000).

The catalytic properties of Pd and Pt supported on woven glass fibers (GF) were investigated in three-phase hydrogenation of NB and 100% yield of AN was attained. The catalytic activity for the best catalysts was two times higher than the activity of commercial Pt/C catalyst traditionally used for liquid–phase hydrogenation. V. Holler, *Chemical & Engineering Technology*, 23(03):251-255(2000).

Some of the noble metal catalysts such as Pd/C, Pt/C, Ru/C, and Rh/C catalysts exhibit 100% selectivity to AN in super critical CO$_2$ (scCO$_2$). The NB conversions in case of Pd/C and Pt/C are 100% as well. F. Zhao et al. *Prepr. Pap.-Am. Chem. Soc., Div. Fuel Chem.* 49 (01):13-14 (2004).

An efficient method for selective formation of aromatic amines by selenium-catalyzed reduction of ANCs with CO/H$_2$O under atmospheric pressure is developed.

NB Conversion = 100%   AN Selectivity = 100% --Scheme-12

The ANCs with different sensitive functional groups on them are quantitatively reduced (i.e. 100% conversions, except for *o*- and *p*-Cl derivatives) by CO/H$_2$O to aryl amines with high selectivity (100%)(Scheme-12). This process has a potential for scaling up to commercial productions. X. Liu, et al. *J. Molecular Catalysis A: Chemical* 212:127–130 (2004). S. Diao et al. *Applied Catalysis A: General*, 286:30-35(2005). Pt-NPs core-polyaryl ether trisacetic acid ammonium chloride dendrimer shell nanocomposites (Pt@Gn-NACl) were prepared and used as catalysts for hydrogenation of NBs to ANs, with 100% NB conversions (Scheme-13).

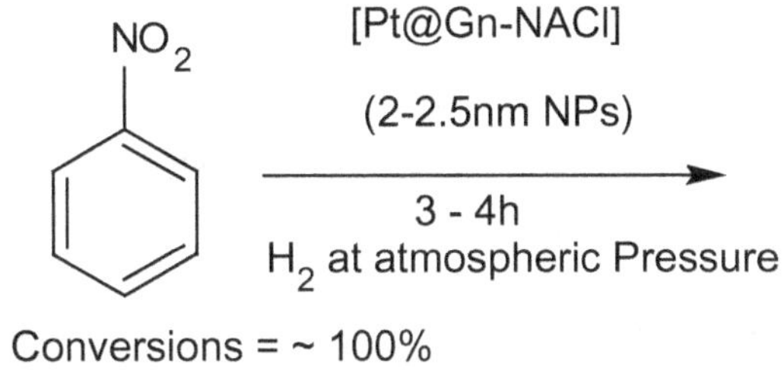

NO_2
[Pt@Gn-NACl]
(2-2.5nm NPs)
3 - 4h
H_2 at atmospheric Pressure
NH_2

Conversions = ~ 100%
--Scheme-13

The resin as a stabilizer for the preparation of Pt-NPs catalysts is appropriate and it helps in 100% NB conversion. P. Yang, et al. *J. Molecular Catalysis A: Chemical:* 260(01-02)04-10(2006). The kinetics of the semi-batch oxidative synthesis of AN under atmospheric pressure using DuPont's NiO/ZrO_2 cataloreactant was studied in a microreactor flow set-up equipped with a calibrated online mass spectrometer.

NO_2
$NiO / ZrO_2/ H_2$
590K
NH_2

Microreactor Flow Syst.   Sel = 100% --Scheme-14

It was found that the reaction temperature is a crucial parameter for the selectivity to AN (Scheme-14). At 590 K, exclusively aniline was formed (100% AN selectivity). www.faqs.org/patents/app/20120215029.

Pd supported on Mg-Al hydrotalcite with different Pd loadings (0.5, 1, 2 and 5 wt. % of Pd) was prepared and tested for the vapor phase hydrogenation of NB at atmospheric pressure in the temperature range of 498–598 K (Scheme-15).

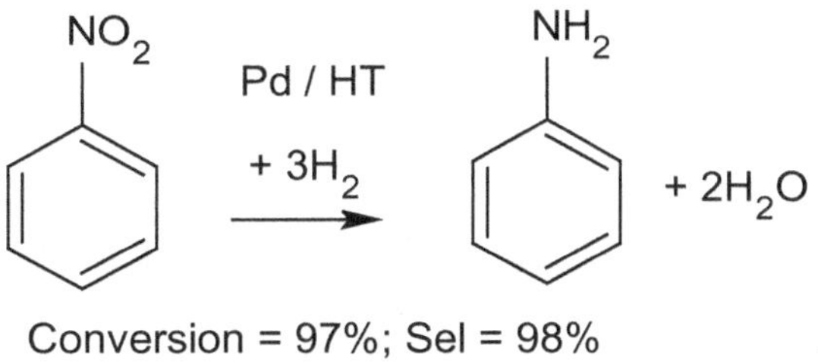

NO_2
Pd / HT
+ $3H_2$
NH_2
+ $2H_2O$

Conversion = 97%; Sel = 98%   --Scheme-15

The catalyst with 0.5wt% Pd supported on hydrotalcite (0.5 wt% Pd/HT) reduces NB to aniline with NB conversion of 97% and aniline with 98% selectivity at 498K. P. Sangeetha, et al. *J. Molecular Catalysis A: Chemical*, 273:244–249(2007).

Reduction of NB by Fe-Cu catalyzed process, electrochemical reduction characteristics & the mechanism of reductions in the process are discussed. W. Xu, et al. i) *J. Hazardous Materials,* 123(01-03):232-241(2005). ii) *J. Environmental Sciences,* 18(02): 379-387(2006), iii) *J. Environ Sciences* (China). 20(08):915-921(2008).

Effectiveness and potential of multiphase reaction systems using dense phase CO_2 are demonstrated for several catalytic hydrogenation reactions, especially NB hydrogenation to AN. As a result, the 100% selective hydrogenation of NB to AN can be achieved at any conversion level up to 100% with conventional supported Ni catalysts in the presence of dense phase CO_2. S. Fujita, In: Supercritical Fluids, chapter 11, ISBN: 978-1-60741-930-3. Editor: Marcel R. Belinsky © 2009 Nova Science Publishers, Inc.

Vapour phase hydrogenation of NB on 1 wt% Pd supported on hydrotalcite, MgO and Al_2O_3 catalysts are reported. The catalyst (Pd/H-Talcite) is more active than on other supports and is found as an effective catalyst for NB hydrogenation. P. Sangeetha, et al. *Applied Catalysis A: General,* 353 (02):160-165(2009).

Iron-based catalyst and formic acid system for the reduction of NAs to ANs in good to excellent yields has been developed under mild conditions. Notably, the process constitutes a rare example of base-free transfer hydrogenations. G. Wienhofer, et al. *J. American Chemical Society,* 133(32):12875-12879(2011).

Pd-NPs-Ionic Liquid Brush (SiO_2-BisILs[PF_6]Pd°) have been used as an efficient catalyst for NB hydrogenation in the absence of solvent under mild conditions, where the NB conversions are 100% and the AN yields are good to quantitative (75 - 100%). The catalyst can be recycled and used again to yield 100% AN (Scheme-16). J. Li, X. Shi et al. *ACS Catalysis,* 01:657–664 (2011).

Conversions = 100%   Yields = 75 - 100%

R = CH_3, Cl, OCH_3, COOH, $CONH_2$, $COCH_3$, Ph, etc.   --Scheme-16

Gas-phase, liquid-phase and two phase hydrogenation of NB is reported. C. Kartusch, et al. *ACS Catalysis,* 02:1394-1403(2012). U. Hartfelder et al. *Catalysis Communications,* 27: 83-87 (2012). J. Pasek, et al. EP 2471768 A1 (2012).

Hydrogenation of NB over Au/MeOx catalysts, A Matter of the Support: M. Makosch, et al. *ChemCatChem,* 04:59-63 (2012). Selective hydrogenation of *m*-DNB to *m*-nitroAN over Ru-SnOx/Al$_2$O$_3$ catalyst is reported *m*-NAN selectivity of 97% at the complete conversion of *m*-DNB. H. Cheng et al. *Catalysts*, 04:276-288(2014).

Vapour-phase NB hydrogenation was studied using different metal (Ru, Pd, Ni, Pt) catalysts supported on Pd/Ca$_3$(PO$_4$)$_3$(OH). The (PdHAP) catalyst was found much superior to other metals with 97% aniline selectivity without any other byproducts. M. Sudhakar et al. *Indian J. chemistry*, 53A:550-552(2014).

Reduction of NB with sulfides catalyzed by black carbons from three types of crop-residue ashes is found suitable for remediation technique. W. Gong, et al. *Environmental Science & Pollution Research*, 21(09):6162–6169(2014).

Recently "Hydrogen-Free" hydrogenation of NB over Cu/SiO$_2$ via coupling with 2-butanol dehydrogenation is reported, for the first time, with both aniline & 2-butanone in 100% yields. M. Li., et al. *Topics in Catalysis*, 58(02): 149-158(2015). Nitrogen-doped diamond electrode shows high performance for electrochemical reduction of NB to AN (96.5%). Q. Zhang, et al. *J Hazardous Materials*, 265:185-190(2014).

Photocatalytic hydrogenation of NB to AN over tungsten oxide-silver (WO$_3$-Ag hybrid nanowires (WO$_3$-Ag NWs) is studied. The synergy of Ag NWs and WO$_3$ NWs plays the most vital role in this enhanced photocatalytic performance with excellent reusability. F. Li, C. Wen, *Materials Letters*, 142:201-203(2015).

An efficient immobilized Ru metal containing ionic liquid (ImmRu-IL) catalyst selectively reduces NB and other NAs through dehydrogenation of dimethylamine borane complex. N. M. Patil, et al. *RSC Advances*, 06(57):52347-52352 (2016).

Lu et al. have demonstrated that TiO$_2$ can act as an excellent cathodic electrocatalyst for reduction of NB, when its crystal shape, exposed facet and oxygen-stoichiometry are finely tailored by the local geometric and electronic structures. C. Liu, et al. *Environ. Science & Technology*, 50(10):5234-5242(2016).

Aniline is manufactured by a new process by liquid phase NB hydrogenation over a noble metal catalyst supported on carbon. The NB conversion to aniline is nearly 100% in a single pass. This information is based on "Economics of Aniline Production from Nitrobenzene," a report published by Intratec (Intratec Solutions, March 1, 2016). http://www.chemengonline.com/aniline-production-nitrobenzene-liquid-phase-intratec-solutions/?printmode=1   http:www.intratec.us/products/aniline-production processes

Most recently, the synthesis of Pd/SBA-15 catalyst employing surface-bonded vinyl as a reductant and its application in the hydrogenation of NAs is reported (Table-2).

Table-2: Pd/SBA-15 Comparison with Other Catalysts by Different Groups:

No	Catalyst	Metal %	Temp. (K)	Press(MPa)	Yield (%)
1	Ni–Fe$_2$O$_4$/carbon	–	423	1.0	~100
2	Au/SiO$_2$-Org	01.00	413	4.0	99
3	Pt–NHC	67.00	303	0.1	99
4	Ru/RGO	03.40	383	2.0	99
5	SS-Pd[b]	00.50	323	–	98
6	Fe$_3$O$_4$–NH$_2$–Pd	08.43	RT	0.1	99
7	Pd/HAM@γ-AlOOH	07.20	RT	0.1	~100
8	Pd@Fe$_3$O$_4$	00.87	RT	0.1	99
9	Pd/PMO–SBA-15	01.50	318	0.1	54
10	Pd/SBA-15	01.70	313	0.1	98

High performance of a cobalt–nitrogen (Co-N) complex for the reduction and reductive coupling of nitro compounds into amines and their derivatives with full NAs conversions and >97% AN selectivity is reported. P. Zhou, et al. *RSC Advances,* *03*(02):(2017). e1601945. doi:10.1126/sciadv.1601945.

NB rReduction with Pd/SBA-15: Reaction Conditions: Pd/SBA-15 (10.0 mg), substrates (1 mmol), ethanol (4 mL), decane (100 mg), 313 K, 1atm H$_2$; rotation rate 1200 rpm. Y. Duan, M. Zgeng et al. *RSC Advances,* 07:3443-3449 (2017).

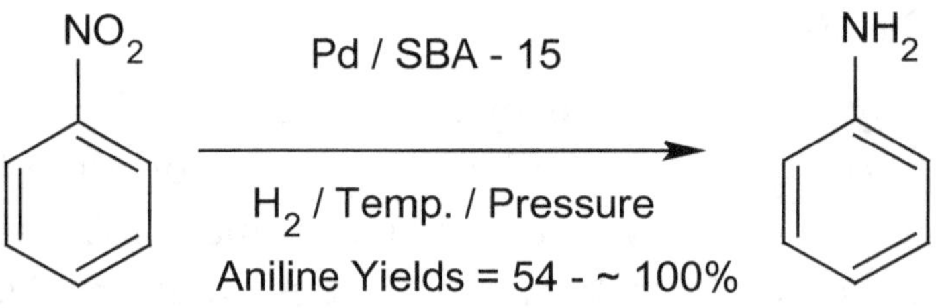

--Scheme-17

The authors have also compared this catalyst with other most successful catalyst systems developed by other groups for by catalytic hydrogenation of NB to AN. The Pd/SBA-15 catalyst gave 98% yield and over 95% aniline selectivity under mild reaction conditions (Scheme-17).

1.11.10: Nickel as a Catalyst:

Amongst non-noble metals Ni is one of the most commonly used metal catalysts in organic synthesis. Especially, here Ni catalysed reductions of NAs to ANs are highlighted.

Ni-NPs were generated by metal disintegration and stabilized by filamentous carbon exhibited excellent performance producing clean AN (~99% yield). N. Mahata, et al. *Applied Catalysis A: General*, 351:204-220(2008). Selective hydrogenation of NB to AN in dense phase carbon dioxide over Ni/γ-Al$_2$O$_3$ is investigated (Scheme-18).

Conversion = 100% ; Sel = 100% --Scheme-18

Significance of molecular interactions is studied and found that these interactions between the catalyst, solid surface and the nitro group are responsible for the best results. Dense phase CO$_2$ was found better than organic solvents as ethanol and n-heptane. The selectivity to the desired product, AN, was almost 100% over the whole NB conversion range of 0–100% (Scheme-17). X. Meng, et al. *J. Catalysis*, 264(01): 01-10(2009). Nano sized silica supported Ni catalyst was found highly selective in the reduction of NB to AN under catalytic hydrogenation conditions (Scheme-19).

NB Conversion = 100%   Aniline Selectivity = 99% --Scheme-19

The catalyst is found highly efficient in reducing NB with100% conversion and with 99% selectivity for AN at 1.0 MPa hydrogen pressure and at 90°C, when the NB:Ni mole ratio was at 305. J. Wang et al. *Ind. Eng. Chem. Res.* 49(10), 4664-4669 (2010). Pd-M/ZnO (M = Ag, Cu, Ni) prepared in single step is found efficient for NB reduction using SBH as a hydrogen source. Under catalytic hydrogenation of NB to AN over SiO$_2$ supported nickel catalysts (Ni/SiO$_2$; Ni =3.7nm), the selectivity of aniline reached 99% with a 100% conversion of NB in 5.5 h at 90°C and 1.0 MPa pressure. J. Wang et al. *Ind. Eng. Chem. Res.*, 49(10):4664–4669(2010). Y. Hu, et al. *Science China Chemistry*, 53(07):1541-1548(2010).

A novel Pd-Ni bimetallic NPs catalyst supported on epigallocatechin-3-gallate (EGCG) grafted collagen fiber (CF), (Pd-Ni/CF-EGCG), has been prepared. The catalyst behavior and the hydrogenation of NB catalyzed by Pd-Ni/CF-EGCG was investigated at 308K and 1.0 MPa hydrogen pressure. X. Liao, et al. *Chin J Catalysis*, 31(12):1465-1472(2010). The hydrogenation of NB in water as a green solvent has been investigated by using Ni/TiO$_2$

catalyst. The catalyst gets deactivated due to Ni(OH)$_2$ formation when it reacts with water. This adversity was overcome by coating Ni/TiO$_2$ catalyst with 11% carbon. It enhanced the catalytic activity and reduced the reaction time (Scheme-20).

Ni/TiO$_2$ = Time 6h   (Ni/TiO$_2$)@C$_{11}$% = 3 - 5h

Conversion = ~73%   Conversion > 97%

Aniline Selectivity = 99.6%   Aniline Selectivity = > 94% --Scheme-20

With Ni/TiO$_2$, the NB conversion reached ~80% at the first run and it dropped to 10% in the second run. On the other hand, with (Ni/TiO$_2$)@C-11% catalyst the conversions could reach >95% for the first three runs and the reaction rate was about 2.3 times higher than that of the Ni/TiO$_2$ catalyst. W. Lin, H. Cheng, et al. *J. Catalysis*, 29:149-154(2012).

Liquid phase hydrogenation of NB over nickel present in elemental state on supports with catalytic activity order of Ni/rutile > Ni/anatase > Ni/TiO$_2$ is reported. A conversion of 99% was observed for Ni/rutile at 140°C and 1.96 MPa H$_2$ pressure.

Conv. = 99%   H$_2$ / 140°C / 1.96Mpa   --Scheme-21

Aniline is the only product (100% selectivity) formed which demonstrates the catalytic hydrogenation of NB proceeds with atom economy (Scheme-21). K. J. A. Raj, et al. *Chinese J. Catalysis*, 33:1299-1305(2012).

Highly-dispersed Ni-NPs over CNTs were successfully prepared and the catalyst exhibited excellent catalytic performance in liquid phase selective hydrogenation of *o*-CNB to *o*-CAN along with the highest yield of 98.1% in 150min. J. Wang, et al. *Catalysis Science & Technology*, 03 (04):982-991(2013).

NiO-NPs supported on multi-walled CNTs (MWCNTs) and AC were prepared and applied as nanocatalyst in hydrogenation of NB to AN (Scheme-22).

NiO - NPs / MWCNTs/AC

H_2 (12bar) / 140°C

AC = Activated Carbon

MWCNTs = Multiwalled Carbon Nanotubes

Aniline Yield = 99.5%, and Anilne Selectivity = 100% --Scheme-22

Excellent yield (99.5%) with 100% selectivity to aniline is accomplished. M. Hashemi, M. M. Khodei et al. *J. Synthesis & Reactivity in Inorganic, Metal-Organic, and Nano-Metal Chemistry*, 46(07): 959-967 (2016).

A highly chemoselective catalytic transfer hydrogenation (CTH) of NAs to corresponding amines is achieved with Fe–Ni bimetallic NPs (Fe–Ni NP's) as the catalyst and $NaBH_4$ at room temperature. D. R. Petkar, et al. *RSC Advances*, 04:8004-8010(2014).

The catalyst preparation methods also play a considerable role in the hydrogenation of NB, Ni/TiO_2, Ni/ZrO_2 prepared by reductive deposition method shows excellent conversion of NB (99%) to AN in vapor phase at atmospheric pressure (Scheme-23).

Ni - NPs/TiO_2 or Ni - NPs/ZrO_2

Catalytic Hydrogenation

NB Conversion = 99%   --Scheme-23

The effect is attributed to the higher active surface area on the Ni-NPs due to a special preparation method. M. Varkolu, et al. *Applied Petrochem Res.*06:15–23(2016).

Preparation of Ni/bentonite and Ni-Mo-P alloy as the catalysts and their applications in the catalytic hydrogenation of NB to AN- are reported. Y. J. Xiliang L. Z.Qin et al. *Chinese J. Chemical Engineering:* 24(09): 1195-1200(2016).

Qin and Liu et al have developed a novel supported Ni-B/SiO_2 amorphous alloy catalyst which gives both the NB conversion and the AN- selectivity at about 100%. Z. Qin, Z. Liu et al. *Chemical Engineering Communications*, 201(03):338-351(2014). Z. Liu et al. *Catalysis Communications*, 85:17-21(2016).

Another similar example is the hydrogenation of NB to AN- over C60-Stabilized Ni catalyst. The NB conversion and the aniline selectivity both reached above 99.9%, within 40min at or below 90°C and 2MPa H_2 pressure. C60 improved the aniline selectivity and enhanced the Ni/C60 stability attributed to its hydrophobicity. Y. Qu, H. Yang et al. *Catalysis Communications*, 97, 83-87(2017).

1.12.00: Manufacturing of Aniline from Other Sources:

1.12.10: Aniline from Chlorobenzene & Ammonia:

Alternatively, Aniline is also manufactured by the ammonolysis of CB or phenol. Kanto Electrochemical Co. manufactures of AN by heating a mixture of CB and aqueous NH_3 at 180-220°C and 60-75 bar in the presence of CuCl and NH_4Cl ("Newland Catalyst"). The aniline selectivity is 91%. Bayer also makes use of Newland catalyst (a mixture of CuCl and NH_4Cl) for the commercial production of AN with 91% selectivity starting with CB and aq. NH_3 as reactants. http://www.lookchem.com/Chempedia/Chemical-Technology/Organic-Chemical-Technology/7791.html.

One of the oldest AN manufacturing processes is reported by Prahl and Mathes in 1935. Amination of CB with ammonia catalyzed by Cu, Ni, Fe salts or acidic catalysts as molybdic acid, tungustic acid etc. at 250-450°C produces AN in 96-100% yield. It is claimed that this process can replace the vapour phase hydrogenation of NB (Scheme-24) (1).

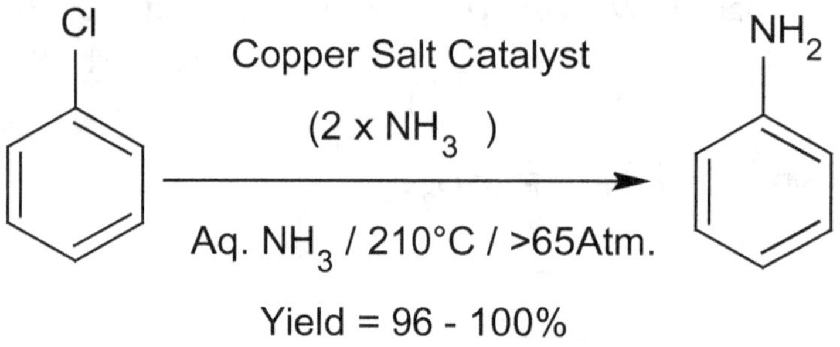

Copper Salt Catalyst

$(2 \times NH_3)$

Aq. NH_3 / 210°C / >65Atm.

Yield = 96 - 100%

--Scheme-24

Similarly, an aniline derivative N-methylaniline is obtained from CB and methylamine (2). Williams et al. from Dow Chemical Co. have reported a commercial method for the production of AN by reacting CB with ammonia in the presence of copper metal and or copper salt (CuCl) or a mixture of Cu/CuCl (3). D. G. Jones from Socony Mobil Oil Company has studied the production of aromatic amines by ammonolysis of aromatic halides in the presence of alumino-silicate catalysts. Here, aniline is manufactured by reacting CB with excess NH_3 and the catalyst. As a catalyst, alumino-silicate catalyst is cation exchanged with Cu or Zn and found better than the former (4). During the first part of twentieth century, commercially aniline was manufactured by amination of CB in the presence of NH_3 and Cu_2O or $CuCl_2$ as a catalyst (5) at high temperatures and pressures (Scheme-25).

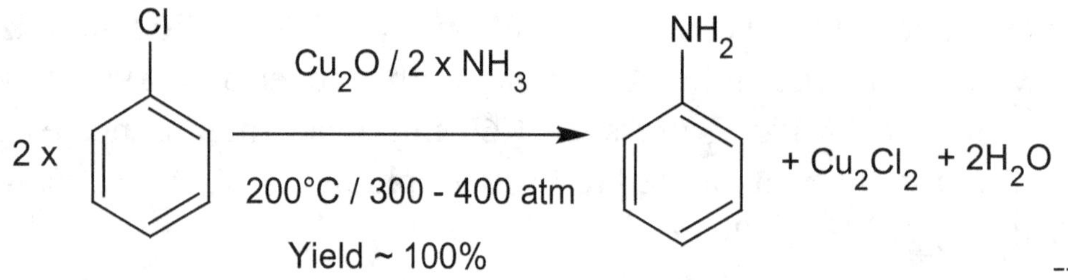

$2 \times$

Cu_2O / $2 \times NH_3$

200°C / 300 - 400 atm

Yield ~ 100%

$+ Cu_2Cl_2 + 2H_2O$

--Scheme-25

An improved process involves reacting CB and aq. NH_3 in the presence of CuCl catalyst at 150-200°C for 1-2 hours (6). Similarly, in a continuous process using CB, aq. NH_3, and CuCl are reacted under high temperature between 200-220°C to obtain AN (7). The use of metal-based complexes for the direct amination of aryl halides with NH_3 has been recently reviewed (8).

References:

1a). W. Prahl, et al. US2001284 (1935): 1b). P. Walter, et al.US 2001284 A (1935).

2). E. C. Hughes, et al. *Industrial Engineering & Chemistry,* 42(05):787-790(1950).

3). Hale and Britton, US1607824 A (1926), H. R. Slagh, US2391848 A (1945), W. H. Williams, et al. US2432551 (1947): 4). D. G. Jones, US 3231616 A(1966).

5a). https://www.scribd.com/doc/219108298/Project-Production-of-Aniline

5b). http://www.transtutors.com/chemistry-homework-help/nitrogen-containing-compounds/preparation-of-aniline.aspx. 5c).Review: S. Dutta, https://www.researchgate.net/figure/254216716_fig2_FIGURE-3-Aniline-production-process-hydrogenation-of-nitrobenzene-Stream-1.

5d). www.intratec.us/products/aniline-production-processes

6). E.C. Hughes and V. Franklin, US 2490813 A (1949).

7). W. Williams, et al. Production of Aromatic Amines, US2432551A (1947).

8). Y. Aubin, et al. *Chemical Society Reviews,* 39:4130-4145 (2010).

1.12.11: New in Amination of Aryl Halides:

Transition metal catalysts have been used successfully to prepare pharmacologically important compounds (9). Metals as Cu and Ni and their salts have been used in the preparation of aniline from CB and phenol. Pd has found a special position as a highly selective catalyst in amination of aryl halides and phenols. For some time, Pd catalyzed CN and CO coupling reactions were encountered with certain limitations. This problem was overcome by using a Pd-ligand which facilitated the coupling reaction between ammonia and CB. This reagent is also used in C-O cross coupling reactions, thus making it possible to make phenols (10). In another case a new robust Pd/phosphine catalyst system for the selective monoarylation of ammonia with different aryl bromides and chlorides has been developed. The catalyst gives full conversion with most substrates with 1–2mol% of Pd source and a fourfold excess of ligand (11a, b). A simple, efficient catalyst system for the Pd-catalyzed amination of aryl chlorides, bromides, and triflates is also reported. Butchwald group has developed novel Pd-ligands for the preparation of ANs with high selectivity (11c). Ammonia is found as an environmentally friendly nitrogen source for primary aryl amine synthesis (12). The direct amination of aryl halides with ammonia and Pd-catalysts is reviewed (13).

References:

9). S. Tasler, et al. *Advanced Synthesis & Catalysis*, 349:2286-2300(2007).

10a). M. C. Willis, *Angewandte Chemie Intl. Ed.*, 46(19):3402-3404(2007).

10b). A. Borzenko et al. *Angewandte Chemie Intl Ed.*, 54(12):3773–3777(2015).

10c). J. F. Hartwig, *Angewandte Chemie Intl. Ed.*, 37(15):2046-2067(1998).

10d). S. M. Crawford, *Chemistry A European J.*, 19(49):16760-16771(2013).

11a). T. Schulz, et al. *Chemistry*, 15(18):4528–4533 (2009).

11b). J. Shi, P. Yang, et al *Dalton Transactions*, 21(07):938-945 (2008).

11b(i). S. L. Buchwald, et al. *Chemical Reviews*, 116(19):12564-12649(2016).

11b(ii). J. P. Wolfe, et al. *J Organic Chemistry*, 65(04):1144-1157 (2000).

11b(iii). J. P. Wolfe, et al. *J Organic Chemistry*, 65(04):1158-1174 (2000).

11b(iv). G. A. *Grasa, et al. J Organic Chemistry, 66(23):7729-7737 (2001).*

11b(v). *L. R. Moore, et al. J Organic Chemistry, 69(23):7919-27(2004).*

12). S. Enthaler, *ChemSusChem*, 3(9), 1024-1029(2010).

13). A. F. Littke, et al. *Angewandte Chemie Intl Edition*, 41(22):4176-4211(2002).

For further details see a separate chapter in this book on Haloanilines.

1.12.12: Aniline by Amination of Phenol:

Phenol is subjected to gas-phase amminolysis to obtain aniline with the Halcon/Scientific Design process at 200 bar and 425°C. $Al_2O_3 \cdot SiO_2$ (possible as zeolites) and oxide mixtures of Mg, B, Al, and Ti are used as catalysts. These can be combined with additional co-catalysts such as Ce, V, or W. The catalyst regeneration required previously is not necessary with the newly developed catalyst. With a large excess of NH_3, the selectivity to aniline is 87-90% at a phenol conversion of 98%. This process has been operated since 1970 by Mitsui Petrochemical in a plant which has since been expanded to 45000 tons per year. A second plant with a capacity of 90000 tons per year was started up by US Steel Corp. (now Aristech) in 1982. In 1977, Mitsui Petrochemical started production of *m*-toluidine, 2000 tons/year, by using the same technology (1). Initially, the process based on aminolysis of CB and phenol was not used for aniline production. Based on 1985 data, USA (570 tons) is the largest producer of aniline followed by West Germany (200 tons). The consumption of AN in the USA alone in 1985 was 350000 tons (1a).

Aniline is usually produced by catalytic hydrogenation of NB, but these processes involve certain limitations as cost, exothermic reaction, health hazards and environmental issues. Still some of the countries are likely to continue using these technologies only because of their set processes, availability of raw materials and marketing strategies. In the old days, alternatives to hydrogenation of NB for commercial production of aniline were sought and one of them was the direct amination of phenol, but the cost of phenol was very high and hence it is the prohibitory factor. In the early days, Halcon had tried to come up with a process for aniline manufacturing using phenol as the feedstock. Based on the results, Halcon had drawn some conclusions as: i) Aniline yields were quantitative at high conversions, ii) Aniline purity was very high and iii) the phenol to aniline based process is more economical than those based on NB hydrogenations. Mitsui Petrochemical Industries Limited was to set up a plant at Chiba, Japan with 44 million pounds capacity per year, because Sunoco Chemical had shut down its US plant (2). Phenol being an important chemical, in nineteenth century it was manufactured commercially by coal tar distillation. Since 1899, BASF started making phenol through sulfonation of benzene and continued till 1960. Thereafter, the main method of manufacturing phenol was by oxidation of cumene. Different processes for production of phenol are described in detail (3). Little later, a process for preparing aniline by the direct amination of phenol with ammonia in the vapor phase in the presence of a solid, heterogeneous (SiO_2-Al_2O_3) or Pd/Al_2O_3 catalyst is reported (4, 5).

The cumene process (Hock Process) is an industrial process for developing phenol and acetone from benzene and propylene (6). In 2003, nearly 7 million tons of phenol was produced by the cumene process. It is believed that manufacturing of aniline by amination of phenol, which in turn is made available through Hock process, is relatively more economical than other processes. It is a gas phase aminolysis at 400°C and 200bar. The process developed by Halcon is being used by Mitsui Petrochemicals, Japan and Aristech, USA, which utilizes Silica/Alumina with Tungusten or Vanadium as a catalyst. Phenols and cyclic alcohols have also been converted into amines by reaction with ammonia. Aniline has been synthesized from phenol, ammonia and hydrogen over Cu/ Al_2O_3/B_2O_3 and SiO_2/B_2O_3 (7). One of the basic processes, the amination of phenols using fixed bed reactors uses phenol and excess of ammonia (molar ratio 1:20) at 370°C, 1.7MPa pressure. Phenol is converted (99%) into aniline with 98-99% selectivity (Scheme-26).

OH → NH$_2$

20 x NH$_3$
Alumina - Silica
Fixed Bed Reactor
370°C / 1. 7 MPa

Conversion = 99%   Selectivity = 99% --Scheme-26

The process is simple, inexpensive catalyst with long life, good quality product, pollution less and hence green technology suitable for large-scale continuous productions (8, 9). Phenol and NH_3 heated over a heterogenous catalyst as fluoride alumina yields aniline with 99.7-99.8% selectivity with highest conversion rates (10). Amination of phenol with aqueous ammonia over Al_2O_3-SiO_2 catalyst offers AN. The single step conversion of benzene into phenol has facilitated this process for bulk production of phenol as feedstock (Scheme-27) (11). Also we have seen in the previous section that Pd-ligands are used in making phenol.

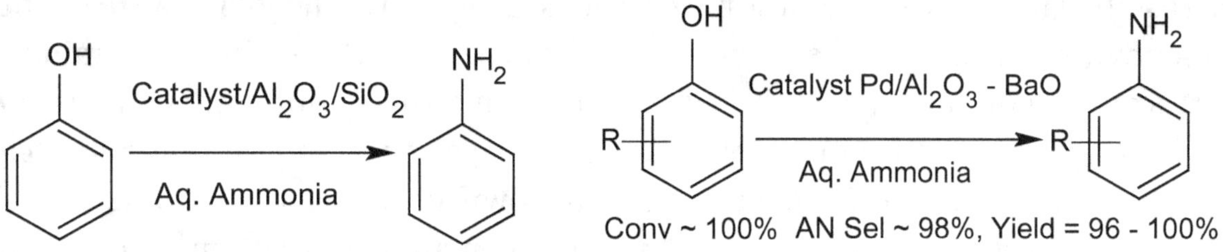

Scheme-27

Similarly, Pd/Al_2O_3-BaO is used as an efficient catalyst for the production of substituted ANs with excellent phenol conversions (up to100%), AN selectivity (>99%) and excellent yields (>97%) (12). Saha et al have reviewed the recent developments in aniline preparations and productions (13).

1.12.13: One-Pot Synthesis of Anilines from Phenol: A one pot, practical synthesis of ANs from phenols through Smiles rearrangement have been reported (Scheme-28) (14, 15, 16).

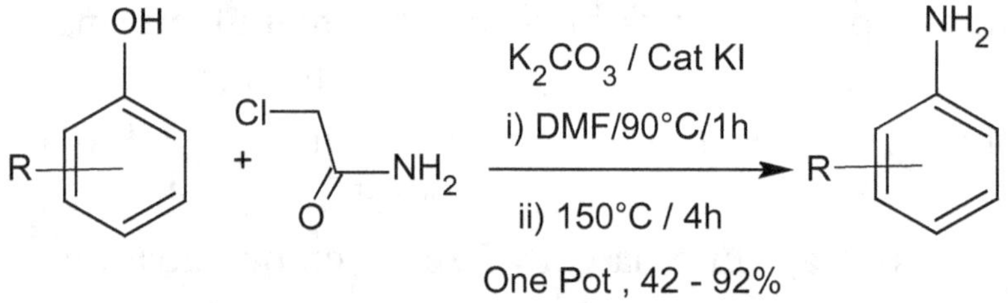

--Scheme-28

It uses $ClCH_2CONH_2/K_2CO_3/DMF$ system with catalytic KI, through Smiles rearrangement as the key step (17). A one-pot synthesis of ANs from phenols has been developed using an *ipso*-oxidative aromatic substitution (iSOAr) process. The products are obtained in good yields under mild and metal-free conditions (18).

References:

1). http://www.lookchem.com/Chempedia/Chemical-Technology/Organic-Chemical-Technology/7791.html.

1a). Industrial Aromatic Chemistry: Raw Materials - Processes - Products, Heinz-Gerhard Franck, Jürgen W. Stadelhofer, Springer Science & Business Media, 06-Dec-2012 - Science - 486 pages, PP 199.

2). Report in *Chemical & Engineering News*, 47(14):44 (1969). http://pubs.acs.org/doi/abs/10.1021/cen-v047n014.p044.

3a). Competition Science Vision (Monthly Magazine), 09(105):136 Pages (2006), Published by Pratiyogita Darpan Group India. 3b). http://www.greener-industry.org.uk/pages/phenol/4PhenolProdMethSum.htm

4). J. L. Russel, US 3578714 (1971). 5a). Y. Ono *J. Catalysis,* 72(01):121-128(1981). 5b). M. Yasuhara, et al. US 4987260 A(1991). 6). https://en.wikipedia.org/wiki/Cumene_process.

7). http://shodhganga.inflibnet.ac.in/bitstream/10603/11822/6/06_chapter%201.pdf

8). M. Becker, et al. EP 0127395 A2(1984). D. Clarence, et al. US 4380669 A(1983).

9). R. S. Downing. et al. *Catalysis Today,* 37(02)**:**121-136(1997).

10). L. A. Cullo, WO1991001293 A1(1991). US US5091579 A(1992). US5214210 A, (1993). https://www.icis.com/resources/news/2004/02/27/560820/aniline-the-builder

11). B. Macdonald, in *ECN Chemscope,* 38-39 (1997).

12). J. Ma et al. *Bull. Korean Chemical Society,* 33(02):387-392(2012).

13). B. Saha, et al., Recent Advancements of Replacing Existing Aniline Production Process With Environmentally Friendly One-Pot Process: An Overview: *Critical Reviews in Environmental Science & Technology,* 43(01):84-120(2013).

14). Y. S. Xie et al. *Tetrahedron Letters* (2013).

15). V. R. Arava et al. *Der Pharma Chemica,* 5(6):12-27 (2013).

16). M. Mizuno, M. Yamano, *Organic Letters,* 7(17):3629-3631(2005).

17). http://www.tcichemicals.com/en/us/support-download/tcimail/application/143-15a.html

18). C. P. Frazier, et al. *Organic & Biomolecular Chemistry,* 14:5520-5524 (2016).

1.12.13a: PHENOL: Recent Developments;

Phenols are one of the most important organic compounds with various applications in different industries, including pharmaceuticals, nylon, plastics, bisphenol-A, etc. Phenol is also used as a raw material for the production of AN, both in the laboratories and at commercial scales. Considering its importance, different options of manufacturing phenol at industrial scale have been explored. Most recently, Dakka et al (US 9260387 B2 (2016) have given a retrospective view of old and new processes of manufacturing phenol at commercial scale.

The total annual worldwide production of phenol and acetone through a cumene feed-based process, in 2009 as its jubilee year, has reached about 10.5 million tons and about 6.5 million tons per year respectively. In 2015, the world demand for phenol was >10 million tons and the global phenol supply were expected to exceed 10.7 million tons in 2016. The global phenol foreign trade exceeded USD 3.6 billion in 2012. Europe was the leading phenol exporter and importer (19). Since its invention, Udris (1942) and Hock (1944), manufacturing of phenol by DSM (1960), Solutia (1990) to most recent efforts by Exxon Mobil have been covered by Plotkin (20b). Direct oxidation of benzene with nitrous oxide (N_2O) at 570-720K over a zeolite catalyst produces phenol in 99% selectivity with 95% benzene conversion. Considering low benzene conversions in other methods, as seen above, this is one of the best methods both in terms of benzene conversion and aniline selectivity (Scheme-29).

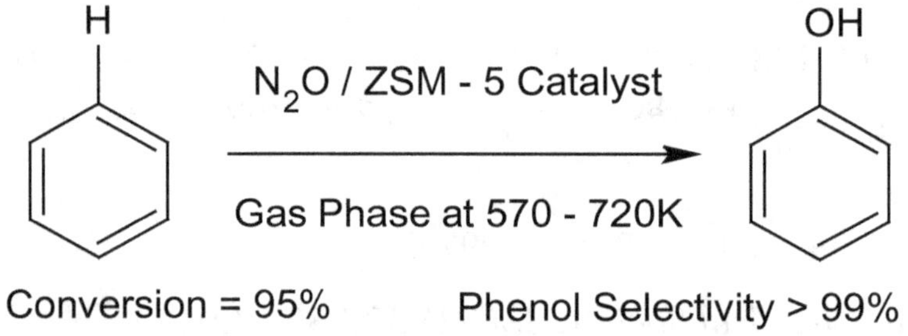

H

N_2O / ZSM - 5 Catalyst

Gas Phase at 570 - 720K

OH

Conversion = 95%   Phenol Selectivity > 99% --Scheme-29

Other H-ZSM-5 or Fe-ZSM-5 give phenol in excellent yields (>99-100%) (20, 21, 22, 23-). Asia was poised to raise its phenol capacity to 6.37 million *tons* by mid-*2016*.

References:

19a). V. Zakoshansky (vlazak@illallc.com), Phenol Process Celebrates Its 60th Anniversary: The Role of Chemical Principles in Technological Breakthroughs, Russian Journal of General Chemistry, 79(10), PP10 (2009). 19b). https://mcgroup.co.uk/news/20140131/global-phenol-supply-exceed-107-mln-tonnes.html

20a). J. M. Dakka, et al. Exxon Mobil Process for Producing Phenol: US 9260387, (Feb. 16, 2016). 20b). https://www.acs.org/content/acs/en/pressroom/cutting-edge-chemistry/what-s-new-in-phenol-production-.html (March 21, 2016)

21). Direct hydroxylation of benzene to phenol by nitrous oxide. a). A. K. Uriarte, M. A. Rodkin et al. *Studies in Surface Science and Catalysis:* Vol. 110, 857-864(1997), b). G.I. Panov· G.A. Sheveleva, et al. *Applied Catalysis A: General,* 82(1), 31-36 (1992). c). L. V. Pirutko et al. *Applied Catalysis A: General* 227, 143–157(2002).

22). S. Niwa et al. *Science,* 295(5552), 105-107 (2002).

23a). https://www.icis.com/resources/news/2000/09/25/122615/oxidation-of-benzene-to-phenol-maintaining-a-high-product-ratio/. 23b). Manufacturing of Aniline from phenol and ammonia: http://chempedia.info/info/151260/.

1.12.13b: Phenols and Anilines from Aryl Halides: Phenol is an ideal raw material for AN as it constitutes one step process, but its high price is a concern using it as a feedstock.

--Scheme-30

Hence, it is necessary to develop economical processes for making phenol at commercial scale. CuI-NPs catalyzed selective synthesis of phenols, ANs and thiophenols from aryl halides is developed in the absence of ligands and organic solvents (Scheme-30) (24, 25, 26). Arpe et al have covered all the different aniline manufacturing processes by NB hydrogenation, aminolysis of CB, phenol and direct amination of benzene (27, 28).

References:

24a). A. Tlili, N. Xia et al. *Angewandte Chemie Intl Ed.* 48(46):8725-8728(2009).

24b). S. Xia, L. Gan, et al. *J. Am. Chemical Society*, 138(41):13493-13496 (2016).

25). P. S. Fier & K. M. Maloney, *Organic Letters*, 18 (09):2244-2247(2016).

26). H.J. Hu et al. *J. Organic Chemistry*, 76(07):2296-2300 (2011).

27a). *Industrial Organic Chemistry*, K. Weissermel & H. Arpe, John Wiley & Sons, 2003 - Science - 491 pages, PP 377-379. 27b).*Ullmann's Encyclopedia of Industrial Chemistry*, VCH Verlagsgesellschaft mbH, Weinheim, A19, 299-311 (1991).

28a). http://www.greener-industry.org.uk/pages/phenol/4PhenolProdMethSum.htm

28b). https://www.acs.org/content/acs/en/pressroom/cutting-edge-chemistry/what-s-new-in-phenol-production-.html. 5c). https://en.wikipedia.org/wiki/Cumene_process

1.12.14: Aniline by Amination of Cyclohexanol:

Direct amination of cyclohexanol yields aniline with water as the only side product making it environmentally cleaner process (Scheme-31).

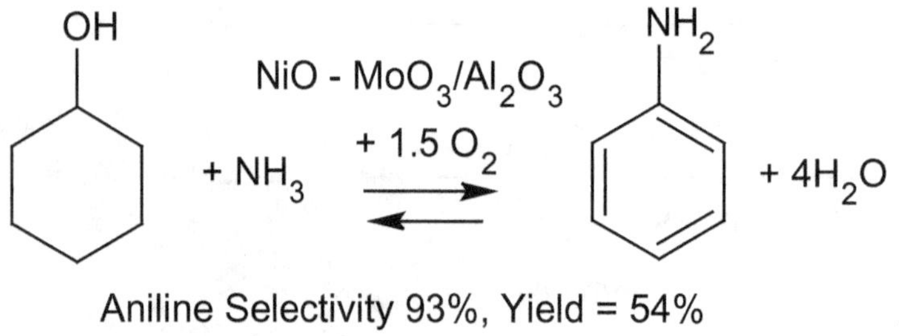

Aniline Selectivity 93%, Yield = 54%　　　--Scheme-31

Aniline is obtained in 90-95% by the amination cyclohexanone or cyclohexanone-cyclohexanol mixture over Ni, Pt metals over Al_2O_3 or Al_2O_3/SiO_2 supports (29). The reaction is catalyzed by $NiO/MoO_3/Al_2O_3$ at 698K and atmospheric pressure (1.013×105 Pa). The Aaniline yield is 54% and the selectivity was 93% (Scheme-32).

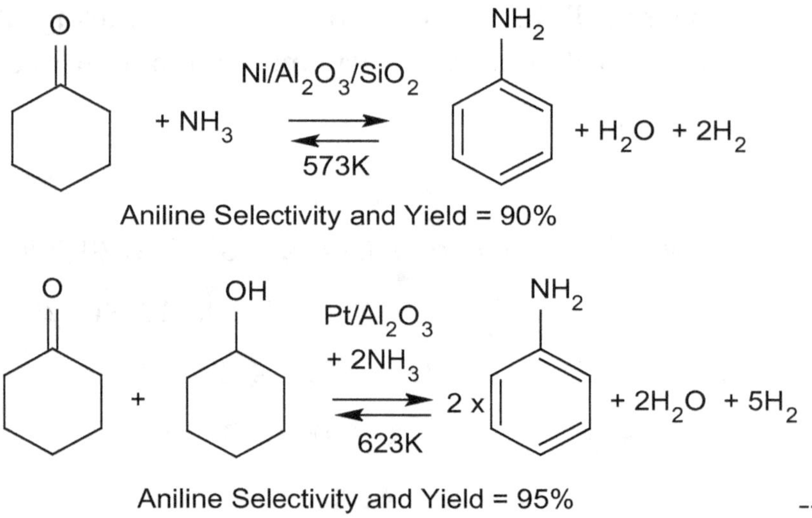

Aniline Selectivity and Yield = 90%

Aniline Selectivity and Yield = 95%　　　--Scheme-32

Becker at al also have obtained AN in 96% selectivity using Cu, NI or group VIII metals (as Pd, Pt, Ru, Rh, etc.) in a gas phase amination of cyclohexanol and a 9 : 1 weight ratio mixture of cyclohexanol/cyclohexanone at 250°C (30).

29). C. Tong et al. *Chinese Science Bulletin*, 47(23):1937-1939(2002).

30). J. Becker, et al. *Applied Catalysis A: General*, 197(02):229-238(2000).

1.12.15: Direct Amination of Benzene (DAB):

Until around 1970 the aminolysis of CB and phenol with ammonia and CuCl, SiO_2-Al_2O_3 catalysts respectively were the commercial routes for AN manufacturing by several companies as Halcon, Mitsui Petrochemicals, Sunoco Chemicals etc. However, phenol cost is a prohibitory factor by this route. As an alternative, benzene is selectively and directly aminated to AN in the presence of ammonia using a Group VIII element as the catalyst supported on a carrier at elevated temperatures and high pressures (1). Novel cataloreactant systems consisting primarily of noble metals as Rh, Ir, Pd, and Ru and a reducible metal oxide, especially NiO as an oxidant, and ZrO_2 or K-TiO_2 as carrier have been discovered. An optimized cataloreactant, Rh/Ni-Mn/K-TiO_2 achieved a stable 10% benzene conversion and >95% selectivity to aniline at 300°C and 300 bar for 2 h (2a, b). Du Pont has developed an interesting new manufacturing process for AN. Benzene and ammonia can be heated over a NiO/Ni catalyst containing promoters including ZrO_2 at 350°C and 300 bar to give a 97% selectivity to AN with a 13%benzene conversion. The process has limitations of catalyst regeneration and low benzene conversions (2c).

1.12.15a: DAB with Ammonia and Metal Catalysts:

Modern reaction routes for the production of aniline are focused on a DAB route. Usually the generated hydrogen is "consumed" by oxidation to water (Scheme-33).

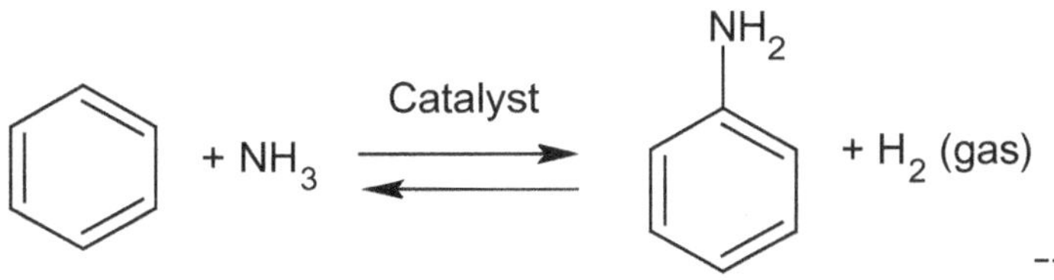

--Scheme-33

This process has certain limitations as it generates i) hydrogen gas which is inflammatory and ii) on longer residence time it shifts the reaction back to benzene. Many other metal catalysts as nickel, iron etc. have been used. Here, the corrosion and the high temperatures as 500-1000K were major limiting factors. Therefore, potential industrial applications are limited due to higher costs in comparison with the classical NB reduction route. Better results can be obtained using metal oxides of iron, nickel, copper, gold, silver etc. DuPont's NiO/ZrO_2 cataloreactant was used in a microreactor flow set-up equipped with a calibrated online mass spectrometer at 590 K produces aniline as the sole product (3, 4). With these metals and metal oxides the aniline selectivity is >90%, but the benzene conversions are about 5% (5). Membrane-Electrode Assembly (MEA) is also used for DAB to aniline with 60% benzene conversion rates (6). Zhai et al have reported a highly efficient amination of benzene to aniline mediated by Br_2 with metal oxide as cataloreactant (7). Amination of benzene by aq. NH_3 over Pt-loaded TiO_2 photocatalyst gives aniline with high selectivity (8).

1.12.15b: Direct Amination of Benzene with Ammonia and H_2O_2:

The existing aniline production processes tend to be energy and time consuming and pollutant producing as well. The "greening" of global chemical manufacturing by minimizing energy consumption and waste production has become a major concern for chemical research and chemical industry. DAB with aq. NH_3 in the presence of hydrogen peroxide is done at 50°C and atmospheric pressure with moderate yields, but higher selectivity for AN than that of phenol. The reaction is catalyzed by metal/metal oxide catalysts such as Ni-Zr-Ce/Al$_2$O$_3$, V-Ni/Al$_2$O$_3$, Mo-Ni/Al$_2$O$_3$ and Mn-Ni/Al$_2$O$_3$ etc. With Ni-Zr-Ce/Al$_2$O$_3$ this process indeed provides a new greener chemical route for the production of aniline with atom efficiency. Some other noble metals and other co-catalysts are also found more suitable, especially with very high (>95%) selectivity to AN (9). Au-NPs on alumina (Au/Al$_2$O$_3$) are found highly efficient catalyst for a single-step amination of benzene to aniline in high yields with aq. NH$_3$ and H$_2$O$_2$ at 50°C for 2hrs (10).

1.12.15c: Direct Amination of Benzene with Hydroxylamine.HCl:

Amination of benzene to aniline that uses ammonia gas as aminating agent, and molecular oxygen or lattice oxygen as oxidant generally requires high temperature and high pressure. However, using hydrogen peroxide as oxidant and aqueous ammonia as aminating agent, or direct amination with hydroxylamine the amination can be carried out under mild conditions. Recent advancements of replacing existing aniline production process with environmental friendly one-pot oxidative amination of benzene to aniline are reviewed in detail. It focuses on oxidants such as molecular oxygen, lattice oxygen and hydrogen peroxide, and aminating agents such as ammonia gas, hydroxylamine or aqueous ammonia. The future study on one-step direct amination of benzene to aniline is also discussed (11). Progress of one-step oxidative amination of benzene to aniline is available at: https://www.researchgate.net/publication/286612190. Amongst other catalysts, sodium metavanidate (NaVO$_3$) and hydroxyl amine gave the satisfactory aniline yield and turnover (64mol%, 48mol AN per mole of V) with a selectivity of 95.6-100%% to AN (Scheme-34) (12).

Mn-MCM-41 mesoporous materials showed significant catalytic activity for single-step amination of benzene in acetic acid–water medium under mild reaction conditions using hydroxylamine as aminating agent. Mn-MCM-41 (Si/Mn = 20) showed the highest benzene conversion (68.5%) and 100% selectivity to AN (13).

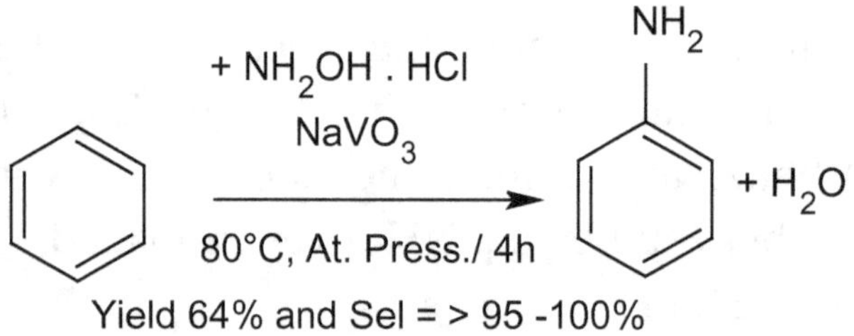

$$\text{+ NH}_2\text{OH . HCl} \xrightarrow[\text{80°C, At. Press./ 4h}]{\text{NaVO}_3} \text{C}_6\text{H}_5\text{NH}_2 \text{ + H}_2\text{O}$$

Yield 64% and Sel = > 95 -100%   --Scheme-34

While scrutinizing the activities of different metal catalysts supported on solid supports, hydroxylamine with Cu-diamines and Cu-triamines–MCM-41 were found most effective as they gave 80% and 86% yields and 100% aniline selectivity respectively (14). A highly efficient, reusable catalyst comprising Cu(II) nanoclusters supported on CuCr$_2$O$_4$ spinel NPs for the oxyamination of benzene to AN (H$_2$O$_2$ + NH$_3$) under mild aqueous reaction conditions is reported (15). Hassan has published a system and process for the production of AN (16). Recent development of one-step production of aniline is reviewed by many groups (17).

References:

1). J. Becker et al. *Catalysis Letters,* 54(03):125-128(1998).

2a). H. Hagemeyer, et al. *Applied Catalysis A: General,* 227(01-02):43-61 (2002).

2b). T. Desrosiers, et al. *Catalysis Today,* 81(03):319-328 (2003).

2c). German Patent: 2,114,255 (1971).

3a). J. Becker, et al. *Catalysis Letters:* 54:125(1998).

3b). M. Kralik et al. Aniline - Catalysis and Chemical Engineering: https://www. researchgate.net/publication/263010315_Aniline_Catalysis_and_Chemical_ Engineering

4a). N. Hoffmann, et al. *Catalysis letters,* 103(01-02):155-159 (2005). 4b). N. Hoffmann, et al. *ChemSusChem,* 01(05):393-396(2008).

5). D. Poojary et al. WO 2000069804 A9 (2001), and US6933409 B1 (2005).

6). A. Mendes et al. US20160032469 A1 (2016).

7). H. Zhai et al. *Chemistry Letters,* 35:1358-1359 (2006).

8a). H. Yuzawa, et al. *J. Physical Chemistry C,* 117(21):11047-11058(2013), 8b). H. Yuzawa, et al. *Chemical Communications.* 46:8854-8856(2010). 8c). http://docslide. net/documents/direct-aromatic-ring-amination-by-aqueous-ammonia-with-a-platinum-loaded-titanium.html

9a). C. Hu, et al. *Industrial Engineering & Chemical Res.* 46(10):3443-3445 (2007). 9b). Y. Xia, et al. *J. Acta Phys Chim Sin,* 21(12):1337-1342(2005). 9c). T. Yu, et al. *Catalysis Science & Technology:* 3159-3167(2014). 9d). C. H. Yang, et al. *Advanced Materials Research,* 550-553:2607-2611(2012).

10). G. Chen, et al. *Advanced Materials Research,* 661:47-52(2013).

11). C. Yang; G. Chen, *J. Chemical Society of Pakistan,* 38(02):282-286 (2016).

12). L. Zhu et al. *J. Catalysis,* 245(02):446-455(2007).

13a). K. M. Parida et al. *Applied Catalysis-A, General,* 351(01):59-67(2008). 13b). K.M. Parida, et al. *Applied Catalysis A: General,* 351(01):59-67(2008). 13c). K. M. Parida, et al. *J. Molecular Catalysis A: Chemical,* 318(01-02), 85-93(2010).

14). R. T. Driessen, et al. *Chemical & Engg Technology,* 40(05):838–846(2017).

15a). S. S. Acharyya, et al. *Chemical Communications:* 50:13311-13314(2014). 15b). P. Kubanek, et al. US: 8642810-B2 (2014).

16). Hassan, et al. US 8153076 B2 (2012).

17a). H. Yuzawa, et al. *J. Physical Chemistry C,* 117(21):11047-11058 (2013).

17b). B. Saha, et al. *Critical Reviews in Environ Science and Technology*, 43(01):84-120(2013). 17c). N. A. Romero, et al. *Science*, 349(6254):1326-1330 (2015).

17d). A. Corma, Huntsman Intl., EP2016/053743(2016), WO2016155948 A1(2016).

17e). W. Zhou and N. Jiao, Chapter 2: Nitrogenation Strategy for the Synthesis of Amines: "Nitrogenation Strategy for the Synthesis of N-Containing Compounds", Springer, Singapore, Pages 09-27(2017).

1.13.00: New Mechanism for Nitrobenzene Reduction:

In the olden days, phenol and aniline were prepared/manufactured by nucleophilic substitution of labile groups on the ANCs. Mechanism of SP^2 carbon nucleophilic substitution and reactivity in ANCs such as phenol, CB is reviewed (1, 2). Although, Haber's NB reduction mechanism has been looked at in detail for a long time, the curiosity of its understandings through different angles based on new techniques is a subject of continuing interest (3). As we know, the reduction of NB to AN- is a well-known reaction and was first performed by Zinin (1842) using inorganic sulfide as a reductant (Zinin reaction), while Bechamp (1854) used iron and hydrochloric acid. Nowadays, the reaction is conducted using stannous chloride (SnCl$_2$), where the reduction has been presumed to proceed with two intermediates, nitrosobenzene (NOB) and phenyl hydroxylamine (PHA) (Scheme-35). Yamabe et al have extensively reviewed the different methodologies and reaction pathways covering a long period of time (4).

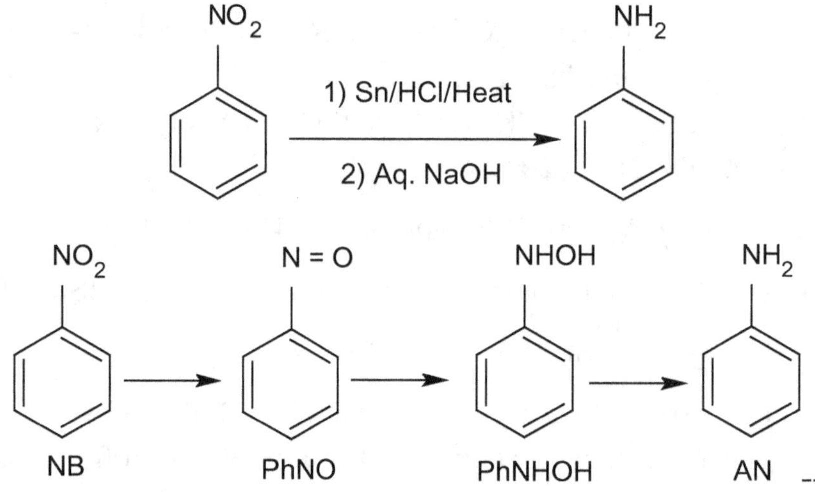

--Scheme-35

Depending upon the reaction solvents, conditions and pH etc different products are formed during NB reductions. In the non-aqueous solvents as DMF and ACN, the NB radical anion is very stable and the reduction to PHA occurs in two steps: i) a single electron addition to form the radical anion, ii) it is followed by a three-electron addition and protonation to form PHA. Incremental addition of a proton donor, such as benzoic acid causes an increase in the first reduction wave at the expense of the second wave until at a ratio of acid to NB of 4:1 is reached. A single four electron reduction wave is observed equivalent to that obtained in aqueous solutions (5). Zuman et al have observed that the sequence of protons and electrons in the four-electron reduction step was proved to be: H^+, e, H^+, e, 2e, $2H^+$. PHA formed in the four-electron reduction is protonated and further reduced to AN. They also had observed some other chemical reactions among the intermediates during hydrogenation of NB (6).

Zilberberg et al. have used the associative and dissociative adsorption of NB as the basis for the mechanistic pathway in NB to AN reduction over the surface of the metallic iron. Based on the quantum chemical analysis, it is suggested that the direct electron donation from the metal surface into the π^* orbital of NB is a decisive factor responsible for subsequent transformation of the nitro group. The detailed molecular mechanism of the NACs reduction by Fe(0) is not yet fully understood. Formally, the overall nitro-to-amino reduction is a six-electron-transfer reduction process.

$$Ar\text{-}NO_2 + 6e\text{-} + 6H^+ \rightarrow Ar\text{-}NH_2 + 2H_2O$$

However, the actual reduction of the nitro to amino group by iron comprises a variety of pathways including coordination by iron centers, transformation of nitro group, condensation of NACs, etc. The first 2e- transfer step in the nitro reduction reaction is proposed to be the formation of nitroso group.

$$Ar\text{-}NO_2 + 2e\text{-} + 4H+ \rightarrow Ar\text{-}N{=}O + 2H_2O$$

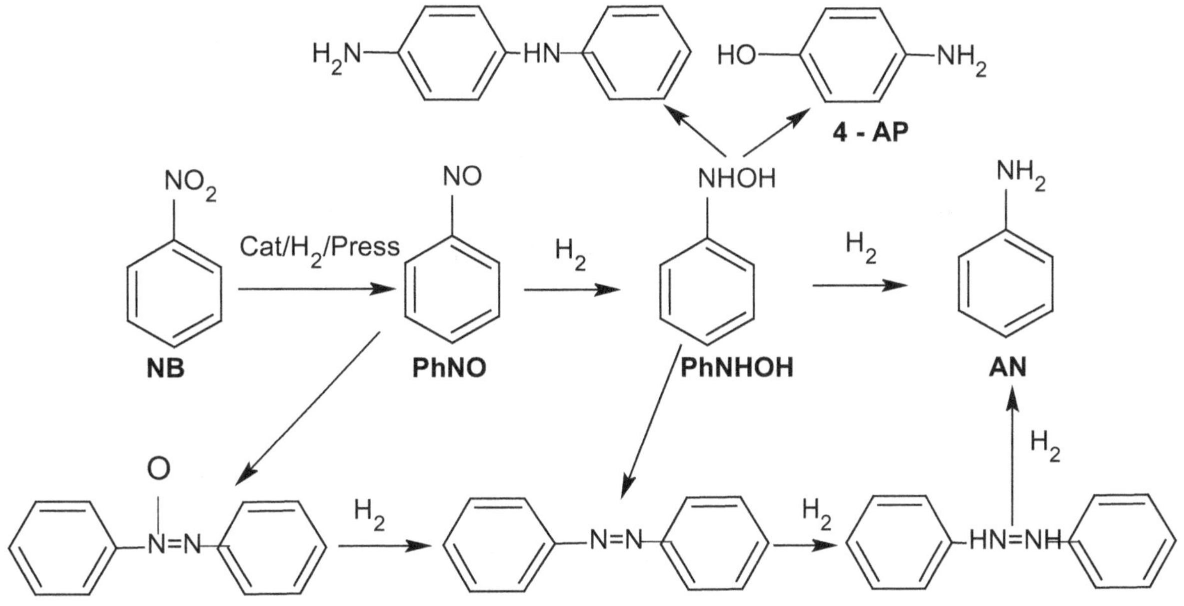

Scheme-36A: General Reaction Pathways in Liquid Phase Hydrogenation of NB

This step is thought to be crucial as it describes the initial reaction mechanism as well as the generation of products for subsequent reactions. This work highlights the actual steps involved at electronic level during the reduction steps (Scheme-36A, B).

Croma (8) and Mokasch (9) have independently proposed different pathways for NB hydrogenations. Mokasch et al have observed that the hydrogenation of NB over Au/MeOx catalysts is dependent on the matter of the support. By varying the support, the hydrogenation of NB occurs either through a direct route in the case of Au/TiO$_2$catalysts or through a condensation route over Au/CeO$_2$ catalysts. Adsorption on the surface of the catalysts is a possible key factor in determining the route. Possible reactions pathways during aniline synthesis are shown in Scheme-36B.

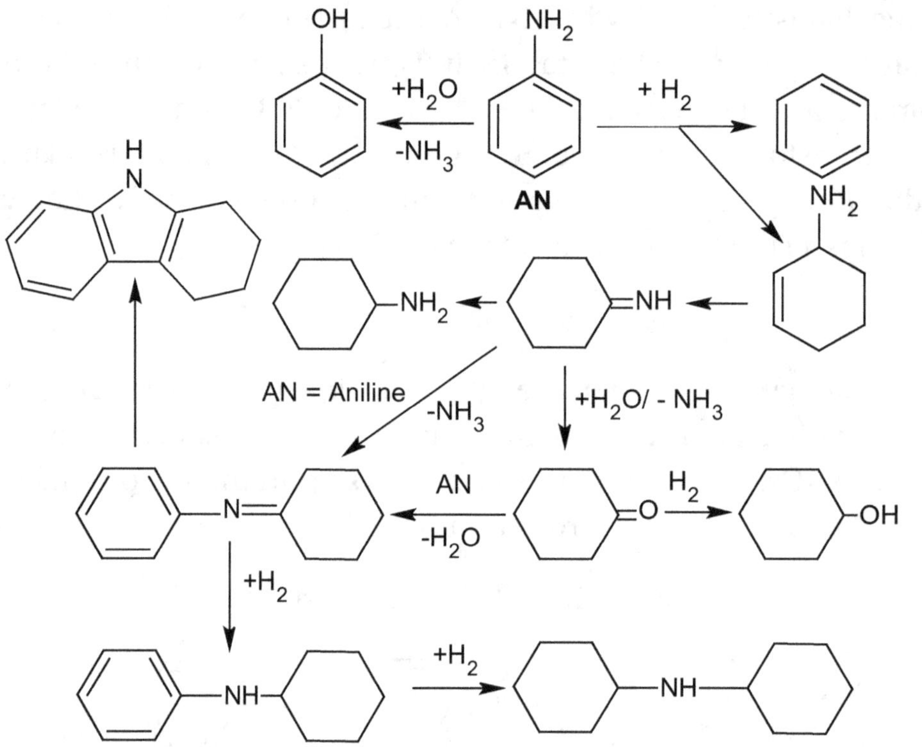

Scheme-36B: General Reaction Pathways in Liquid Phase Hydrogenation of NB

Recently, the Haber mechanism (Scheme-37) in hydrogenating NB to AN- is shown to be incorrect and a new mechanism is proposed by Gelder et al.

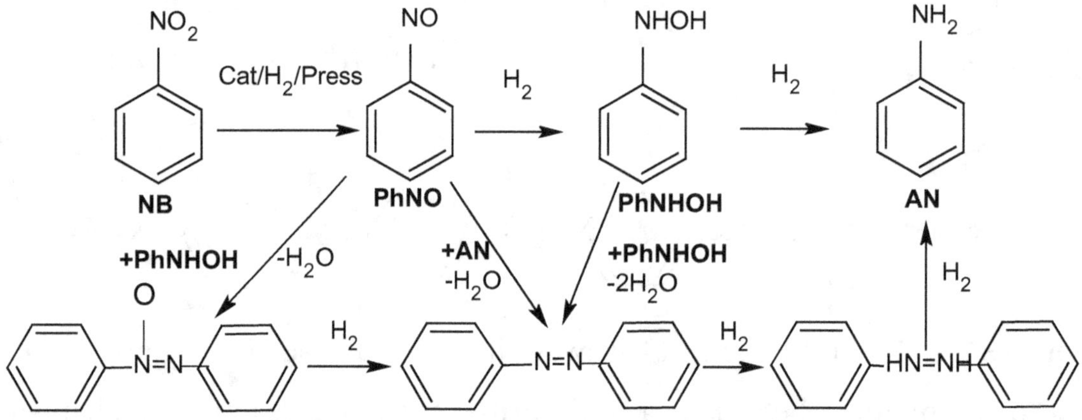

Scheme-37: Haber Reaction Mechanism

Gelder et al claimed that the lower rate of hydrogenation to AN from NOB and the different kinetic isotope effects make it clear that nitrosobenzene (NOB) cannot be an intermediate in the hydrogenation of NB. Therefore the Haber mechanism is wrong and must be updated (10).

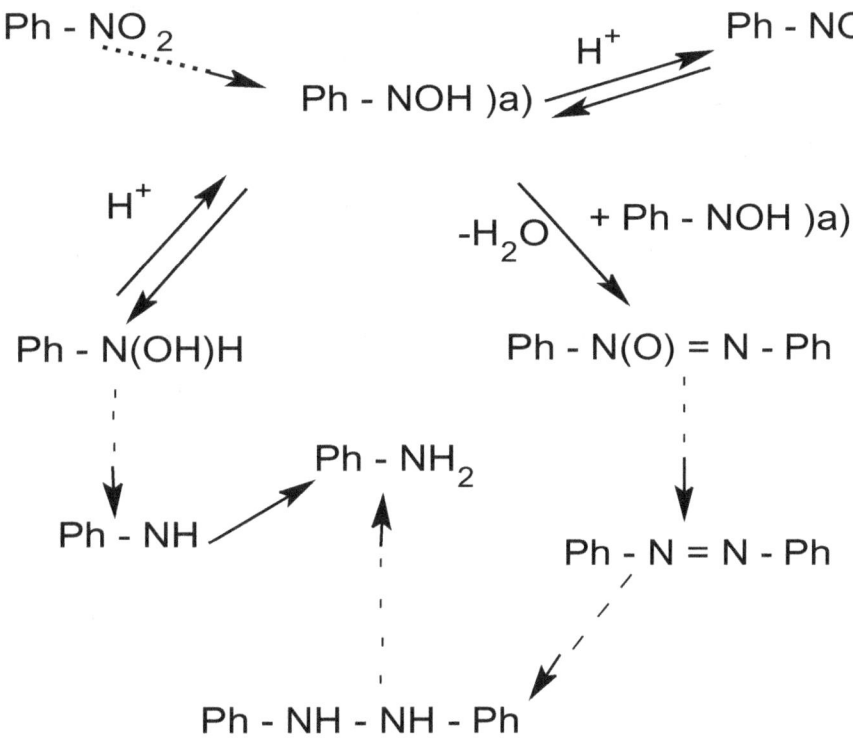

Scheme-38: Gelder et al. Proposed New Reaction Mechanism

A detailed analysis of the surface reaction mechanisms for NB and NOB has led Gelder et al. to propose a new mechanism (Scheme-38). We are unable due to the space limitations to lay out the full detailed mechanism for both the compounds hence dotted arrows indicate multiple steps. The analysis suggests that there is a common surface intermediate, namely Ph-N(OH)(a). In the NB mechanism, Ph-N(OH)(a) reacts with adsorbed hydrogen whereas in the NOB mechanism, Ph-N(OH)(a) reacts with itself to eliminate water and produce azoxybenzene. In NB hydrogenation each step involves the addition of hydrogen hence the kinetic isotope effect is seen. However, in the NOB mechanism there is a coupling reaction and a hydrogenolysis reaction either of which could lead to an inverse kinetic isotope effect. The different processes occurring in the two mechanisms would be consistent with the different rates observed and the fact that NOB has a slower rate of AN formation. The NB and NOB hydrogenation reactions were repeated with a pre-activated Raney-Nickel catalyst (Aldrich) and similar reaction profiles were obtained. When the hydrogen pressure was reduced to 1 bar, trace levels of NOB were detected in the reaction mixture. This is predictable from Scheme-2, as under limited hydrogen supply it is possible for Ph-N(OH)(a) to dehydrogenate to give NOB as a byproduct of NB hydrogenation.

This new understanding of the mechanism has implications for both catalyst and the reactor design. To obtain a high activity and selectivity, it is essential that the hydrogen flux at the surface is maintained with good access to the reaction site and no diffusion limitations. The presence of typical by-products such as azobenzene and azoxybenzene may indicate that there are local regions on the catalyst surface where the hydrogen flux is insufficient to inhibit Ph–N(OH)(a) coupling, even though the system overall may not be in a diffusion controlled regime (10, 11, 12).

The new insights into the reaction pathways lead to more knowledge about the mechanism of that particular reaction. Some other groups, after Gelder in 2005, have proposed new mechanisms for catalytic hydrogenations of NB to AN and they are as briefly discussed below. Kralik supports the Gelder pathways of mechanism in NB to AN hydrogenation (14). Recent studies in this regard, after Gelder's proposed NB reduction mechanism, Sheng et al have got insights into the mechanism and kinetics of NB reduction to AN using the Pt(111) model catalyst. The density functional theory (DFT) with the inclusion of Vander Waals interactions have been used for fundamentally understanding the mechanisms at atomic and molecular levels. It was found that the double H-induced dissociation of N-O bond was the preferential path for the activation of nitro group, having a much lower reaction barrier than that of the direct dissociation and single H-induced dissociation paths. The overall mechanisms have been identified as shown in Scheme-39.

$$C6H5NO2^* \rightarrow C6H5NOOH^* \rightarrow C6H5N (OH)2^* \rightarrow C6H5NOH^*$$

$$\rightarrow C6H5NHOH^* \rightarrow C6H5NH^* \rightarrow C6H5NH2^*$$

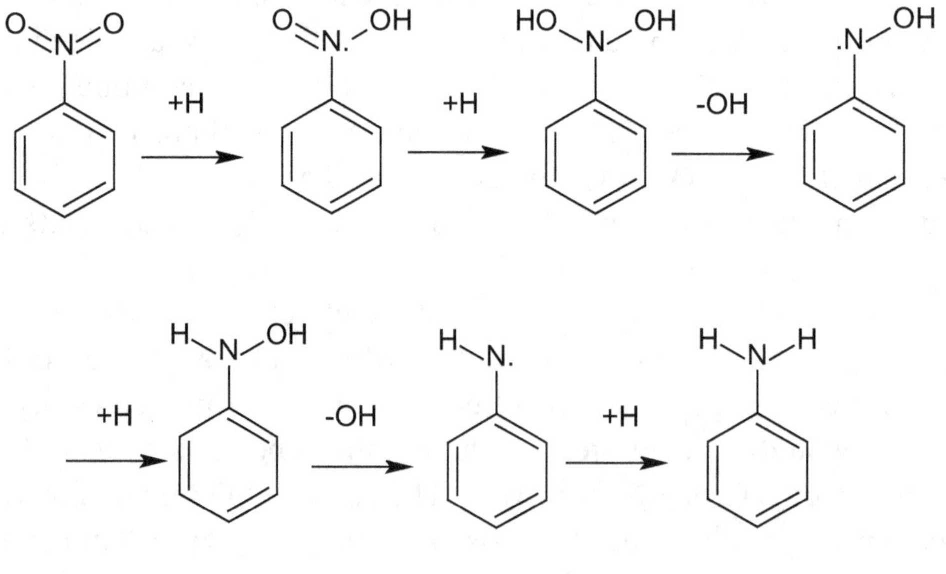

Scheme-39: Illustration of overall mechanism of NB reduction to AN on Pt (1 1 1).

The authors have realized six transition states involving five intermediates between NB and AN during the reduction process (15).

Direct *vs.* indirect pathway for NB reduction reaction on a Ni catalyst surface as a density functional study is reported (16). The DFT calculations are performed to understand and address the previous experimental results that showed the reduction of NB to AN prefers direct over indirect reaction pathways irrespective of the catalyst surface. NB to AN conversion occurs *via* the PHA intermediate (direct pathway) or *via* the azoxybenzene intermediate (indirect pathway). The study indicates that the parallel adsorption behavior of the molecules over a catalyst surface is preferable over vertical adsorption behavior. Based on the reaction energies and activation barrier of the various elementary steps involved in direct or indirect reaction pathways, the authors have found that the direct reduction pathway of NB over the Ni(111) catalyst surface is more favorable than the indirect reaction pathway (Scheme-40).

Direct Reduction Pathway

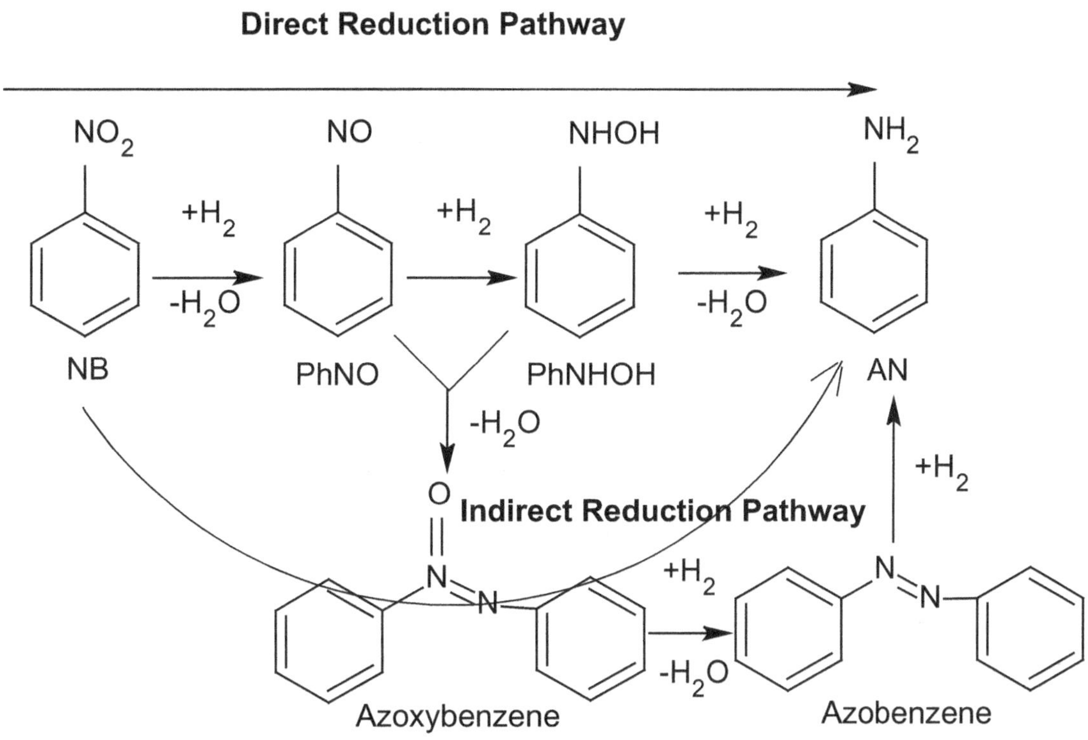

Scheme-40: Direct and Indirect pathways in Reduction of NB to AN

After Gelder's new mechanism (2005), a new insight into the hydrogenation mechanism of NB to AN on Pd_3/Pt (111) is investigated using a density functional theory (DFT) study. Four different pathways are postulated. The hydrogenation mechanism of NB on Pd_3/Pt (111) bimetallic surface preferentially follows the direct hydrogenation route and fits the Jackson reaction mechanism. DFT calculations are also used to investigate the adsorption and hydrogenation mechanism of NB to AN on Pd_3/Pt(111) bimetallic surface. It gives insights into the hydrogenation mechanism of NB to AN reduction and supports the Tada theory of direct reduction pathway (17). Ag clusters (0.24wt% and 1.5 nm) photodeposited on TiO_2 particles in a highly dispersed state, dramatically enhanced both the activity (84% conversion after 1 h irradiation) for the reduction of NB to AN with 100% selectivity in CH_3OH as a solvent (18).

We have seen earlier that Zilberberg et al have proposed a six electron/six proton mechanism in NB to AN reduction. Here, the mechanism of the carbon catalyzed reduction of NB by hydrazine as a source of hydrogen is studied. Hydrazine is a two-electron reducing agent as shown by trapping the diimide intermediate with norbornene. The NB reduction is a four-electron process proceeding first to PHA which was observed in low concentrations in the reacting system by NMR. PHA is further reduced to aniline in a second step (Scheme-41).

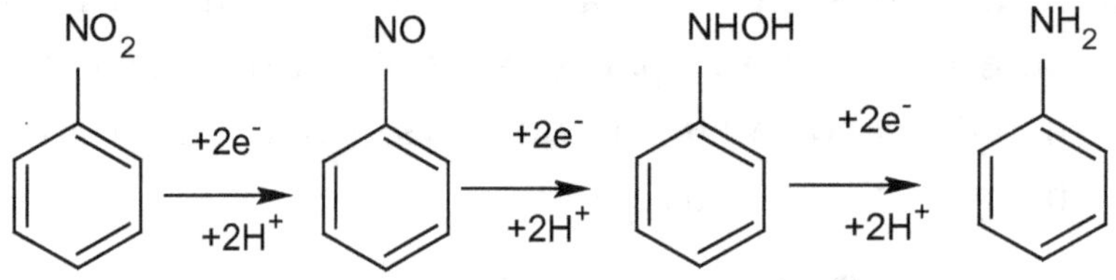

Scheme-41: NB Reduction Using Hydrazine/Carbon

The two-electron intermediate, NOB, gave different products than NB under the set conditions and could not be trapped, supporting the expected four-electron pathway.

By serving as an adsorbent and collecting hydrazine on their surfaces, carbons make it possible to execute a four-electron reduction using a two-electron reducing agent. It was unambiguously concluded that the hydrazine reaction at the carbon surface is the rate determining step (19).

The electroreduction of p-nitrothiophenol (PNTP) on gold and silver electrodes has been investigated using density functional theory (DFT). Zhao et al have recently demonstrated that the NAs undergo electrolytic hydrogenations involving six electrons, and six protons (20). Jensen et al also have proposed a six electron/six proton absorption mechanism during NB reduction over CdS quantum dots under visible light photo-irradiations (Scheme-42) (21).

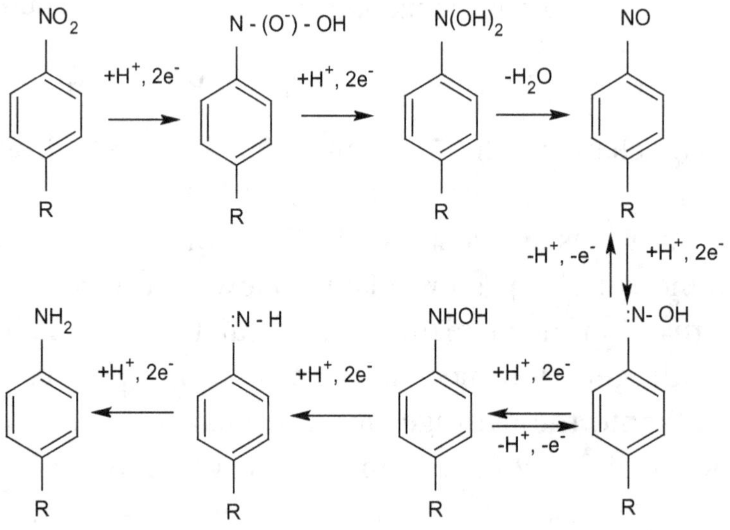

Scheme-42: Basic steps in Electroreduction of NACs to Anilines

The scientific world is like an ocean and you are trying to get few drops from it. I get the same feelings while writing books because so much information is available in the public domain about any subject one would like to delve into. Whatever we have covered so far about aniline and its analogs, what we believe, is fairly good but certainly not everything under the sun is covered. We have tried to put additional information, obviously relevant to aniline and its analogs only, but under different headings as Haloanilines, Aminophenols, Homogenous and Heterogenous Catalytic Hydrogenations of NBs and other NAs. Also the reader may find some of the information under a particular heading which also would be there under another heading because they are relevant both the places. For example, water serves as a hydrogen source and also as a solvent in the selective hydrogenations of HNBs, NAs etc. as i) hydrogenation with metal catalysts, ii) as well as using nanocatalysts or nanoparticles on solid supports under a sub-title "Catalysts on Solid Supports".

References:

1). J. F. Bunnett, *Q. Reviews Chemical Society,* 12:01-16(1958).

2). C. Moreau, et al. *J. Mole. Catalysis A*: Chemical, 161(01-02):141-147 (2000).

3). J. W. Larsen et al. *Carbon,* 38 (2000) 655–661(2000).

4). S. Yamabe & S. Yamzaki, *J. Physical Organic Chemistry,* 29:361-367(2016).

5). W. H. Smith & A. J. Bard, *J. Am. Chemical Society,* 97(18):5203-5210 (1975).

6). P. Zuman et al. *Electroanalysis J,,* 04(08):783-794(1992).

7a). I. Zilberberg, et al. *Intl. J. Molecular Science:* 03:801-813 (2002).

7b). J. Klausen, et al. *Environmental Science Technology:* 29:2396-2404 (1995).

8). A. Corma et al. *Angewandte Chemie,* 119(38):7404-7407(2007).

9). M. Mokasch, et al. *ChemCatChem,* 04(01):59-63(2012).

10a). E. A. Gelder et al. *Chemical Communications*: 522-524(2005).

11). E. A. Gelder et al. *Catalysis Letters,* 84(03-04):205-208(2002). 12a).Hydrogenation of NB over Metal Catalysts: Ph. D. Thesis by E. A. Gelder, Univeristy of Glasgow: 2005. 12b). http://theses.gla.ac.uk/1045/1/2005gelder1phd.pdf

13). E. A. Gelder, et al. Competitive Hydrogenation of NB, Nitrosobenzene and Azobenzene. In: Catalysis of Organic Reactions, 21st Conference, R. Schmidt (Ed.), *CRC Press,* 167-176(2006).

14a). M. Kralik et al. *J. Chemistry & Chemical Engineering:* 06:1074-1082 (2012). 14b). M. Králik, et al. *Aniline - Catalysis and Chemical Engineering,* Editor: J. Markos,

In Proceedings of the 41st *International Conference of Slovak Society of Chemical, Engineering,* Tatranské Matliare, Slovakia, 723–733(2014), PP 726 - 728.

15). T. Sheng et al. *Chemical Engineering J.,* 293:337-344 (2016).

16). A. Mahata et al. *Phys. Chem. Chem. Phys.,* 16, 26365-26374 (2014).

17). L. Zhang, et al. *RSC Advances,* 05(43):34319-34326 (2015).

18a). H. Tada et al. *Langmuir:* 20(19):7898-900 (2004).

18b). H. Tada, et al. *ChemPhysChem,* 06(08):1537-543 (2005).

19a). J. W. Larsen et al. *Carbon,* 38:655–661(2000).

19b). Y. P. Li et al. *J. Hazardous Materials,* 148 (01-02):158–163 (2007).

20). L. B. Zhao et al. *J. Physical Chemistry C* 119(9), 4949–4958 (2015).

21). S. C. Jensen et al. *J. Am. Chem. Soc.,* 138 (5), 1591–1600 (2016).

Chapter-2

AMINOPHENOLS

2.01.00: Introduction:

Aminophenols (APs: 2, 3, 4 or o-, m-, p-) are important chemicals and intermediates for many commercial products in the field of pharmaceuticals, agro products, electrical, paints and dyes etc. Amongst them 4-aminophenol (4-AP) or p-aminophenol (PAP) finds more applications than the other two APs. In the following sections, we will look at the synthesis and manufacturing methods of these aminophenols (APs). Throughout the text they are referred to their abbreviations. For example, the 4- or p-aminophenol is referred to as 4-AP or PAP, and so is the case with the other two derivatives as 2-AP and 3-AP.

2.02.00: Synthesis of o-, m-, and p-Aminophenols:

In the following sections we will be seeing the reports especially focusing on the synthesis and production of 4-AP or PAP using different catalysts and methodologies. First, we will look at different approaches making all the three types (o-, m-, p- or 2-, 3-, 4-) of APs.

2.02.10: 2-Aminophenol: (2-AP):

2-Aminophenol or o-Aminophenol (2-AP or o-AP) is an organic compound with the formula $C_6H_4(OH)NH_2$ and is an important raw material in the production of many dyes and pigments and bicyclic heterocycles. The following information briefly highlights its synthetic/manufacturing methods and its applications in making other commercially important products. Synthesis of 2-aminophenols and heterocycles from them are reported by Ru-catalyzed C–H mono- and dihydroxylation (1). Along with its isomer 4-AP, it (2-AP) is an amphoteric molecule and a reduction product of 2-NP. It is industrially manufactured by reducing the corresponding 2-NP by hydrogenation in the presence of various catalysts including iron. It is a useful product for the synthesis of dyes and heterocyclic compounds such as benzoxazoles, which can be used in pharmaceutical industry in making drugs such as flunoxaprofen and antifungal agents with antioxidant, anti-allergic, anti-tumoral and anti-parasitic activities (2). Chlorzoxazone is a muscle relaxant and is synthesized from 4-chloro-2-aminophenol and ethyl chloroformate in the presence of a base (3).

The synthesis of 2-amino-4-chlorophenol (a) and 5-nitro-2-aminophenol and some other derivatives (b) is known in the literature (4).

2.02.11: 3-Aminophenol (3-AP):

3-Aminophenol is an organic compound with formula $C_6H_4(NH_2)(OH)$. It is an aromatic amine and aromatic alcohol. It is the meta-isomer of 2-AP and 4-AP. Its synthesis from 3-hydroxyphenol or *m*-catechol by reacting with NH_4Cl and 10% NH_4OH is reported (5). An interesting observation is made that APs are used in the synthesis of chiral alcohols. Novel chiral molecules were prepared using 1, 3-APs in a Friedel–Crafts reaction and followed by an optical resolution. The catalytic activity of the APs was studied for the addition of diethylzinc to benzaldehyde with high enantioselectivity (94% ee). The same ligand was also used with other aldehydes, to give optically active alcohols in good chemical yields and ee values up to 99% (6).

2.02.12: 2-, 3-, 4-NP Reductions Using Catalysts on Solid Supports:

Based on literature search it has revealed that most of the reported work makes use of solid supported metal catalysts for the synthesis of APs. The following examples strongly support this observation.

2.02.12a: 2-AP: The combined action of immobilized PHA mutase on a support and zinc in a flow-through system catalyzes the conversion of NACs to the corresponding 2-APs, including a novel analog of chloramphenicol. With NB as a model substrate, a continuous reaction system that consists of an initial reduction of NB to PHA using zinc followed by enzymatic conversion to *o*-AP is presented (7).

2.02.12b: 3-AP: A method for the manufacturing of 3-AP is reported by dehydrogenating 3-amino-2-cyclohexene-1-one with a solid supported Pd or Pd-Pt catalyst in a solvent in the presence of a base and its use directly without purification to produce 3,4'-oxydianiline, which is a derivative of 3-AP. The compounds obtained from 3-AP are valuable intermediates for drugs, dyestuffs and herbicides (8). Different ways of synthesizing 3-APs are described and are available at: http://www.molbase.com/en/synthesis_591-27-5-moldata-41407.html

2.02.12c: 4-AP: The results of a study on the thermal stability of gold catalyst supported on mesoporous titania-nanofibres are presented (Scheme-1). The reduction of 4-NP to 4-AP by SBH was evaluated as a probe reaction. After calcination at high temperature, the activity of the catalysts supported on mesoporous TiO_2 nanofibre was well maintained, and the particle size of Au-NPs had hardly changed.

NO$_2$

Au - TiO$_2$ Mesoporous NanoFibres

Solvent / Ambient Temp. / Pressure

NH$_2$

OH

OH

p-(OH)NB

p-(OH)AN --Scheme-1

This might be attributed to the peculiar crystallographic structure and mesoporous nanoarchitecture of the mesoporous TiO$_2$ whisker (9). The transformation of Au$_3$M/SiO$_2$ (M = Ni, Co, Fe) into Au–MO$_x$/SiO$_2$ catalysts for the reduction of 4-NP was found necessary as the former was not active in the catalytic transfer reduction of 4-NP to 4-AP. The Au-M-NPs were deposited on SiO$_2$ to obtain Au-M/SiO$_2$, which were not active in the catalytic reduction of 4-NP unless they were converted into Au–MO$_x$/SiO$_2$ after appropriate thermal treatment in air (10). Ni/graphene nanostructures were synthesized, characterized and tested for their electron-enhanced catalytic activity for hydrogenation of 4-NP to 4-AP. The 4-NP conversion was 100% with 90% selectivity to 4-AP (Scheme-2).

NO$_2$

Ni/Graphene Nanostructures

H$_2$ / Press/ Temp.

NH$_2$

OH

OH

Conversion = 100%

Sel = 90% --Scheme-2

The catalyst is stable, recyclable and has the value of non-noble metal/graphene nanocomposites in the development of catalysts for green chemistry (11). Fe-Ni/Fe$_3$O$_4$ embedded nanostructures were successfully synthesized, which exhibited excellent catalytic performance for the reduction of p-nitrophenyl compounds while a kinetic law was observed. The reduction of 4-NP to 4-AP was achieved well (12). A simple, economic, and facile approach for the synthesis of novel reduced graphene oxide-zinc tungstate-iron oxide (rGO-ZnWO$_4$-Fe$_3$O$_4$) nanocomposites by a one-pot microwave method and its efficiency as a catalyst in reducing 4-NP to 4-AP using SBH is tested. The catalyst has excellent stability and reusability and an efficient candidate in research and industrial applications (13). Amongst solid supports, Y-Al$_2$O$_3$ is the first choice of many groups as it has been found beneficial in the selective hydrogenations of NAs. Nandanwar et al. have reported the synthesis of CuO/Y-Al$_2$O$_3$ by microemulsion process and its use in the catalytic reduction of 4-NP to 4-AP in the presence of SBH. CuO/γ-Al$_2$O$_3$ catalysts were prepared by dispersing highly stable CuO-NPs on γ-alumina by mechanical stirring and the reduction of 2, 3, and 4-NPs were accomplished in good yields (14). The catalytic hydrogenation of 4-NP to 4-AP was investigated over Ni/TiO$_2$ catalysts prepared by a liquid-phase chemical reduction method and is much superior to Raney-Ni catalyst. The catalytic activity of anatase titania supported nickel catalyst Ni/TiO$_2$ (A) is higher than that of rutile titania supported nickel catalyst Ni/TiO$_2$(R), because of their structures. 4-NP conversions were almost quantitative at 99.9% conversion rates (15).

References:

1). X. Yang, et al. *Organic Letters*, 15(10): 2334-2337(2013).

2a). M. S. Mayo, et al. *J. Organic Chemistry*, 79(13):6310–6314(2014). 2b). D. R. Mileski, et al. *J. Pharma Science*, 54(02):295–298(1965), 2c). Z. He, J. C. Spain, *J. Industrial Microbiology & Biotechnology*, 23:138–142(1999). 2d). M. Aslam, et al. *Medicinal Chemistry Research*, 25(01):109-115(2016). 2e). C. Silva, et al. *J. Advanced Research*, 02(01):01-08(2011). 2f). E. Sener, et al. *Quant. Structure.-Activity Relations*,. 10:223-228 (1991).

3). I. Itoh, et al. US 4743595 A (1988). L. Bogogna et al. US 9567308 B1 (2017).

4a). H. E. Fierz-David, *Fundamental Processes of Dye Chemistry*: 110-111(1949). 4b). T. Deligeorgiev, *Dyes and Pigments*, 23(01):85-90(1993).

5). M. Heidelberger, *Advanced Laboratory Manual Organic Chemistry*, 28-29(1923).

6a). X. Yang, et al. *Tetrahedron: Asymmetry*, 18(10):1257-1263(2007). 6b). P. Singh, *Am. J. Material Science*, 04(02):74-83(2014). 6c). J. Raj, et al. Res. J. Chemical Sciences, 05(09):01-10(2015).

7). H. R. Luckarift, et al. *Chemical Communications*, 00(03):383–384 (2005).

8a). W. Muller, US: 4212823A (1980). 8b). S. E. Jacobson, US 5202488A (1993).

9). X. Ma, et al. i). *Chinese J. Catalysis*, 33(09):1463-1469 (2012), ii). X. Ma, et al. *Chinese J. Catalysis*, 33(09):1480-1485(2012).

10). C. Lin, et al. *Catalysis Letters*, 144(06):1001-1008(2014).

11). Y. Wu, et al. *J. Physical Chemistry C*, 118 (12):6307–6313(2014).

12). Dandan Wu, et al. *Inorganic Chemistry*, 56 (09):5152–5157(2017).

13). K. B. Denthaje, et al. *Ind. Engg. & Chemical Res.*, 55(27): 7267–7272(2016).

14). S. U. Nandanwar, et al. *Chinese J Catalysis*, 33(9–10):1532-1541(2012).

15). R. Chen, et al. *Chinese J. Chemical Engineering*, 14(5) 665-669 (2006).

2.03.00: 4-Aminophenol (4-AP) or *p*-Aminophenol (PAP):

Aromatic nitro compounds (ANCs) or nitro aromatic compounds (NACs) are widely generated as byproducts in various industries, including in the production of pigments, pesticides and medicines. 4-Nitrophenol (4-NP) is among the most common ANCs, and is harmful to the environment. 4-AP the reduction product of 4-NP is an important intermediate for the manufacture of polymeric products, dyes, agrochemicals, and pharmaceuticals. Various methods to synthesize 4-AP have been reported, such as multi-step iron-acid reduction of 4-NP, catalytic reduction of nitrobenzene in acidic medium, and electrochemical synthesis. Among these methods, the catalytic reduction is an alternative green process for 4-AP production. In most of the cases, the noble metals have been used in the catalytic reactions for their high catalytic activities. However, the high cost and scarcity in nature limit their practical applications. Therefore, non-noble metal catalysts have attracted attention because of their availability and lower costs. 4-AP or PAP is prepared by the reduction of 4-NP, which in turn can be prepared by nitration of phenol itself.

$$C_6H_5OH + HNO_3 + H_2SO_4 = OH\text{-}C_6H_4\text{-}NO_2$$

$$HO\text{-}C_6H_4\text{-}NO_2 + 6H_2 \text{ (Molecular } H_2 \text{ or Hydrogen Donor)} = HO\text{-}C_6H_4\text{-}NH_2 + 2H_2O$$

Alternatively, the partial hydrogenation of NB affords N-hydroxyaminobenzene (HAB) or phenylhydroxylamine (PHA), which in the presence of an acid rearranges to 4-AP.

$$C_6H_5NO_2 + 2\,H_2 = C_6H_5NHOH + H_2O$$

$$C_6H_5NHOH = HOC_6H_4NH_2$$

PAP is a building block compound for many end products. Prominently, it is the final intermediate for paracetamol at commercial scale, where 4-AP is treated with acetic anhydride to give paracetamol. Along with haloanilines (HANs), the wide range applications of APs as raw materials for many useful end products place them amongst the top important AN derivatives, For example, Deerfield Beach, FL, Feb. 03, 2016 (GLOBE NEWSWIRE) – Zion Research has released a new report titled "Acetaminophen (Paracetamol) Market for Pharmaceuticals, Dye Industry and Chemical Industry - Global Industry Perspective, Comprehensive Analysis, and Forecast, 2014-2020" According to the report, global acetaminophen market was valued at around USD 801.3 million in 2014 and is expected to reach USD 999.4 million in 2020, growing at a CAGR of around 3.8% between 2015 and 2020. In terms of volume, the global acetaminophen market stood at above 149.3 kilo tons in 2014. http://www.marketresearchstore.com/report/acetaminophen-market-z45985. https://www.bizjournals.com/prnewswire/press_releases/2016/03/29/MN57575.

A stepwise reduction and acetylation process for the production of pure N-acetyl-p-aminophenol (APAP) or Paracetamol from *p*-NP in an aqueous system that avoids the need for strong acids or excess acetic anhydride is present in the system until the hydrogenation reaction has reached substantial completion (1). The huge demand for just one derivative (Paracetamol) of 4-AP gives an idea about the importance of 4-AP and other APs in different industries.

2.03.10: Synthesis of 4-AP by Catalytic Hydrogenation of NB:

A conventional method for the synthesis of 4-AP is a two-step reaction involving iron–acid reduction of 4-NP. This method causes serious effluent disposal problems due to the stoichiometric use of iron–acid, which leads to the formation of Fe–FeO sludge (1.2 kg/kg of product) in the process, which cannot be recycled. Rode et al. have reported the preparation of 4-AP via single-step catalytic hydrogenation of NB in acid medium over Pt catalyst and the process conditions have been optimized (Scheme-3).

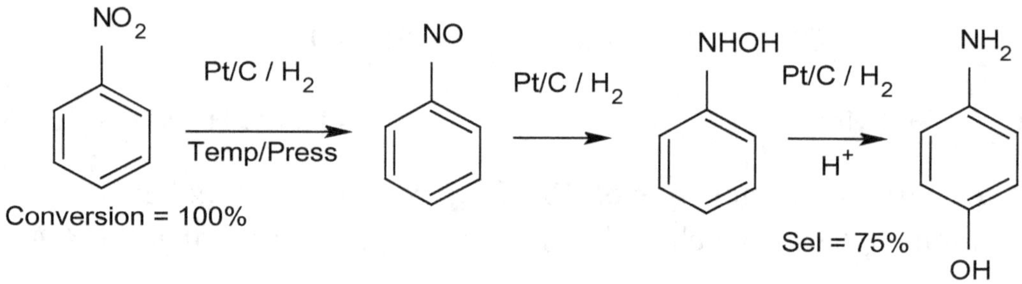

--Scheme-3

Complete conversion of NB was achieved with 75% selectivity to 4-AP under the best set of conditions. This study also highlights the effect of various process parameters such as temperature, hydrogen pressure, and the substrate and acid concentration on the rate of reaction and selectivity to 4-AP (2). Synthesis of 4-AP in two steps involving catalytic hydrogenation of NB to PHA at 303K/H2(0.69MPa) and its Bamberger rearrangement in acid medium to 4-AP is reported. These aspects were studied separately in a batch reactor using a well-characterized 3% Pt/C catalyst. The reaction is characterized by complete conversion of NB and 90% selectivity to PHA and little AN, even at lower temperatures (Scheme-4).

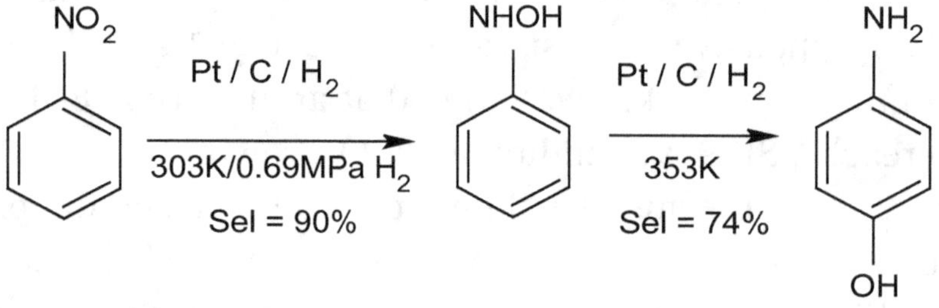

--Scheme-4

The second step of PHA rearrangement to PAP could be achieved under a hydrogen atmosphere at elevated temperature of 353K to give 4-AP with 74% selectivity (3).

Recently, the kinetics of Bamberger rearrangement of PHA to 4-AP in acetonitrile/TFA system is studied, where a substrate acid complex as *p*-selectivity driver is reported (4). In another case, the effect of a number of reaction parameters such as hydrogen pressure, reaction temperature, stirring rate, and the amounts of NB, catalyst, and the surfactant present in the reaction mixture had on the rate and selectivity of the hydrogenation was examined. Optimization of these parameters led to the formation of 4-AP at a selectivity (4-AP/AN) of 5.4 with a productivity of over 80000g of 4-AP/g-Pt/h (5). There are some other reports of synthesis of 4-AP by catalytic liquid phase hydrogenation of NB in the presence of metal/support catalyst solid acid at 189°C/4h. Under the optimal reaction conditions, NB conversion and 4-AP selectivity reached 61.0% and 77.8% respectively (6). An efficient hydrogenation catalyst in sulfuric acid for the conversion of NB to 4-AP using N-doped carbon with encapsulated molybdenum carbide using a one-pot preparation process is reported. The outer layer of N-doped carbon exhibits high activity and excellent selectivity with molybdenum carbide as the catalyst in the hydrogenation of NB to 4-AP in acid (7). Pt/C catalyst (1%) was chosen for optimization of reaction conditions and kinetic studies because of its higher catalytic activity compared to that of other heterogeneous metal catalysts. The catalytic activity and initial rate of reaction was found to increase with increase in the polarity of solvent (Scheme-5).

--Scheme-5

Kinetic studies and other parameters were studied and a simple Langmuir–Hinschelwood (L–H)-type model was found to represent the kinetics of hydrogenation of 4-NP to 4-AP satisfactorily (8).

Compared with commercial Raney-Ni, catalytic properties (activity, selectivity, and stability) of the nano-sized Ni are superior, which is attributed to a combination effect of the small particle size (nano-size Ni) and high-density surface defects (Scheme-6).

--Scheme-6

Based on prior understanding and in order to enhance the catalytic activity of the non-noble metal, supported Ni-Pt nanocatalysts were synthesized (~10 nm). The incorporation of Pt to the pristine Ni catalysts led to the enhanced activity towards the reduction of 4-NP to 4-AP, because the modification of Pt affected the electronic structures of Ni and acted as a promoter. The optimal Ni-Pt catalysts with an appropriate amount of Pt showed excellent activity in the reduction of 4-NP (9, 10). The catalytic hydrogenation of 4-NP to 4-AP was investigated in a laboratory-scale batch-slurry reactor.

Spherical Ni-NPs with different sizes and different structures were prepared starting from nickel oxalate and using hydrazine as a reductant in the presence of citric acid, cetyltrimethylammonium bromide, Tween 40, and D-Sorbitol as organic modifiers, which affected the size and structure of the resultant Ni-NPs (11).

--Scheme-7

All of the Ni-NPs showed higher catalytic activity and selectivity than the conventional Raney-Ni catalyst in the hydrogenation of 4-NP to 4-AP (Scheme-7). The synthesis of 4-AP from NB is studied by examining the effects of different reaction media, promoters, supports, and preparation methods of catalysts. The results show that the aqueous solution of sulfuric acid is the most suitable reaction medium. $AlBr_3$ is an effective promoter and Pt/modified over Al_2O_3 is a promising catalyst for this reaction (12).

An improved process for the selective preparation of 4-AP from NB is reported (13a) by the catalytic hydrogenation of NB in an aqueous acidic reaction medium containing a dimethylalkylamine oxide and found that the hydrogenation rate and the selectivity of the reaction is increased. Similarly, a process for the production of unsubstituted and lower alkyl substituted 4-APs over a Pt catalyst and a sulfur compound is reported (Scheme-8). The hydroxylamine intermediate is heated to 70°C, thereby effecting rearrangement to the corresponding PAP (13b).

--Scheme-8

An intimately bi-functional catalyst composed of Pt-NPs encapsulated in crystals of HZSM-5 zeolite has been studied for one-pot hydrogenation of NB to 4-AP without the use of inorganic acid (14).

A single-step conversion of NB to 4-AP through catalytic hydrogenation in acidic medium is well known, however the main shortcoming of this route is the use of sulfuric acid for the rearrangement of the PHA intermediate.

--Scheme-9

Wang et al. have prepared a $S_2O_8^{2-}/ZrO_2$ (PSZ) solid acid and $Pt\text{-}S_2O_8^{2-}/ZrO_2$ (Pt-PSZ) bifunctional catalysts and used for the synthesis of 4-AP in non-acid medium (Scheme-9). PSZ solid acid exhibits high activity for PHA rearrangements. However, the PAP yields were very low 23-39% (15). A new insight is gained during the preparation of 4-AP via hydrogenation of NB in a single liquid phase ($CH_3CN-H_2O-CF_3COOH$; Acetonitrile+H_2O+TFA)) has been carried out in the presence of precious metal catalyst (Pt/C) and sulfolane as a promoter. The solvent and the promoter help enhance the selectivity of 4-AP. CH_3CN decreases the hydrogenation activity compared to other solvents. CF_3COOH promotes the formation of the desired product both via Bamberger rearrangement in solution as well as by a surface catalyzed reaction, while H_2O is responsible for 4-AP formation in both reactions. Indeed, the formation of 4-AP may occur both in solution via acid catalyzed Bamberger rearrangement and on the catalyst surface by the formation of a surface Pt-nitrenium complex, which undergoes surface nucleophilic attack by H_2O. No other intermediates are observed (16).

A nanocomposite catalyst containing palladium–nickel boride–silica and reduced graphene oxide ($Pd@Ni_xB-SiO_2$/RGO, abbreviated as Pd@NSG was successfully fabricated and its enhanced hydrogen spillover mechanism and high catalytic performance towards reduction of 4-NP to 4-AP are discussed (17). Pd-NPs were mounted on a ceramic membrane support in which the support surface was silanized with amino-functional silane and was used for catalyst immobilization and were tested in the liquid-phase hydrogenation of 4-NP to 4-AP. The newly prepared catalyst was found superior to Pd-NPs deposited on the ceramic membrane support without silanization (18). Similarly, Gold (Au-NPs, 5.6nm) on mesoporous silica (GMS) effectively reduce 4-NP to 4-AP. The nanoreactor framework catalyst is very robust, readily separable, reusable, and has potential for practical applications (19).

2.03.11: Borate reduction of nitrophenols: A process for the direct production of PAP and N-acetyl-PAP from p-NP using a borate ion additive during hydrogenation to eliminate undesirable by-products and color formation is described (Scheme-10).

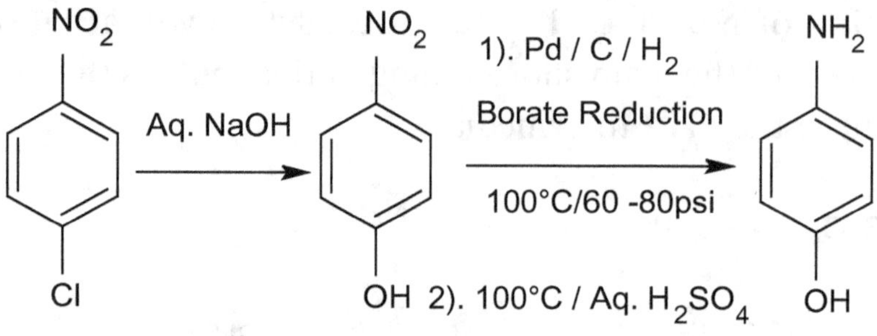

--Scheme-10

An improved process for making PAP involves alkaline hydrolysis of 4-CNB and subsequent reduction of *p*-NP to PAP with Pd/C catalyzed hydrogenation in the presence of borate ion, which helps eliminate undesirable hydrolysis products (20).

2.04.00: Ionic Liquids:

We are aware of the fact that ionic liquids (ILs) play important roles in organic synthesis. Here, Ir/C and Bronsted acid functionalized ionic liquids are used as an efficient catalytic system for hydrogenation of NB to 4-AP (21).

2.05.00: CO+H$_2$O, CO$_2$+H$_2$O and scCO$_2$:

There are different hydrogen sources other than molecular hydrogen for the catalytic transfer hydrogenation (CTH) of NAs, and one of them is CO+H$_2$O, CO$_2$+H$_2$O, and scCO$_2$. These systems in the presence of a metal catalysts act as hydrogen sources for reducing different functional groups including nitro groups in NAs. The reader can visit the individual subject headings in chapter-5 for more information. Here biphasic H$_2$O/CO$_2$ system as a versatile reaction medium for organic synthesis (as cycloaddition, reduction, hydroformylation, etc.) is reviewed. In our context, this system is used earlier by some groups to make PAP from PHA (22). Some groups have recently studied the synthesis of PAP through Bamberger rearrangement of PHA under pressurized CO$_2$+H$_2$ system, with PHA conversion at 100% and the selectivity for PAP is found to be at 80% at 100°C for 1h under 4MPa CO$_2$. This is the first time, the Bamberger rearrangement of PHA was realized in a CO$_2$+H$_2$O system. The process fully avoids the need of inorganic strong acid and is environmentally benign (23). The synthesis of PAP through NB hydrogenation under acidic environment can be done either using transition metals (Fe,Ni, Zn, Sn etc.) or supported mono or bimetals in pressurized CO$_2$/H$_2$O system and in the presence of Pt-Pb/SiO$_2$ as a catalyst. PAP was obtained with 82% selectivity when the reaction was carried out at 110°C under 5MPa CO$_2$ and 0.2MPa H$_2$ (24).

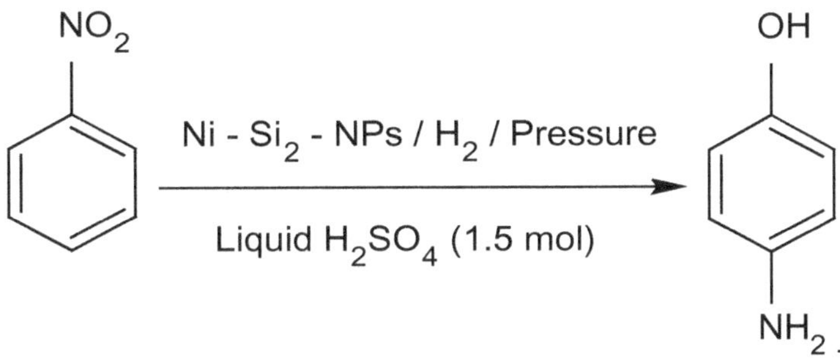

$$\underset{\text{NO}_2}{\text{[benzene ring]}} \xrightarrow[\text{Liquid } H_2SO_4 \text{ (1.5 mol)}]{\text{Ni - Si}_2 \text{ - NPs / H}_2 \text{ / Pressure}} \underset{\text{NH}_2}{\overset{\text{OH}}{\text{[benzene ring]}}}$$ --Scheme-11

Here is an example where a Ni-silicides-NPs (Ni-Si$_{2-}$NPs$_)$) are used as a substitute for noble metals for hydrogenation of NB to PAP in sulfuric acid (Scheme-11) (25). A green process for the production of PAP from NB (conversion: 98.5%) hydrogenation in CO_2/ H_2O or $scCO_2$-H_2O over Pt/Al$_2O_3$ catalyst, is found as an effective and green medium for production of PAP. During the hydrogenation of NB the acidity and CO formed *in situ* were demonstrated to promote the formation of PAP (26).

2.06.00: Hydrazine Hydrate:

Highly efficient and selective reduction of substituted NAs, including 4-NP, with hydrazine over supported Rh-NPs is investigated under mild conditions. The corresponding ANs, as 4-AP, were obtained quantitatively (27). An efficient catalyst made of bismuth, iron and graphene oxide composite was developed for the reduction of NAs (quantitative conversions) with hydrazine hydrate to afford corresponding ANs. The bismuth and graphene remarkably promoted the activity of iron oxide catalyst, so that a mixture of ferric chloride, bismuth nitrate and graphene oxide catalyzed the complete conversion of 4-NP into 4-AP in a continuous flow batch reactor at 90-110°C for 1h (28).

2.07.00: Synthesis of 4-AP by NaBH$_4$ (SBH) Reduction of 4-NP:

Natural products are used for the preparation of M-NPs to be used in the selective hydrogenation of 4-NP to 4-AP. The following examples illustrate the contributions from different groups in this context.

2.07.10: Non-Noble Metal Catalysts:

One-pot green synthesis of Ni-NPs was carried out successfully in aqueous medium by using hydrazine hydrate as a reducing agent. Greener stabilizing agent starch is used as a stabilizer, which also acts as a particle protector toward oxidation (29).

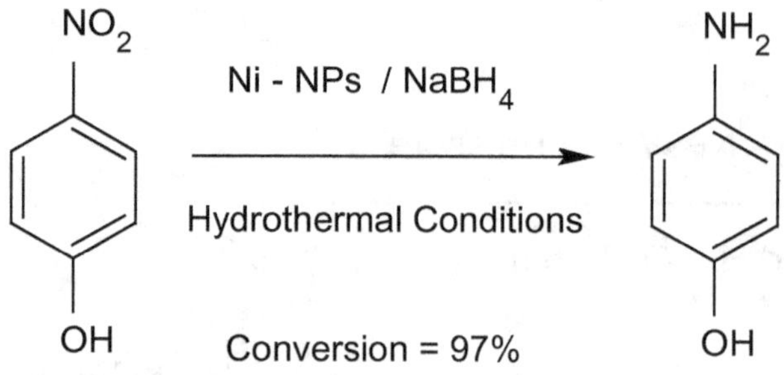

--Scheme-12

The as-prepared Ni-NPs were used as a catalyst in the reduction of 4-NP to 4-AP (97%) under hydrothermal condition in the presence of SBH as a source of hydrogen (Scheme-12). Similarly, Co-NPs as reusable catalysts for reduction of 4-NP under mild conditions with sodium borohydride (SBH, $NaBH_4$) as hydrogen donor is reported (30). A simple sol-gel method was used to synthesize magnetic Fe_2O_3-Cu_2O-TiO_2 nanocomposites and have been tested in the catalytic reduction of 4-NP to 4- AP by using SBH as a reductant (31).

2.07.11: Noble Metal Catalysts:

Here, Ag and Au-NPs are prepared by using a photochemical green synthesis on Calcium-Alginate, which helps stabilize the M-NPs on its surface and also acts as a reductant (Scheme-13) (32).

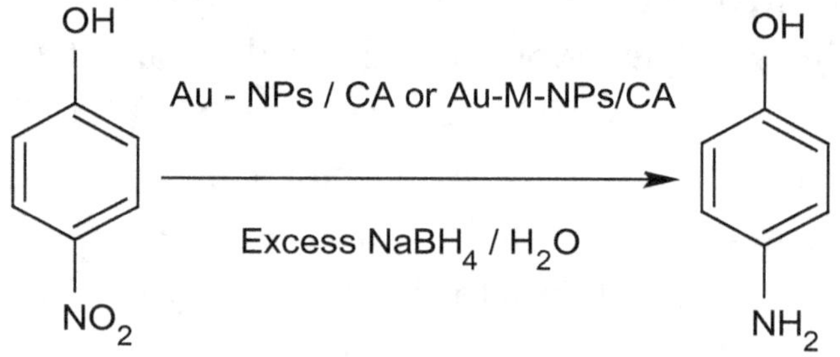

--Scheme-13

The as-prepared new solid-phase biopolymer-based catalysts are very efficient, stable, easy to prepare, eco-friendly, cost-effective, and they have the potential for industrial applications. Based on the importance of metals on solid supports, the reduction 4-NP to 4-AP by SBH is catalyzed by both monometallic and bimetallic NPs on poly-(amido) aminedendrimers is presented. The resulting dendrimer encapsulated NPs (DENs) as monodisperse with alloys of Pt/Cu, Pd/Cu, Pd/Au, Pt/Au, and Au/Cu were synthesized and evaluated as catalysts for 4-NP reduction with SBH and it was found that their catalytic properties are dependent on the adsorbate's binding energy (33) Superior catalytic performance was observed in a test reaction of 4-NP and 4-nitroaniline (4-NA) reduction in the presence of freshly prepared ice cold aqueous solution of SBH at room temperature (Scheme-14).

--Scheme-14

Hybrid Au-NPs-reduced graphene oxide nanosheets (Au-rGO) are proved as active catalysts for highly efficient reduction of NAs using SBH as a hydrogen donor at room temperature for 30min. This is a simple, one-step synthesis of hybrid Au-rGO nanosheets through electrostatic self-assembly, which are found good for multiple applications (34). In another case, Ultra small core–shell Ag-NPs have been synthesized by an up scaled modification of the polyol (PAA) process and compared with those prepared with Glutothione (GSH) and bovine serum. With PAA as a stabilizer and SBH as a hydrogen donor the Ag-NPs are better than other stabilizers and effective in the reduction of 4-NP to 4-AP (35). Biodegradable amphiphilic copolymer composed of poly-(2-ethyl-2-oxazoline) and poly-(ε-caprolactone) have received much more attention in the last decades due their potential applications in the fields related to environmental protection, medicine, agriculture, and the chemical processes. Here, Ag-NPs (10-15nm) were prepared via reduction of silver nitrate ($AgNO_3$) using biodegradable amphiphilic copolymers in aqueous solution. The catalyst shows high catalytic activity toward the reduction of 4-NP in the presence of an excess amount of SBH (36).

References:

1). a). J. Huber, US 4264525 A (1981). b). J. B. Warner, US 4670589 A (1987). c). J. H. VanNess et al. US4670589 (1987).

2). C. V. Rode, et al. *Organic Process Research & Dev.*, 03(06):465-470(1999).

3a). C. V. Rode, et al. *Chemical Engineering Science*, 56(04):1299-1304(2001). 3b). C. V. Rode, et al. *Industrial Engineering & Chemical Research*, 50(09):5478–5484(2011). 3c). E. Bamberger, *Berichte*: 27:1347 and 1548(1894).

4). N. Fonzo et al. *Applied Catalysis A: General*, 516:58-69(2016).

5a). S. K. Tanielyan , et al. *Org. Process Research & Development*, 11(04):681-688(2007). 5b). US: 4264525 (1981).

6). Y Lu, et al. *Chemical Engineering Journal*, 229:105-110(2013).

7). T. Wang, et al. *Chemical Communications*, 52:10672-10675(2016).

8). M. J. Vaidya, et al. *Organic Process Research & Dev.*, 07(02):202–208(2003).

9). H. Shang, K. Pan, et al. *Nanomaterials*, 06:103-124(2016).

10a). Y. Du, et al. *Applied Catalysis A: General*, 277(01-02): 259-264(2004). 10b). S. Wang, et al. *J. Chemical Technology & Biotechnology*: 83:1466–1471 (2008).

11). A. Wang, et al. *Langmuir*, 25 (21):12736–12741(2009).

12). Z. Liu, *J. Natural Gas Chemistry*, 08(04):305-310(1999).

13a). US:4307249(1981). 13b). US: 4415753 (1983).

14). J. Gu, et al. *Catalysis Communications*, 97:98–101(2017).

15). S. Wang et al. *J. Chemical Technol & Biotechnology*, 83(11):1466-1471(2008).

16). Quartarone, et al. *Applied Catalysis A: General*, 475:169-178(2014).

17). R. Krishna, et al. *RSC Advances*, 05(74):60658-60666(2015).

18). R. Chen, *Industrial Engineering & Chemical Res..*, 50 (8):4405–4411(2011).

19). L. Chen, *Industrial Engineering & Chemical Res..*, 50(24):13642–13649(2011).

20). D. C. Ruopp & M. A. Thorn, US4264526 (1981).

21). H. Wang, et al. *RSC Advances*, 07(50):31663-31670(2017).

22). M. A. Pigaleva, et al. *RSC Advances*, 05(125):103573-103608(2015).

23a).i). CN102001954-A(2011). ii).CN102001954-B(2013). 23b). S. Liu, *Industrial Engineering & Chemical Research*, 53(20):8372–8375(2014).

24). T. Zhang, et al. *Chinese Chemical Letters*, 28(02):307-311(2017).

25). Z. Dong, et al. *Applied Catalysis A: General*, 520:151-156(2016).

26a). L. Zhao, et al. *J. CO2 Utilization*, 18:229-236(2017). *26b). T. Zhang, et al. Organic Process Res. & Development*, 19(12):2050–2054(2015). 26c).T. Zhang, et al. *Chinese Chemical Letters*, 28(2): 307-311 (2017). 26d).R. Robles, et al. *ChemSusChem*, 04:1035-1048(2011).

27). P. Luo, et al. *Catalysis Science & Technology*, 02:301-304(2012).

28). H. Sun, et al. *Chemical Engineering Journal*, 314:328-335(2017).

29). N. V. Suramwar, Synthetic Communications, 43(01):57-62(2013).

30). A. Mandal, et al. *Bulletin of Material Science*, 40(02):321-328(2017).

31). P. Babji, et al. *Intl J. Chemical Studies*, 04(05):123-127(2016).

32). S. Saha, et al. *Langmuir*, 26(04):2885–2893(2010).

33a). Z. D. Pozun, et al. *J. Physical Chemistry C*, 117:7598–7604 (2013). 33b). R. Vadakkekara et al., *Colloids and Surfaces A: Physicochemical and Engineering Aspects*, 399:11-17(2012).

34). Y. Choi, et al. *J. Material Chemistry* 21:15431-15436(2011).

35a). C. Kastner & A. F. Thünemani, *Langmuir*, 32 (29):7383–7391(2016). 35b). S. Gu, et al. *J. Physical Chemistry C*, 118(32):18618–18625(2014).

36). S. Jafari, *Green Chemistry Letters & Reviews:* 09(01):20-26(2016).

2.08.00: 4-NP Reductions Using Different Techniques:

2.08.10: Electrochemical Reductions:

The electrochemical reduction of NB has been studied in aqueous organic solvents containing sulfuric acid. It is found that the yield of 4-AP and the current density for the reaction are dependent upon the solvent, cathode metal, electrode potential, concentration of acid, and the rate of stirring of the catholyte (37). The yield of 4-AP is >75% at a current density of 150 mA cm^{-2}. Ordered mesoporous carbons (OMCs) modified glassy carbon electrode (GCE) (OMCs/GCE) was employed to investigate the electrochemical behavior of o-NP (2-NP), m-NP (3-NP) and p-NP (4-NP) in ambient–N$_2$ phosphate buffer saline. Compared with bare GCE, the OMCs/GCE exhibited higher electrocatalytic activity towards NP isomers (38). The electrochemical degradation of 4-NP under different conditions was investigated and 4-AP was obtained in 92% yield on a stainless steel cathode and Ti/Pt anode through cyclic voltammetry (39). Performance of three different anodes in electrochemical degradation of 4-NP is investigated. Among electrodes investigated, the IrO$_2$-PbO$_2$/Ti electrode resulted in 98% of COD removal in 30 min comparatively at a less energy consumption depicting its higher performance efficiency in 4-NP degradation (40). 4-NP is obtained from phenol by nitrating phenol chemically and then it is reduced to 4-AP electrochemically. Since 4-AP is fixed as a 4-AP sulfate, it is directly used for producing Paracetamol (41). Polyaniline nanofibres (PANINFs) have been prepared by electrochemical polymerization of aniline monomers in acidic aqueous media without using any templates and surfactants. The subsequent treatment of such nanofibres with AgNO$_3$ aqueous solution leads to in-situ chemical reduction of Ag+ on them to form Ag-NPs decorated PANINFs (Ag-NPs/PANINFs) nanocomposites. The catalytic activity and electrochemical properties of these nanocomposites were tested toward reduction of 4-NP to 4-AP by SBH and found that these nanocomposites exhibit excellent catalytic activity (42a). Sterically hindered 2-APs (42b) and the synthesis of quinones from 4-APs (42c) by electrochemical reduction of NAs have been reported.

2.08.11: Microwave Irradiations (MW):

Microwave is an important technique useful in many applications. Fe$_3$O$_4$ Nanostructures were synthesized with a simple, economic and one pot synthetic protocol and the nanostructures were coated with tetraethyl orthosilicate for stabilization purposes. The synthesized SiO$_2$/Fe$_3$O$_4$ nanostructures with fine semi-spherical textures showed high heterogeneous catalytic activity for the reduction of 4-NP (99.5% conversions) to 4-AP using SBH under microwave radiations. The catalyst is used as a heterogeneous catalyst in environmentally and industrially important reactions (43). A simple, economic, and facile approach for the synthesis of novel reduced graphene oxide-zinc tungstate-iron oxide (rGO-ZnWO$_4$-Fe$_3$O$_4$) nanocomposites by a one-pot microwave method and its efficiency as a catalyst in reducing 4-NP to 4-AP using SBH can be achieved (44). Bimetallic nanoparticles (BM-NPs) as efficient catalysts in the "Facile and Green Microwave Synthesis", has gained importance (45). Microwave assisted Ni-NPs are prepared with Ag and Au, and used in the reduction of 4-NP to 4-AP with SBH as a hydrogen donor (46).

2.08.12: Photocatalytic Reduction of 4-NP to 4-AP:

Some of the techniques, instruments as photoirradiation, unltrasonication, microwave irradiations, etc. do help in solving the problems and add value to the process of reduction of NAs to ANs. Here is an example where an unprecedented photoactivity of Ag-NPs photodeposited on nanocrystalline TiO$_2$ are found useful in the efficient reduction of 4-NP to 4-AP at room temperature. The use of Na$_2$SO$_3$ as a harmless scavenger agent for the reduction of a nitroaromatic endocrine disruptor yields a valuable reducing reagent (47). Solar-light driven photocatalytic conversion of 4-NP (Conversion > 92%) to 4-AP on CdS nanosheets and nanorods in ethylenediamine as solvent is performed. The better catalytic activity of both morphologies can be attributed to quantum size effect and good optical absorbance (48). Visible light assisted reduction of 4-NP to 4-AP on Ag/TiO$_2$ photocatalysts synthesized by hybrid templates resulted in 98% 4-NP conversion in two minutes of irradiations (49). Photocatalytic reduction of 4-NP with SBH is abundantly reported in the literature. A green photochemical approach towards the synthesis of carbon nanofibre- and graphene-supported Ag-NPs and their use in the catalytic reduction of 4-NP is tried. Carbon–Ag-NPs hybrids are demonstrated to be notably effective catalysts for the reduction of 4-NP to 4-AP with SBH (50). Silver and gold NPs have been grown on calcium alginate gel beads using a green photochemical approach and used in the green synthesis of 4-AP by catalytic reduction of 4-NP. The catalytic efficiency of alginate-based Ag catalyst was much more compared to that of the Au catalyst (51). The surface charge of TiO$_2$ affects the adsorption rates of target molecules, thereby influencing the associated photocatalytic degradation rates of these molecules. This study describes the specific photocatalytic reduction of 4-NP to 4-AP (100%) with arginine-modified TiO$_2$ (Arg–TiO$_2$-NPs). The terminal amine groups of the arginine monolayer create a positive TiO$_2$ surface charge over a wide range of pH values. At high pH, the degradation rate of 4-NP increased due to improved target adsorption (52). The selective photocatalytic reduction of 4-NP to 4-AP in TiO$_2$ suspensions prepared in aliphatic alcohols (methanol, ethanol, 1-propanol, 2-propanol, 1-butanol, i-butanol) was investigated. The photoreduction rate is significantly affected by the solvent parameters such as viscosity, polarity and polarisability (53).

2.08.13: Ultrasonocation:

Ultrasonication assisted synthesis of palladium-nickel/iron oxide core-shell nanoalloys as effective catalyst for Suzuki-Miyaura and 4-NP reduction reactions are presented. These magnetic NPs can be separated from the reaction mixture by external magnetic field. The catalyst is found efficient and this strategy is simple, economical and promising for industrial applications (54). The catalytic reduction of 4-NP to 4-AP occurs at the surface of a specially prepared highly efficient Cu-NPs, with PAA in water at room temperature (55). Ultrasound assisted synthesis of Au-NPs in an ecofriendly manner using *Calothrix* algae is studied. The production of the Au-NPs in the reaction mixture is significantly accelerated by inducing ultrasound irradiation and tested in reducing 4-NP to 4-AP. The advantage of using ultrasound relates to the ecofriendly and rapid synthesis of Au-NPs, which have various biotechnological applications including reductions of NPs (56). New insights for the mechanism of nitro to amine groups conversions have been revealed on both N-doped graphene and Ag-NPs catalysts based on the paper assisted ultrasonic spray ionization mass spectrometry. Some of the observations made are: 1) water molecules, not $NaBH_4$, are the hydrogen source for the reduced amino groups, 2) $NaBH_4$ could contribute to the ionization of H^+ from water facilitating its adsorption on nitro groups, 3) six different intermediates have been detected to depict the whole catalytic process and no condensed roots are involved in, and 4) this reduction process is spontaneous to some extent and even without catalysts it is not totally stopped as observed from the spectral measurements. X. Kong, et al. https://www.researchgate. net/publication/317859259_Insights_into_the_Reductionof4-Nitrophenol_to_4-Aminophenol_on_Catalysts.

References:

37). J. Marquez & D. Pletcher, *J. Applied Electrochemistry,* 10(05):567-573(1980).

38). T. Zhang, et al. *Electrochimica Acta,* 106:127–134(2013).

39a). P. Jiang, et al. *J. Environmental Science* (China): 22(04):500-506(2010). 39b). M. N. Behnajady, et al. *J. Hazardous Materials,* 154(01-03):778-786(2008). 39c). P. Canizares, et al. *Industrial Engg & Chemical Research,* 43(9):1944-1951(2004). 39d). F. R. Zaggout, et al. *J. Environmental Management,* 86(01):291-296(2008).

40). P. Murugaesan, *Environmental Technology.* 36(20):2618-27(2015).

41). P. A. Anantharaman, *Bulletin of Electrochemistry,* 01(05): 471(1985).

42a). G. Chang, et al. *Catalysis Science & Technology*: 02:800–806(2012). 42b). Z. Chen & Q. Wang, *Organic Letters,* 2015, *17* (24):6130–6133 (2015). 42c). R Harman, J. Cason, *J. Organic Chemistry,* 17(07):1058–1062(1952).42d). Electrochemistry for the Environment: Christos Comninellis, Guohua Chen, Springer Science & Business Media, 15-Oct-2009 - Science - 563 pages, PP 532.

43). M. T. Shah, et al. *Microsyst Technology*, pp01-14 (2017).

44). M. J. S. Mohamed, et al. *Ind. Eng. Chem. Res.*, 55 (27):7267–7272(2016).

45). M. Blosi, et al. *Materials* (Basel), 09(07): 550 (2016).

46a). S. Joseph, et al. *Research J. Recent Science*: 03(ISC-2013):185-191 (2014) 46b). M. A. Bhosale, et al. *RSC Advances*, 05(65):52817-52823(2015). 46c). H.S. Park, et al. *Asian J. Chemistry*. 27(04,):1240-1242(2015). 46d).K. Hsu, et al. *Nanoscale Research Letters*: 09(01): 484-496(2014). 46e). J. Li, et al. *Asian J. Chemistry*, 26(07): 2153-2155(2014).

47a). A. H. Gordillo, et al. *J. Photochemistry and Photobiology A: Chemistry*, 257:44-49(2013). 47b) A. Gardillo, et al. *Applied Catalysis B: Environmental*, 144:507-513(2014). 47c). A. Gordillo, et al. *RSC Advances*, 05(20):15194-15197 (2015).

48). A. Khan, et al. *Inorganic Chemistry Communications*, 79:99-103(2017).

49). M. M. Mohamed, et al. *Applied Catalysis B: Environ*, 142-143:432-441(2013).

50). J. Maria et al. *RSC Advances*, 03:18323-18331(2013).

51). S. Saha, et al. *Langmuir*, 26(04):2885-2893(2010).

52). W. Ahn, et al. *Applied Catalysis B: Environmental*, 74(01-02):103-110(2007).

53). I. Surina, et al. *J. Photochem & Photobio A: Chem*, 107(01-03):233-237(1997).

54). N. Ghanbari, *UltrasonicsSonochemistry*, In Press, Available online 13 May 2017

55). R. Kaur et al. *Chemistry An Asian J.* 09(01):189-198(2014).

56). B. Kumar, et al. *Advances in Natural Sciences: Nanoscience and Nanotechnology*, Published 24 May 2016. http://iopscience.iop.org/article/10.1088/2043-6262/7/2/025013.

2.09.00: Bimetallic Catalysts:

Bi-metallic effect in the reduction of 4-NP to 4-AP is seen by using Cu-Sn and Au-Cu alloys. The severe plastic deformation of metals leads to the formation of nano textured surfaces as well as the retention of significant strain energy, characteristics which are known to promote catalytic activity. Cu-Sn based alloys with plastically deformed surfaces of copper and copper-based alloys are used in the reduction of 4-NP to 4-AP by SBH (57). Taking advantage of the knowledge on bimetallic catalyzed hydrolysis of SBH for hydrogen generation, another concept of metal-organic framework (MOF) has been used for hydrogen generation from SBH. Dopamine-directed in-situ and one-step synthesis of Au@Ag core–shell NPs immobilized as MOF and used as a synergistic catalyst in the reduction of 4-NP to-4-AP using SBH as a hydrogen source. The bimetallic catalyst system is highly efficient and stable as demonstrated by multiple recycling experiments (58). An Ag/Fe$_3$O$_4$@SiO$_2$-APTES nanostructure was prepared by encapsulation of Ag in magnetically modified silica nanostructures, which was able to avoid undesirable phenomena such as agglomeration of Ag-NPs and their leaching. The catalyst (1mg Ag contents) is used for the reduction of 4-NP exhibiting excellent conversion yields (99%) in a short time (10min) under mild and green environmental conditions, and capable of almost 100% catalyst recovery (Scheme-13).

--Scheme-15

The reaction was fast as it was completed in 2-25 minutes depending upon the catalyst amount as 02g-2mins, 0.1g-10mins, and 0.01g-25mins etc.(Scheme-15). A mesoporous organosilanes layer was able to facilitate the contact of organic reactant with the catalyst surface while the incorporated magnetite allowed easy separation of the catalyst for reuse (59). In-situ synthesis of Ag-NPs@Ag(I)-AMTD metal-organic gel composite and its catalytic properties for the reduction of 4-NP to 4-AP with aqueous SBH are investigated. This material exhibited a remarkable and durable activity for the catalytic reduction of 4-NP to 4-AP and other NAs by SBH in aqueous solution. This method affords a facile means of embedding the active Ag-NPs within the metal-organic gel matrix and offers the necessary stability of the resulting silver nanostructures for catalytic transformation (60).

References:

57a). E. Menumero, et al. *Catalysis Science & Technol.*, 06(14):5737-5745(2016).

57b). S. Thota, *Chemical Communications*, 52(32):5593-5596(2016).

58). P. Huang, W. Ma, *Chemistry An Asian J*, 11(19): 2705–2709(2016).

59). R. Ahmadi, M. Jafarzadeh, et al. *Monatshefte für Chemie - Chemical Monthly*, 148(08):1423–1431(2017).

60). Y. Cheng, et al. *Colloids and Surfaces A: Physicochemical and Engineering Aspects*, 522:43-50(2017).

2.10.00: Recent Developments in the Synthesis of 4-AP from 4-NP Using Metal Catalyst over Solid Supports:

Some of the most recent publications, especially about the synthesis of 4-AP from 4-NP are given below along with relevant references for the reader's ready reference.

The metal catalysts are supported on inorganic solids, natural products, resins etc.

1). Ag-NCs on electron-rich PPy-MAA-composite, for reduction of 4-NP. S. Giri, et al. *Applied Catalysis B: Environmental*, 209:669-678(2017).

2). Protein-directed Au-NPs with excellent catalytic activity for 4-NP reduction with $NaBH_4$ (SBH). K. Liua, *Materials Science and Engineering*: C, 78: 429-434(2017).

3). Cu/graphene with high catalytic activity prepared by glucose blowing for reduction of 4-NP with $NaBH_4$. L. Jin, et al. *J. Cleaner Production*, Vol; 161: 655-662(2017).

4). New insights for the mechanism of reduction of 4-NP to 4-AP with $NaBH_4$ have been revealed. X. Kong, et al. *Chem Physics Letters*, 684:148-152(2017).

5). Catalytic reduction of 4-NP by palladium-resin composites. N. Jadbabaei, et al. *Applied Catalysis A: General*, 543:209-217(2017).

6). Fabrication of highly stable metal oxide hollow nanospheres for 4-NP reduction. G. Wu, et al. *ACS Applied Materials & Interfaces*, 09(21):18207-18214(2017).

7).Fe_3O_4/Fe-Ni Nanostructures selective catalytic reduction of *p*-nitrophenyl compounds. D. Wu, et al. *Inorganic Chemistry*, 56(09):5152-5157(2017).

8). Enhanced stability and catalytic activity of bismuth nanoparticles modified with porous silica. K. Chen, et al. *J. Physics and Chemistry of Solids*, 110: 0 9-14(2017).

9). Cu/graphene with high catalytic activity prepared by glucose blowing for reduction of 4-NP. L. Jin, et al. *J. Cleaner Production*, 161:655-662(2017).

10). Insights into the reduction of 4-NP to 4-AP on catalysts. X. Kong, *Chemical Physics Letters*, 684:148-152(2017).

11). Catalytic reduction of 4-NP by palladium-resin composites. H. Zhang, et al. *Applied Catalysis A: General*, 543:209-217(2017).

12).$Ni(OH)_2$/Graphene composite to Ni@Graphene core-shell for NP reduction. J. Wu, et al. *Carbon*, 117:192-200(2017).

13a). Magnetically-recyclable Ni@h-BN composites for efficient hydrolysis of ammonia borane: Y. Wu, et al. *Intl J. Hydrogen Energy*, 42(25):16003-16011(2017).

13b). Synergetic Catalysis of Non-noble Bimetallic Cu–Co Nanoparticles Embedded in SiO_2 Nanospheres in Hydrolytic Dehydrogenation of Ammonia Borane. Q. Yao, et al. *J. Phys. Chem. C*, 119 (25):14167–14174(2015).

14). Ag-Pt-NPs supported on magnetic graphene oxide nanosheets for catalytic reduction of 4-NP. M. Kohantorabi, *Appl Organometallic Chemistry,* 2017,78, e3806.

15). In situ mosaic Co-based N-doped mesoporous carbon for highly selective hydrogenation of NACs. F. Zhang, et al. *J. Catalysis,* 348:212-222(2017).

16). Microwave-irradiated preparation of reduced graphene oxide-Ni nanostructures for reduction of 4-NP. H. Qiu, *Applied Surface Science,* 407:509-517(2017).

17). Facile preparation of Ag/Ni(OH)$_2$ composites with enhanced catalytic activity for reduction of 4-NP. F. Bao, *RSC Advances,* 07(23):14283-14289(2017).

18). Catalytic reduction of 4-NPusing Au-NPs biosynthesized by cell-free extracts of Aspergillus sp. W. Shen, et al. *J Hazardous Materials,* 321:299-306(2017).

19). Difunctional Cu-doped carbon dots: catalytic activity for the reduction reaction of 4-NP. J. Du, et al. *RSC Advances,* 07(54):33929-33936(2017).

20). Catalytic reduction of 4-NP over Ni-Pd nanodimers supported on nitrogen-doped reduced graphene oxide. O. L. Liu, et al. *J. Hazardous Materials,* 320:96-104(2016).

21). Nickel phosphide nanostructures on nanofibrous membrane for reduction of 4-NP: K. Liu, et al. *Applied Catalysis B: Environmental,* 196:223-231(2016).

22). MOF derived porous Co@C hexagonal-shaped prisms with high catalytic performance. H. Li, *J. Materials Research,* 31(19):3069-3077(2016).

23). Ag nanoparticles on graphene oxide/TiO$_2$ nanocomposite for the reduction of 4-NP to 4-AP. B. Jaleh, et al. *Ceramics International, 42* (7), 8587-8596(2016).

24). Ni-NPs coated with carbon nanosheets as a highly active heterogeneous hydrogenation catalyst. G. Wu, et al. *Catalysis Communications, 79:*63-67(2016).

25). ZnO-loaded Co0.85Se nanocomposites (NCs) for decomposition of hydrazine hydrate for reduction of 4-NP. T. Xu, et al. *Appl Catal A: General,* 515:83-90(2016).

26). Green synthesis of Pd/RGO/Fe$_3$O$_4$ nanocomposites using Withania coagulants leaf extract for the reduction of 4-NP. M. Atarod, et al. *J Colloid and Interface Science,* 465:249-258(2016).

27). Synthesis of 4-AP from 4-NP reduction over Pd@ZIF-8. H. Jiang, et al. *Reaction Kinetics, Mechanisms and Catalysis,* 117:307-317(2016).

28). Fabrication of Au/CNT hollow fiber membrane for 4-NP to 4-AP reduction. Q. Zhang, et al *RSC Advances,* 06:41114-41121(2016).

29). Amorphous Ni-B/carbon nanohybrids: synthesis and catalytic hydrogenation of 4-NP to 4-AP. W. Liu, et al. *RSC Advances,* 06(97):94451-94458(2016).

30). Preparation of mesoporous TiO$_2$-C composites as an advanced Ni catalyst support for reduction of 4-NP. W. Gao, et al. *New J. Chem,* 40(5):4200-4205(2016).

31). Ni-NPs on carbon black for reduction of nitrophenols under mild conditions. J. Xia, et al. *Applied Catalysis B: Environmental*, 180:408-415(2016).

32a). NiWO$_4$-ZnO-NRGO ternary nanocomposite as an efficient photocatalyst for reduction of 4-NP: 32b). M. M. J. Sadiq, et al. *J. Physics and Chem. of Solids:* 109:124-133(2017*). 32c). Industrial & Engg Chem. Res.*, 55(27):7267-7272(2016).

33). CeO$_2$/Pd nanocomposites by pulsed laser ablation in liquids for the reduction of 4-NP to 4-AP. R. Ma, et al. *Ceramics International*, 41:12432-12438(2015).

34). A green approach for efficient 4-NP hydrogenation catalyzed by a Pd-based nanocatalyst. D. Zhang, et al. *Catalysis Communications*, 66:95-99(2015).

35). Advances in NP reduction by gold- and other transition metal nanoparticles. P. Zhao, et al. *Coordination Chemistry Reviews*, 287:114-136(2015).

36). Ni-NPs/rGO hybrids for catalytic reduction of 4-NP. Y. Tian, et al. *Colloids and Surfaces A: Physicochemical and Engineering Aspects,* 464:96-103(2015).

37). Pd-NPs on the surface of Fe$_3$O$_4$ @dextran particles in the 4-NP reduction reaction: R. S. Luciano, et al. *RSC Advances*, 05:8289-8296(2015).

38). Graphene-metal/metal oxide nanohybrids in heterogeneous catalysis. Y. Cheng, et al. *Catalysis Science & Technology*, 05:3903-3916(2015).

39). Cu/rGO/Fe$_3$O$_4$ nanocomposite using Euphorbia wallichii leaf extract for the reduction of 4-NP. M. Atarod, et al. *RSC Advances*, 05:91532-91543(2015).

40). Supported gold nanoparticles@agarose film by thiols and their synergy in efficient catalysis. N. Gogoi, et al. *RSC Advances.* 05:101860-101870(2015).

41). In Situ-Generated Co°-Co$_3$O$_4$/N-Doped carbon nanotubes for hydrogenation of nitroarenes. Z. Wei, et al. *ACS Catalysis*, 05(08):4783-4789(2015).

42). A MOF-derived nickel based N-doped mesoporous carbon catalyst for the reduction of 4-NP-to 4-AP. W. Zuo, et al. *RSC Advances*, 06(14): 11749-11753 (2016).

43). Cu-NPs using extract of the leaves of *Euphorbia esula L* for ligand-free reduction of 4-NP to 4-AP. M. Nasrollahzadeh, *RSC Advances*, 04(88): 47313-47318 (2014).

2.10.10: Recent Developments in 4-AP Synthesis with Metal-NPs:

In the recent times the metal-nanoparticles (M-NPs) have gained a considerable importance for their catalytic efficiency and selectivity in organic synthesis. There are some other reports from different groups on reduction of toxic 4-NP to 4-AP and other NPs in producing non-toxic APs with M-NPs over or without solid supports, and they are as listed below along with relevant references.

1). The Pd-Ag nanowires as efficient catalysts for the reduction of common and toxic NPs (2, 3 and 4-NPs) pollutants: J. Isagani & B. Janairo, *Intl. J. Philippine Science & Technology*, 08(02):41-43(2015).

2). Synergistic effects of Au-Pd nanoalloys and reducible supports on the catalytic reduction of 4-NP with SBH N. Bingwa, *Langmuir*, Article ASAP.

3). Magnetic polyaniline-chitosan nanocomposites decorated with Pd-NPs for reduction of 4-NP. M. M. Ayad. et al. *Molecular Catalysis*, 439:72-80(2017).

4). Rh-Ag/rGO Nanocatalyst: synthesis and superior catalytic performances for the reduction of 4-NP with SBH, i). C. Wang, *J. Mater. Sci.*, 52(16):9465–9476(2017). ii). C. Wang, et al. *J. Mater Sci.*, 52: 9465(2017).

5). Metal-free catalyst for the hydrogenation reduction of 4-NP to 4-AP is studied. J. Liu. et al. *J. Colloid and Interface Science*, 497:102-107(2017).

6). Biogenic synthesis of gold nanoparticles by yeast *Magnusiomyces ingens* LH-F1 for catalytic reduction of nitrophenols with SBH. X. Zhang. *Colloids and Surfaces A: Physicochemical and Engineering Aspects*, 280-285(2016).

7). Nanoscale zero-valent iron/SBH and its catalytic activity for reduction of PNP: S. Bae. *Applied Catalysis B: Environmental*, 182:541-549(2016).

8). Cucurbit [7]uril-stabilized gold nanoparticles as catalysts of the nitro compound reduction reaction with SBH E. Blanco, *RSC Advances*, 06(89): 86309-86315(2016).

9). Bimetallic yolk–shell Ni@PtNi nanocrystals supported on rGO and their excellent catalytic properties for 4-NP reduction. L. Mei, *New J. Chemistry*, 40(03): 2315-2320(2016).

10). Catalytic reduction of 4-NP using Ag-NPs with adjustable activity is investigated. C. Kästner & A. F. Thüneman, *Langmuir*, 32(29),:7383–7391(2016).

11). Au/Pd nanoalloys immobilized in spherical polyelectrolyte brushes with SBH. S. Gu, et al. *Phys. Chem. Chem. Phys.*, 17(42): 28137-28143(2015).

12). Au(0)/SiO$_2$ and synergism in the catalytic reduction of 4-NP with SBH P. Mahamallik & A. Pal, *RSC Advances*, 05(95):78006-78016(2015).

13). Hollow porous Au-NPs with tunable particle size for the reduction of 4-NP with NaBH$_4$ is reported. M. Guo, et al. *J. Hazardous Materials*, 310:89-97(2016).

14). Pt–Au dendrimer-like NPs supported on polydopamine-functionalized graphene for 4-NP (100% conversion after 6 runs) reduction with SBH. W. Ye. *Applied Catalysis B: Environmental*, 181:371-378(2016).

15). Metal oxide hollow nanospheres for 4-NP reduction to 4-AP: G. Wu, et al. *ACS Appl. Mater. Interfaces*, 09(21):18207-18214(2017).

16). A bioinorganic nanohybrid catalyst was synthesized by combining esterase with Pt-NPs. This hybrid catalyst can be successfully used in the multistep synthesis of acetaminophen (Paracetamol) in one pot and high yield. These results demonstrated that the nano-biohybrid catalyst offers advantages in the synthesis of fine chemicals with industrial applications. B. H. San, et al. *ACS Applied Materials & Interfaces*, 08(44): (2016).

17). Gold on methyl methacrylate (PMMA) gave the highest catalytic performance among polymer supported Au-NPs reported so far for the reduction of 4-NP to 4-AP using excess $NaBH_4$. K. Kuroda, et al. *J. Molecular Catalysis A: Chemical*, 298(01-02):07-11(2009).

CHAPTER-3

HALOANILINES (HANS)

3.00.00: Introduction:

Haloanilines occupy an important place in the synthetic organic chemistry both at academic and industrial levels for their wide range of applications in many industrial products and they are obtained by the catalytic reduction or hydrogenation of halonitrobenzenes (HNBs). HNBs are obtained by the nitration of halobenzenes, which gives a mixture of 2-, 3- and 4- or *o*-, *m*-, *p*- HNBs and also 1, 3-, 2, 4-, and 3, 4-Dinitro-HBs (DNHBs or HDNBs). Usually, the halo derivatives subjected to reductions are -Cl, -Br, and -I, which serve as precursors for other industrially important products such as dyes, paints, antioxidants in rubber industry, polyesters and also in many agro and pharmaceutical products. HANs are prepared by well-established and highly efficient metal catalyzed hydrogenation of respective HNBs in the presence of molecular hydrogen or other hydrogen sources.

Although, the substituents on NAs could be anything from alkyl groups to reducible functionalities like halogens, nitriles, or benzyl groups etc., some groups as halogens and O- and N-benzyls are susceptible for hydrogenolysis resulting into their cleavage from aryl C-X bond (X= removable group as halo or benzyl). Other sensitive groups as CN, CHO, COOR, CH=CH$_2$, etc. present in HNBs are also susceptible for undergoing hydrogenation to give the corresponding aniline derivatives. There are two main challenges during hydrogenations of HNBs substituted with sensitive groups and they are: i) preventing hydrodehalogenation of in-situ formed HANs and ii) chemoselective reduction of nitro group without affecting the other sensitive groups. Especially, considering the importance of HANs as intermediates or raw materials for their derivatives as pharmaceutical products (most of the fluoroquinolines as ciprofloxacin, norfloxacin, moxifloxacin, gatifloxicin etc. and many more), it is of paramount importance to find a solution to catalytic hydrogenation of HNBs to HANs without hydrodehalogenation, which is a serious challenge especially at complete conversion of the nitro substrates. It is reported that over a novel system of platinum/iron-oxide (Pt/γ-Fe$_2$O$_3$) nanocomposite catalysts, the hydrodehalogenation of HANs was completely suppressed even at complete conversion of HNBs. With the help of tools and techniques it is proposed that the electron transfer from Pt-NPs to oxygen vacancies in the activated Pt/γ-Fe$_2$O$_3$ catalysts may play an important role in completely suppressing the hydrodehalogenation of HANs in the hydrogenation of HNBs. Chao et al. have reviewed this matter in detail with emphasis on the great efforts and remarkable contributions of different authors during the long exploration for the solutions of this hard obstacle. Stress is placed on supported metal catalysts and polymer-protected metal nanoclusters or colloidal catalysts (1).

3.01.00: Synthesis of (*o*-, *m*-, *p*-,) Chloroanilines (CANs):

Just like other derivatives, chloroanilines are also reffered to as *o*-, *m*-, *p*- or 2-,3-,4-chloroanilines.

3.01.10: 2-Chloronitrobenzene (2-CNB): 2- or *o*-CNB itself is not a valuable product, but is a precursor to other useful compounds. The compound is particularly useful because both of its reactive sites can be utilized to create further compounds those are mutually ortho-derivatives. For example, 2-chloroaniline is a precursor to 3, 3'-dichlorobenzidine, which is a precursor to many dyes and pesticides. 2-CNB can be reduced to 2-CAN with catalytic hydrogenation process in the presence of transition metal catalysts. However, here the Bechamp NB reduction process using Fe/HCl as a catalytic system is relavant (2a, 3).

3.01.11: 3-Chloronitrobenzene (*m*- or 3-CNB): Since the nitration of chlorobenzene (CB) synthetic route does not efficiently produce 3-isomer, the route most commonly used by the chemists is the chlorination of NB. This reaction must be carried out with a sublimed iron (III) catalyst at 33-45°C. 3-CNB can be reduced to 3-CAN with Fe/HCl mixture by the Bechamp reduction procedure. 3-CAN is a useful compound and sometimes it is referred to as an Orange GC Base (2b, 3).

3.01.12: 4-Chloronitrobenzene (*p*- or 4-CNB): 4-CNB or *p*-CNB is a pale yellow solid and is an organic compound with the formula $ClC_6H_6NO_2$. *p*-CNB is an intermediate in the preparation of a variety of derivatives as i) 1,3-, 2,4-dinitroCB and 3,4-dichloroNB through its nitration, ii) with iron metal it gives 4-CAN, which is a common intermediate in the production of a number of industrially useful compounds, including common antioxidants found in rubber. The electron-withdrawing nature of nitro-group makes the benzene ring susceptible to nucleophilic substitution of the halides in HNBs. Thus, the strong nucleophiles as hydroxide, methoxide, amines and amide displace chloride to give respectively 4-NP, 4-nitroanisole, 4-NAN. Another major use of 4-CNB is its condensation with aniline to produce 4-nitrodiphenylamine (2c, 3).

3.02.00: Synthesis and Manufacturing of HANs:

Many other aryl amines and their derivatives, as AN analogs, as 4-nitroaniline (4-NA), *o* and *p*-phenylene diamines etc. have been synthesized and used in different industries as precursors to other products (2b). Some of the catalysts hydrogenate the benzene ring to cyclohexene or completely hydrogenate it to cyclohexane, as we have seen earlier in case of complete hydrogenation of NB to cyclohexylamine (4). However, with suitable catalysts and optimized experimental conditions it is possible to selectively hydrodehalogenation of some of the polyhaloanilines to obtain mono- or di-HANs, which leads to 3 or 3, 5-CANs. For example, liquid phase catalytic hydrogenation with Pd/Charcoal (Pd/C) as a catalyst 3,5-dichloranilines are obtained in very good yields (≥95%) by selective hydrodechlorination of pentahalogenated compounds as starting materials, along with tri and tetra derivatives. This selective hydrodechlorination in 2 and 4 positions (in relation with NH_2 or OH group) is effective in aqueous acidic medium containing HCl and traces of ions such as I^-, Br^-, Sn^{2+}, Ag^+, Bi^{3+}, Pb^{2+}, Hg^+, Tl^+ etc. It is also effective in an organic medium (chlorobenzenes as solvent) containing HCl and a Lewis acid like $AICI_3$, ZnI_2, $SnCl_2$. It is obvious that only protonated forms of substatres are chemisorbed and selectively dechlorinated in 2 and 4 positions (5).

Halogenated aromatic amines have multiple applications in the manufacture of a variety of organic products with commercial applications. Hence, their chemoslective synthetic procedures in high yields and selectivity are of paramount importance in the cost-effective production of commercial products. The most abundantly produced and used HANs are the chloroanilines (CANs), which are highly valuable organic intermediates used in different industries as agro-products such as herbicides and pesticides, paints and dyes, polyesters and in the manufacturing of many pharmaceutical products, cosmetics, and many more products (6). In the old days, usually CANs were synthesized or manufactured by gas phase or liquid pahse catalytic hydrogenation of the respective CNBs (7, 8).

References:

1).X. Chao, et al. *Current Organic Chemistry,* 16(02):280-296(2012).

2a). https://www.revolvy.com/topic/2-Nitrochlorobenzene&uid=1575. 2b).https://www. revolvy.com/topic/3-Nitrochlorobenzene&uid=1575. 2c).https://www.revolvy.com/ main/show.php?cmd=list&qf=all.

3). G. Booth, "Aromatic Nitro Compounds" in Ullmann's Encyclopedia of Industrial Chemistry, Wiley-VCH: Weinheim, 2005. doi:10.1002/14356007.a17_411

4). M. L. Buil, et al. *Organometallics, 29* (19):4375–4383(2010).

5). G. Cordier et al. *Studies in Surface Science and Catalysis,* 41:19-31(1988).

6). A. Boehnecke, et al., CICADS-Report-48, WHO, Geneva, pp. 78(2003).

7). Gas Phase: F. Cardenas-Lizana, a). *ACS Catalysis,* 03(06):1386–1396(2013). b). *Applied Catalysis A: General,* 473:41-50((2014). c). X. Wang, et al. *J. Phys. Chem. C,* 117(02):994–1005(2013).

8). Liquid Phase: a). Y. W. Chen, et al. *Modern Research in Catalysis*, 2013, 2, 25-34(2013). b). Book: *Hydrogenation with Low-Cost Transition Metals*, Jacinto Sa, Anna Srebowata, CRC Press, 04-Nov-2015 - Science - 203 pages.

3.03.00: Catalytic Hydrogenation of HNBs or HNAs to HANs:

(HNBs = Halonitrobenzenes, HNAs = Halonitroarenes, HANs = Haloanilines)

During past few dacades various supported metal catalysts have been used for liquid-phase hydrogenation of *o*-, *m*-, *p*-CNBs. In some cases the rsults are quite satisfactory, but their commercial applications have got limitations because of environmental concerns. These concerns were due to the use of organic solvents in the hydrogenation process using high pressures and high temperstures and ultimately leading to generation of huge amounts of wastes, their expensive disposals and environmental pollutions along with high costs associated with each step involved therein. As a conscious scientific society, there are always constant efforts in resolving the commercial problems associated with benefits to human life. Keeping in mind these issues, a lot of progress has been made within last couple of decades in developing new, efficient methodologies for the reduction HNBs to HANs. The fruitful work published in the public domain is summarized below to highlight the recent developemnts in the methodologies and tehcnologies in the chemoselective catalytic hydrogenation of HNBs to the corresponding HANs using differnet pathways, procedures, methods for the synthesis of different (*o*-, *m*-, *p*-) HANs. In general the research is focused on developing efficient metal based catalysts as Transition Metals, Mono-Metal-NPs (M-NPS), Bi-Metal–NPs, (BM-NPs). These metal nanoparticles (M-NPs) are prepared by a special method in each case and are mounted on solid supports. Based on the kinetic studies and insights into mechanistical pathways during the process developments, the emphasis is on fine or nanosized metal particles specially prepared on solid supports, so that the synergistic effects between the M-NPs and the surface structures help in achieving the desired catalyst activities and the selectivity of the end products (HANs). Also care is taken to see the catalysts under given conditions give 100% substrate conversion and quantitative yields. Interesting enough, there are some reports with almost 100% conversions of CNBs and 100% selectivity and yields of the desired *o*-, *m*-, or *p*-CANs. In some cases very negligible or absolutely no dehalogentaion of CANs had detected. Let us have a look at the work on selective catalytic hydrogenations of *o*-, *m*-, and *p*-CNBs.

Reduction of NAs is accomplished using hydrogenation processes. There are two ways of hydrogenations, i) Catlytic Hydrogenation (CH) usually using Metal Based Catalysts and Molecular Hydrogen and ii) Catalytic Transfer Hydrogenation (CTH) which makes use of Metal Catalysts and Different Hydrogen Sources Other Than Molecular Hydrogen. First we will look at the CH using molecular hydrogen over different metal catalysts and then cover the other hydrogen sources.

3.03.10: Catalytic Hydrogenation of *o-* or 2-CNB

3.103.10a: Non-Noble Metal Catalysts:

2-, or *o*-Chloroaniline (*o*-CAN) is one of the first three (*o-*, *m-*, *p-*) and many other CANs and finds applications in polymer, rubber, pharmaceutical and dye industries. Some of the most significant and new developments in synthesizing *o*-CAN are presented below. Pt/C-Ni-NPs are selective towards HNB hydrogenations. Similarly, a three dimensional flower-like Co-Ni/C bimetallic catalysts composed of interlaced carbon flakes with highly dispersed M-NPs (the Co 50:Ni 50 /C) were synthesized and tetsed for the selective hydrogenation of *o*-CNB to *o*-CAN (Scheme-1).

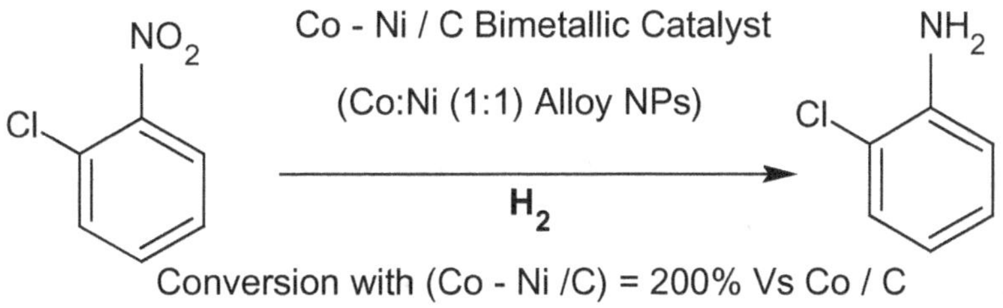

--Scheme-1

The conversion of *o*-CNB over the bimetallic Co-Ni/C catalysts was increased by up to 200% compared with that over a mono-metallic Co/C catalyst, which was obviously due to synergistic interactions between the Co and Ni species (9). Liquid-phase catalytic hydrogenation of 3, 4-DCNB over Pt/C catalyst under gradient-free flow conditions in the presence of pyridine, along with dehalogenation is investigated (10). Monodispersed Pt-NPs were prepared by reduction of Pt-acetylacetonate in octadecene with the presence of $Fe(CO)_5$. The catalyst exhibited high activity and selectively for hydrogenation of *o*-HNBs to the corresponding *o*-HANs under mild reaction conditions (11).

The γ-Fe_2O_3/MC catalyst obtained by in-house synthesis exhibits catalytic activity and selectivity for the hydrogenation of Cl^-, Br^- and I^- functionalized NBs without any obvious dehalogenation. The hydrogenation reactions had a product yield and selectivity for the corresponding HAN of 100% when using hydrazine hydrate as the reducing agent (12). The reduction of *o*-CNB in soils by Fe(0) at ambient temperature and pressure was studied (Scheme-2).

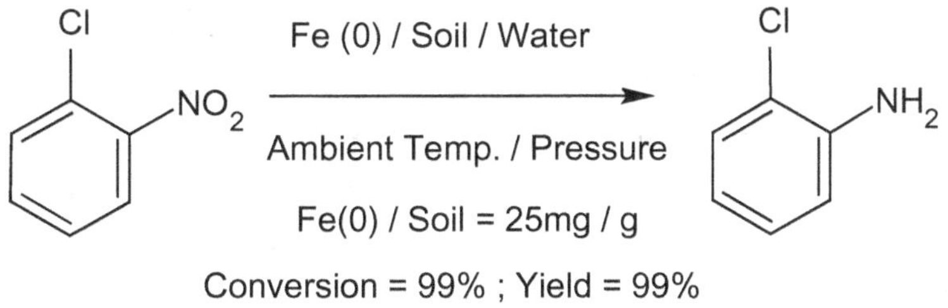

--Scheme-2

The effects of several parameters of the reaction system as well as the changes of the reaction products with time were investigated. The *o*-CNB was initially reduced by Fe°

to o-chloronitrosobenzene (o-CNOB), and finally to o-CAN. After 4 hours of reaction the efficiency of the reduction reached 99% (13).

A type of Pd/α-Fe$_2$O$_3$ catalyst was synthesized by a convenient UV light-induced reduction in the presence of Fe^{3+} ions, and tested for the hydrogenation of o-CNB to o-CAN under mild conditions as 50°C, 2 h, 1 MPa hydrogen pressure to furnish 100% o-CNB conversion and 91.4% selectivity of o-CAN (14).

Cobalt-doped Fe$_3$O$_4$-NPs is prepared and the catalyst is used for the hydrogenation of CNBs to CANs (Scheme-3). The reaction proceeds at low temperatures in aqueous phase and at atmospheric pressure, resulting in approximately 100% yield and selectivity.

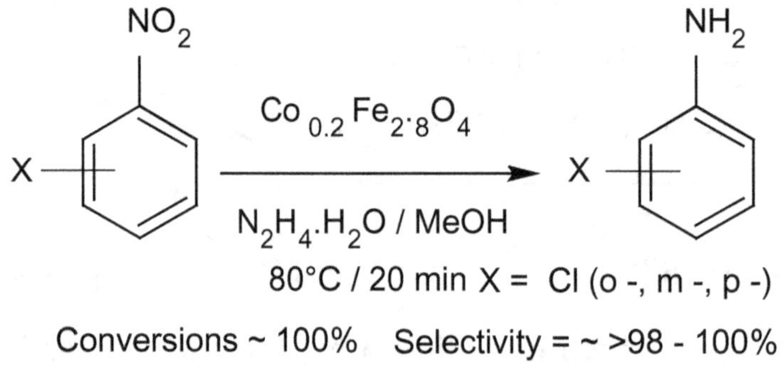

Conversions ~ 100%   Selectivity = ~ >98 - 100% --Scheme-3

The o-CNB, m-CNB and p-CNB conversions were 100%, while the selelctivity were 99.4%, 98.9% and 99.6% respectively. The additive effect of the metal was seen as higher concentrations of Co and lower iron contents led to lower CAN selelctivity up to about 97-96%. Actually, the individual Fe° and Co° also gave high selelctivty but the nitro conversions were low, especially in case of iron (15).

Shi et al. have studied the interactions of CuO active components with γ-Al$_2$O$_3$, CeO$_2$, and CeO$_2$/γ-Al$_2$O$_3$ supports. They found that CuO clusters could be stabilized based on the strong Cu–Ce interaction. The stabilized CuO clusters play a key role in NO, CO, reductions (16). Similarly, hydrogenation of o-CNB to o-CAN over Ni/TiO$_2$ catalyst prepared by sol-gel method gives 99% o-CNB conversions and 99% o-CAN selectivties under suitable reaction conditions (17).

3.03.10b: Catalytic Hydrogenation Using Noble-Metal Catalysts:

Pt/C catalyst is one of the most important and common catalysts in hydrogenation of o-CNB to 2, 2′-dichlorohydrazobenzene (18). Noble metals dispersed in PVP have shown good promise for the selective reduction of HNBs to HANs. Here, the selective hydrogenation of o-CNB to o-CAN was accomplished without dehalogenation over finely dispersed PVP-stabilized Ru colloids (PVP–Ru). It was observed that some metal cations added to the system increased the activity while the selectivity remained constant (19). A PVP-Pt catalyst was synthesized via chemical reduction of Pt ions with hydrazine hydrate in a PVP/n-butanol/H$_2$PtCl$_6$ aqueous solution and tested in the liquid-phase hydrogenation of m-CNB to m-CAN under mild conditions as 303K and 0.1 MPa H$_2$

pressure. The as-prepared catalyst exhibited higher activity and selectivity than prepared via conventional ethanol reduction with the same Pt load. The catalytic performance of PVP-Pt catalyst was remarkably improved by addition of 0.2 wt %Sn^{4+}. The modification mechanism may be related with the interaction of Sn^{4+} with nitro group of m-CNB and -NH_2 in m-CAN (20).

The method of preparation of a catalyst has its impact on the performance of the catalyst, as the catalysts of noble metals on activated carbon are subjected to the action of a sulfoxide together with hydrazine or its derivatives and are found useful in the reduction of o-CNB to o-CAN (Yield= 98.4%). In contrast, Pt on activated carbon (Pt/AC) catalyst was treated first with DMSO and then with hydrazinehydrate solution. This catalyst reduced 2-CNB to 2-CAN, with a reproducible yield of 99.6%. Also, the 3-chloro-4-methyl aniline (99.8%) and 2-CAN (99.6%) were obtained when the relevant nitro substrtaes were treated with hydrazine at 90°C in the presence of above prepared catalyst. The high yields of CANs were attributed to the preparation method of the Pt catalyst using DMSO and hydrazine (21). The reduction of m-CNB to m-CAN without or very little hydrodehalogenation is a matter of interest for a long tine now (22). Mali et al. have published a brief review of nano-catalysts application in the reduction of substituted nitro-compounds into corresponding ANs is reported. The influence of the metal nature, catalyst's supports and catalysts particle size in the reaction's selectivity has been highlighted. Nano-structured advantages of catalysts are reported evincing future challenges (23). Some of the examples presented above and below demonstrate the fruits of the hard work of many research group in developing appropriate catalyst systems for chemoslective hydrogenations of HNBs with desired results and without or negligible dehalogenation. Liquid phase hydrogenation of CNB isomers (x-CNB x = 2, 3, 4) to the corresponding x-CANs under mild reaction conditions over Pd and Pt catalysts supported on cationic resins (Dowex-D) has been studied. The substrate conversions upto 90% and the selectivities from 85% (2-CAN and 3-CAN) to 95% (4-CAN) were achieved in diethyl ether-methanol as solvent (24). The use of magnesium fluoride as a support for ruthenium (Ru/MgF_2) has enabled to obtain a catalyst of high activity and selectivity (100%) from hydrogenation of o-CNB to o-CAN in the liquid-phase at 353 K under 4 MPa H_2 in aqeous methanol (H_2O:MeOH = 1:2) (25) Magnetic nanocomposite (Pt/γ-Fe_2O_3) catalysts with high activity and selectivity for selective hydrogenation of o-CNB are reported. There was a complete conversion of o-CNB to o-CAN without any hydrodehalogenation of o-CAN (26). Polyurea (PU) spheres with size of 2-10 μm were derived through the polymerization of CO_2 with 1,4-butanediamine, and a series of PU-supported metal nanocatalysts including Pt/PU, Au/PU, Pd/PU (~3.0nm) were prepared by the reduction of metal ions by SBH. The Pt/PU catalyst was tested in the catalytic hydrogenation of o-CNB, and a high selectivity of 99.5% toward o-CAN at complete conversion of o-CNB was obtained at room temperature (27). Pt°-NPs in the size range of 0 to 10 nm were prepared in-situ by impregnation of $H_2PtCl_6.6H_2O$ into the nanopores of modified and activated montmorillonite followed by reduction with different reducing agents. The catalyst prepared by reduction with hydrazine was found highly active and efficient in the chemoselective reduction of o-CNB (100% conversion) with >99% selectivity towards o-CAN at 45°C, 10bar H_2 for 15min. There was very negligible dehydrohalogenation due to C-Cl bond cleavage (28). Some groups have prepared the M-NPs using natural products resources. For example, here the selective hydrogenation of o-CNB is perfomed over biosynthesized Ru–Pt bimetallic nanocatalysts, which were synthesized using the *Diospyros kaki* plant leaf extract and then immobilized on the carbon substrates to make supported nanocatalysts. The biogenic $Ru_{0.5}Pt_{0.5}$/Vulcan XC catalyst displayed outstanding catalytic performance in selective hydrogenation of o-CNB under mild condition without adding solvent (29). The pathways during hydrogenation of o-CNB over BM-NPs are shown in Scheme-4. Magnetically recyclable Pt/C-Ni catalysts consisting of Pt-NPs on the surface of flower-like C-Ni-NCs was synthesized and tested for the selective hydrogenation of o-CNB to o-CAN at 30°C and atmospheric pressure. The Pt/C(Ni) catalysts show higher selectivity towards o-CAN than the Pt/C catalyst due to the synergistic effect between Pt and Ni species (30). Liu et al. have studied the hydrogenation of o-CNB to o-CAN over a PVP-stabilized Pt/Ru colloid with 99.0% selectivity to o-CAN at 100% conversion of o-CNB (31). The synthetic pathways for o-CNB over bimetallic NPs are illustrated in Scheme-4.

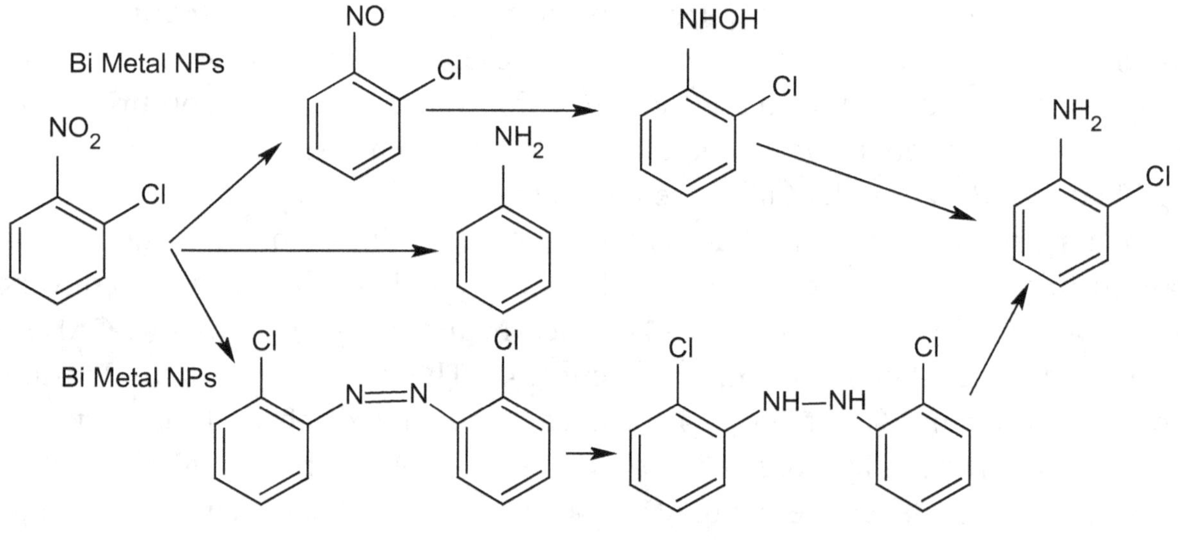

Scheme-4: Pathways of o-CNB Hydrogenation over Bimetallic NPs

Core@shell structured bimetallic-NPs are of immense interest due to their unique electronic, optical and catalytic properties. However, their synthesis is non-trivial. Au/Pd and Pt/Pd core@shell NPs have been synthesized and tested for the selective production of industrially valuable CANs from CNBs (32). Therefore, Spitale et al. have synthesized and analyzed the Au/Pd nanoalloys at atomic level. Grand canonical simulations using different sampling procedures were used to study the growth mechanism of Pd atoms on Au seeds of different shape (33). Rh-Ni (3:1) is a highly synergistic bimetallic catalyst system for the reduction NAs with high selcitivity (>93%) 1atm H_2 pressure at room temperature is accomplished (34) (Scheme-5).

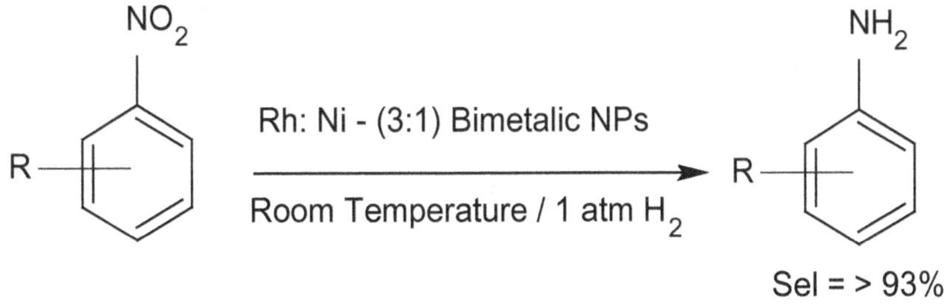

Sel = > 93% --Scheme-5

3.03.10c: Solvent-Free Synthesis of o-CAN over Noble-Metal Catalysts:

The Pt-NPs were prepared by in-situ reduction and immobilized on or stabilized with polyethylene glycols (PEGs). The Pt-NPs/PEG catalyst was found suitable for a highly selective reduction of o-CNB to o-CAN with selectivity >98% and 99.7% yield at complete conversion of o-CNB (Scheme-6) (35).

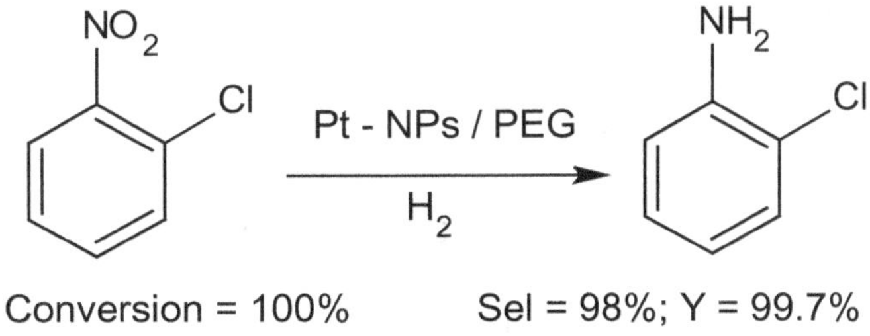

Conversion = 100%   Sel = 98%; Y = 99.7% --Scheme-6

A simple and efficient method used to prepare large and uniform Pd-NPs supported on oxygen-poor AC and the catalyst was used for the solvent-free selective hydrogenation of o-CNB to o-CAN, with selectivity to o-CAN up to 99.8 %. This catalyst shows great potential for industrial applications, especially in views of the green and efficient synthesis of o-CAN (36). In another case, the solvent-free selective hydrogenation of o-CNB to o-CAN over Pt-NPs/Al$_2$O$_3$ is studied. Addition of other metal cations to the system improved both the o-CAN selectivity and o-CNB conversions (Schem-7).

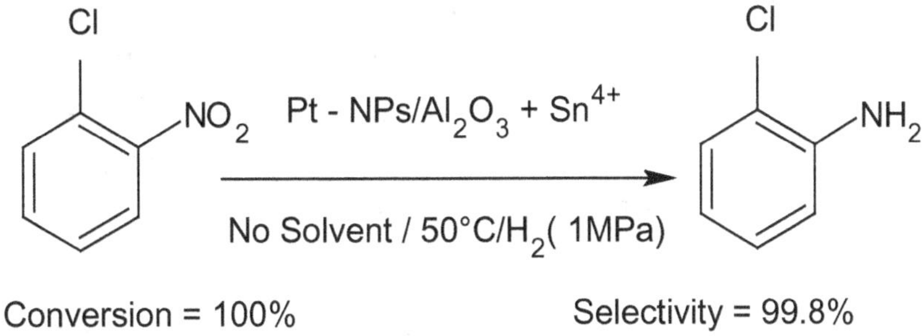

Conversion = 100%   Selectivity = 99.8%   --Scheme-7

Especially, 99.8% selectivity to o-CAN was attained at ~100% conversion of o-CNB with the addition of a metal cation (Sn^{4+}) to reaction system (37). A solvent free but highly selective hydrogenation of o-CNB and m-CNB, with selectivity to the corresponding CANs of ≥99.4% and complete conversion of the substrates was realized over a robust Pt/Fe$_3$O$_4$ catalyst (Scheme-8).

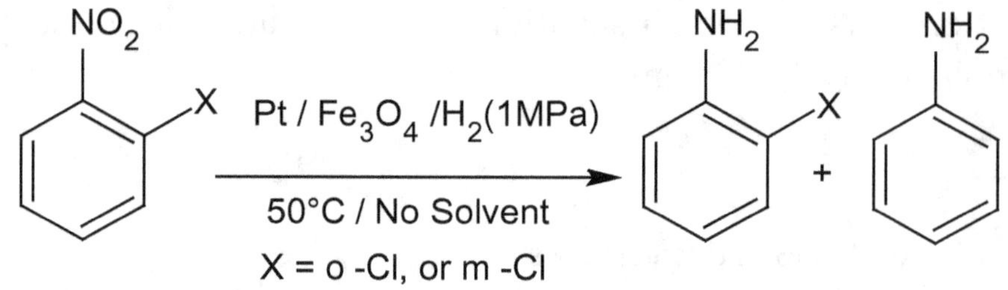

Conversion = 100%   Sel = 99.4%   0.5% --Scheme-8

The catalyst was prepared by adsorbing Pt-NPs on a Fe$_3$O$_4$ support (38). Liu et al. have reported the selective hydrogenation of o-CNB using supported Pt-NPs without solvent. Colloidal Pt-NPs (2.2 nm) were immobilized on γ-iron oxide (γ-Fe$_2$O$_3$) powder to make a supported catalyst (Pt-NPs/γ-Fe$_2$O$_3$) with controlled Pt-NPs size and size distribution (Scheme-9). This catalyst was used to catalyze the selective hydrogenation of o-CNB to o-CAN without using any solvent.

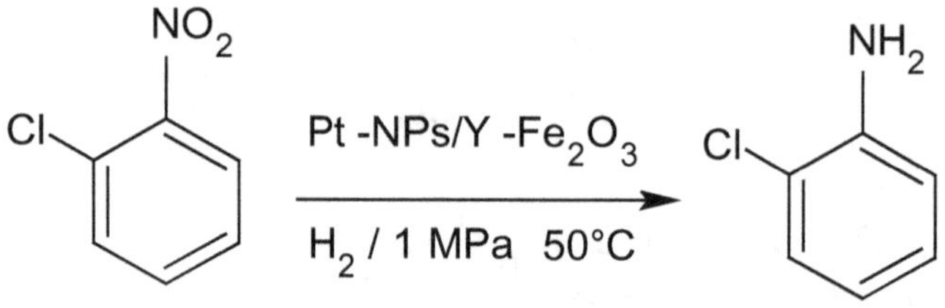

99.95% 2 -CNB Conversion and 99.9% 2 -CAN Selectivity --Scheme-9

The results were extraordinary as >99.9% selectivity for o-CAN and 99.95% conversion of o-CNB was achieved (39).

Selective hydrogenation of o-CNB over SnO$_2$ supported Pt-Ru bimetallic nanocatalysts without solvent is presented. Pt-SnO$_2$ heteroaggregate nanocatalysts were synthesized by in-situ transformation of Pt@Sn core–shell NPs and their catalytic performance for hydrogenation of various NAs was investigated. Based on the Pt@Sn-NPs core@shell-like structures and Pt-SnO$_2$ heteroaggregate nanostructure's spatial arrangements, the Pt-SnO$_2$/Al$_2$O$_3$ nanostructures were proved better for hydrogenation of various NAs relative to individual Pt/Al$_2$O$_3$ nanocatalysts. Mechanistically, the Pt-SnO$_2$ nanocatalysts can slightly facilitate the adsorption of H$_2$ and o-CNB and strongly weaken the binding of Pt/o-CAN, resulting in more available reactants and easier release of products from the catalyst surfaces. It is observed that the enhanced catalytic performance may originate from a cooperative interaction between Pt and SnO$_2$ (40).

3.04.00: Catalytic Hydrogenation of m-HNBs to m-HANs (Cl, Br, I):

Monometallic and bimetallic catalysts on or without solid supports have been used for chemoslective hydrogenation of m-HNBs (Cl, Br, I). Further, experimentally it is also proved that the type of supports, the method of M-NPs preparations and the addition of an additive as a base or an ion helps enhancing the substrate conversions and end product selectivities.

3.04.10: Catalytic Hydrogenations of m-HNBs over Noble-Metal Catalysts:

Supported monometallic (Pt, Ni,) and bimetallic (Ni–Pt) catalysts were prepared for the selective liquid phase hydrogenation of m-CNB to m-CAN. It was found that the use of sodium carbonate as an additive substantially reduced the extent of dehydrohalogenation in the case of monometallic 1% Pt/C catalyst, which gave the highest selectivity of 96% to m-CAN. Ni–Pt bimetallic catalyst although showed almost complete selectivity (>99%) to m-CAN, its activity was several fold lower than that of 1% Pt/C–Na$_2$CO$_3$ system. Compared with Ni monometallic catalyst, bimetallic Ni–Pt showed higher activity and selectivity due to the presence of electron rich surface and metallic Pt stabilized by Ni having lower ionization potential compared with Pt (41). The Pt/Y-Fe$_2$O$_3$ catalyst has attracted a great attention in the recnt time, which is visible from its applications in the hydrogenation of different HANs (Cl, Br, I–ANs). Here the authors have presented a case Pt/γ-Fe$_2$O$_3$ catalyst which exhibited excellent catalytic activity at 2MPa H$_2$ pressure towards chemoslective hydrogenations of o,m,p-BNBs to the corresponding o,m,p-BANs with selectivities > 99.9% (Scheme-10).

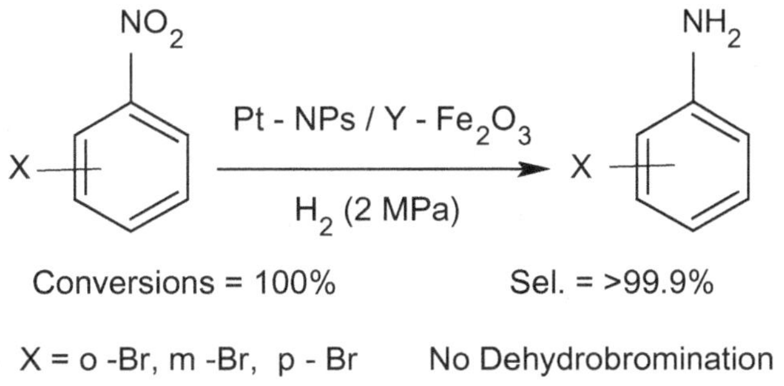

Conversions = 100%   Sel. = >99.9%

X = o -Br, m -Br, p - Br   No Dehydrobromination --Scheme-10

For the first time, there was no hydrodebromination of BANs observed even at the complete conversion of the nitro substrates (42).

The PVP-Pt catalyst was prepared and tested for the m-CNB hydrogenation to m-CAN. The catalyst activity and the selectivity were significantly improved compared with the ethanol reflux method. The addition of a small amount of a rare earth ion, Pr^{3+}-modified PVP-Pt/Pr (Pt: Pr molar ratio of 1:0.14) further improved the activity of the catalyst (43). Li et al. have further probed into the selective liquid phase hydrogenation of m-CNB to m-CAN over PVP/Pt and Pt/Sn catalysts under mild conditions as 303K and 0.1MPa H$_2$. The catalytic performance of PVP-Pt catalyst was remarkably improved by addition of 0.2 wt % Sn^{4+}. The modification mechanism may be related with the interaction of Sn^{4+} with nitro group of m-CNB and the NH$_2$ in m-CAN ultimately giving 99.3% selectivity for m-CAN (20). The promoting effect of CNTs confinement on the catalytic performance in the hydrogenation of m-CNB over amorphous Ni-B/CNTs catalysts: is reported (44).

Au-NPs deposited on α-Fe$_2$O$_3$ substrate were prepared by different procedures including the deposition–precipitation in presence of urea or NaOH, and by the traditional wet impregnation followed by calcination. The catalyst was tested for the chemoselective hydrogenation of o-, p- and m-CNBs at ambient temperature on 0.5%-Au/α-Fe$_2$O$_3$ catalysts. Especially m-CAN was obtained in 97% yield at 99.7% conversion, the highest amongst the three isomers (45). Magnetic-Pt (Pt-NPs (3.1nm)/Fe$_3$O$_4$) catalyst was found highly selective for hydrogenation of m-CNB. The activity of Pt was dependent on its particle size and the favorable support from Fe$_3$O$_4$ (46).

3.04.11: Hydrogenation of m-HNBs over Non-Noble Metal Catalysts:

A Fe-modified Co–B amorphous alloy supported on carbon nanotubes (Fe–Co–B/CNTs) for the selective hydrogenation of m-CNB with about 97% selectivity to m-CAN is achieved (47). There are not many reports on selective hydrogenation of bromonitrobenzenes (BNBs) and especially those on iodonitrobenzenes (INBs) without dehalogenation, but some work is already covered under different sections in this chapter and also one can vist the review of the published work (48).

References:

09). Y. Xie, et al. *Chinese J. Catalysis,* 33(11-12):1883-1888(2012).

10). V. G. Dorokhov, et al. *Russian Chemical Bulletin,* 65(08):2040-2045(2016).

11). R. Xie, et al. *Science China Chemistry,* 58(06):1051-1055(2015).

12). M. Tian, et al. *Green Chemistry,* 19(06):1548-1554(2017).

13a). Y. Chen, et al. *Chinese J. Environmental Engineering:* (2007-01). 13b). H. Hu, et al. *Environmental Science & Pollution Research,* 21(07):5132-5140(2014).

14). W. Jiang, et al. *Applied Catalysis A: General,* 520:65-72(2016).

15). B. Yang, et al. *Nano Research,* 09(07):1879-1890(2016).

16). C. Shi, et al. *Chinese Journal of Catalysis,* 33(09):1455–1462(2012).

17). C. Jixiang et al. *Chem. Engg. Journal,* 148:164-172(2009).

18). C. Jiang, et al. J Zhejiang Univ Sci B. 06(05):378–381(2005).

19). M. Liu, et al. *J. Molecular Catalysis A: Chemical,* 138(02-03):295-3-3(1999).

20). F. Li, *Russian J. Physical Chem. A,* 89(05):766-770(2015).

21). G. Lippert, et al. US 3897499 A (1975).

22a). F. C. Trager, US 2772313:-A (1956). 22b). J. R. Kosak, US 3145231 A(1964).

23). M. Mali, et al. *Synth Catalysis:* 02:02.08(2017).

24). M. Kralik et al. *Chemical Papers*, 68(12):1690–1700(2014).

25). M. Pietrowski, et al. *Catalysis Letters*, 128(01-02):31-35(2009).

26). J. Zhang, et al. *J. Catalysis*, 229(01):114-118(2005).

27). L. Hao, et al. *J. Colloid Interface Sci.*, 15:424:44-48(2014).

28). D. K Dutta & D. Dutta, i) WO2015011716-A1 (2915): ii). US9284259-B2(2016).

29). Z. Zhang, et al. *Ind. Eng. Chem. Res.*, 55(26):7061-7068(2016).

30). Y. Xie, et al. *Catalysis Communications*, 28:69-72(2012).

31). M. Liu, et al. *J. Catal.* 278(01):01-07(2011).

32). C. J. Serpell, *Nature Chemistry*, 03:478–483(2011).

33). A. Spitale, et al. Phys Chem Chem Phys., 17(42): 28060–28067 (2015).

34). S. Cai, et al. *ACS Catalysis*, 03:608–612(2013).

35). H. Cheng, et al. *J. Colloid & Interface* Sciences, 336(02):675-678(2009).

36). Q. Zhang, et al. *Reac Kinet Mech Cat*, 114:629–638(2015).

37). Q. Bai, et al. *Progress Natural Science: Materials Intl*, 25(03):179-184(2015).

38). C. Lian, et al. *Chemical Communications*, 48(25): 3124-3126 (2012).

39). M. Liu, et.al. *Applied Catalysis A: General*, 439–440:192–196(2012).

40a). M. Liu, *Chemical Engineering J.*: 232:89–95(2013). 40b).M. Liu, et al. *ACS Catalysis*, 07(03):1583–1591(2017).

41a). R. B. Mane, *Ind. Eng. Chem. Res.*, 51(48):15564–15572(2012). 41b). *C. V. Rode, et al. Ind. Eng. Chem. Res.*, 33(07): 1645–1653(1994).

42a). X. Wang, et al. *J. Molecular Catal: A-Chemical*, 273(01-02):160-168(2007). 42b). A Review: X. Chao, et al. *Current Organic Chemistry*, 16(02):280-296(2012)

43). F. Li, et al. *Reaction Kinetics, Mechanisms & Catal*, 116(02):479-489(2015).

44). F. Li, et al. *The Canadian J. Chem. Engg.*, Recetly Published Article, 2017.

45). C. H. Campos, et al. *Applied Catalysis A: General*, 482(127-136(2014).

46). W. Du, *et al. Ind. Engineering & Chemical Research*, 53(12):4589–4594(2014).

47). F. Li, et al. *Reaction Kinetics, Mechanisms & Catal*, 120(02):651-662(2017).

48). M. Pietrowski, *Current Organic Synthesis*, 09(04):470-487(2012).

3.05.00: Hydrogenation of *p*-Chloronitrobenzene (*p*-CNB):

3.05.10: Catalytic Hydrogenation of *p*-CNB with Molecular Hydrogen: The M-NPs are used in the reduction of NAs under different set of conditions as catalytic hydrogenations, photocatalyzed reductions, reductions using hydrazine, formic acid and other ammonium salts etc as hydrogen donors. Here, the emphasis is laid on a single metal based and bimetallic NPs used in reductions of NACs. NPs, as the name itself indicates, are very small particles usually measuring about few tens to 100nm in size. Because of the finite size they have much bigger surface area per weight compared to large particles and hence they are more active than large particles. In recent years, the researchers have made a very good progress in the field of NPs in terms of their different synthetic methods for preparing different sizes and their myriad applications, as in industrial coatings, in energy and electronics as conductors/semiconductors, polymer industry especially for making synthetic skin comprising Ni-NPs and a polymer, in medicines especially as anti-cancer, anti-bacterial, as drug and gene delivering agents, anti-ageing drugs, X-ray and MRI technologies, medicine preparations and last but not least as highly effective catalysts in chemistry. The recent developments in this field have been reviewed by some groups (50). Here, we have compiled the reactions used for enhancing the reaction rates, increasing the yields and selectivities, reducing the cycle times and costs, and also reducing the environmental loads. This will be evident from some of the reactions, where M-NPs as catalysts have enabled performing reactions with simple, low cost raw materials, at ambient temperatures and atmospheric pressures, with or without solvent, recyclable catalyst systems making them economically feasible and environmentally friendly processes even at commercial scales. In our current interests the synthesis of aryl amines and especially AN and its analogs, NPs have enabled researchers to come up with very simple processes based on strategies listed above. For example, besides catalytic hydrogenation of HNBs with molecular hydrogen, the synthesis of AN and its analogs by catalytic transfer hydrogenation of NB with transition non-noble metals as iron, zinc, tin etc. in association with formic acid or ammonium formate, ammonium chloride and hydrazine hydrate as the hydrogen sources have been used to deliver aryl amines in very high to excellent yields.

3.05.11: Catalytic Hydrogenation of *p*-CNB on Solid Supports:

Because of its noble nature, gold has superior ability of hydrogen adsorption and hence a great potential for competitive selectivity and activity in hydrogenation reactions. Preparation of gold-containing binary metal clusters (M = Cu–Au, Ag–Au, and Au-Ru) by co-deposition-precipitation method and for hydrogenation of CNBs is successfully investigated. The M-Au-Alloys (M with enriched surface) were loaded on TiO_2 and were tested for the hydrogenation of *p*-CNB in a batch reactor at 373 K and 1.1 MPa H_2 pressure (Scheme-11).

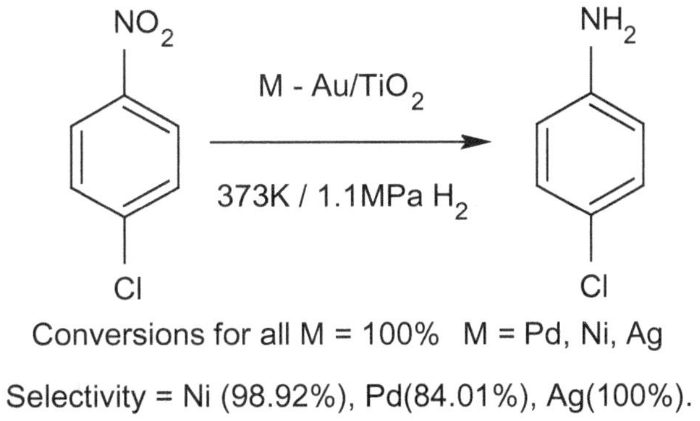

Conversions for all M = 100%   M = Pd, Ni, Ag

Selectivity = Ni (98.92%), Pd(84.01%), Ag(100%). --Scheme-11

Considering the surface enrichment and electronic density the Pd-Au/TiO$_2$ had the highest reaction rate and its conversion of p-CNB reached 100% within 20 min. Ni, Pd, Ag, all on TiO$_2$ also gave 100% p-CNB conversions and 98.92, 84.01 and 100% p-CAN selectivity respectively. Cardenas-Lizana et al. also have observed that Pd-promoted selective gas phase hydrogenation of p-CNB with Au-NPs supported over alumina and titania was 100% selective for p-CAN. This means the dehalogenation of p-CAN was not detected at all (51).

3.05.12: Hydrogenation on Metal Oxides as Solid Supports:

Earlier, we have seen under o-HNB and m-HNB that the chemoselective hydrogenation of o,m,p-HNBs over specially prepared Pt/Y-Fe$_2$O$_3$ to respective o,m,p-HANs with 99.9% selectivity and 0% AN due to total inhibition of dehalogenation of HANs, even at complete conversion of the nitro substrates are observed (Scheme-12).

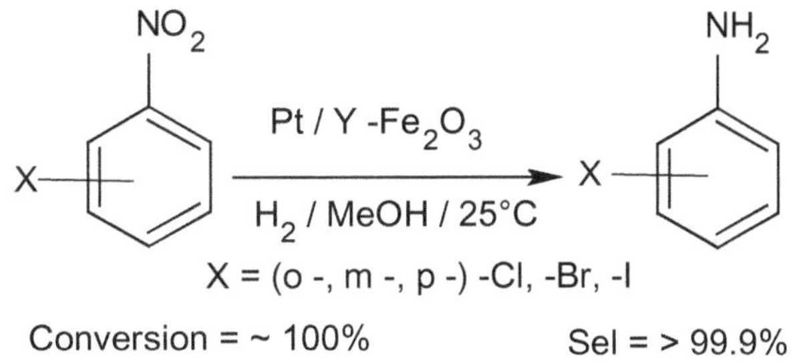

X = (o -, m -, p -) -Cl, -Br, -I

Conversion = ~ 100%   Sel = > 99.9% --Scheme-12

Even, m-BAN was obtained with 99% selectivity at 100% conversion. See references under o-, and m-HNB hydrogenations above. Pt-Au/TiO$_2$ gives 100% selectivity to p-CAN from selective hydrogenation of p-CNB at 60°C and 4MPa H$_2$ (52). Chemoselective hydrogenation of NAs (-Br, -Cl) to ANs in 99.9% yields in the liquid-phase hydrogenation over Pt-NPs confined within a hyper cross-linked polystyrene (HPS) polymeric matrix is studied. This is achieved even for HNAs through boosting NPs efficiency by confinement within highly porous polymeric framework (53).

Bimetallic nano particles on solid supports perform better than monometals in delivering HANs in quantitative selectivity (Scheme-13).

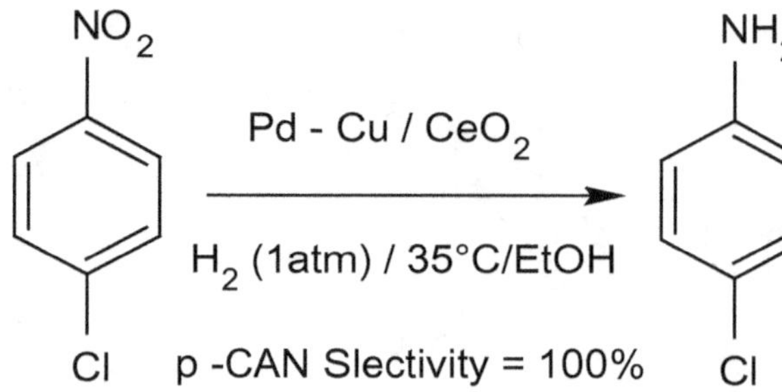

Scheme-13

NO₂ → NH₂ with Pd - Cu / CeO₂, H₂ (1atm) / 35°C/EtOH, p-CAN Slectivity = 100%

Here is a very good example of a first time prepared catalyst $Pd_{0.01}Ru_{0.01}Ce_{0.98}O_{2-\delta}$, for a highly active and selective liquid phase hydrogenation of p-CNB under ambient conditions with 100% selectivity to p-CAN (54).

Wang et al have studied the use of nonreducible (Al_2O_3) and reducible ($Ce_{0.62}Zr_{0.38}O_2$, CZ) carriers to support nanoscale Au in gas phase p-CNB hydrogenation. Reaction over Au/Al_2O_3 generated p-CAN as the sole product, whereas Au/CZ catalyzed nitro-group reduction and also dechlorination of p-CAN to AN (55). Cardenas-Lizana et al. also have demonstrated that the Au/Al_2O_3 in the continuous gas phase (423 K) hydrogenation of p-CNB has been achieved in exclusive selectivity to p-CAN (56). In selective gas phase hydrogenation of p-CNB over Pd (1-10%wt) catalysts with respect to defining the role of the supports as activated carbon (AC) and non-reducible (SiO_2 and Al_2O_3) and reducible (ZnO) oxides have been examined at 1 atm H_2, and 453 K. Pd/AC generated p-CAN and AN, while Pd on SiO_2 and Al_2O_3 exhibited hydrodechlorination character generating AN and NB, and Pd/ZnO promoted the sole (100%) formation of p-CAN at all levels of conversions of p-CNB, absolutely with no hydrodechlorination

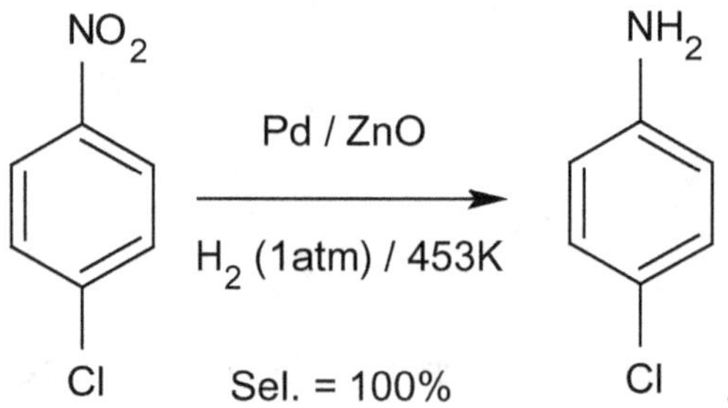

Scheme-14

NO₂ → NH₂ with Pd / ZnO, H₂ (1atm) / 453K, Sel. = 100%

Reaction selectivity is linked to Pd particle size and electron density with the formation of $Pd^{\delta+}$ on AC and the occurrence of $Pd^{\delta-}$ on SiO_2 and Al_2O_3 (Scheme-14).

The reaction exclusivity to p-CAN over Pd/ZnO is attributed to the formation of Pd-Zn alloy (demonstrated by XPS), which selectively activates the $-NO_2$ group. This is the first report that demonstrates 100% selectivity towards p-CNB transformation to p-CAN over supported Pd (57). Substituted NAs, including o- and p-CNBs have been hydrogenated using Pd-quinoline complex. The o- and p-CAN were obtained in 92% and 95% yields when the hydrogenation was carried out in DMF and aq. NaOH at 35°C and 1atm H_2 pressure (58). As seen above, F. Cardenas-Lizana et al. have made very good contributions to the chemoselective hydrogenations of organic functional groups including the nitro groups in NAs. Some of their additional work is presented below with titles and references, especially the hydrogenations of *p*-CNB to *p*-CAN.

F. Cardinas-Lizana, et al. Clean production of CANs by selective gas phase hydrogenation over supported Ni. *Appl. Catalysis A: General*, 334:199–206(2008).

F. Cardinas-Lizana et al. Exclusive production of *p*-CAN (100% Sel) from *p*-CNB over Au/TiO₂ and Au/Al₂O₃. *ChemSusChem*, 01(03): 215-212(2008).

F. Cardinas-Lizana, et al. Pd-promoted selective gas phase hydrogenation of *p*-CNB over alumina supported Au. *Journal of Catalysis*, 262:235–243(2009). F. Cardinas-Lizana, et al. *Gold Bulletin*, 42(02):124-132(2009).

F. Cardinas-Lizana, et al. β-Molybdenum nitride response in the gas phase hydrogenation of *p*-CNB. *Catalysis Science & Technology*, 01:794–801(2011).

F. Cardinas-Lizana, et al. Selective gas phase hydrogenation of *p*-CNB over Pd catalysts: Role of the support. *ACS Catalysis*, 03:1386–1396(2013).

F. Cardinas-Lizana, et al. Catalyst deactivation in *p*-CNB hydrogenation over supported gold. *Chemical Engineering Journal*, 25(05):695-704(2014).

3.05.13: Hydrogenation of *p*-CNB on C/CNTs as Catalyst Supports:

Thermally reduced Au-NPs prepared by the carbonization of ordered mesoporous carbon as a heterogeneous catalyst was effective in transforming *p*-CNB and 4-NP to the corresponding amines in high yields and selectivity (59). Nitrogen-doped CNTs as highly selective, noble metal-free catalysts exhibited a very high activity towards hydrogenation of *p*-INB and high selectivity to *p*-IAN molecule in a slurry type parallel reactors at 20-110°C and under 5-60 bar hydrogen pressure (Scheme-15) (60).

--Scheme-15

In comparison with reference catalyst Pt/C (70.6% sel), the NCNT catalysts showed higher selectivity (~96%) to *p*-IAN at 100°C/40bar H₂ under identical conditions.

Ru-NPs are supported on nitrogen-doped porous carbon (NPC) derived from MOF-ZIF-8 for the first time and tested in the selective hydrogenation of *p*-CNB and *p*-BNB to corresponding HANs at 1.5 MPa with very low level of dehalogenation (61). Ru – NPs/CNTs gives 100% selectivity to *p*-CAN from selective hydrogenation of *p*-CNB at

60°C and 4MPa H$_2$ (62). Au-NPs supported by multi-wall carbon nanotubes (MWCNT) selectively hydrogenate substituted NAs, including o-, m-, p- CNBs at 90°C for 0.3-4h which gave o-, m- p-CANs in 78%, 81% and 82% respectively (63). Ir-NPs immobilized on MWCNT were synthesized and found excellent in selective hydrogenation of HNBs at room temperature and balloon hydrogen pressure in a methanol/water mixture, especially showing a selectivity of 99.9% to p-CAN at complete conversion of p-CNB. The high activity and selectivity of the catalyst is attributed to the interactions such as competitive adsorption and hydrogen bonding between solvent water and reactants (64). Other groups also have reported similar results with 100% substrate (o, m, p -Cl, F, NO$_2$, etc.) conversions and corresponding HANs chemoselectivity using different metal catalysts, especially when Pd-NPs size is larger than 25nm (65). Recent developments in the last decade in the chemoselective hydrogenation of HNBs have been reviewed (66).

An activated carbon supported Pt-Fe bimetallic catalyst (Pt-Fe/AC) was prepared and tested for selective hydrogenation of p-CNB to p-CAN. Compared with Pt/AC catalyst, the Pt-Fe/AC catalyst exhibited higher activity and full inhibition of dechlorination with 100% selectivity towards p-CAN at 100% conversion of p-CNB at 30°C and 1.0 MPa H$_2$, in ethanol for 150mins (Scheme-16).

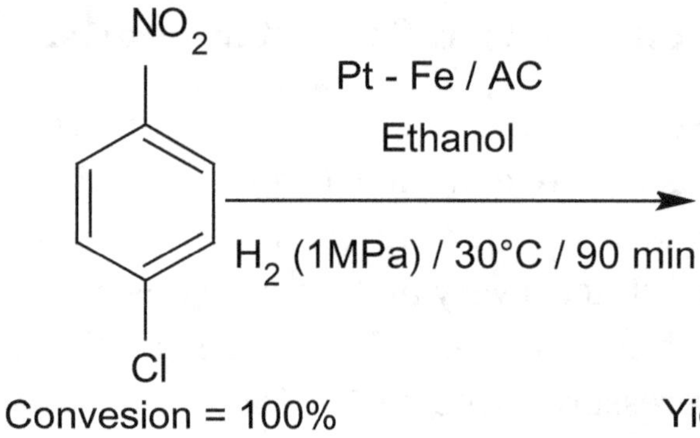

Scheme-16

X-ray photoelectron spectroscopy (XPS) revealed the electron transfer from Pt-NPs to Fe$_2$O$_3$ leading to the electron-deficient state of the Pt-NPs, which is responsible for improved performance of the Pt–Fe/AC catalyst in this case (67). Ultrasmall-sized Pt-NPs (~1 nm) supported on CNTs with nitrogen doping and oxygen functional groups were synthesized and applied in the catalytic hydrogenation of NAs. The advanced identical location transmission electron microscopy (IL-TEM) method was applied to probe the structure evolution of the Pt/CNT catalysts in the reaction. The results indicate that Pt-NPs supported on CNTs with a high amount of nitrogen doping (Pt/H-NCNTs) afford 2-fold activity to that of Pt-NPs supported on CNTs with oxygen functional groups (Pt/oCNTs) (Scheme-17).

X = Cl, Br, I   Sel. = >99 -100%   --Scheme-17

Compared with Pt/oCNTs, Pt/H-NCNTs exhibited a higher selectivity (>99%) in chemoselective hydrogenation of HNBs to HANs due to the electron-rich chemical state of Pt-NPs and the strong metal–support interactions on H-NCNTs are capable of stabilizing the Pt-NPs, stability, recyclability and ~100% selectivity (68). The in-house synthesized Pd-NPs over CNTs (Pd/CNTs) and graphene (Pd/G) gave excellent performance in the liquid phase chemoselective hydrogenation of p-CNB to p-CAN at room temperature.

p -CAN Slectivity = 100%   --Scheme-18

Both the catalyst gave 100% selectivity to p-CAN (Scheme-18) (69). Similarly, chemoselective hydrogenations of HNBs have been reported using noble metal catalyst on solid supports as carbon nano fibres (CNFs) (Scheme-19).

X = Cl, Br, I, OH, OBn, COR, COOR, $CONH_2$ etc.   --Scheme-19

High yields and selectivities are obtained for most of HANs (70). The Co/CoO-NPs coated with graphene layers (Co/CoO@Carbon) have been developed and found excellent in chemoselective hydrogenation of some challenging NAs with reducible functionalities to the corresponding ANs (71a). In another case $Co°Co_3CO_4$@N-doped CNTs gave excellent catalytic activity and perfect selectivity (>99%) for 21 examples (71b).

3.05.14: Catalytic Hydrogenation of HNBs on Polymer Support:

Poly-(N-vinyl-2-pyrrolidone) (PVP)-stabilized Iridium nanoparticles (Ir-NPs) showed high chemoselectivity for hydrogenation of NAs having –CHO, -C=O, –CN, and –Cl functional groups to afford the corresponding ANs under atmospheric hydrogen at room temperature in aqueous–organic biphasic medium (72). A synergistic effect of the polymer-anchored bimetallic Pd-Ru catalysts can lead to a remarkable increase in the selectivity for p-CAN in the selective hydrogenation of p-CNB under atmospheric pressure and in the presence of a small amount of base (73). A Novel synthesis of Au-NPs as catalysts using chloroauric acid ($HAuCl_4$) as gold precursor on polymer/metal oxide hybrid materials (Au/P[VBTACl]-M (M= Al, Ti or Zr) and their use as heterogeneous catalysts in liquid phase hydrogenation of p-CNB is reported. The selectivity for all the catalytic systems was >99% toward the p-CAN. Au-NPs/Al_2O_3 could be used four times with 98-100% yields of p-CANs. The reactivity of the Au catalysts toward the p-CNB hydrogenation reaction is attributed to the different particle size distributions of Au-NPs in the hybrid supports (74). The p-CNB is such a pollutant which is widely used as intermediates for chemical syntheses of drugs, herbicides, dyes, etc. A feasible degradation pathway for p-CNB is bio-reduction with H_2 as the electron donor converting it into p-CAN and AN. Flux analysis indicated that the reduction of p-CNB to p-CAN could consume fewer electrons than that of nitrate and sulfate. The HFMBfR (hollow-fiber membrane biofilm reactor) had high average H_2 utilization efficiencies (maximum 98.2%) at different steady states in this experiment (75).

3.05.15: Hydrogenation of HNBs over Mono-Metallic Catalysts:

Pt nanocatalysts as prepared by solovolysis and stabilized by surfactants can be used for heterogeneous hydrogenation catalysts in the selective high-pressure transformation of 3, 4-DCNB to corresponding ANs. The catalytic performance of the new systems was better than conventional Pt/C systems (Scheme-20) (76).

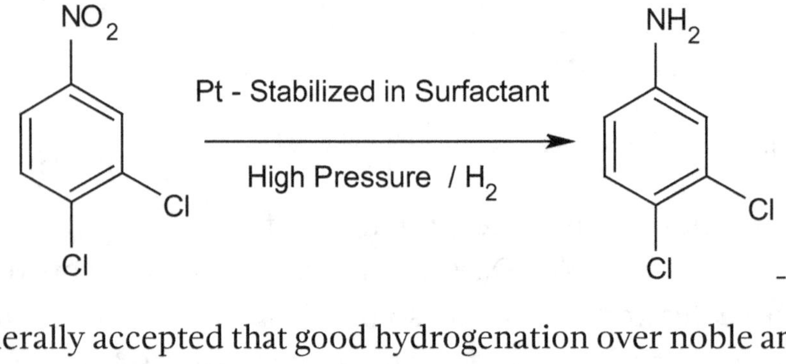

--Scheme-20

It is generally accepted that good hydrogenation over noble and nonnoble metal based catalysts such as Pt, Ru, or Ni are not chemoselective for hydrogenation of nitro groups in substituted aromatic molecules. Corma et al. have proved transforming nonchemoselective into highly chemoselective metal catalysts by controlling the coordination of metal surface atoms with selected support. Thus, for highly chemoselective and general hydrogenation

Pt, Ru, and Ni catalysts can be prepared by generating nanosized crystals of the metals on the surface of a TiO_2 and decorating the crystal faces by means of a simple catalyst activation method.

Chemoselectivity increased from <1% to 95% --Scheme-21

With these modifications, the authors were able to enhance the catalyst acativity by almost 2 orders of magnitude and increasing the chemoselectivity from less than 1% to more than 95% (Scheme-21) (77). Ligand-free Pt-NPs (2.2nm) encapsulated in hollow porous carbon shells are found as highly active heterogenous catalysts in the reduction of NAs under mild hydrogenation conditions as 0.1-0.2MPa H_2 (78).

Highly efficient and durable platinum nanocatalysts stabilized by thiol-terminated poly(N-isopropyl acrylomide (Pt-NPs@PNIPAM)) for selective hydrogenation of HNBs to HANs is reported with excellent results. A highly chemoselective hydrogenation of HNBs is achieved using Pt@PNIPAM and Pt@PNIPAM-SH to obtain 99.9% selectivity for p-CAN from at 100% p-CNB conversion. Other HANs (o-, m-. p-Cl, Br, I-NBs) were also reduced to corresponding HANs with high to excellent selectivities (~93-99.8%) at 100% conversions for all of them (79). Pd-NPs with 25 nm supported on AC were synthesized and were used for the selective hydrogenation of p-CNBs with selectivity of >99.90% to p-CANs. Finally, the industrial applications of the proposed catalyst were evaluated in several pilot factories (80). The poisoned Au-NPs at edge/corners of sites of anatase are synthesized, which form unique sites for selective adsorption and activation of nitro groups, thus leading to high activity and selectivity. This strategy for preparation of supported gold catalysts opens a new door for the design of highly efficient heterogeneous catalysts in the future (81). The catalytic continuous gas phase hydrogenation of p-CNB has been investigated over a series of oxide (Al_2O_3, TiO_2, Fe_2O_3 and CeO_2) supported Au (1 mol %) catalysts at 423K/H_2=1atm. p-CAN was generated as the sole reaction product (100%) over all the Au catalysts with no evidence of C-Cl and/or C-NO_2 bond scission and/or aromatic ring reduction (82).

In this study, sub-nano (<3 nm) Ir-NPs and Pd-NPs were prepared and their catalytic properties for hydrogenation of HNBs were evaluated. Results show that high selectivity (>99%) fo p-CAN from p-CNB was achieved over small Ir-NPs, which is attributed to hydrogen consumption rates and reaction rates on the metal surfaces and the distances between the oxygen and Ir (short) and Pd (long) (83). A reliable methodology is developed to perform the HNBs and NAs hydrogenation process, which involves a one pot synthesis, simple procedure and short reaction times in aqueous medium without any hydrogen pressure using PVP-Pd-NPs (Scheme-22).

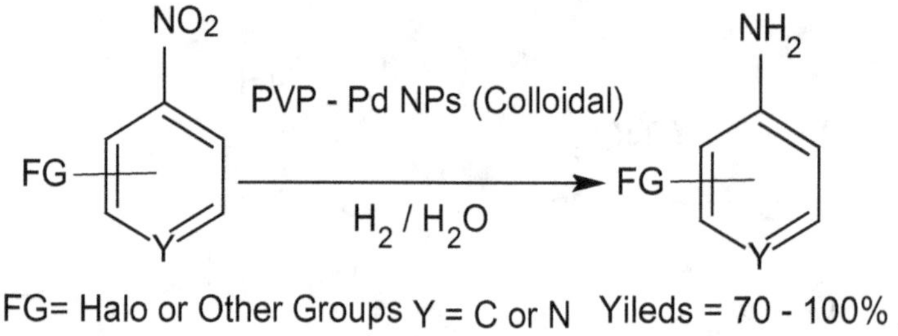

FG= Halo or Other Groups Y = C or N   Yileds = 70 - 100% --Scheme-22

A highly chemoselective reduction is done at room temperature under mild condition with good to quantitative yields (70-100%) of the substituted aryl amines, leaving many sensitive functional groups unaffected (84). In another case, a chemoselective and highly efficient hydrogenation of NAs using an economical and environmentally friendly H_2/MoO_2Cl_2 system has been developed. The substituted anilines (-Cl, -CH=CH$_2$, -CN, etc) are obtained in 100% yields (85).

3.05.16: Hydrogenations of HNBs over Bimetallic-Catalysts:

A highly selective hydrogenation of CNBs to CANs catalyzed by a recyclable, water-soluble Ru/Pt bimetallic catalyst in an aqueous-organic biphasic system was studied. Here the Pt acted as a promoter and increased the catalyst activity for the p-CNB hydrogenation (99.9% conversion) at 25°C/1.0Mpa H_2, with 99.4% p-CAN selectivity. The Ru/Pt catalyst showed high activity and selectivity for the hydrogenation of other CNBs and DCNBs with substitutions in different positions (86).

Pt-Au/TiO$_2$ catalysts were prepared by deposition-precipitation with urea, and the influence of addition of a second metal Pt and reaction conditions on hydrogenation of p-CNB to p-CAN were investigated. The results showed that the Au/TiO$_2$ with minute amount of Pt exhibited high catalytic activity for hydrogenation of p-CNB to p-CAN in 100% yield in 1 h at 50°C,1.0 MPa H_2 especially when the Pt loading was 0.02% (Scheme-23).

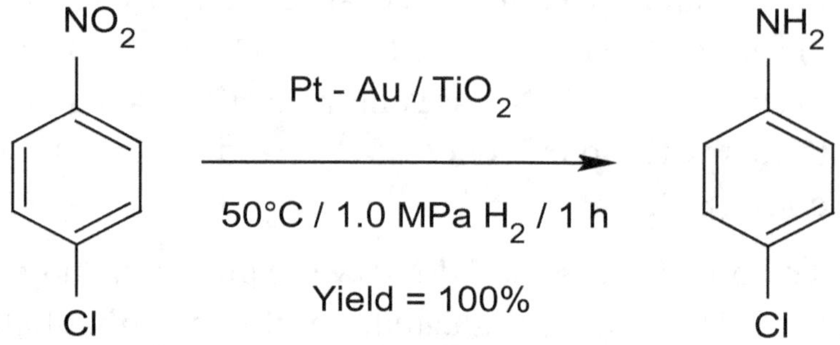

--Scheme-23

The addition of micro Pt entities to Pt-Au/Al$_2$O$_3$ exceptionally enhances the catalytic activity while maintaining the selectivity towards p-CAN (100%) at 100% conversion of p-CNB at 333 K (87). Enhanced catalytic hydrogenation activity and selectivity of

Pt-M$_x$O$_y$/Al$_2$O$_3$ (M = Ni, Fe, Co) heteroaggregate catalysts by *in–situ* transformation of Pt-M alloy NPs were found superior to Pt/Al$_2$O$_3$ in the hydrogenation of *p*-CNB to *p*-CAN. The activity is due to its (*p*-CNB's) strong intercations with Pt-MxOy (88). Selective hydrogenation of *p*-CNB over a Fe-promoted Pt/AC catalyst is investigated. The bimetallic catalysts as Fe-Pt/AC exhibited excellent performance in producing CANs with high selectivity (>99%) in comparison to the Pt/AC (89). Core@shell structured bimetallic-NPs (BM-NPs) are currently of immense interest due to their unique electronic, optical and catalytic properties. However, their synthesis is non-trivial. The bimetallic Au-Pt and Pt-Pd core@shell NPs have been synthesized by a new supramolecular route giving striking atomic-resolution images of the core@shell architecture and the unique catalytic properties. The newly structured BM- NPs have demonstrated a remarkable improvement in the selective production of industrially valuable CANs from CNBs (90).

We have seen earlier that the versatility of Ni-NPs in the selective hydrogenation of HNBs. In a particular case the selectivity of main product (*p*-CAN) was greater than 99% over La-NiB catalysts (91). The liquid-phase hydrogenation of (*o-*, *m-*, *p-*)-CNBs with Pt-M/CNTs (M = La, Pr, Ce, Sm, Nd) catalysts gave >99% conversions for all the CNBs, while yields for *o-*, *m-*, and *p-* CANs were 97, 98 and 97.5% respectively. The electron deficient species of the second metal promote the rate constant on Pt surface by activating the nitrogrn –oxygen bond. Pt-Ce/CNT exhibited the best catalytic activity and stability at 1 wt% and gave the highest yield for *p*-CAN (97.5 mol %), while Pt-Sm/CNT gave the best yields for *m*-CAN (98 mol %) and *o*-CAN (96.9 mol %) (92). The Pt-Ni@mSiO$_2$ and Pt–NiO@mSiO$_2$ nanocatalysts show significantly improved relative activity and selectivity for *p*-CNB reductions compared to that of control Pt@mSiO$_2$ nanocatalysts. The catalysts are stable and their enhanced catalytic performance is ascribed to the metal–metal interaction for the Pt-Ni@mSiO$_2$ catalysts and metal–oxide interaction for the Pt–NiO@mSiO$_2$ catalysts (93). Nickel-iron mixed oxide prepared from a nickel-iron hydrotalcite precursor was found to be a highly efficient catalyst for the chemoselective reduction of NAs under mild reaction conditions (94). Also, bimetallic Pd-M (M = Ni, Co, Fe) NPs were synthesized using butyllithium as a reductant and were used to prepare Pd–M$_x$O$_y$/KIT-6 catalysts which were used for the efficient and selective hydrogenation of *p*-CNB (95).

Pietrowski has extensively reviewed the recent developments during last decade in heterogeneous selective hydrogenation of HNBs to HANs. The first sections of the review contains a concise report on applications of halogenated anilines, methods for their preparation and reaction pathways for hydrogenation of HNBs. Subsequent sections discuss hydrogenation properties of platinum group metals, new catalysts based on gold and silver as well as nickel catalysts. Particular attention is paid to positive effects of the use of new unconventional supports and catalyst preparation methods with an emphasis on important factors affecting high activity and selectivity of heterogeneous catalysts for selective reduction of HNBs (96).

References:

50a). P. Tartaj, et al., *J. Physics D: Applied Physics.* 36: (R182–R197)(2003), 50b). O. V. Salata, *J. Nanobiotechnology*, 02:03(2004).

51a). F. Cardinas-Lizana, et al. *J Cataysis:* 262:235-243(2009). 51b) F. Cardinas-Lizana et al. *Phys. Chem. Chem. Phys.,* 17(42): 28088-28095(2015).

51c). Y. Tsu & Y. Chen, *AIMS Materials Science,* 04(03):738-754(2017).

52). D. He, et al. *Green Chemistry,* 14(01): 111-116 (2012).

53a). F, Cardenas Lianza, et al. *J. Catalysi:* 301:103–111(2013). 53b). F. Cardinas-Lizana, et al. *Catalysis Today,* 173(01):53-61(2011).

54). R. Mistri, et al. *J. Molecular.Catalysis A: Chemical,* 376:111-119(2013).

55). X. Wang, *J. Physical Chemistry C,* 117(02):994-1005(2013).

56). F. Cardinas-Lizana, et al. *Chemical Engineering J.,* 255:695-704(2014).

57). F. Cardinas-Lizana, et al. *ACS Catalysis.,* 03(06):1386–1396(2013).

58a). C. R. Saha, et al. *Proc Indian Natl Science Academy,* 03:538=549(1985). 58b). A. Corma, *Science,* 313(5785):332-334(2006).

59). H. Fu, et al. *J. Catalysis,* 344313-324(2016).

60). a) H. Lu, EP2837612-A1: b) WO2009080204 A1: & c) WO2009149849 A1.

61a). X. Li, et al. *Dalton Transanctions,* 45(39):15595-15602(2016). 61b). M. Oubenali, et al. *ChemSusChem,* 04(07): 950–956(2011).

62). C. Antonetti, et al. *Applied Catalysis A: General,* 421-422:99-107(2012).

63a). M. Cano, et al. *Materials Today Communications:* 03:104-113 (2015).63b). V. Pandarus et al. *Catal. Sci. Technol,* 01:1616-1623(2011).

64a). H.B. Li, et al. *Synthesis & Recativity in Inorg, Metal-Org, and Nano-Metal Chem,* 46(10):1499-1505(2016). 64b). Y. Motoyama, et al. *Chemistry An Asian J.* 09:71-74(2014). 65). J. Lyu, et al. *J. Physical Chemistry C,* 118:2594-2601(2014).

66a). H. U. Blaser, et al. *ChemCatChem:* 01:210-221(2009). 66b). P. Serna & A. Corma, *ACS Catalysis,* 05(12):7114–7121(2015).

67a). M. Gu, et al. *Chinese Journal of Applied Chemistry,* 32(10):1164-1169(2015). 67b). H. Chen, et al. *RSC Advances,* 07(46): 29143-29148 (2017).

68). W. Shi, et al. *ACS Catalysis,* 06(11):7844-7854(2016).

69). A. B. Dongil, et al. *Catalysis Communications:* 7555-7559(2016).

70). M. Takasaki, et al. *Organic Letters,* 10(8):1601-1604(2008).

71a). B. Chen, et al. *ChemCatChem*, 08(06):1132-1138(2016). 71b). Z. Wei, et al. *ACS Catalysis*, 05(08):4783-4789(2015).

72). M. J. Sharif, et al. *Chemistry Letters*, 42(09):1023-1025(2013).

73). Z. Yu, et al. *J. Chem. Soc, Chem Commun*, 1155-1156(1995).

74a). C. H. Campos, *J. Chemistry*, Volume 2017 (2017): Article ID 7941853, 9 pages. 74b). *Y. Yamane, et al. Organic Letters*, 11(22):5162–5165(2009).

75a). S. Xia, et al. *J Hazard Materials*, 192(02):593-598(2011). 75b).H. Li, S. Xia et al. Biodegradation: 25(02):205-215 (2014).

76). H. Bönnemann, *New J Chemistry*, 22:713-718(1998).

77). A. Corma et al. *J. American Chemical Society: 130* (27), pp 8748–8753(2008).

78).S. Ikeda, *Angewandte Chemie*, 118(42):7221–7224(2006).

79). W. Yu, et al. *RSC Advances*, 07:751-757(2017).

80). J. Lyu, et al. *J. Physical Chemistry C*, 118:2594–2601 (2014).

81). L.Wang, *ACS Catalysis*, 06(07):4110-4116(2016).

82). F. Cardenas-Lizano, et al. *Gold Bulletin*, 12(02):124-132(2009).

83). L. Ma, et al. *Chinese J.Chemical Engineering*, 25(03):306-312 (2017).

84). P. M. Uberman, et al. *Green Chemistry*: 19(03):739-748 (2017).

85). P. M. Reis & B. Royo, *Tetrahedron Letters*, 50:949–952(2009).

86). Y. Zhou, et al. *China Petrol Process & Petrochem Technol.*, 17(02):26-31(2015).

87). D. He, et al. *Green Chemistry*, 14(01):111-116 (2012).

88). X. Wang, et al. *J. Physical Chemistry C*, 117(14):7294–7302(2013).

89). H. Chen, et al. *RSC Advances*, 07:29143-29148(2017).

90). C. J. Serpell, et al. *Nature Chemistry*, 03:478-483 (2011).

91). Y. Liu, et al. *J. Nanoparticle Res.*, 08(02):223-234(2006).

92). J. Yang et al. *Indian J. Chemistry*, 48A: 1358-1363(2009).

93a). H. Liu, et al. *RSC Advances*, 05(26): 20238-20247(2015). 93b). H. Liu, et al. *Catalysis Science & Technology*, 05(01):405-414 (2015).

94). Q. Shi, et al. *Advanced Synthesis & Catalysis*, 349(11-12):1877-1881(2007).

95). X. Zhang, *Catalysis Letters*, 145(03):784-793(2015).

96a). M. Pietrowski, *Current Organic Synthesis*, 09(04):1470-1487(2012). 96b).G. Avgouropoulos, et al. *Environmental Catalysis Over Gold-Based Materials:* Royal Society of Chemistry, 23-Jul-2013 - Science - 227 pages.

3.06.00: Catalytic Hydrogenation of HNBs without Dehalogenation:

Hydrodehalogenation is a serious problem in the hydrogenation of halonitribenzenes (HNBs) or halonitroarenes (HNAs). Initially, the catalysts used were not selective towards nitro group only and also the reaction times were quite long till the HNBs or HNAs are completely converted to corresponding ANs. During this long residential time the in-situ formed HANs also underwent hydrodehalogenation and some other sensitive functional groups on the same molecules were affected as well. As a result, the yields and selectivity of HANs were lower than expected. Actually, there are some reports where Pd reduces HNBs to HANs and AN and sometime even to give AN plus NB. This is attributed to the highly active catalyst under set experimental conditions that continues hydrogenation even after complete conversion of the nitro substrate and causes hydrodehalogenation of the in-situ formed HAN. Also, earlier we have seen that NB is completely (100mol% conversion) reduced to cyclohexylamine (97a). But later on some other recent reports have obtained almost quantitative conversions of HNBs and about 100% yields and selectivity of ANs. These developments have been recently summarized (97b, 97c).

The following examples illustrate all the above factors plus the effect of an additive or promoter. Without dehalogenation inhibitor (DCDA) the dehalogenation of o-CNB was almost 50% giving only 54.5% of o-CAN, while in the presence of an inhibitor the yields were more than 99% (98). Amongst others, formamidine acetate is found as the most effective inhibitor (at 0.012 mol %) of dehalogenation during hydrogenation process, in aq. alcohol at 60°C/10MPa, of 2, 4-dinitrochlorobenzene, as it gives almost same yields of chlorodiamine (97.5% vs 99%) in one third of the time taken by DCDA (99). p-BNB and p-INB underwent dehalogenation under these experimental conditions to give the corresponding anilines in lower yields, 83% and 23% respectively. The susceptibility of iodo derivative is understood based on its labile nature for displacement. Stratz compared Raney-Ni without inhibitor and with dicyandiamide as an inhibitor in the hydrogenation of o-CNB in methanol below 120°C and 1MPa H_2. Proportion of dechlorination decreased from 44.3% without inhibitor to 0.3% in the presence of the inhibitor, and the yield of o-CAN increased from 54.5% without inhibitor to 99.8% with the inhibitor (Schemes -24-26).

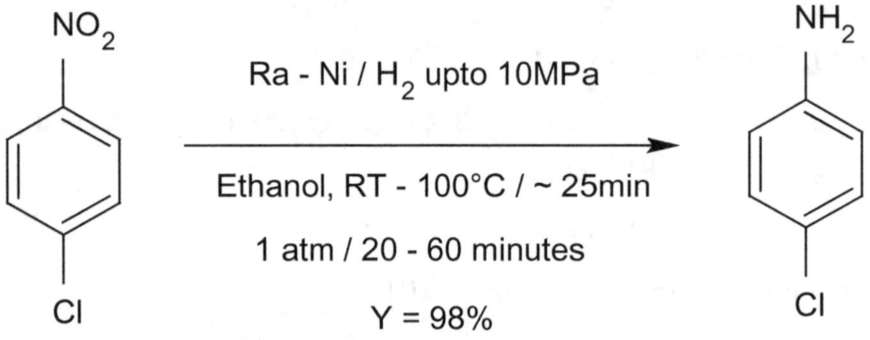

--Scheme-24

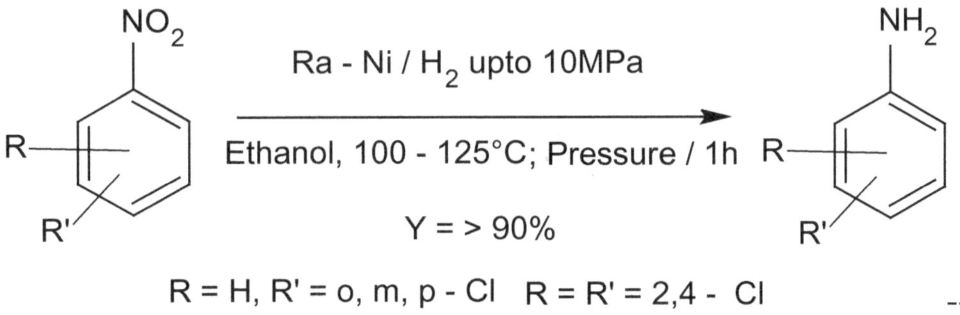

R = H, R' = o, m, p - Cl   R = R' = 2,4 - Cl   --Scheme-25

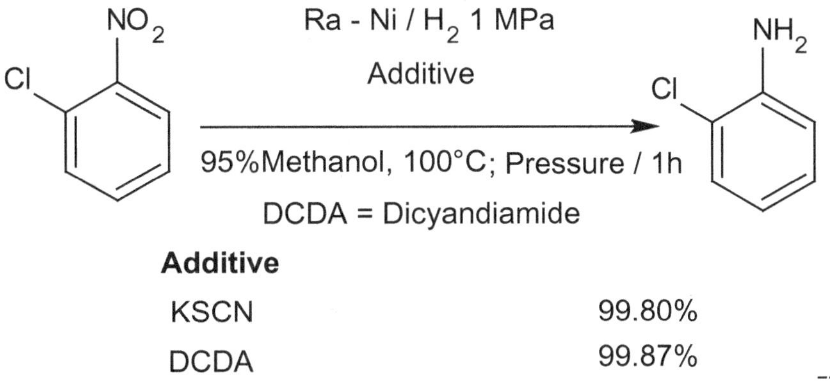

DCDA = Dicyandiamide

Additive

KSCN	99.80%
DCDA	99.87%

--Scheme-26

The dechlorination decreased further to 0.2%, and the yield of *o*-CAN increased to 99.3% over a Raney-Ni promoted with chromium and iron. Similarly, dehalogenation over the chromium and iron-promoted Raney-Ni was only 0.6% with *m*-CNB, and 0.2% with 3, 4-diCNB. Under optimized conditions and Pt with DCDA as dehalogenation inhibitor. Particularly, *p*-CAN was obtained in 99% from *p*-CNB. Ref: S. Nishimura, PP 340-350 (see below). Hydrogenation of *o*-CNB in the presence of triethylphosphite and DCDA gave 96.7% and 96.8% yields respectively, compared to 54% yield when only Pd/C was used as a catalyst (98). Pt catalysts are found to be highly selective in the hydrogenation process of HNBs to HANs and especially the Pt sulfides have found a unique place in hydrogenation of HNBs. Dechlorination is not detected at all with any of the PGM sulfides, including Pd, whereas debromination was observed with a different degree with respect metal sulfide used, with PdS being the worst case and PtS being the best one.

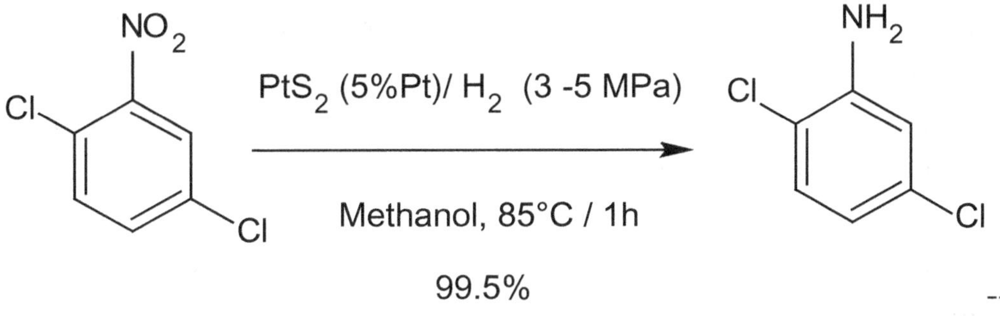

99.5%   --Scheme-27

NO$_2$ 　PtS$_2$ (5%Pt)/ H$_2$ (3 -5 MPa) 　NH$_2$

Methanol, 100 - 130°C / 1h

99.5%

Br 　Br 　--Scheme-28

A compilation of recent data on various Pt catalysts is reported. Especially in this case, PtS$_2$ is found highly chemoselective with p-CNB and p-BNB with 99.5% yields in both the cases (Scheme-27-28)(100). Ir and Re catalysts are used rarely but are found very selective in hydrogenation of HNBs with lowest possible or no dehalogenation levels (Scheme-29).

NO$_2$ 　Re$_2$S$_7$ (5%)/ H$_2$ (13 MPa) 　NH$_2$

Methylcellosolvel, 60°C / 0.5h

100%

Br 　Br 　--Scheme-29

For example, 5% Ir/C gave 98.8% of 3,4-dichloroaniline from the corresponding precursor and was found better than from 5% Pt/C (97.99%). Similarly Re sulfide (Re$_2$S$_7$) gave 100% of p-BAN starting from p-BNB (101-102). Re$_2$O$_3$ Rhenium oxide (Re(III) oxide) also gave the same selectivity at 124°C and 22MPa (103).

Some of the additive are highly selective and help achieve impossible things. For example, triethylphosphite (TEP) does help 3% Pd/C catalyzed hydrogenation, at commercial scale, for the manufacturing of p-CAN and 2,5-DCAN from the respective raw materials at 100°C in IPA or water (104). HCl-acidified attapulgite-supported Pt-Nanocomposites (Pt-NCs) catalyst exhibited excellent catalytic activity and selectivity for the selective hydrogenation of p-CNB, m-CNB and o-CNB. The selectivity to respective CANs could be up to 100% at complete conversion of CNBs under 2.0MPa and 50°C (104c). However, some of the remarkable methodologies for making HANs, with high atom efficiency and environmental friendly attributes, are limited due to the formation of hydrodehalogenated products form the already formed HANs, especially at the complete conversion of HNBs. Therefore, many effective strategies have been developed in order to inhibit the hydrodehalogenation in the selective hydrogenation of HNBs to HANs. There are many examples of chemoselective hydrogenation of HNBs, as discussed above, without or very negligible dehalogenation of HANs.

Considering the growing importance of solid supports for M-NPs, recently iron oxide (Y-Fe$_2$O$_3$) has drawn an attention as independently reported by Niu and Komatsu et al. Both the groups have achieved 100% chemoselectivity for HANs using Pt-Zn/SiO$_2$ catalyst. Especially, Xiao et al. have contributed a lot through development of Pt-NPs supported on Fe$_2$O$_3$ (Pt/Fe$_2$O$_3$ nanocomposites catalyst), which completely suppressed the hydrodehalogenation of HANs at the complete conversion of HNB. Mechanistically, the electron transfer from Pt-NPs to oxygen vacancies in the activated Pt/Y-Fe$_2$O$_3$ catalysts may be playing an important role in completely suppressing the hydrodehalogenation of HANs. Recently Xiao et al have reviewed all these great efforts and remarkable contributions of different research groups during the long exploration for the solutions of this hard obstacle, with an emphasis on the developments of supported metal catalysts and polymer-protected metal nanoclusters or colloid catalysts (105). Recently, Yu et al. for the first time have developed a highly selective hydrogenation of HNBs to HANs under mild conditions catalyzed by well-dispersed Pt-NPs protected by thiol-terminated poly(N-isopropyl acrylomide) (PNIPAM-SH). The polymer acts as a protector of Pt-NPs and also helps inhibit the highly active Pt catalyst from producing undesired hydrodehalogenation products through anchoring the thiol groups to the surface of Pt-NPs over a period of time (106). The metal–support interaction in the partially reduced Pt/α-Fe$_2$O$_3$, Ir/γ-Fe$_2$O$_3$, Rh/γ-Fe$_2$O$_3$, and Ru/γ-Fe$_2$O$_3$ catalysts was investigated by IR-CO probe and X-ray photoelectron spectroscopy (XPS). Excellent catalytic properties over nanocomposite catalysts on M/Y-Fe$_2$O$_3$, espeecially on Pt/Y-Fe$_2$O$_3$ for selective hydrogenation of HNBs were already investigated (107). A partially reduced Pd/γ-Fe$_2$O$_3$ nanocomposite catalyst (Pd/γ-Fe$_2$O$_3$-PR) was prepared using PdO and ferric hydroxide colloidal particles as starting materials (Scheme-30).

--Scheme-30

This catalyst exhibited excellent catalytic properties for the selective hydrogenation of p-CNB to p-CAN (99.2%) at complete conversion of the substrate and intermediates. The hydrodechlorination of p-CAN was fully suppressed over Pd/γ-Fe$_2$O$_3$-PR (108). The Pt/γ-Fe$_2$O$_3$ nanocomposite catalyst exhibited high catalytic activities and superior selectivities to o-BAN, m-BAN and p-BAN, with the selectivities >99.9% for all the three BANs, in the hydrogenation of o-BNB, m-BNB and p-BNB respectively (Scheme-31). Considering the labile nature of Br, such selectivity is of great importance to the bromoanilines (BANs).

$$\text{Br.} \xrightarrow{\text{NO}_2} \xrightarrow[\text{H}_2]{\text{Pt / Y -Fe}_2\text{O}_3} \text{Br.} \xrightarrow{\text{NH}_2} \not\longrightarrow \xrightarrow{\text{NH}_2}$$

(o -, m - , p -)   Selectivity = 99.9%   --Scheme-31

The hydrodebromination of bromoanilines (BANs) over the Pt/γ-Fe₂O₃nanocomposite catalyst was fully suppressed even at complete conversion of the substrates for the first time (109). Through our observation while looking at the literature on selective hydrogenation of HNBs, it is found that Pt catalyst is best of all in inhibiting hydrodehalogenation of HNBs during hydrogenation, without any inhibitors but certainly enhances the inhibiting effects by addition of Cr or Ni as additives or promoters. Metal sulfides also are helpful in inhibiting hydrodehalogenation. Pt-sulphide on C(Pt-S/C) in methanol at 85°C/2.5 h under 800psi could reduce 3,5-DCNB to 3,5-DCAN in 99.5 yield (110). Also see references 100,101 102 for selective hydrogenationof HNBs using metal sulfides.

Kosak et al have studied the effect of various inhibitors for the selective hydrogenation of HNBs to HANs. Especially, Pt-Morpholine catalyst system showed the best results because morpholine acts as an inorganic acid acceptor to control the amount of carbon-halogen cleavage, it acts (i) as a true suppressor to inhibit the dehalogenation reaction and (ii) as an acid acceptor when some dehalogenation occurs (111). For details on reduction of m-CNB to m-CAN and other HNBs reductions see Shodhganga, Chaper-IV. Supported Pt –NPs with controlled particle size on different iron oxide supports (γ-Fe₂O₃, Fe₃O₄, Fe₂O₃) were used in the liquid phase hydrogenation of o-CNB to o-CAN under mild reaction conditions (298 K, 0.1 MPa). The Pt/γ-Fe₂O₃ exhibited higher catalytic selectivity (99.9%) and activity than other similar works (112). Liquid phase hydrogenation of CNB isomers (x-CNB x = 2, 3, 4) to the corresponding CANs (x-CAN) under mild conditions (0.6 MPa, 25°C, diethyl ether-methanol as solvents) over Pr or Pt catalysts on functionalized Dowex-D gave 80% CNB conversions and up to 95% CAN selectivity (113).

The challenges in chemoselective hydrogenation of HNBs are very high as the chances of dehalogenations are also very high. But, with the advanced knowledge about the behaviours of the HNBs and the requirements of the nanocatalysts has helped develop the metal nanacatalyst which meet desired expectations. The following examples are illustrative of such new developments and achieving the desired results in quantitative substrate conversions, AN yields and its selctivities,

The hydrodebromination of bromoanilines (BANs) over the Pt/γ-Fe₂O₃nanocomposite catalyst was fully suppressed even at complete conversion of the substrates (BNBs) for the first time (114). Similar results were obtained, without hydrodechlorination during the hydrogenation of o-CNB to o-CAN using Pt/Y-Fe₂O₃. The results are listed earlier in this chapter under o-CNB reductions (115). Not only mono-HNBs but also the 2, 4-di-HNBs have been successfully reduced to corresponding HANs using this catalyst with selectivity up to 99.9%. For example, a partially reduced Pt/γ-Fe$_2$O$_3$ magnetic nanocomposite catalyst (Pt/γ-Fe$_2$O$_3$-PR) exhibited excellent catalytic properties in the selective hydrogenation of 2, 4-dinitrochlorobenzene and iodonitrobenzenes. The selectivity to 4-chloro-m-phenylenediamine (4-CPDA), meta-iodoaniline (m-IAN), and p-iodoaniline (p-IAN) reached 99.9%, 99.8%, and 99.4%, respectively, at complete conversion of the nitro substrates. The hydrodehalogenation of 4-CPDA and IANs was fully suppressed for the first time over Pt/γ-Fe$_2$O$_3$-PR. Because of the modified surface of the supports, the electron donation from Pt to the nitro group of the substrate is believed to be the cause of superior selectivity to the HANs (116).

Process for reducing CNBs catalyzed by Pt-NPs stabilized on modified montmorillonite clay is reported (117). The supported Pt°-NPs show efficient catalytic activity for the selective reduction of CNBs. As a typical example, at a H$_2$ pressure of 10 bars, temperature 45°C for a period of 15 min, the Pt°-NPs (prepared by reduction with hydrazine) exhibit conversion of o-CNB up to 100% and selectivity >99% to o-CANs with very negligible amount of C-Cl bond cleavage. Chemoselective hydrogenation of HNBs over Pt/C catalysts in scCO$_2$ to produce HANs with excellent selectivity and inhibition of dehalogenation is reported (118). The method of preparation of nanocatalyst has very good effect on their activity, and the yield and selectivity HANs, as demonstrated by the following example. Polyethylene glycol (PEG) stabilized platinum nanoparticles (SPPNs) were immobilized on solid supports such as γ-Al$_2$O$_3$, SBA-15, TiO$_2$ and active carbon, and tested in the hydrogenation of p-CNB in scCO$_2$, to obtain p-CAN with excellent selectivity (>99.3%) in the whole range of conversion. Such high selectivity to corresponding haloanilines (HANs) (>99.1%) was also obtained in the hydrogenation of o-CNB, m-CNB, p-BNB, m-INB and 2-chloro-6-nitrotoluene. The dehalogenation and the accumulation of intermediates were fully inhibited simultaneously in scCO$_2$ (119). Selective reduction of HNBs using [Rh]-thioligand species, i.e., [Rh]/DHTANa, as catalyst in an aqueous bi-phase system is reported. When NB and 3-CNB were independently reduced using [Rh]/DHTANa (A) and H$_2$ (4MPa), AN, and 3-CAN were obtained in high selectivity (99%) and high yields (99%) with conversions for both at >99%. RhTPPTS (triphenylphosphine-3, 3′, 3″-trisulfonic acid trisodium salt) also gave ANs in >99% yields. In case of 4-INB (85% conversion) in aq. biphasic system of toluene: H$_2$O at 60°C and 4MPA H$_2$ for 24h, 4-IAN was obtained without dehalogenation. Interesting enough, when the temperature was increased it gave a complete conversion and 4-IAN in 97% yield. Aniline was a side product in small amount. The importance of the process is in light of high conversion and high yields with little hydrodehalogenation of a product with highly sensitive 4-iodo group (120). In an interesting study using the well-known Pt-NPs/Y-Fe$_2$O$_3$ catalyst absolutely no hydrodehalogenation was observed in catalytic hydrogenation of o-, m, p-BNBs (121). How the catalyst preparation method enhances the HAN selectivity while inhibiting the dehalogenation is demonstrated by synthesizing Pd-on-Si catalysts prepared *via* galvanic displacement for the selective hydrogenation of *p*-CNB (122). Thermoregulated PEG1000-DIL/toluene biphasic system was applied in the synthesis of HANs by reduction of HNBs. The palladium catalyzed reduction with hydrogen, and the $FeCl_3 \cdot 6H_2O$ and $Fe_5HO_8 \cdot 4H_2O$ catalyzed reduction with hydrazine hydrate were investigated, where the latter gave highest catalytic activity in the given reaction media with excellent yields of up to 99% (123). The direct redox reaction (galvanic displacement) between Pd^{2+} and substrate Si was used to deposit Pd on Si, and the Pd–Si catalysts enabled a chemoselective hydrogenation of *p*-CNB with the selectivity for *p*-CAN higher than 99.9% at complete conversion of *p*-CNB.

3.06.10: Effect of Bimetallic Nanocomposites and Solid Supports:

Bimetallic nanocmposite chemoselective hydrogenation of *o*-CNB is achieved using copper catalysts. Among different catalysts employed, CuO–NiO/CS exhibited high conversion (96.6%), high product selectivity (98%), and excellent stability (124). A specially prepared $Pt/Co(OH)_2$ catalyst gives excellent results, both in terms of substrate conversions and CAN selectivity. $Co(OH)_2$ coated Pt-NPs (1.8nm) are proved to be a highly efficient catalyst for the chemoslective hydrogenation of HNBs (as CNBs, BNBs, INBs) with 99.9% selectivity for the HANs obtained. $Co(OH)_2$ was confirmed to prohibit the dehalogenation effectively, and the $Pt/Co(OH)_2$ catalyst could be recycled for several times (125). The selective gas phase (1 atm, 453 K) hydrogenation of *p*-CNB over a series of laboratory synthesized and commercial Pd (1-10% wt.) supported on activated carbon (AC), non-reducible (SiO_2 and Al_2O_3) and reducible (ZnO) oxides has been examined. This is the first report that demonstrates 100% selectivity for *p*-CNB to *p*-CAN over supported Pd (126). Porous manganese oxide (OMS-2) and platinum supported on OMS-2 catalysts have been shown to facilitate the selective hydrogenation of the nitro group in CNBs to give CANs with no dehalogenation. Complete conversion was obtained within 2 h at 25°C, and the selectivity to CAN remained at 99.0% even at 100°C (127). All the three above examples show quantitative selectivity for HANs, so obviously there is no dehalogenation of HANs.

The method through which the deposition of Cu adatoms to Pd/C is done influences the selectivity of HANs in hydrogenation reaction. The two structures, obtained by two different deposition methods, had different selectivity in the hydrogenation of *p*-CNB, as Cu adatoms suppressed the formation of aniline to 0.1%, while deposition of bulk Cu had no significant influence on the dechlorination side reaction (128). Mo_2C-supported Au and Au–Pd catalysts (nominal Au/Pd = 10 and 30) obtained for the first time from colloidal NPs (4-5nm) stabilized by polyvinyl alcohol (PVA) were tested in the gas phase hydrogenation of NB, *p*-CNB and *p*-nitrobenzonitrile and delivered 100% selectivity to the target amine in each case. Inclusion of Pd served to increase selective hydrogenation rates where Au–Pd/Mo$_2$C out performed Au–Pd/Al$_2$O$_3$, a response that is attributed to increased surface hydrogen. Equivalent Au/Al$_2$O$_3$ and Au–Pd/Al$_2$O$_3$ were prepared and served as benchmarks (129). Ru-nanocomposites or nanocatalysts (Ru-NCs) supported fullerenes (C60) used for the catalytic hydrogenation of NB to AN, are optimized (130). A new Co$_3$O$_4$/NGr@CNT as a fixed bed catalyst is prepared for using in a continuous flow reactor. Under optimized conditions, no dehalogenation side products could be detected. This remarkable selectivity and mechanical stability render Co$_3$O$_4$/NGr@CNT as a catalyst particularly relevant for application in continuous processes based on a packed bed reactor (131).

References:

97a). X. Lu et al. *RSC Advances,* 06(19):15354-15361 (2016). 97b). N. M. Patil, et al. *RSC Advances,* 05(105):86529-86535(2015). 97c). R. K. Rai, et al. *Inorganic Chemistry,* 53(06):2904–2909(2014).

98). A. M. Stratz, in: J. R. Kosak (Ed.), *Catalysis of Organic Reactions,* Marcel-Dekker, NewYork, 1984, p. 335.

99). P. Baumeister, et al in Hetero Catalysis and Fine Chemicals-II, Guisnet, et al. ed. Elsevier Sci. Amasterdam (1991), P321.

100). Shigeo Nishimura, in Handbook of Heterogenous Catalytic Hydrogenation for Organic Synthesis, John Wiley & Sons, 2001, PP344-345

101). US126127 102) H. S. Broadbent, *J. American Chemical Society*: 76:1519(1954). 103). H. S. Broadbent, et al. *J. Organic Chemistry*: 27:4400(1962).

104a). R. J. Gait, *British Patent*: 1498722(1978). 104b). D. D. May, US 5068436 A (1991). 104c). H. Ma, et al. *Catalysis Communications*: 10:1363-1366(2009).

105a). C. Xiao, *Current Organic Chemistry*; 16(02):280-296(2012). 105b). W. Du, et al. *Ind. Eng. Chem. Res.,* 53(12):4589–4594(2014). 105c). H. Niu, et al. *Ind. Eng. Chem. Res.,* 2016, 55 (31):8527–8533. 105d). B. Yang, et al. Nano Research, 09(07):1879-1890(2016). 105e). T. Komatsu, et al. *ACS Catalysis*: 06:642-646(2016). 105f). H. Baek, et al. *Synfacts* 2016; 12(06): 0656(2016). 105g). Hydrogenation with Low-Cost Transition Metals: Jacinto Sa, Anna Srebowata, CRC Press, 04-Nov-2015 - Science - 203 pages, PP55.

106). W. Yu, *RSC Advances*: 07:751-757(2017).

107). M. Liang, et al. *J. Catalysis*: 255:335-342(2008).

108). H. Liu, et al. *J. Molecular Catalysis A: Chemical,* 308(01-02):79-86(2009). 109a). H. Wang, et al. *J. Molecular Catalysis A: Chemical,* 273(01-02):160-168(2007). 109b). M. Liang, et al. *J. Catalysis,* 255(02):335-342(2008).

110). Book: Catalytic Hydrogenation in Organic Syntheses: Rylander, Elsevier, 02-Dec-2012 - Science - 336 pages, PP 124-125.

111). J. R. Kosak *Annales of N. Y, Academy of Science:* 172-175(1970).

112). Y. Fang, et a. *RSC Advances,*4(23):11788-11793(2014).

113). M. Kralik, et al. *Chemical Papers,* 68(02):146-150(2016).

114). X. Wang, et al. *J. Molecular Catalysis A: Chemical,* 279(01-02):160-168(2007).

115). J. Zhang, et al. *J. Catalysis,* 229(01):114-118(2005).

116). M. Liang, et al. *Journal of Catalysis,* 255(02):335-342(2008).

117). D. K.Dutta, & D. Dutta, US9284259-B2(2016).

118). S. Ichikawa, et al. *Chemical Communications* (Camb): (07):924-926(2005).

119). H. Cheng, et al. *J. Colloid Interface & Science:* 415:01-06(2014).

120). V. D. Rathod, *Catalysis Communications:* 84:52–55(2016).

121). X. Wang, et al. *J. Molecular Catalysis A-Chemical,* 273(01-02):160-168(2007).

122). Q. Wei, Y. Shi, et al. *Chemical Communications,* 52(14):3026-3029(2016).

123). H. Zhi, et al. *Chemical Journal of Chinese Universities,* 34(03):573-578(2013).

124). K, N, Rao, et al. *Catalysis Communications,* 11(02):142-145(2009).

125). H. Cheng, et al. *J. Colloid Interface Science,* 377(01):322-327(2012).

126). F Cardenas-Lizana, et al. *ACS Catalysis,* 03(06):1386-1396(2013).

127). I. J. McManus, *Faraday Discussions:* 188:451-466 (2016).

128). T. Mallat, A. Baiker, *Applied Catalysis A: General,* 200:03-22 (2000).

129). X. Wang, et al. *Catalysis Science & Technololgy,* 06(18):6932-6941(2016).

130). K. Hassan, et al. *Iranian J. Chem & Chemical Engineering:* 34(01):21-32(2015)

131). T. Baramov, P. Loos, *Adv. Synthesis & Catalysis,* 358(18):2903–2911(2016).

3.07.00: Catalytic Hydrogenation of HNBs with Nickel Catalyst:

Just like Iron, Nickel is one of the most commonly used metal catalysts either as a metal, metal salt, metal nanoparticles, and in a bimetallic form with other transition metals and also with or without a solid support. Its importance as an efficient catalyst is also because of its lower cost than noble metals, a crucial factor for developing a cost-effective process for manufacturing chemicals at commercial scales. The nanoalloys (a mixture of two or more metals) possess a series of useful properties such as magnetic, semiconducting, field-emission materials that are being applied in the catalytic processes and the creation of optical, electronic, and magnetic materials (132). Earlier, Meng et al. have studied the catalytic hydrogenation of NB to AN in dense phase CO_2 over Ni/Y-Al_2O_3 catalyst at 353K and 4MPa H_2 pressure (133).

A series of nanosized Ni–B catalysts were prepared by reduction of nickel acetate with aq. $NaBH_4$ in alcohol and along with Raney-Ni (Ra-Ni) were tested for liquid phase hydrogenation of p-CNB.

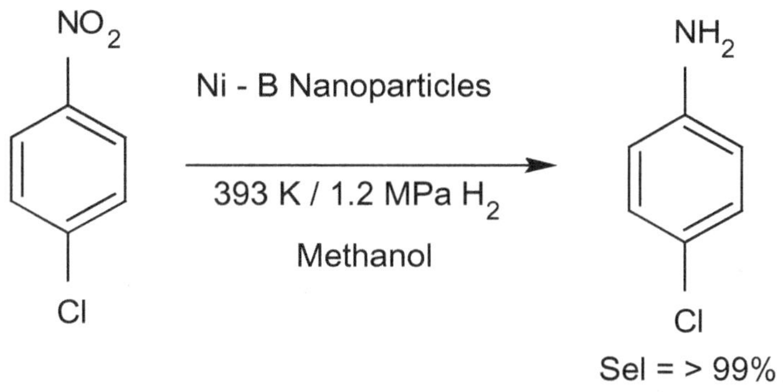

Sel = > 99% --Scheme-32

The Ni–B catalyst was passivated by B, therefore it was more stable than Raney-Ni and did not catch fire after exposure to air. The selectivity of main product (p-CAN) was greater than 99% on all of the Ni–B catalysts, which was far more than Raney-Ni catalyst (Scheme-32)(134).

Liquid-phase hydrogenation of p-CNB was performed in a well-stirred, high-pressure batch reactor system. La-Ni-B was much more active than Ni-B and the Raney-Ni catalyst. The selectivity of the main product (p-CAN) was >99% on La-Ni-B catalysts. The effect of the lanthanum promoter can be attributed to the electronic modification of Ni by La (135). Another example is of p-CNB hydrogenation to p-CAN with 0.02% Pd-Ni-B catalyst with 100% p-CNB conversion and very high selectivity for p-CAN attributed to both ensemble and electronic effects. Electron-enriched Ni could activate the polar-NO_2 groups of p-CNB and depress the dehalogenation of p-CAN (136). Ni-Co-B in the presence of W has been reported to be a good catalyst for the hydrogenation of p-CNB, W being a promoter (Scheme-33).

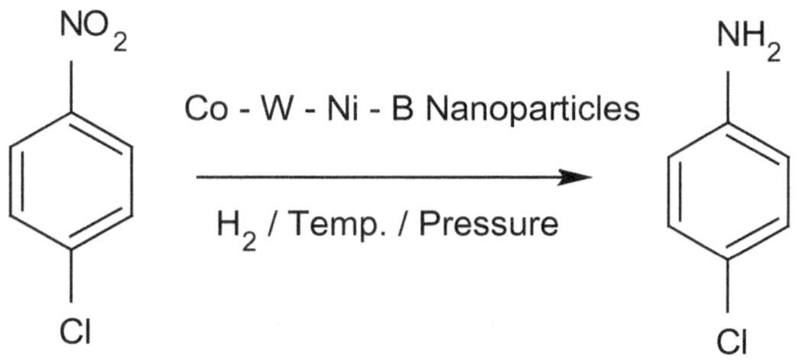

--Scheme-33

Tungsten (W) species not only acted as a spacer to prevent Ni-Co-B particles from aggregating but also donated partial electrons to Ni and Co. Here is another example of hydrogenation of p-CNB over Ni-Co-B bimetallic catalyst further modified with Mo, La,

Fe and W to form nanoalloys. The different behaviors of Mo-, La-, Fe- and W-modified Ni-Co-B catalysts displayed different results in the *p*-CNB hydrogenation (Scheme-34)-(137).

NO$_2$ → M- Ni - Co - B Nanoalloy, H$_2$ / Temp. / Pressure, M = Mo, Fe, W, La → NH$_2$ + Other Side Products (with Cl substituents)

--Scheme-34

A series of W-, Ru-, and Mo-promoted Ni–B amorphous nanocatalysts and also addition of Ni, Co, W, and B to Co-Ni-B has a positive effect on the *p*-CNB conversion and selectivity for p–CAN (Scheme-35) (138).

NO$_2$ → M - Ni - B Nanoparticles, H$_2$ / Temp. / Pressure, M = Ru, Mo, W → NH$_2$ (with Cl substituents)

Mo - Ni - B Gave Best Results   Sel = High --Scheme-35

In the former case, the catalysts also exhibited high selectivity to *p*-CAN and Mo–Ni–B showed the highest activity. In another case, the hydrogenation of *p*-CNB on nanosized modified Ni-Mo-B catalysts is studied.

NO$_2$ → Ni-Mo-B - Alloy (0.4:1:3:0.1), 333 - 353K/ 1.2 MPa H$_2$ → NH$_2$ (with Cl substituents)

Ni:Mo:B:Additive = 0.4:1:3:0.1   Additives: Pd, La, Co, Fe --Scheme-36

The Ni-Mo-B catalysts were prepared chemical reduction with SBH. Nanoalloys have good selectivity and activity in the reduction of HNAs (Scheme-36) (139).

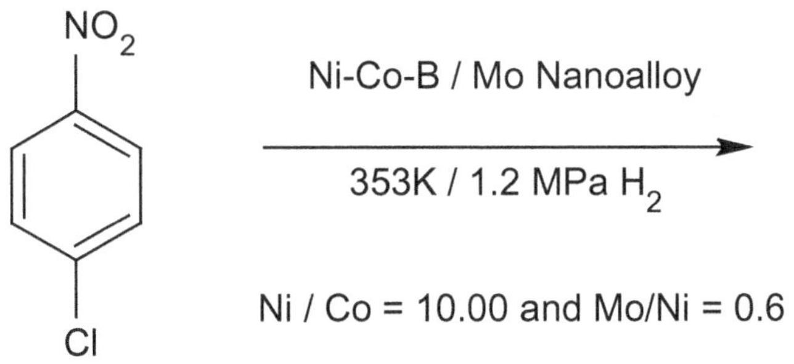

NO$_2$ 　　　Ni-Co-B / Mo Nanoalloy 　　　NH$_2$

$$\xrightarrow{\text{353K / 1.2 MPa H}_2}$$

Ni / Co = 10.00 and Mo/Ni = 0.6 　　　Cl 　--Scheme-37

Similarly, the combination of other metals/atoms with Ni has led to high catalysts activity and HAN selectivity as tested on hydrogenation of *p*-CNB to *p*-CAN (140). Su et al. have found that in Mo-Ni-B nanoalloys, the effect of Mo content was crucial. The catalyst activity increased with an increase of Mo content until a Mo/Ni atomic ratio of 0.6 was reached (141). As seen above, the Ni-Mo-B alloy catalyst has been reported to be a good catalyst for the hydrogenation of *p*-CNB to *p*-CAN. Further, the authors have investigated the effect of additives (Pd, La, Fe and Co) on the catalytic properties of Ni-Mo-B in the hydrogenation of *p*-CNB (Scheme-37). The catalysts were prepared and characterized by different techniques. The catalysts, with additives, were tested for liquid-phase hydrogenation of *p*-CNB at 1.2 MPa H$_2$ pressure and 333–353K. The reaction activity is in the order of Pd-NiMoB > La-NiMoB > Co-NiMoB > NiMoB > Fe-NiMoB. Except Pd, the other additives were in the form of hydroxide and acted as a spacer to prevent Ni-Mo-B catalysts from sintering and aggregation (142). Amorphous nanosized Ni-B catalysts have been reported to be a good catalyst for the liquid-phase hydrogenation of CNBs. This study was conducted to investigate the effect of lanthanum on the catalytic properties of the Ni-B catalyst in the liquid-phase hydrogenation of *p*-CNB (Sch-38).

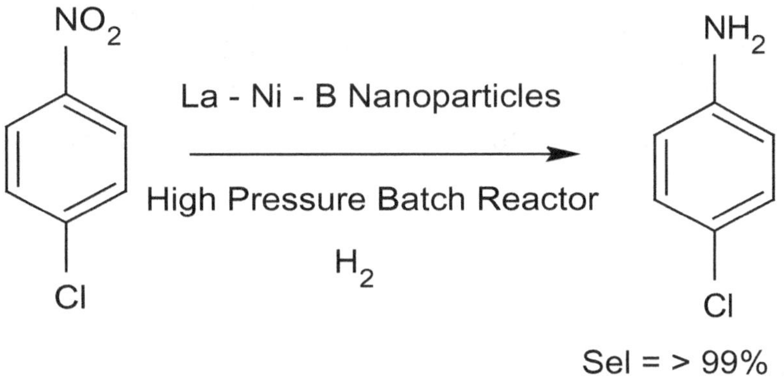

NO$_2$ 　　La - Ni - B Nanoparticles 　　NH$_2$

$$\xrightarrow{\text{High Pressure Batch Reactor}}$$

H$_2$ 　　　　Cl

Sel = > 99% --Scheme-38

La-Ni-B was found much more active than Ni-B and the Raney-Ni catalyst. The selectivity of the main product *p*-CAN was >99% on La-Ni-B catalysts (143).

A series of Ni–P–B catalysts were prepared by liquid-phase reduction and the catalysts were tested by liquid-phase hydrogenation of *p*-CNB at 393K and 1.2MPa hydrogen pressure in a batch reactor. Boron in the Ni–P–B powder, donated electrons to the nickel metal and phosphorus accepted electrons from the nickel metal. The catalyst with the lower concentration of nickel on the surface seemed to have higher activity (Scheme-39).

Sel = > 99% --Scheme-39

The reaction conditions also have a pronounced effect on the catalytic activity. Using methanol, compared to ethanol, as the reaction medium significantly increased the conversion of *p*-CNB. The selectivity of *p*-CAN was >99% on the Ni–P–B Catalyst and was found better than Raney-Ni catalyst (144). Ni°-nanoparticles of approximately 5 nm were generated and supported on montmorillonite and were found as efficient and selective heterogeneous catalyst for hydrogenation of HNBs to corresponding HANs with conversion 78–100% and selectivity 96–99.4% (145). The Pt/C-Ni nanocomposite catalysts show higher selectivity towards *o*-CAN than the Pt/C catalyst due to the electronically synergistic effect between Pt and Ni species from *o*-CNB hydrogenation at 30°C and 0.1MPa hydrogen. The catalyst can be recovered and reused (146).

Chemoselective hydrogenation of CNBs to CANs in scCO$_2$ over Ni/TiO$_2$ can be achieved at any conversion levels up to 100% at 35°C. No accumulation of any intermediates and very little dehalogenation at any conversion level is observed (147). Ni-B/Al$_2$O$_3$ was found superior to Ni-B/SiO$_2$ with conversions 40% and 74.5%, and selectivity 96.8% and 94.3% respectively. Ni-B over CNTs and AC (activated carbon) gave higher conversions (over 85%) and high selectivity of 95.5% and 90.4%, respectively were achieved. Y. Fang, http://en.cnki.com.cn/Article_en/CJFDTOTAL-GYCH200610013.htm Pt doped with another metal and supported on C reduces *p*-CNB to *p*-CAN with 99% selelctivty but at very high temperature (500°C). M. Hua, et al. http://en.cnki.com.cn/Article_en/CJFDTOTAL-GYCH200704011.htm

Some examples of selective hydrogenation of HNBs over C or CNTs, without dehalogenation, are discussed above.

3.08.00: Hydrogenation of HNBs in Batch and Flow Reactors:

The batch processes in chemical/pharmaceutical industries have been finding difficulties in particular cases. Therefore, in recent times the continuous flow systems are being explored. The selective hydrogenation of functionalized NAs and especially those of HNBs pose major challenges from both academic as well as industrial viewpoints. Here is an example of selective hydrogenation of HNBs to HANs using batch and flow system as used by Bayer Pharma, Germany (148). The commonly used Fe, Zn and HCl method produced HANs with the metal oxide waste. Chemoselective hydrogenation of 4-INB as a model experiment was found successful by the current methodology in batch processes. The optimized conditions were transferred from batch to a continuous flow process. The HANs were obtained in good to quantitative yields (78-100%), and very high to quantitative selectivity (98-100%). Finally, the optimized flow conditions were applied to transformations which represent important steps in the syntheses of the active pharmaceutical ingredients Clofazimine and Vismodegib (149a). A Review on recent devevelopments in selective heterogenous hydrogenations of HNBs to HANs is published (149b).

An alloy of Ni-Cu (1:1) proved inactive when the experiments were conducted in an autoclave in a flow-type catalytic apparatus at atmospheric pressure, the yield of p-CAN did not exceed 25%. However, in a stucdy it is found out that the industrial catalyst Ni-Kieselguhr with 50% Ni catalyzes the reduction of p-CNB in an autoclave under a pressure of 50–100 atm to p-CAN with yields up to 99% under optimum conditions (Scheme-40(150).

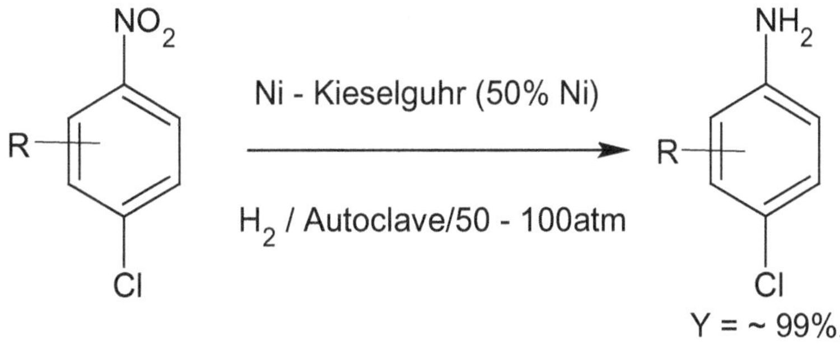

Y = ~ 99%--Scheme-40

3.09.00: Catalytic Hydrogenation of HNBs in a Biphase System:

In a Ru/Pt bimetallic catalyst, the Pt acted as a promoter and hence increased the catalyst activity for the p-CNB hydrogenation (99.9% conversion) at 25°C amd 1.0Mpa H_2, with 99.4% selectivity to p-CAN. The Ru/Pt catalyst also showed high activity and selectivity for the hydrogenation of other chloro- and dichloro-NBs with substitutions in different positions (151). There are some other reports which present the NAs reductions in multi-pahse systems (152).

3.10.00: Catalytic Hydrogenation of HNBs in Ionic Liquids:

Ionic liquids (ILs), for the first time, were found as an excellent medium for the heterogeneously catalyzed hydrogenation of HNBs to corresponding HANs. In the search of best possible or at least better solutions to hydrogenation problems than existing methodologies, ionic liquids have shown fairly satisfactory results. Three types of heterogenous catalysts of Raney-Ni, Pt/C and Pd/C were tested. The ionic liquids give rise to higher selectivity and lower dehalogenation in the hydrogenating process compared with that observed in conventional organic solvents (153).

As a part of Aniline and Its Analogs, we will cover major work in this context under chemoslective reductions of NAs (NB with various substituents at 2, 3, 4- position in its benzene ring). However, here are few examples where the HNBs are multinuclear substrates. For example, the chemoselective reduction of 5-nitro-2-chloro-2',4'-dimethylbenzenesulfonanilide on a bimetallic catalyst supported on y-alumina (Pd-Ru/γ-Al₂O₃) in ionic liquids is presented (Scheme-41).

--Scheme-41

The use of ionic liquids [[EPy]Br or [BPy]Br] as reaction medium for hydrogenation of 5-nitro-2-chloro-n-(2',4'-dimethyl)benzenesulonamide (NCD) to 5-amino-2-chloro-n-(2',4'-dimethyl) benzenesulfonamide (ACD) has led to obtain high ACD selectivity (~100%) on Pd-Ru(1:1)/γ-Al₂O₃ catalysts at 338K under 0.35MPa of hydrogen pressure (154). An alternative methodology toward highly selective catalysis of HNBs to corresponding HANs in ionic liquids is reported (155). Later, Huang et al. have found out that the ionic liquids-modified Ni/Al₂O₃ catalysts are found as an excellent media for the heterogeneously catalyzed selective hydrogenation of HNBs to corresponding HANs (Scheme-42).

R = o,m,p - Halides; Good Selectivity; Lower Dechlorination --Scheme-42

It gives rise to higher selectivity and lower dehalogenation in the hydrogenating process compared with that observed in conventional nickel catalyst (156). A CNT supported catalyst containing cobalt/cobalt oxide (Co/Co₃O₄) NPs encapsulated within a shell of nitrogen-doped graphene layers (Co₃O₄/NGr@CNT) was prepared. It shows excellent chemoselectivity in the hydrogenation of the showcase substrate the 4-INB. In contrast to traditional AC supported catalysts, the CNT support allow for the application of the newly prepared Co$_3$O$_4$/NGr@CNT as a fixed bed catalyst in a continuous flow reactor. Under optimized conditions, the dehalogenation side products were not detected at all (157).

3.11.00: Catalytic Hydrogenation of HNBs with a Non-Metal Catalyst:

Usually, we are accustomed to see reports on metal catalyzed hydrogenations of organic functional groups. However, here is an example, where a nonmetal catalyst for molecular hydrogen activation is used with a comparable catalytic hydrogenation capability of noble metal satalysts.

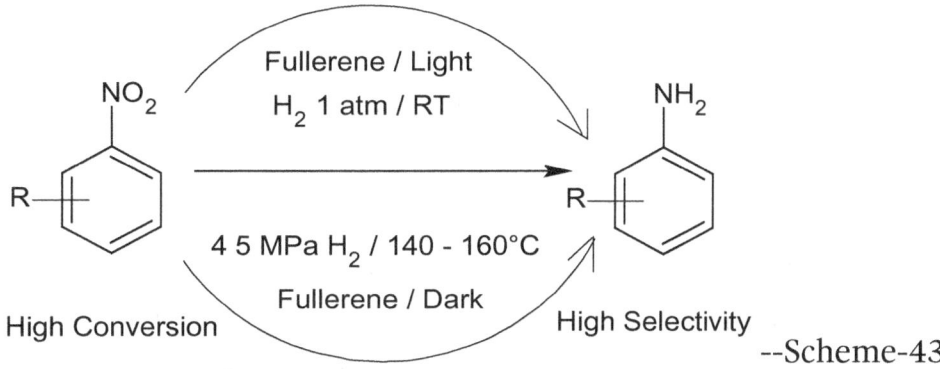

--Scheme-43

Fullerene can activate molecular hydrogen and is a novel nonmetal catalyst used in the hydrogenation of NACs to arylamines with high conversions and selectivity (Scheme-43), under mild conditions as: i) room temperature, H$_2$ (1atm), and light irradiation or ii) at 120–160°C and 4–5 MPa H$_2$ pressure without light irradiation, which is comparable to the case with a noble metal catalyst (158).

3.12.00: Catalytic Hydrogenation of 2, 3, 4-HNBs:

Long time ago, using a novel method of converting HNBs, notably CNBs, to their corresponding HANs in >90% yields by hydrogen reduction with a minimum dehalogenation was reported using Ru-NPs/Al$_2$O$_3$ catalyst prepared by depositing metallic Rh on activated alumina having a particle size of less than 200mesh (159). 3,4-DCNB, 2,5-DCNB and 4-Cl-2-NT have been successfully reduced to the corresponding ANs in high yields (94.5 -97.5%) and 99.9% purity by catalytic hydrogenation over PT/C (or Darco) with about 2-3% dehalogenated products (160). A highly efficient and selective hydrogenation of CNBs to CANs using water-soluble bimetallic catalysts Ru-DPPES and Ru/Pt-DPPES is investigated. The variety of CNBs used in this study were 2-CNB, 3-CNB, 4-CNB, 2,4-DCNB, 3,4-DCNB, 1,2-DCNB with selectivityies upto >99-100% and the conversions also >99% t0 ~ 100% (161). As seen under *o, m, p*-HNB hydrogenation above, with Ag-NPs/Y-Fe$_2$O$_3$ as a catalyst the 2, 4-dinitro-CB gave 4-chloro, *m*-phenylene diamine with 99.9% selectivity at 100% conversion. Exactly the same results were obtained for i) 3,4-DCNB (100% conv) to give 3,4-DCAN (99.9% sel), ii) 3-chloro-2-methyl-AN (99.9sel) at 100% conversion of 2-chloro-6-nitrotoluene., iii) methyl-4-chloro-3-aminobenzoate (99.9% with100% conversion of methyl 4-chloro-3-nitrobenzoate and many oher examples. All the reductions are carried out between 300-333K and 0.1MPa H$_2$ (162). A process for reducing CNBs using Ru with a minor amount of Pt as the catalyst (Ru:Pt= 75:1 to 30:1) is used for the chemoselective hydrogenation of HNBs to the corresponding CANs with high yields (96-97%) (163).

References:

132a). D. J. Collins, et al. *Ind. Eng. Chem. Prod Res. Dev.*, 21(02):279–281(1982). 132b). M. Blanca, et al. *Ind. Eng. Chem. Res.*, 50(13):7705-7721(2011).

133a). X. Meng, et al. *J. Catalysis*, 264(01):01-10(2009). 133b). X. Meng, et al. *J. Catalysis*: 264:01-10(2009).

133c). B. M. Bhanage & M. Arai, "Transformation and Utilization of Carbon Dioxide", Springer Science & Business Media, 27-Jan-2014 - Science - 388 pages, PP 374.

134). Y. Liu, et al. *J. Nanoparticle Research*, 08(02):223-234(2006).

135). Y. Liu & Y. Chen, *Ind. Eng. Chem. Res.*, 45(09):2973–2980(2006).

136). Y Chen, et al *J. Nanoparticles*: Volume 2013 (2013), Article ID 132180, 10 pages. http://dx.doi.org/10.1155/2013/132180

137a). B. Zhao, et al. *J. Non-Crystalline Solids*, 356(18-19):839-847(2010). 137b). B. Zhao, et al. *Ind. Eng. Chem. Res.*, 49(04):1669–1676(2010).

138). L. Chen & Y. Chen, *Ind. Eng. Chem. Res.*, 45(26):8866–8873(2006).

139a). Y. Chen & D. Lee, *Catalysis Surveys from Asia*, 16(04):198-209(2012). 139b). Y. Chen & N. Sasirekha, *Ind. Eng. Chem. Res.*, 48(13):6248-6255(2009). 140). Q. Shi, et al. *Adv. Synthesis & Catalysis*, 349(11-12,):1877–1881(2007). 141a). M. Lin , et al. *Ind. Eng. Chem. Res.*, 48(15):7037–7043(2009). 141b). J. F. Su, et al. *Ind. Eng. Chem. Res.*, 50(03):1580-1587(2011).

142). Y. Chen, & D. Lee, *Catalysis Surveys from Asia*, 16(04):198-209(2012).

143a). Y. Liu, & Y. Chen, *Ind. Eng. Chem. Res.*, 45(09): 2973–2980(2006). 143b): Y. Chen, *J. Non-Crystalline Solids*, 355(22-23):1193-1201(2009).

144). Y. Liu, *Ind. Eng. Chem. Res.*, 45(01):62–69(2006).

145). D. Dutta, et al. *Applied Catalysis A: General*, 487:158-164(2014).

146).Y. Xie, et al. *Catalysis Communications*, 28: 69–72(2012).

147a). M. Meng, et.al.*Journal of Catalysis*, 269(01): 131-139. 147b). H. H. Seltzman & B. D. Berrang, *Tetrahedron Letters*, 34(19):3083-3086(1993).

148). P. Loos, et al. *Org. Process Research & Development*, 20(2):452–464(2016).

149). M. Pietrowski, *Current Organic Synthesis*: 09:470-487(2012).

150). N. P. Sokolova, et al. *Russian Chem Bulletin*, 15(11):1824-1829, 1966(1984).

151). Y. Zhou, et al. *China Petrol Procce & Petrochem Technol.*, 17(02):26-31(2015).

152a). G. Evdokimova, *Applied Catalysis A: General*, 271:129–136(2004). 152b).G. D. Yadav, et al. *Industrial Engineering & Chemical Research*, 46:2951-2961(2007).

153). D. Xu, *J. Molecular Catalysis A: Chemical*, 235(01-02):137-142(2005).

154a). Y. Zhao, et al. *Modern Applied Science*: 04:155-161(2010). 154b). K. Ju, et al. *Microbiol Mol Biol Rev*, 74(02):250-272(2010).

155). D. Dian et al. *J. Molecular Catalysis A: Chemical*, 236(01-02):137-142(2005).

156). X. Huang, et al. *Advanced Materials Research*, 233-235:2904-2908(2011).

157a). T. Baramov, et al. *Adv Synthesis & Catalysis*, 358(18):2903-2911(2016).

157b). H. Alex, et al. *ChemCatChem*, 09(16):3210-3217(2017).

158). B. Li, et al. *J. American Chemical Society*, 131(45):16380-16382(2009).

159a). F. C. Trager, US 2772313-A(1956). 159b).T. Keil, US 3067253-A(1962).

160a). L. Spiegler, US 3073865-A(1963). 160b). U. Jersak, US 4053527-A(1977).

161a). Y. Zhou, et al. *China Petrol Processing & Petrochem Technol.*, 17(02):26-31(2015).
161b). C. Xiao, et al. *Current Organic Chemistry*, 16(02):280-296 (2012).

162a). C. Evangelisti, et al. *J. Molecular Catalysis A: Chemical*, 366:288-293(2013). 162b). Y. Wang, et al. EP 1826180-A1(2007). 162c). US 20090124834-A1(2009). 162d). Y, Wang, et al. US 7947191-B2(2011). 162e). *P. Serna & A. Corma, ACS Catalysis*, 05(12):7114–7121(2015). 163). J. R. Kosak, US 4760187-A(1988).

3.13.00: Hydrogenation of *p*-CNB with Different Hydrogen Sources:

We have seen above the catalytic hydrogenations of NAs over different metal catalysts using molecular hydrogen. Now we will focus on the reductions of NAs using hydrogen sources other than molecular hydrogen. The other hydrogen sources are as alcohols, formic acid and formates, glucose, hydrazine, $CO+H_2O$, CO_2+H2O, $scCO_2$ and water itself over different catalysts. These reductions are generally referred to as Catalytic Transfer Hydrogenations, "CTH".

3.13.10: CTH of HNAs with Alcohols:

Some of the reductions are carried out with photocatalytic irradiations in the presence of an alcohol (IPA, EtOH, MeOH etc.) as a solvent and especially as a hydrogen donor or hydrogen source. Gnerally, when the authors say that it is the selective hydrogenation of NAs, with an exception, it is the exclusive reduction of the nitro group without affecting other functional groups, which include halogens, -OH, -CH$_3$, -CHO, -COOR, -NO$_2$, -NH$_2$, -COCH$_3$, -CN etc, of which one or two are present in the same NA that is subjected to hydrogenation process. In case of alcohols as hydrogen donors, they are used without any activation or some time by activation with photocatalysis or ultrasonication.

A highly efficient and selective process is developed for the hydrogenation of NAs over phoroactivated rutile TiO$_2$ using alcohol as a hydrogen donor at room temperature and atmospheric pressure. The corresponding ANs are produced in high yields compared to common anatase and P25TiO$_2$ (Scheme-44).

--Scheme-44

The Ti^{3+} atoms located at the oxygen vacancies on the rutile surface behave as the adsorption site for NAs and the trapping site for photoformed conduction band electrons, which enhance the selective formation of ANs. The process is chemoselective for nitro groups, so it is tolerant for other reducible substituents (164). The catalytic activities of a series of in-situ generated homogeneous Ru systems and ligands have been successfully prepared and tested in transfer hydrogenation of NAs to ANs. Combination of [RuCl$_2$(p-cymene)]$_2$ and tridentate phenanthroline based ligand 2-(6-methoxypyridin-2-yl)-1,10-phenanthroline (phenpy-OMe) exhibited the highest catalytic activity for this reaction using 2-propanol or isopropylalcohol (IPA) as a hydrogen source under mild conditions and in excellent yields. It is a highly chemoselective process towards nitro groups in the presence of sensitive functionalities (165). Hydrogenation properties of CNB have been studied over Pt/CNTs and Pt-M/CNTs catalysts (M = Mn, Fe, Co, Ni and Cu) in ethanol at 303 K and normal pressure. Both catalytic activities and the yields of p-CAN are improved over Pt-M/CNTs catalyst, but the Pt-Fe/CNTs exhibits the best catalytic activity (TOF is 0.47 s^{-1}), and Pt-Mn/CNTs catalyst gives the highest yield of p-CAN (98.5%) (166). Now, with new techniques it is possible to visualize the actual events happening on the catalyst and support surfaces. In this particular case and in most of the other cases the reasons for the catalytic activity and the products selectivity are identified. For example, the liquid-phase hydrogenation of p-CNB was carried out at 373k and 1.2 MPa hydrogen pressure in methanol with 500 rpm stirring speed and in the presence of 2 mmol Ni catalyst. Based on the electron transfer between elemental nickel and boron, Ni-CoB(1:0.1) had the most d-band electrons which were responsible for the highest activities in hydrogenation of nitro groups (167). A functionalized plasmonic Au/TiO$_2$ photocatalyst with an Ag co-catalyst was successfully prepared by the combination of two types of photodeposition methods.The catalyst quantitatively reduced NB with IPA as hydrogen donor to AN under UV irradiation (168).

A highly efficient and selective heterogeneous photocatalytic system for nitro reduction to aryl amines was established using CdS, Ni$_2$P and Na$_2$S/Na$_2$SO$_3$ as a photosensitizer, a cocatalyst and a sacrificial electron donor (an alcohol) in aqueous solution, respectively. Two competing pathways for photocatalytic H$_2$ production and nitro reduction were found. Also, the reduction of NAs to ANs was confirmed to proceed through both the direct and condensation routes (169). Zhang et al. have reported photocatalytic reduction of p-CNB on illuminated TiO$_2$-NPs in the absence of oxygen and in the presence of the sacrificial electron donors as alcohols under UV light irradiation. TiO$_2$ used were A101, P25, and R201 (Scheme-45).

--Scheme-45

The sacrificial electron donors consisted of IPA, methanol, ethanol, formic acid, etc. p-CAN yield could reach up to 99.20% in 12 h when P25 used, while A101 and R201 gave lower yields. The mixture of IPA and formic acid (v:v = 90:10) was selected as reductive medium because of less by-product and higher p-CAN yield (170). A novel electron-transfer system for the chemoselective reduction of NAs is developed using a Sm(0) and 1,1'-Dioctyl-4,4'-bipyridinium dibromide catalyst (Scheme-46)(171).

Ligand = 1,1'-Dioctyl, 4,4'-bipyridinium dibromide   --Scheme-46

R = alkene, azide, amide, sulfonamide, tBoc, halide, silylether, benzylether etc.

The catalytic system gave 79–99% yield with high selectivity over a number of other functional groups. In another case, the reduction of various NACs into their corresponding amines (21 examples, including o- and p-CNBs, p-FNB, p-BrNB, and p-INB), is achieved by zeolite-supported Cu-NPs (Cu-NPs/Zeolite) with good selectivity and excellent yields in IPA as a sustainable reducing agent (172).

3.13.11: CTH of HNBs/HNAs with Boranes and Borohydrides:

We have seen under boranes in other chapter that boranes and borohydrides are very good sources of hydrogen and are commonly used both in reseach laboratories and in industries for appropriate reduction processes, including reductions of NAs and HNAs (173). A series of m-, and p-CNBs were reduced with $NaBH_4$ (SBH) in the presence of Cu-NPs catalysts to afford the m- and p-CANs in high yields (95%). Other substituents present on the same molecules were as $CH=CH_2$, COOR, CHO, $COCH_3$, NO_2 etc. (Scheme-47).

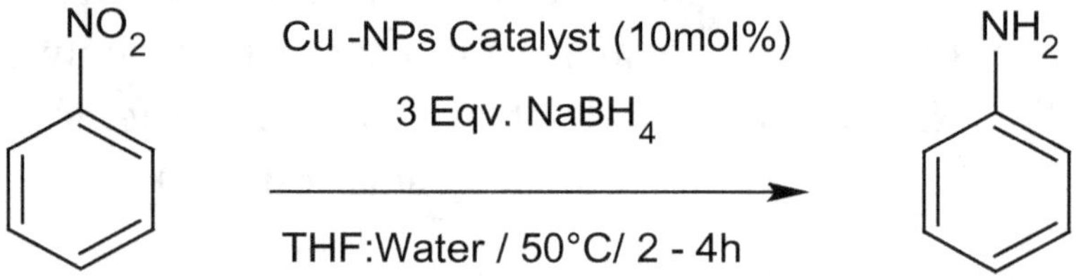

Cat = Cu salts or Oxide : Y= 71 - 85%; No Cat : Y =< 3%

Cat. = Cu - NPs : R = H: Y = 98%(GC), 95% (isol), R = m, p -Cl: Y = 95%--Scheme-47

Copper being a non-noble metal this procedure can lead to an economical process with industrial applications (174).

Copper (II) bromide as a procatalyst is used for *in-situ* preparation of active Cu-NPs for the efficient reduction of NAs using SBH (Scheme-48).

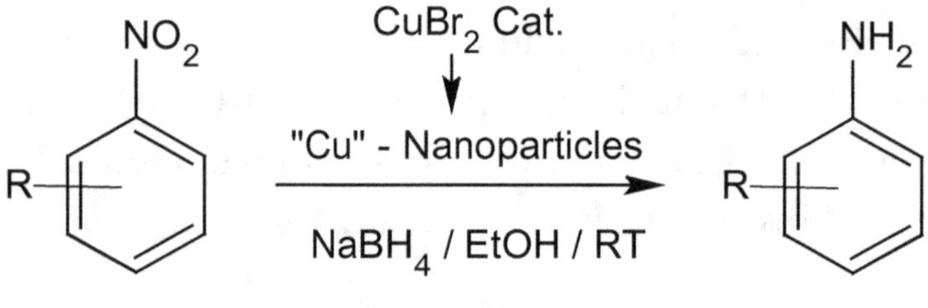

R = Cl, I, COOH, OBn, CH_2NO_2 etc.   --Scheme-48

During reduction, Cl, I, COOH, aliphatic nitro and OCH_2Ph groups remain intact displaying the chemoselectivity of the process (175).

Supported bimetallic Pt–Ni-NPs (BM-NPs) of tunable compositions have been prepared and tested for the reduction of NAs in aqueous SBH and were found to be superior to the activities of monometallic Pt-NPs (176). SBH in the presence of charcoal (0.4–0.8 g) reduces varieties of NAs in H_2O-THF (1:0.5 mL) at 50–60°C to afford high to excellent yields of corresponding aryl amines (177). Au-NPs deposited on nanocrystalline MgO are found very efficient and highly chemoselelctive catalyst for the reduction of NAs in aqueous medium at room temperature using SBH as a hydrogen source (178).

Hydrolytically stable borane–amines can be activated in-situ through Pd catalysis and perform reductions not possible otherwise. Pd catalyzed activation of aminoboranes in methanol enhances the reduction of NAs to ANs (179).

R = H, CH$_3$, CN, OH, F, OCH$_3$, COOMe etc.

R = H = 99%

Y = 95 - > 99% --Scheme-49

Hence, borane–trimethylamine is an efficient hydrogen-transfer reagent for the open vessel reduction of NAs to ANs (Scheme-49).

Graphene modified Ni$_{30}$Pd$_{70}$ nanoalloy can cause tandem dehydrogenation of ammonia-borane (AB, H$_3$N-BH$_3$) and can catalyze the reduction of NAs, including HNBs to the corresponding ANs and HANs in excellent yields (>99%) in methanol at room temperature (RT). G-NiPd showed composition-dependent catalysis on the tandem reaction with G-Ni$_{30}$Pd$_{70}$ being the most active and completed the reaction in 5–30 min with the conversion yields reaching up to 100% (Scheme-50).

R = H, 4 - OH, 4 -CH$_3$, 4 - Br, 3 -NH$_2$, 2 - NH$_2$ Yields >99%, --Scheme-50

Especially p-BNB was reduced to p-BAN in >99% yield within 5 minutes in aqueous methanol at RT. Considering the labile nature of bromine, this is really an excellent system both for chemoselectivity and yields without causing dehydrobromination This is one of the best example of hydrogenation of HNBs without dehydrohalogenation.(180). The mesoporous titania-supported Au-NPs assemblies (Au/MTA) catalyze the activation of SBH and 1,1,3, 3-tetramethyl disiloxane (TMDS) compounds, which act as transfer hydrogenation agents for the reduction of NAs to the corresponding ANs and these catalytic systems (/MTA-NaBH$_4$ and Au/MTA-TMDS) show promise for the efficient synthesis of aromatic amines at industrial scales (181). Aqueous SBH in the selective reduction of CNBs catalyzed by Au-NPs anchored onto the magnetic poly-(ionic-liquid) polymer (MNP@PIL@Au) as a robust and recoverable catalyst is described. The reaction was carried out for various NAs in water under mild conditions to obtain ANs in high yields (182). Iron-catalyzed, general and operationally simple formal hydrogenation using Fe(OTf)$_3$ and SBH is found highly chemoselective towards nitro group in the presence of other functionalities as halides, ester and amide etc.(Scheme-51).

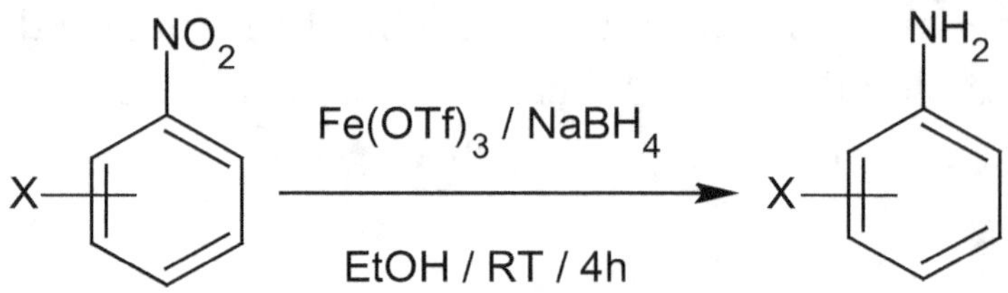

X = 2 -Cl (70%), 4 -Cl (80%), 4 -Br (51), 2 -F(73%), 4 -F(87%). --Scheme-51

The synthesis of the analgesic benzocaine from the corresponding NA showcases the utility of this methodology (183).

3.13.12: CTH HNAs with CO+H_2O, CO_2+H_2O and scCO$_2$:

The dehalogenation of in-situ formed CANs during catalytic hydrogenations of CNBs is a challenge for synthetic organic chemists all along. Over a period of time different alternatives with better selectivity for CNAs is achieved. According to Meng et al. supported Ni catalysts, prepared using the incipient wetness impregnation was studied in the hydrogenation of o-CNB in compressed CO_2 and in ethanol at 35°C. The total conversion and selectivity to o-CAN was significantly enhanced in scCO$_2$ compared with that in ethanol, and a high selectivity of about 99% was attained at a nearly complete conversion. Meng et al. have further demonstrated that the selective hydrogenation of NB can be achieved over Ni/γ-Al_2O_3 in dense phase CO_2 with 99% selectivity to AN over the whole conversion range of 0-100%, and the molecular interactions of CO_2 with the reaction species are significant for the improvement in the selectivity (184). Ichikawa et al. have reported that the rate of hydrogenation of CNB over Pt/C was markedly enhanced and the dechlorination reaction was significantly suppressed in scCO$_2$. The selectivity to CAN was >99% at 100% conversion. The rate of the hydrogenation of nitro groups is markedly increased in scCO$_2$ compared to the neat reaction, due to modification of the catalyst surface with CO generated from CO_2 hydrogenation (185). Ichikawa et al have obtained o-CAN in 99.7% selectivity from o-CNB by hydrogenation over Pt/C in scCO$_2$ (Scheme-52).

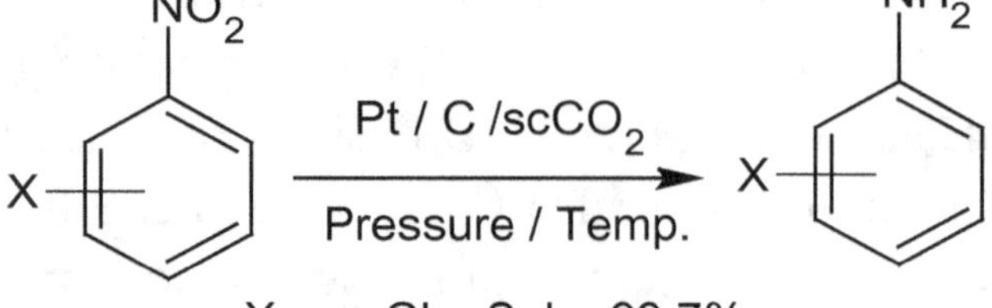

X = o -Cl = Sel = 99.7%

Other Halo Groups, Convs = 100%; Sel =99% --Scheme-52

Other HNBs were also reduced to corresponding HANs with 99% selectivity at 100% conversions of the substrates (186). Hydrogenation of p-CNB by Ni–B nanocatalyst in CO_2-expanded methanol gave higher conversions of p-CNB than in methanol without dissolved CO_2. The conversion enhancement was attributed to the increased hydrogen solubility and mass transfer rate resulting from the methanol volume expansion (187). A PVP-Stabilized Ru colloid afforded over 99.9% selectivity in the hydrogenation of o-CNB to o-CAN (188). Hydrogenation of o-CNB over Pd/C and in scCO$_2$ at 308K and 8-14MPa pressure afforded o-CAN in lower (93%) yield than PVP-Ru catalyst (189).

Water as a clean solvent and promoter in the organic synthesis has attracted more attention than in the past. Herein the effect of water was studied for the hydrogenation of o-CNB over Pt/C and Pd/C catalysts in ethanol, n-heptane and compressed CO_2. Under experimental conditions the o-CNB conversions in i) water, ii) water+n-heptane and iii) water + scCO$_2$ were 100% in all the solvents and o-CAN selectivities were 98.6%, 99.5% and 99% respectively (190). A new method for reduction of NAs employing metallic iron in pressurized CO_2-H_2O medium has been developed. Particularly, p-CNB can be reduced to p-CAN in 99% yield under the optimized conditions (Scheme-53)

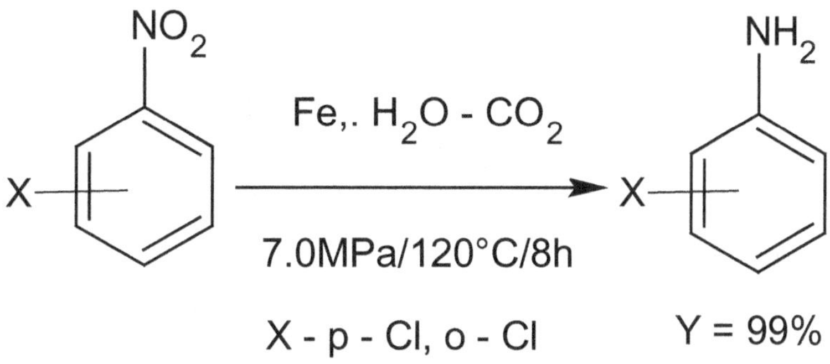

$$X - p - Cl, o - Cl \qquad Y = 99\%$$ --Scheme-53

The method is environmentally benign, easily scaled-up and applicable for the preparation of substituted ANs (191). Polyethylene glycol stabilized Pt-NPs were immobilized on solid supports such as γ-Al$_2$O$_3$, SBA-15, TiO$_2$ and active carbon, forming supported polyethylene glycol stabilized Pt-NPs (SPPNs). In the hydrogenation of p-CNB in scCO$_2$, the SPPN showed high selectivity to p-CAN (>99.3%) in the whole range of conversion. Such high selectivity to corresponding HANs (>99.1%) was also obtained in the hydrogenation of o-CNB, m-CNB, p-BNB, 2-Cl-6-NT and m-INB. The dehalogenation and the accumulation of intermediates were simultaneously and fully inhibited in scCO$_2$ (192).

3.13.13: CTH of HNAs with Formic Acid and Formates:

Formic acid is one of the most efficient and readily available alternative hydrogen sources for the selective reduction of NAs and HNAs. The following examples illustrate the general applicability of this simple hydrogen source. The role of transition-metal-loaded silicon NPs for the photocatalytic reduction of NAs in the presence of formic acid under visible light irradiation is investigated. Formic acid assumes the role of both as a hydrogen source and a sacrificial reagent for the introduction of electrons into the generated holes of semiconductors (Scheme-54).

--Scheme-54

The reactions smoothly proceed under mild conditions without gaseous hydrogen. In particular, palladium-loaded silicon (Pd/Si) was the most suitable catalyst for the conversion of NAs to ANs, compared to Pt/Si, Ru/Si, and Pd/C (193).

Azo compounds are conveniently reduced to either partially reduced hydrazo compounds or completely reduced to ANs by employing Raney-Ni in the presence of hydrazinium monoformate depending upon reaction conditions. The reduction process is selective, rapid and high yielding with tolerance to other functional groups as –COOH, halogens etc. Gowda et al have many pubications on the selective reduction of azo and NACs to corresponding ANs including HANs in good to excellent yields using different catalysts and formates. The CTH is accomplished by i) Zn-HCOOEt$_3$, ii) Zn-HCOOHN-NH$_2$, iii) Polymer Supported zinc-formate etc (194). The deprotection of protected amine groups using zinc dust and ammonium formate is accomplished without affecting many other reducible or hydrogenolysable substituents such as halogens, methoxy, ester etc (195). Yu et al. also have studied the effects of chitosan-supported formate and Mg and found it as a highly efficient catalyst for the reduction of azo compounds to corresponding ANs (Scheme-55).

--Scheme-55

This method was found to be highly facile with selectivity over several other functional groups, such as halogen and other groups (196). Transfer hydrogenation of NACs using recyclable polymer-supported formate as hydrogen donor and Pd-C as a catalyst produces corresponding amines in excellent yields (90–98%) (197a). ANCs were reduced to respective amines in high yields by using 5% Pt/C with ammonium formate or formic acid as hydrogen donor. It was observed that the former was more efficient donor than the later. Further it has been found that the reduction of nitro groups occurs without hydrogenolysis of halogens and the reducible substituents remain unchanged under the reaction conditions (197b).

A novel process for the selective reduction of HNAs as 5-chloro-2-(2,4-dichloro-phenoxy) aniline (selectivity upto 99.77%) from its precursor nitro derivative(99.91%

Conv) with 3-5%Pt/C and 5mol% $HCOONH_4$ under different set of conditions is accomplished. The reduction went smoothly without the hydrogenolysis of the C-X (where, X represents F-, Cl-, Br- and I) and the C-O-C bonds, in the presence of platinum based catalysts and ammonium formate (198). A highly efficient and chemoselective photocatalytic reduction of NAs using TiO_2/polyethylene glycol 400-water (TiO_2/PEG–H_2O) is reported. Sunlight and violet LED (400 nm) irradiation efficiently reduced NAs using oxalic acid or ammonium formate as a hydrogen donor. The catalyst is stable after several runs and it is attributed to the deposition of PEG on TiO_2 (199).

3.13.14: CTH of HNAs with Hydrazine Hydrate (HH):

Catalytic transfer hydrogenation (CTH) is a chemical reaction in which the hydrogen needed for the reduction of an organic functional group is supplied from an inorganic or organic molecule in place of molecular hydrogen. Gnereally, the CTH agents act as a hydrogen donor and they are HH, alcohols, FA, cyclohexene, etc. CTH is known since a ling time and it was first reviewed in 1974 (200).

3.13.14a: CTH with HH & Fe/Fe Salts or Oxides as Efficient Ctalaysts:

Recently, for economical purposes of the processes, it has become imperative to design non-precious-metal-based nanocatalysts with high catalytic efficiency for the hydrogenation of NAs to the corresponding aryl amines under mild reaction conditions. In this study, γ-Fe_2O_3-NPs-supported hollow mesoporous carbon microsphere (h-MCM) nanocatalysts (γ-Fe_2O_3/h-MCM) were prepared, and their catalytic performance for the hydrogenation of NAs using HH as the reducing agen is studiedt. HH only generates N_2 and H_2O as harmless by-products. This study provides a useful platform based on a cost-effective, magnetically recyclable γ-Fe_2O_3-based nanocatalyst for the highly efficient hydrogenation of NAs (201). An efficient chemoselective hydrogenation of HNBs using low-cost catalysts is an important research area of applied catalysis. Here the auothors have prepared a Fe-metal organic gel (Fe-MOG) and it was used in the preparation of a low-cost catalyst (γ-Fe_2O_3/MC) (Scheme-56).

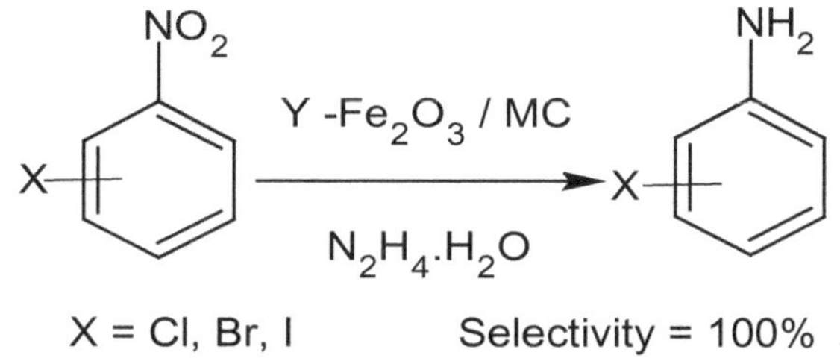

X = Cl, Br, I   Selectivity = 100%   --Scheme-56

The catalyst (γ-Fe_2O_3/MC) had high catalytic activity and selectivity for the hydrogenation of HNBs (Cl^-, Br^- and I^-) without any obvious dehalogenation. The hydrogenation reactions had a product yield and selectivity for the corresponding HANs of 100% when using HH as the reducing agent. The whole hydrogenation reaction process was environmental friendly because of its harmless byproducts (H_2O and N_2) (201b).

Immobilized iron metal-containing Ionic liquid-catalyzed chemoselective transfer hydrogenation of NAs including CNBs into corresponding ANs is accomplished in almost quantitative nitro conversions and aniline selectivity (Scheme-57).

Conversion > 99%   Selectivity > 99% --Scheme-57

Both the conversions and the selelctivty are excellent (> 99%) (202).

In the synthesis of p-CAN by catalytic reduction of p-CNB with HH in the presence of $FeCl_3 \cdot 6H_2O$-$AlCl_3 \cdot 6H_2O$ complex it is found that the catalyst had a splendid catalytic activity and selectivity. The optimal reaction condition in the solvent of alcohol was at 70°C for 2h (Scheme-58).

--Scheme-58

The conversion of p-CNB was 100% and the selectivity of p-CAN was over 99%. *Dyestuffs and Coloration.* 2004-05: Y. Luo, et al. http://en.cnki.com.cn/Article_en/ CJFDTOTAL-GONG200405010.htm

In another case, a novel γ-Fe_2O_3-NPs modified N-doped porous carbon materials (γ-Fe_2O_3/mCN) were prepared and was used as a cost-effective catalyst for the hydrogenation of NB using $N_2H_4 \cdot H_2O$ as the reductant under mild reaction conditions (Scheme-59).

R = Cl, NH_2, CH_2OH, etc.   Selectivity > 99.9% --Scheme-59

The non-noble metal catalyst exhibited high catalytic performance with selectivity of 100% to AN. During the catalytic hydrogenation of NACs with reducible groups as halogens, and amino groups an excellent selectivity of 100% was achieved. X. Cui, *New J. Chem.*, 2017, Advance Article Catalytic hydrogenation of NAs, including HNBs, using metal catalysts and HH as a hydrogen donor is summarized in detail by Shodhganga Chapter-I. The γ-Fe_2O_3 NPs well dispersed in porous carbon were fabricated *via* a Fe-based MOF-templated pyrolysis, and the resultant product exhibited excellent catalytic activity, chemoselectivity and magnetic recyclability for the hydrogenation of diverse NAs under mild conditions using HH as a hydrogen donor (203). Alumina-supported Fe_3O_4-NPs and hydrazine-mediated heterogeneously catalyzed reductions of functionalized NAs to ANs under batch and continuous-flow conditions can be accomplished (204).

3.13.14b: CTH of HNAs with HH & Other Non-Noble Metal Catalysts:

Some of the catalysts supported on solids supports are highly active because of they have oxygen vacancies on their surface, which help absorb hydrogen and also direct chemoselelctive reductions of nitro groups in NAs.

Copper: *In-situ* preparation of Cu-NPs from a copper acetate precursor and its application as an efficient catalyst for the selective reduction of NAs with HH under combined microwave and ultrasound irradiation are described in detail (Scheme-60).

EG = Ethylene Glycol   11 Examples, Yields = 89 - 96%

CMUI = Combined Microwave : Ultrasonic Irradiations   --Scheme-60

The results reveal the synergetic effect of microwave and ultrasound on the synthesis of Cu-NPs and the formation of various aryl amines (205).

Cobalt: CoO-NPs (size 2-3.5 nm) were successfully impregnated on an alumina–silica (mixed oxide) support having a high surface area and were tested for the hydrazine-mediated transfer hydrogenation of NAs using 2 mol% of the catalyst in ethanol at 60°C. The reaction was smooth and chemoselelctive towards substatrates in the presence of other sensitive functional groups such as halide and others. It was found that this inexpensive catalyst on a gram scale reaction was found to be robust and recyclable up to eight runs (206).

Nickel: Long time ago, the isomeric mono-HNBs (F, Cl, Br, I)) can be converted to the corresponding HANs in good yield by refluxing methanolic HH in the presence of Raney-Ni (207). Tlie following HANs also have been reported: *o-*, *m-*, ancl *p*-FANs, *o-*, *m-*, and *p*-CANs, and *o-*, *m-*, and *p*-BANs. Especially, *o*-BAN (yield =90.1%) and *m*-IAN (yield

= 73%). A recyclable highly dispersed Ni/SiO$_2$ catalyst was prepared by atomic layer deposition and used for the chemoselective reduction of NAs using HH as a hydrogen donor. Different kinds of NAs were converted to the corresponding ANs in high yields. The high activity of the catalysts could be a result of the highly dispersed Ni-NPs (208). Purely aqueous-phase chemoselective reduction of a wide range of aromatic substrates including o-CNB to the respective amines has been achieved in the presence of inexpensive non-noble metals like Ni and Co-NPs. The catalysts have high tolerance to other highly reducible groups present in close proximity to the targeted nitro groups using HH as a reducing agent at room temperature (209).

Bi-metallic catalysts are found more efficient than monometallics both in catalytic hydrogenation (CH) and catalytic transfer hydrogenations (CTH). Considering the industrial importance of HANs, a heterogenous non-noble metal nickel-molybdenum oxide catalysts supported with 1,10-phenanthroline on ordered mesoporous silica SBA-15 (Ni-MoO$_3$/CN@SBA-15) were prepared for the first time (Scheme-61).

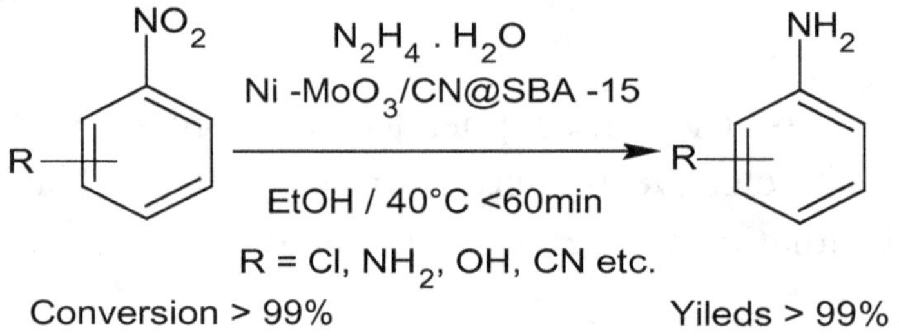

$$\text{NO}_2 \quad \xrightarrow[\text{EtOH / 40°C <60min}]{\substack{\text{N}_2\text{H}_4 \cdot \text{H}_2\text{O} \\ \text{Ni -MoO}_3/\text{CN@SBA -15}}} \quad \text{NH}_2$$

R = Cl, NH$_2$, OH, CN etc.

Conversion > 99%　　　　　　　　　　Yileds > 99% --Scheme-61

The catalyst exhibited unprecedented synergistic effect of metal Ni and MoO$_3$ species, and the catalytic activity and chemoselectivity for the reduction of various substituted NAs to the corresponding aromatic amines in ethanol with HH as a hydrogen donor under mild conditions. Excellent yields of >99% within very short reaction periods ($\leq$60 min) were obtained with a recyclable catalyst (210).

3.13.14c: CTH of HNAs with HH and Noble Metal Catalysts:

The small size polymeric PEG35k-Pd-NPs are key attractions for catalysis due to their large surface to volume ratio, non-toxicity, inexpensive, thermal stability, and recoverability (Scheme-62).

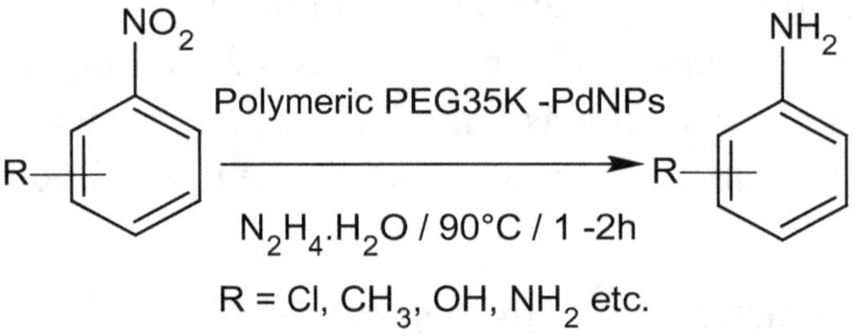

$$\text{NO}_2 \quad \xrightarrow[\text{N}_2\text{H}_4 \cdot \text{H}_2\text{O / 90°C / 1 -2h}]{\text{Polymeric PEG35K -PdNPs}} \quad \text{NH}_2$$

R = Cl, CH$_3$, OH, NH$_2$ etc.　　　　　　--Scheme-62

Polymeric PEG35k-Pd-NPs in the absence of phosphine ligands are insensitive to the air and moisture and act as an active chemoselective heterogeneous catalyst for the reduction of NAs (211). In another case, the 2-CAN and 3-chloro-4-methyl aniline were obtained in 99.9% and 99.8%.respectively when the corresponding c-CNB and 3Chloro-4-methyl-NB were reduced with HH at 90°C in DMSO over 5%Pd/C (212).

3.13.14d: CTH of HNAs with HH and Without Catalyst:

We know that the NAs can be reduced to corresponding ANs under CTH using different hydrogen sources. Here is an example of reduction of various azoarenes to aminoarenes with HH in refluxing ethanol without using of any specialized catalyst. The reaction is fast and cost effective, and yields are high to excellent (85–95%). Substituents such as –OH, –OCH$_3$, –COOH, –Cl, and –Br are unaffected. The method affords an elegant route to the preparation of aminoarenes (213). Rollas have extensively reviewed some of the examples of aromatic and heteroaromatic azo compounds to aryl amines usinh HH in the absence of a catalyst (214).

References:

164a). Y.Shiraishi, *ACS Catalysis*, 02(12):2475–2481(2012). 164b). Y. Shiraishi, et al. *ACS Catalysis,* 03(10): 2318–2326(2013).

165). B. Paul, et al. *RSC Advances*, 06(102):100532-100545(2016).

166).X. Han, et al. *J. Molecular Catalysis A: Chemical*, 277(01-02):210-214(2007).

167). J. Shen, et al. *J. Molecular Catalysis A: Chemical*, 273(01-02)265-276(2007).

168). A.Tanaka, et al. *Chemical Communications*, 49(25): 2551-2553(2013).

169). W. Gao, et al. *Chemical Communications*, 51(67): 13217-13220(2015).

170a). T. Zhang, et al. *Dyes and Pigments*, 68(02-03):95-100(2006). 170b). J. L. Ferry, et al. *Langmuir*, 14(13):3551–3555 (1998).

171). C. Yu , et al. *J. Organic Chemistry,* 66(03):919–924(2001).

172a). S. Thirumeni, et al. *ChemCatChem*, 04(12):1917–1921(2012). 172b). H. Lu, et al. EP 2837612-A1(2015).

173). A Review: R. C. Wade, J. Molecular Catalysis, 18(03):273-297(1983).

174a). Z. Duan, et al. *Bull. Korean Chem. Society*, 33(12):4003-4006(2012). 174b). M. Gawande, et al. *Chemical Reviews,* 116(06):3722-3811(2016).

175). H. K. Kadam & S. G. Tilve, *RSC Advances*: 02:6057-6060(2012).

176). S. K. Ghosh, et al. *Applied Catalysis A: General*, 288(1–2):61–66(2004).

177). B. Zeynizadeh, et al. *Synthesis Communications*, 36(18):2699-2704(2006).

178). K. Layek, et al. *Green Chemistry*, 14(11): 3164-3174 (2012).

179). M. Couturier, *Tetrahedron Letters*: 42:285–2288(2001).

180a). H. Goksu, et al. *ACS Catal.*, 04(06):1777-1782(2014). 180b). W. B. Smith, *J. Hetrocyclic Chemistry*, 24(03):745-748(1997).

181a). S. Fountoulaki, et al. *ACS Catalysis*, 04(10):3504–3511(2014). 181b).J. Rahaim, R. E. Maleczka (Jr.), *Organic Letters*, 07: 5087-5090(2005).

182). F. M. Moghaddam, *Applied Organometallic Chem:* Recent Article, May, 2017.

183). A. J. MacNair, et al. Organic & Biomolecular Chemistry: 12:5082-5088(2014).

184a). X. Meng, et al. *J. Catalysis*: 264:1-10(2009).

184b). X. Meng, et al. *J. Catalysis*: 269:131-139 (2010).

185). S. Ichikawa, et al. *Chemical Communications*: 07 924-926(2005).

186a). S. Ichikawa et al. Adv. Synthesis & Catalysis, 366(11-12):2643-2652(2014). 186b). M. Chatterjee, et al. *Adv. Synthesis & Catalysis*: 354:2009-2018(2012). 186c). M. Chatterjee, et al. *Green Chem.*, 14(12):3415-3422 (2012). 186d). X. Han, et al. *Chem. Soc. Reviews*: 41:1428-1436(2012).

187). Y. Chen & C. Tan, *The J. Supercritical Fluids*, 41(02):272-278(2007).

188). M. Liu, et al. *J. Molecular Catalysis A:* 138:295(1999).

189). C. Xi, et al. *J. Molecular Catalysis A: Chemical*, 282:80-84(2008).

190). H. Cheng, et al. Applied Catalysis A: General, 455:08–15(2013).

191). G. Gao, et al. *Green Chemistry*, 10:439-441(2008).

192). H. Cheng, et al.*J. Colloid & Interface Science:* 415:01-06(2014).

193). K.Tsutsumi, *ACS Catalysis*, 06(07):4394–4398(2016).

194a). G. R. Srinivasa, et al. *Synth Commun*, 34(02): 223-231(2004), 194b). H. S. Prasad, et al. *Synth Commun*, 33(04):717-724(2003), 194c). H. S. Prasad, et al. *Synth. Commun*, 34(01):01-10(2004).

195a). G. R. Srinivasa, *Synthetic Communications*, 34(02):1831-1837(2004). 195b). G. R. Srinivasa, *Synthetic Communications*, 35(09):1161-1165(2005).

196). X. H. Yu, et al. *Synthetic Communications*, 44(05):707-713(2014).

197a). K. Abiraj, et al. *Synthetic Communications*, 30(20):3639-3644(2000). 197b). C. Gowda, B. Mahesh, *Synthetic Communications*, 30(20): 3639-3644 (2000).

198). M. Subramaniyan, et al. EP 2209763 A2: (text from WO2009090669A2, (2010).

199). M. Ramdar, *Chemical Papers*, 71(06):155-1163(2017).

200a). G. Brieger & T. J. Nestrick, *Chemical Reviews*, 74(0 5): 567-580(1974). 200b). J. Toubiana, et al. *Modern Research in Catalysis:* 03:68-88(2014).

201a). M. Tian, et al. *Inorg Chem. Front.*, 03(10):1332-1340(2016). 201b). M. Tian, et al. *Green Chemistry*, 19(06):1548-1554(2017). 201c).US-20090124834-A1(2009).

202). N.M. Patil, et al. *ACS Sustainable Chem. Engg*, 04(2):429–436(2016).

203). Y. Li, et al. *Chem. Commun.*, 52(22):4199-4202(2016).

204). C. O. Kappe, et al. *ChemSusChem*, 07(11):3122–3131(2014).

205). H. Feng, et al. *Sustainable Chemical Processes*, 02:14(2014).

206). P. Linga Reddy, *Chemistry An Asian J*, 12(07):785-791(2017).

207). B. E. Leggeter, et al. *Canadian J. Chemistry:* 38:2363-2366(1960).

208). C. Jiang, et al. *ACS Catalysis*, 05(08):4814-4818(2015).

209). R. K. Rai, et al. *Inorganic Chemistry*, 53(06):2904-2909(2014).

210). H. Huang, *Green Chemistry:* 19:809-815(2017).

211). V. Yadav, Synth Commun, 42(02):213-222(2012).

212). G. Lippert, US 3897499 A(1975).

213). M. A. Pasha et al., *Synth Commun*, 35(07):897-900 (2005).

214). S. Rollas, *Marmara Pharmaceutical J.*: 14: 41-46(2010).

3.14.00: Hydrogenation of HNAs with Metals/Metal-Salts and Acids:

This section and the following sections as silanes, water are part of CTH using hydrogen sources other than molecular hydrogen under "Haloanilines" as a main heading or section. In later sections, Metals and Metal salts, Silanes, Sulphur, Water are covered as individual hydrogen donors with catalysts for NA reductions.

Iron and inorganic acids have been used since a long time ago for the reduction of NAs. Becamp in 1854 was the first one to reduce NB to AN by using Fe/aq.HCl as a reducing agent. Many other groups, including AN manufactuers, follow this approach (215). A method of reducing an aromatic nitro ($-NO_2$) group to the corresponding amine by treatment with iron or zinc metal and a halogen-containing aliphatic carboxylic acid provides amines in good yield and free from unwanted anilides (216). Many HNBs with other substituents are redcued to corresponding ANs with iron, zinc, tin in the presence of an acid or NH_4Cl, and ethanol or EtOAc, I). Fe/AcOH/RT/2h (217a): to corresponding aniline in good yield (-Br-; 70%), and with ii) Fe/NH_4Cl/EtOH/70°C/1h (217b) in high yield (2, 6-dichloro-4-fluoro; 94%), and other combinations used were as iii) Zn/NH_4Cl/EtOH/90°C/6h, iv). $SnCl_2$-H_2O/EtOAc/ON (217c,d). A robust and green protocol is developed for the reduction of functionalized NAs and Hetero-NAs using an inexpensive zinc dust in water with nanomicelles derived from available designer surfactant TPGS-750-M (Scheme-63).

$$R\text{-Ar-}NO_2 \xrightarrow[\text{H}_2\text{O / RT}]{\substack{\text{Zn dust / NH}_4\text{Cl} \\ \text{TPGS -750 -M}}} R\text{- Ar -}NH_2$$

Ar = Aryl, Heteroaryl, R = Halo, CHO, COOR etc. --Scheme-63

This mild process takes place at room temperature and tolerates a wide range of functionalities. Highly selective reductions can also be achieved in the presence of common protecting groups (218). A practical and chemoselective reduction of NAs to ANs using activated iron generated by Fe/HCl or $Zn/FeSO_4$ is described. A variety of functional groups such as alkyne, ketone, nitrile, and aromatic halides are well tolerated under these conditions (219). An efficient $Fe/CaCl_2$ system enables the reduction of NAs and reductive cleavage of azo compounds by CTH. The catalytic system is chemoselective to nitro groups in the presence of sensitive functional groups including halides with excellent yields. The simple experimental procedure and easy purification makes this protocol advantageous (220).

4-Benzyloxy-3-CAN has been frequently used as a building block in the construction of potential anti-cancer, anti-diabetes and anti-viral agents. However, there are no reports on commercial production of this product except the preparations at small scale. These methods utilize the standard reduction processes as (1) reduction of 2-chloro-4-NP to 3-chloro-4-hydroxyAN using Zn/NH_4Cl in $MeOH/H_2O$, followed by Boc protection of the amine, benzylation of the hydroxy group and deprotection of the Boc group, (2) reduction of 4-benzyloxy-3-CNB by either Raney-Ni or Pd/C catalyzed hydrogenation, and (3) reduction of 4-benzyloxy-3-CNB mediated by Fe powder in acetic acid or NH_4Cl solution (Scheme-64).

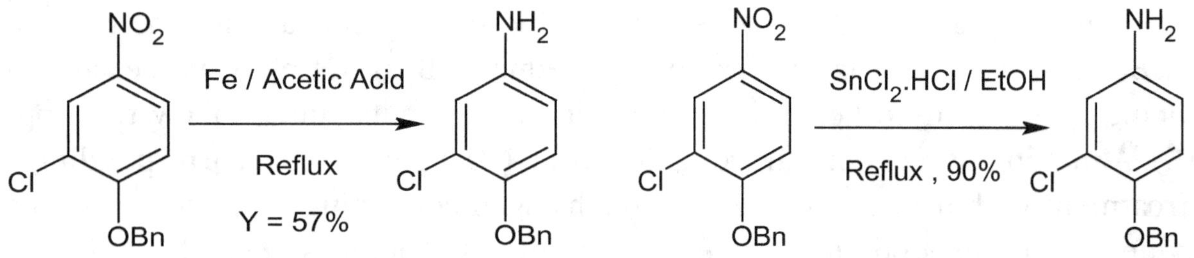

Scheme-64

Since these known methods were not adequate for large scale production, a convenient, safe, large-scale synthesis of the title compound 4-benzyloxy-3-CAN is described. The commercially available 4-benzyloxy-3-CNB is reduced conveniently using $SnCl_2$ to afford

4-benzyloxy-3-CAN in high yield, high purity, and free of tin residues. This process is suitable for kilogram-scale synthesis of the title compound (221). Treatment of waste water pollutant is a serious concern. Many methods for the reduction of toxic NAs have been developed. Here a unique bio-process is used for *p*-CNB reduction in drinking water using a continuous stirring hollow membrane biofilm reactor (222). NB reduction with Fe/HCl is known since 1854.

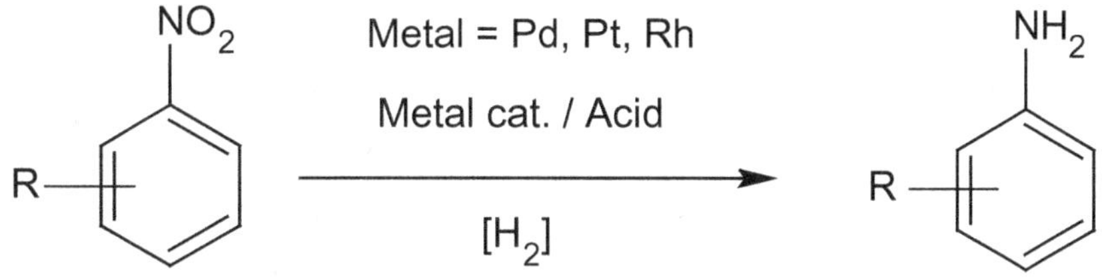

$R = F, Cl, Br, I$

Acid = Formic = Only Fluoronitrobenzene Reduced

Acid = Phosphonic, Phosphorus = All Halo Nitrobenzenes Reduced --Scheme-65

Here is a study in which the aromatic nitro-compounds were reduced to amines in high yields by using Pd, Pt or Rh metal catalyst with formic, phosphinic, or phosphorous acids (Scheme-65). With FA, FNB was reduced but not those containing Cl, Br, or I. With the other acids, NAs containing any of the halogens were reduced with retention of the halogen (223). 4-CNB (4-Cl-NB) was rapidly reduced to 4-CAN in dissimilatory Fe (IH)-reducing enrichment culture. The crystalline magnetite (Fe_3O_4) produced by the Fe (III)-reduction by bacterium and the synthetic magnetite also reduced 4-CNB (224).

3.15.00: Hydrogenation of HNAs with Photocatalysis:

The effects of surface modification of nanocrystalline TiO_2 with specific chelating agents on photocatalytic degradation of NB was investigated in order to design a selective and effective catalyst for removal of NACs from contaminated wastestreams. Aminoacids and other compouds especially Arginine, Lauryl sulfate, and Salicylic acid were found to bind to TiO_2 via their oxygen-containing functional groups thereby modifying its properties for the reduction of NAs. Modification of the TiO_2 surface with arginine in methanol resulted in enhanced NB adsorption and photodecomposition and found better when compared to unmodified TiO_2. In essence, the surface modification of nanocrystalline TiO_2 with electron-donating chelating agents is an effective route to enhance photodecomposition of NACs (225). A detailed and comparative mechanistic study of the photoelectrochemical dehalogenation of the four *p*-HNBs, (X = F, CI, Br, or I) is reported (Scheme-66) Except F^-, all other halogens (Cl, Br, I,) were dehalogenated under the experimental conditions, for which the mechanistic pathways have been discussed (226).

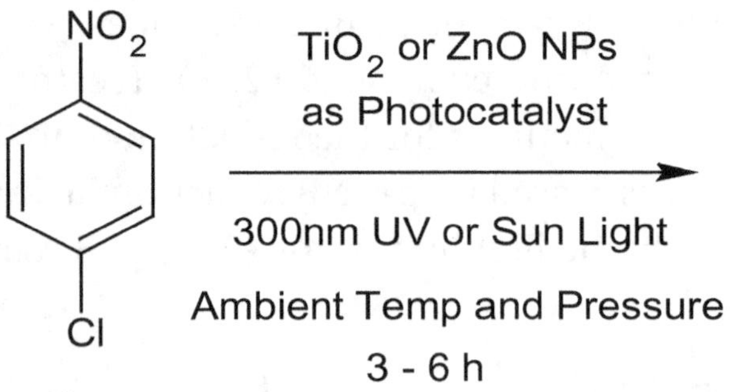

--Scheme-66

The reduction reaction can be carried out at normal temperature in the presence of ultraviolet light or sunlight and nano photocatalyst as TiO_2 or ZnO. The invention has higher reduction conversions and yields. Blue light irradiation of heterogeneous photocatalysts $PbBiO_2X$ (X = Cl, Br) in the presence of triethanolamine as an electron donor process leads to hydrogen evolution and selective reduction of NACs to their corresponding ANs is achieved (227). Chemoselective reduction of NBs having other reducible groups over TiO_2 photocatalyst under metal-free conditions is achieved in aqueous system at room temperature and atmospheric pressure (Scheme-67) (228).

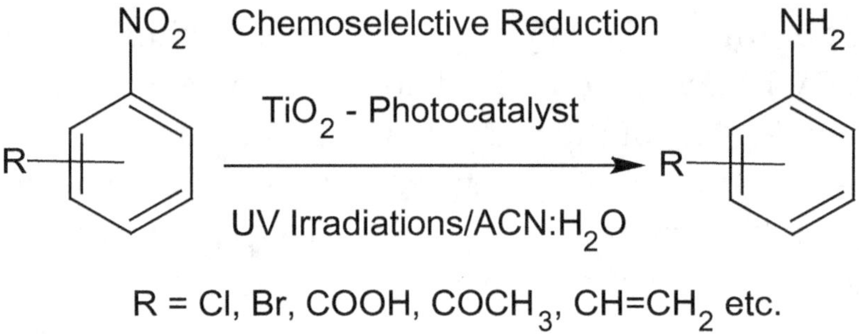

R = Cl, Br, COOH, COCH$_3$, CH=CH$_2$ etc.   --Scheme-67

The reductions are performed without a precious metal or reducing gas and the highest rate was obtained by adding a small amount of water to acetonitrile. Other reducible groups, including Cl-, Br-, were not affected at all and only the nitro group of these compounds was chemoselectively reduced to corresponding amino compounds with high yields (229).

3.16.00: CTH of HNAs with Silanes:

The chemoslective and efficient reduction of ANCs to the corresponding aryl amines with silanes catalyzed by high valent oxo-rhenium complexes can be achieved in the presence of wide range of functional groups (Scheme-68). The catalyst was also regioselective, i) in the reduction of DNBs to the corresponding mononitroanilines and ii) the reduction of an aromatic nitro group in the presence of an aliphatic nitro group (230).

R = Halo, Ester, Lactone, Amide, Bn etc.

Metal Complex = PhMe$_2$SiHReIO$_2$(PPh$_3$)$_2$ PhMe$_2$SiHReOCl$_3$(PPh$_3$)$_2$ --Scheme-68

We have seen under boranes that Au/MTA-TMDS amd Pd/PMSH also chemoselectively reduces *p*-CNB to *p*-CAN and NACs in high yield and selectivity (181). Dehalogenation, usually both in aliphatic and aromatic halides, is observed when silanes are used as reducing agents (231). Therefore, there are not many reports/publications on using slilanes in the selelctive reduction of HNBs. A Japaneese company has published a review on Silicon Based Reducing Agents. https://www.azmax.co.jp/data/cnt_catalog_chemical/pdf/attach_20110517_112436.pdf.   http://www.acros.com/myBrochure/Organosilanes_Brochure_EEM.pdf.

3.17.00: CTH of HNAs with Sulfides:

The metal catalyzed hydrogenation of HNBs gives dehalogenated ANs. Particularly, transition metals like Pt, Pd, Ni, Rh, Copper-chromite catalysts are known to cause dehalogenation of HANs, formed in the hydrogenation of HNBs. Already, we know that the strong alkalies in large amounts cause dehalogenation. Through long history of research and efforts, it has been proved that the addition of a small amount of an inorganic or organic base to the reduction system helps alleviate or eliminate dehalogenation problems. It is believed that the bases may be poisoning the catalysts and hence inhibiting the dehalogenation process. There are some other alternatives to alleviate or eliminate such problems. One of them is the use of another reagent or new catalysts that can chemoslectively reduce HNBs to HANs. The following examples, illustrate these facts.

Although the quantitative production of HANs without dehalogenated products has attracted much attention in the recent past, this matter was looked into quite some time ago as well. The synthesis and production of N-isopropyl-p-CAN and the use of said compound as a selective post-emergence herbicide for narrow-leaf grasses using Pt group metal sulfides for the hydrogenations of the corresponding CNBs is reported long time ago. The dehalogenation of formed HAN was completely inhibited or very negligible amount was detected. Briefly, the present invention comprises effecting nondechlorinating hydrogenation reactions and/or reductive alkylations in the presence of a catalyst comprising a sulfide of a metal selected from the group consisting of Pd, Pt, Rh, Ru and Co (232). Sulfided Pt on carbon catalyst has been used for the selectiove hydrogenation of HNBs without dehalogenation of the formed HANs. Non sulfided Pt catalysts cannot be used for this purpose since they would cause reduction and dehalogenation at the same time (233). A simple solution to a multi-step reaction set can be engineered to produce selectively the desired industrially important products by minimizing both the by-product formation and separation stages. Yadav et al. have investigated in detail the reduction of p-CNB with sodium sulphide under different modes of phase transfer catalysis (PTC), such as liquid-liquid (L-L), liquid-solid (L-S reaction), and liquid-liquid-liquid (L-L-L) processes. This selectivity engineered PTC reaction has been investigated from mechanistic view point and the rationale of selectivity is delineated (234). The use of sodium hydrosulphide as a selective reducing agent for ANCs has been investigated with respect to experimental details, including solvent, yield and ease of purification of product during studies on the x,y-dinitrobiphenyls (x= 2–4, y= 2–4). The hydrosulphide-induced reduction is superior to previously reported methods. Its advantages and possible applications are discussed (235). The modification of diphenyl sulfide to Pd/C catalyst was done by following a procedure and and it was tested in the selective hydrogenation of p-CNB. The Pd-Ph$_2$S/C exhibited a good selectivity of p-CAN in the hydrogenation of p-CNB (236). The reduction of NB (237) and other NAs (238) by sodium sulfide and sodium hydrodisulfide in aqueous media at 50°C has been examined. The reduction proceeds through PHA formation, which consequently gets reduced to AN (Scheme-69).

$$R - Ar - NO_2 \xrightarrow[\text{EtOH / Reflux}]{Na_2S.H_2O} R - Ar - NH_2$$

--Scheme-69

Uniform flowerlike Ni$_7$S$_6$ NCs composed of nanopetals with sharp tips have been synthesized and were tested in the selective hydrogenation of CNBs, for the first time. Compared with other nickel sulfide samples with different morphologies and structures, flowerlike Ni$_7$S$_6$ samples show much higher activity and selectivity in the hydrogenations of NB and CNBs with different chlorine substituent sites (239). Reduction of CNBs and p-nitrotoluene (p-NT) by aqueous ammonium sulphide as a triphase (Liquid-Liquid-Solid) catalysis by anion exchange resin is investigated. About 66 and 57 folds enhancement in rate of reduction of p-CNB and p-NT respectively was obtained with 20% (W/V) loading of the catalyst (240).

3.18.00: CTH of HNAs in Water:

The promoting role of minor amount of water plays a very important role in solvent-free hydrogenation of HNBs. For Pd/C catalyst, minor amount of water reduces the induction time, increases the reaction rate and reaction TOFs. Water might enhance the diffusion, adsorption and dissociation of H$_2$ on Pd catalysts (241). Facile and effective preparation of a series of cobalt-doped Fe$_3$O$_4$NPs via chemical coprecipitation in an aqueous solution is studied. The catalyst allowed the hydrogenation of CNBs to CANs to proceed at low temperatures in absolute water and at atmospheric pressure, resulting in approximately 100% yield and selectivity to CANs (242). The effects of common dissolved anions on the reduction of p-CNB by zero-valent iron (ZVI) in ground water have been investigated. Patil et al have developed a novel, chemoselective, non-hazardous and mild reaction protocol for efficient reduction of NAs to aromatic amines using "iron activated water". Water functions as a terminal hydrogen source without any external catalyst, acid, salts or base and in the absence of a solvent. In the course of the reaction, zero valent iron was oxidized to magnetite. Particularly, p-NT was reduced to p-toluidine (p-AT) in 99% yield and 99% selelctivity at 99% p-NT conversion. Notably, the reduction of substituted HNBs proceeded smoothly to the respective aromatic amines (p-Cl, 88%, p-Br 92%, p-F, 88%) without any dehelogenation (243).

Water functions as a direct hydrogen donor in $scCO_2$ with a novel, ecofriendly, cheap and efficient $Zn-H_2O-CO_2$ system for chemoselective reduction of NBs to ANs in high yields (80% –97% isolated). This process brings together the very important green chemistry technologies–the use of carbon dioxide as a solvent and the use of water as a hydrogen donor (244).

The selective reduction of nitro group in presence of other reducible functional groups as carbonyl, halides etc. is developed using $Co_2(CO)_8$ in water. The catalyst was highly active and selective for chloro and bromo ANs as they were obtained in good to excellent yields at 100% conversions of the CNBs. 1-Chloro-2, 4-DNB gave 100% yield of 2, 4-diamino-1-chloro benzene (Scheme-70).

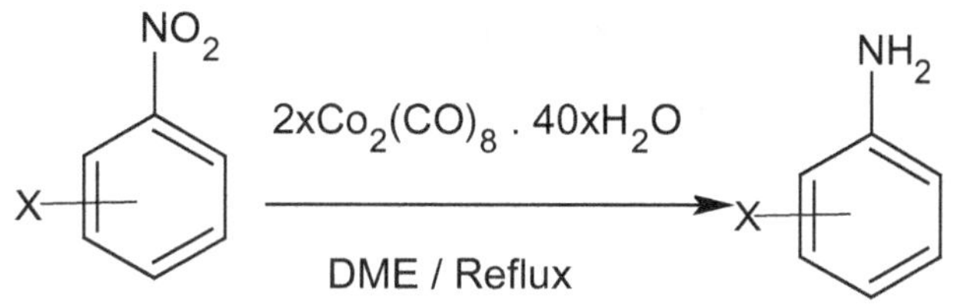

Conv. =100%   Yields =o - Br = 58%, p -Br = 88%, p - Cl = 99%

2,4 - Diamino - 1 - Cl - Benzene = Yield = 100%   --Scheme-70

Many other NAs are reduced with excellent yields and selectivity to the corresponding ANs (245).

3.19.00: Industrial Production of Haloanilines (HANs):

Quite sometime ago a process for the commercial production of p-CAN from NB with a minimum of unwanted by-products using aqueous hydrochloric acid as the source of chlorine is reported. NB is reduced with molecular hydrogen in the presence of a Pt catalyst and aq. HCl, yielding p-CAN as the principal product and limited quantities of p-AP and AN, as valuable secondary products (246). Industrially proven and cost efficient heterogeneous hydrogenation to give highly functionalized sensitive aromatic amines using a new catalytic system with unprecedented chemoselectivity are of current requirements. A highly chemoselective reduction of NAs with modified platinum catalyst is reported (Scheme-71). H. Steiner, M. Benz, High Pressure Kilolab, Solvias, info@ solvias.com www.solvias.com.

R , R' = H or Substituents, 2 -I, 4 -Cl : Y = >90%

Cat = 5% Pt/C with H_3PO_2 - $VO(acac)_2$ / Solvent /Press. 5 bar   --Scheme-71

Using this process technology, several production and pilot processes have been developed and introduced at plants of various customers in the pharmaceutical, agro and fine chemicals industry. Excellent yields and almost quantitative selectivities for ANs are obtained. www.solvias.com/sites/default/files/ORCS_Nitro_US_31.1.03.pdf

The gas-phase continuous hydrogenation of p-CNB over 1 mol% Au/TiO_2 and Au/Al_2O_3 was compared for the first time. Both catalysts exhibit 100% selectivity in terms of $-NO_2$ group reduction, resulting in the sole formation of p-CAN. Au/TiO_2 exhibited a narrower particle size (1–10 nm) distribution than Au/Al_2O_3 (1–20 nm). Au/TiO_2 delivered a higher specific hydrogenation rate (by a factor of up to four), a response that is attributed to Au particle size and contribution of the support to p-CNB activation. CNB isomer reactivity sequence was established and was found to be as, o>p>m, which is attributed to resonance stabilization effects. The results presented here establish a basis for the development of a sustainable alternative route for an exclusive production of CANs from CNBs over Au/TiO_2 and Au/Al_2O_3 (247) The same group has presented another case with Cu-Al (3:1) to achieve 100% selelctivity for p-CAN at 393-523K and atmospheric H_2 pressure. Addition of Ni to the catalyst (Cu-Ni-Al) exhibited two-fold increase in the hydrogenation rate when compared to Cu-Al, while the exclusive nitro-group reduction was maintained (248). Recently, a highly efficient and selective hydrogenation of CNBs to CANs by H_2 over confined Ag-NPs (5-6nm) into mesoporous silica SBA-15 (Ag-NPs/SBA-15) is reported. The catalyst has exhibited excellent activity and selectivity for hydrogenation of HNBs to HANs, and its robust catalytic performance displays its potential application for the production of CANs in the fine chemicals industry (249).

AN, substituted ANs, including HANs have been synthesized on a commercial scale in dye industries although some of them are not recommended due to their carcinogenic properties. Chromic Phenomena: Technological Applications of Colour Chemistry, Peter Bamfield, Michael G. Hutchings, Royal Society of Chemistry, 2010 - Science - 562 pages,

PP 154- onwards. 3-CAN market report is published in 2016 indicating the market and highlights yearly production, sales, demand, analysis and forecast till 2021, https://www.absolutereports.com/news/3-chloroaniline-market-2016-production-sales-supply-demand-analysis-forecast-2021/. A. Pingaley, sales@absolutereports.com

The selective hydrogenation of HNBs is a chemical reaction of great importance in the fine chemical productions. The metal particle size is one of the key factors in controlling the selectivity in HNBs hydrogenations. Through careful studies, it is found that the hydrogenation reactions and dechlorinations are governed by the M-NPs size. Pd-NPs with different sizes (from 2.1 to 30 nm) supported on activated carbon were synthesized and tested for selective hydrogenation of HNBs. The selectivity was over 99.90% when the Pd-NPs are bigger than 25 nm. Finally, the industrial applications of the proposed catalyst were evaluated in several pilot factories. This study provides useful information on controlling the selectivity of other similar reactions catalyzed by noble-metal nanocatalysts (250).

The Ag-NPs/SBA-15 catalyst shows very good activity and selectivity for hydrogenation of CNBs to the corresponding CANs by H_2 compared with that of the SiO_2 supported Ag catalysts. The robust catalytic performance of the as-prepared Ag-NPs/SBA-15 catalyst displays its potential application for the production of CANs in the fine chemicals industry (251). As an alternative to expensive and rare noble metals like Rh, Ru, Pd, Pt, non-noble metals are always sought after in catalytic hydrogenation of NAs to produce industrially important substituted ANs. Iron (Fe) is one such non-noble and abundantly available metal catalyst that can be used for cost-effective production of chemicals at industrial scale. In this context iron oxide based catalyst is developed. Pyrolysis of iron-phenathroline on carbon furnishes a unique structure in which the active Fe_2O_3 particles are surrounded by nitrogen-doped carbon layer. Using this noble, economically feasible catalyst a number of structurally diverse NAs were hydrogenated to the corresponding ANs in good to excellent yields, especially under industrially viable conditions (252).

3.20.00: Fluoroanilines:

Generally the activity of a compound is dependent on the type of bond between two atoms. For example, unactivated aryl halides do not undergo Swarts-type organic halide-inorganic fluoride exchange reactions, which are common in alkyl halides. It is found that the activation by at least two nitro groups is necessary for such reactions with aryl halides and fluoride ions. For example, the replacement of chlorine by fluorine in 2, 4-dinitrochlorobenzene with anhydrous KF at 190-205°C has been reported long time ago (253).

The fluorocompounds have found importance in medicinal chemistry, as some of the most successful drugs as, 5-Fluorouracyl, Fluoronitrezepam (Roche), Atorvastatin

(Lipitor, Pfizer), Lansaprazole, Norfloxacin, Ciprofloxacin, Gatifloxacin, Fluticasone (GSK) etc. are fluoro compounds. However, synthesis of fluoro compounds is not as simple as other halides (254). 2-Fluoroadenosine is synthesized as an anti-cancer agent (255). Aromatic fluorine compounds as p-FANs are made by catalytic reduction of NB in HF. It is also obtained by rearrangement of N-phenylhydroxylamine in anhydrous HF (256). The p-FNB was synthesized by reacting p-CNB with KF. Furher p-FNB can be reduced to p-FAN by any one of the hydrogenation methods, preferably by catalytic hydrogenation with molecular hydrogen. The reduction of NACs by various metals (tin, lead, bismuth) in liquid hydrogen fluoride under an inert atmosphere leads to fluoroaromatic amines, in accordance with the Bamberger reaction (257). Alternatively, p-FAN was synthesized by reduction of p-FNB. The yield of p-FAN was 91.7% under the optimum reaction conditions, such as the molar ratio of p-FNB to active KF 1.0:1.1 at 175°C for 1 h in sulfolane as a solvent instead of DMSO (Scheme-72).

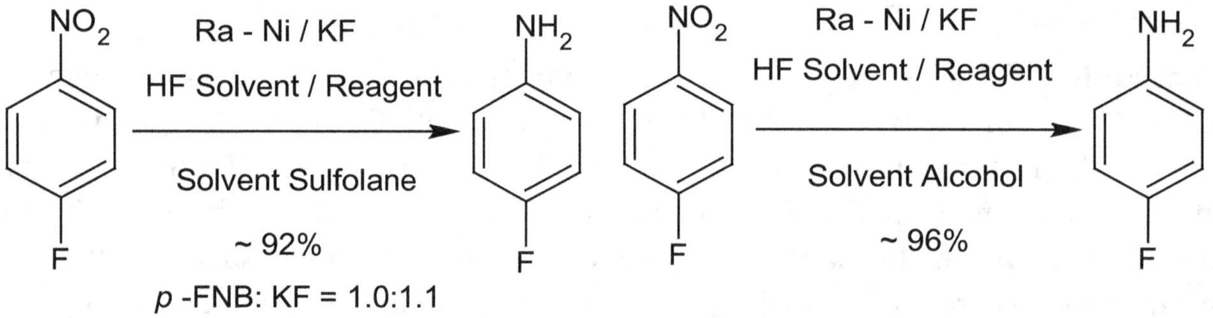

Scheme-72

By application of the catalysis of Raney-Ni p-FAN was obtained from p-FNB in alcohol with the yield 95.8%. With Raney-NI and sulfur solvent as thiourea or sulfolane 100% conversion and > 86- 99% p-FAN yield was obtained (258). The corresponding FANs have been produced by i) reacting chloro-dintroarenes with KF with up to 90% yields and then ii) catalytic reduction using PtO_2 and Pd/C catalysts. In good to excellent yields (68-99.4%). (259). http://en.cnki.com.cn/Article_en/CJFDTOTAL-HGSC200701001.htm

Especially, p-FANs and o-, m-, p-DFANs are also prepared by the reduction of respective azides with anhydrous hydrogen fluoride (260). A method for the preparation of substituted FANs from substituted 1-Cl, F-NBs in high yield is developed. For example, 3, 5-DFAN was prepared from 1-chloro-3, 5-DFNB in 89% yield using the reported method (261). An efficient synthetic method with a potential of industrial applications is developed for producing 3, 5-dichloro-2, 4-DFAN (262).

Particularly, p-Azido tetrafluoroaniline, which is a new photoaffinity reagent, was synthesized in five steps from pentafluoronitrobenzene (A) in 65% overall yield. Compound A was converted into 4-azidotetrafluoronitrobenzene (B) with NaN_3 in 93% yield.

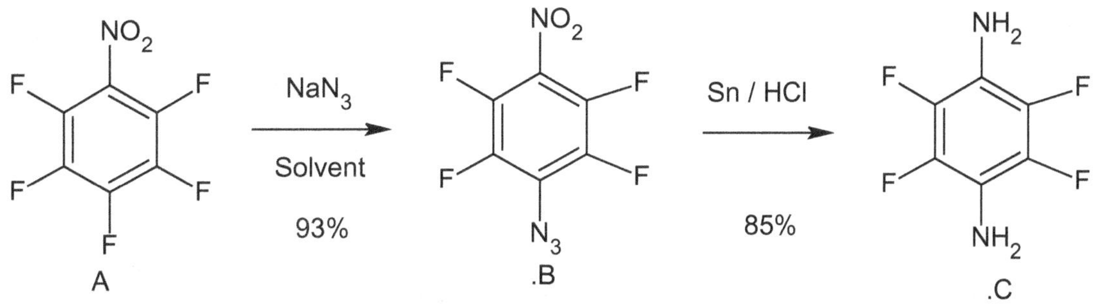

Scheme-73

It was used without further purification to form 1, 4-diaminotetrafluorobenzene (C) by Sn/HCl reduction in 85% yield (Scheme-73) (263). A method of preparing 2,4-DFAN (~70% yield) by reacting 2,4,5-trichloroNB with a fluorinating agent in the presence of a solid-liquid phase interface and a PTC to form 2,4-difluoro-5-chloroNB, followed by hydrogenation over a catalyst is reported (264). Considering the importance of HANs in pharmaceutical and agriculture industries, it is equally important to have efficient, economical and safe commercial processes for their manufacturing. For this purpose, one of the most important factors is the type of catalyst and its efficacy in yielding the desired products in quantitative yields with complete substrate conversions. Here, Zhao et al have used noble metal/non-noble metal oxide NPs catalysts for the chemoselective hydrogenation of 2, 4-DFNB. Carbon-supported Pd/SnO_2 catalysts were synthesized by the chemical reduction method, and their catalytic activity was evaluated for the hydrogenation reaction of DFNB to the corresponding DFAN showing a remarkable synergistic effect of the Pd and SnO_2 NPs. In the $Pd/SnO_2/C$ catalysts Pd atoms underwent modifications in their electronic structure through the use of SnO_2 which led to the suppression of the hydrogenolysis of the C-F bond and the acceleration of nitrosobenzene (DFNSB) conversion (265).

References:

215). S. E. Hazlet, et al. *J. Am. Chem. Soc,* 66(10):1781-1782(1944).

216). P. Buckland, et al. EP0347136-A2(1989).

217a). WO2014149164, b).WO2015129926, c).WO2014149164,d). WO2002016361

218). S. M. Kelly & B. H. Lipshutz, *Organic Letters,* 16(01):98–101(2014).

219). Y. Liu, et al. Adv Synth & Catal, 347(02-03):217-219(2005).

220). S. Chandrappa, et al. *Synlett:* 3019-3022(2010).

221). H. Chen, et al. ARKIVOC: (xiv) 01-06(2008).

222). S. Q. Xia, et al. *Proceeds Water Environ Federation,* Biofilms: 288-298(2010).

223). I. D. Entwistle, *J Chem Soc, Perkin-Trans:* 01:443-444(1977).

224). J. Zeyer, *Appl Environ Microbiol.* 59(12):4350(1993).

225). O. V. Makarova, et al. *Environ. Sci. Technol.,* 34(22):4797-4803(2000).

226a). R. G. Compton, et al. J. Chem. Soc.: Perkin Transactions-2, 00(07): 1581-1587(1994). 226b). T. Zhang, et al. ChemYQ-1634862(2006): 226c).CN-200410072383 (2006).

227). S. Fuldner, et al. *Green Chemistry,* 13(03):640-643(2011).

228). K. Imamura, et al. *Tetrahedron,* 70(36):6134-6139(2014).

229). Y. Shiraishi, *ACS Catalysis,* 02(12): 2475–2481(2012).

230). R. G. de Noronha, et al., *J. Organic Chemistry,* 74(18):6960-6964(2009).

231a). C. Chatgilialoglu, *J. Organic Chemistry,* 53:36413642(1988). 231b), *Tetrahedron Letters*: 30:2733-2734(1989). 231c). J. Giese, *Organic Synthesis:* 70:164-165(1991). 231d). R. J. Rahaim, et al. *Tetrahedron Letters*: 43:8823-8826(2002).

232a). Dovell & Greenfield, US 3350450-A(1967). 232b). H. Greenfield & F.S. Dovell, *J. Org. Chem.,* 1967, *32* (11):3670–3671(1967).

233a). Baessler & Habig, US3929891-A(1975). 233b). F.S. Dovell, et al. *J. Am. Chem. Soc*: 87:2767-2768(1965).

234). G. D. Yadav, et al. *J. Molecular Catalysis A: Chemical,* 200:117-129(2003).

235). J. P. Idoux , *J. Chemical Society C,* 00(03): 435-437(1970).

236). Q. Zhang, et al. *Chin.J.Chemical.Engineering,* 22 (10):1111-1116(2014).

237). O. J. Cope, et al. *Canadian Journal of Chemistry,* 40(12):2317-2328(1962).

238a). WO2014149164: 238b) WO2011014535.

239). F. Cao, et al. *J. Materials Chemistry,* 20(06):1078-1085(2010).

240a). L. Jeeru, *J. Ind. Chem. Engineers,* 58(03):279-296(2016). 240b). S. K. Maity, et al. *Chemical Engineering J.,* 141(01-03):187-193(2008).

241). J. Lyu, et al. *Chinese Chemical Letters*, 25(02):205-208(2014).

242). B. yang, et al. *Nano Research,* 09(07):1879-1890(2016).

243a). C. Le, et al. *Water Science Technology,* 63(7):1485-1490(2011). 243b). R. D. Patil & Y. Sassona, *Organic Chemistry: Current Research,* 04(04):154-158(2015).

244). H. Jiang, et al. *Chinese J Chemistry,* 26(08):1407-1419(2008).

245). H. Lee, et al. *Bull Korean Chemical Society,* 25(11):1717-1719(2004).

246). W. C. Bradbury, US 3265735-A(1966).

247). F. Cárdenas-Lizana et al., *ChemSusChem,* 01(03): 215-221(2008).

248). F. Cardenas-Lizana et al. *ChemCatChem*, 04905):668–673(2012).

249). L. Yang, et al. *RSC Advances*, 06(38):31871-31875(2016).

250). J. Lyu, et al. *J. Physical Chemistry C*, 118(05):2594–2601(2014).

251). L. yang, et al. *RSC Advances*, 206(38):31871-31875(2016).

252). R. V. Jagadeesh, et al. *Science,* 342(6162):1073-1076(2013).

253). B. C. Finger & C. W. Kruse, *J. Am. Chem. Soc.,* 78 (23):6034–6037(1956).

254). R. Duschinsky, et al. *J. American Chemical Soc*iety: 79, 4559-4560(1957).

255). J. A. Montgomery, et al. *J. Am. Chemical Soc*iety: 79(16):4559–4559 (1957).

256). D. A. Fidler et al. *J. Org. Chem.,* 26 (10):4014–4017(1961).

257). M. Tordeux, et al. *J. Fluorine Chemistry*, 74(02):251-254(1995).

258). R. Bailliard, et al. US 5126485 A(1992).

259). I. Takemoto, EP0237899 A1(1987).

260a). Mulvey, et al. US-4145364 (1979): 260b). Gay et al. US 3965183 (1976): 260c). J. R. Patton, EP-0193671(1986):

261). J. O. Smith et al. US-7176334(2007):

262a). CN102617360-B(2013).262b).J.Spencer,etal.*Tetrahedron,*64:10195--10200(2008).

263). K.A. Chehade, H. P. Spielmann, *J. Organic Chemistry*, 65(16):4949-53(2000).

264). R. J. Tull, et al. EP0001825-A1(1979).

265). J. Zhao, L. Ma, et al. *Chinese Chemistry Letters*, 25(08):1137-1140(2014).

CHAPTER-4

SUBSTITUTED ANILINE ANALOGS-I

4.00.00: Catalytic Hydrogenation using Molecular Hydrogen:

Introduction: Analog: In chemistry, a structural analog, also known as a chemical analog or simply an analog, is a compound having a structure similar to that of another one, but differing from it in respect of a certain component. It can differ from the basic molecule in one or more atoms, functional groups, or substructures. A structural analog can be formed by replacement of an atom or atoms or functional group in the base molecule by other substituents. Despite a high chemical similarity, structural analogs are not necessarily functional analogs and can have very different physical, chemical, biochemical, or pharmacological properties. https://en.wikipedia.org/wiki/Structural_analog

In previous chapters we have covered in detail about Aminophenols (APs) and Chloroanilines (CANs) as commercially important AN analogs. In this chapter, other substituted aniline analogs have been briefly covered based on the reports from different research groups and their new findings. Some of the newly developed methodologies for synthesizing these analogs have a great potential to go into the production plants with special features as ease of operations, cost-effectiveness, non-hazardous nature, environmental friendly, quantitative conversions, about 100% yields, and approximately 100% chemoselectivities etc. Since it would be difficult to cover the detailed history of 200 years, here we have tried to give a glimpse of the most recent developments in synthesizing AN analogs. Therefore, some of the important analogs are presented only in the form of their structures and chemical names. These analogs could be made directly from aryl halides, reductions of nitrogen containing compounds especially NAs, aryl azides, aromatic azo and hydrazo compounds etc. using molecular hydrogen or different hydrogen sources. These aspects are covered under different headings depending on what kind of information it contains. For example, reductions of NAs using hydrazine hydrate or ammonium formate can be catalyzed by some metal or metal salts as such or these catalysts could be metals in different forms (a mono-metal, bimetals, metal-NPs) mounted on some kind of solid support as carbon, SiO_2, TiO_2, ZrO_2 etc. So, it is possible that the use of these supported metal catalysts would be found under different sources of hydrogen, including molecular hydrogen or hydrogen gas.

In this section, we will be highlighting the developments in selective hydrogenations of nitro groups in multi-substituted NACs using only molecular hydrogen as the reducing agent. These NACs would be other than aminophenols (APs) and haloanilines (HANs), although some of the substituted NACs would be containing hydroxyl or halo groups as well along with other substituents, such as -CH$_3$, -CH$_2$CH$_3$, (CH3)$_2$CH-, -CH=CH$_2$, OBn, CH$_3$O-, -COOH, -COOR, -COR, -CN, -NO$_2$, NH$_2$, aryl or heteroaryl rings etc. We will also highlight the bi or multinuclear AN analogs including a brief summary of their synthesis. The synthesis and manufacturing of substituted AN analogs using hydrogen sources other than molecular hydrogen are covered in the next chapter.

4.01.00: Catalyic Hydrogenation of Substituted NACs in Alcohol as a Solvent:

Since long time ago (1932), different catalysts have been used for the reduction of NB to AN. Covert et al. have found that the method of preparation of the Ni catalyst plays a significant role in the synthesis of AN from NB. The Ni catalyst prepared from Nickel nitrate and ammonium carbonate is better than sodium bicarbonate or carbonate or potassium hydroxide. The as-prepared Ni-catalyst supported on Kieselguhr catalyzed NB hydrogenation in ethanol to give AN in 95-100% yields (Scheme-1). The substituted NACs may or may not have the hydroxyl and halo groups in them.

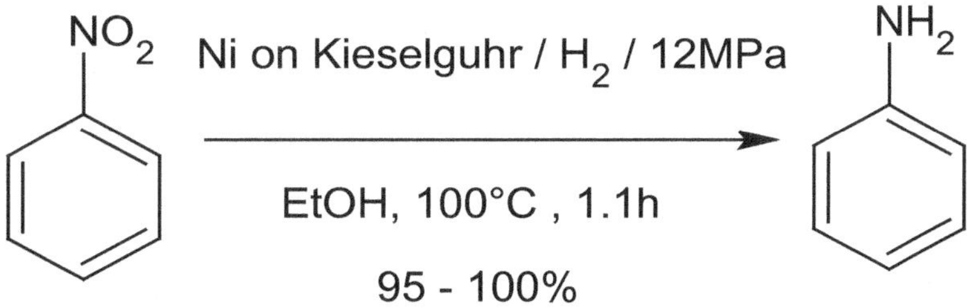

--Scheme-1

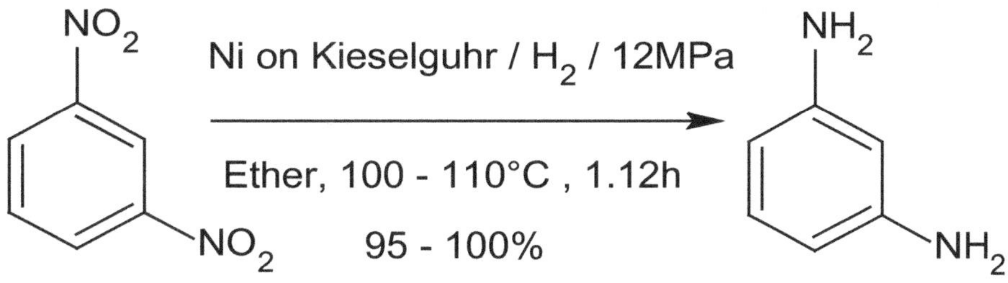

--Schemes-2

Under identical conditions, *m*-dinitrobenzene (*m*-DNB) was also reduced to *m*-phenylenediamine in 95-100% yield (1) (Scheme-2).

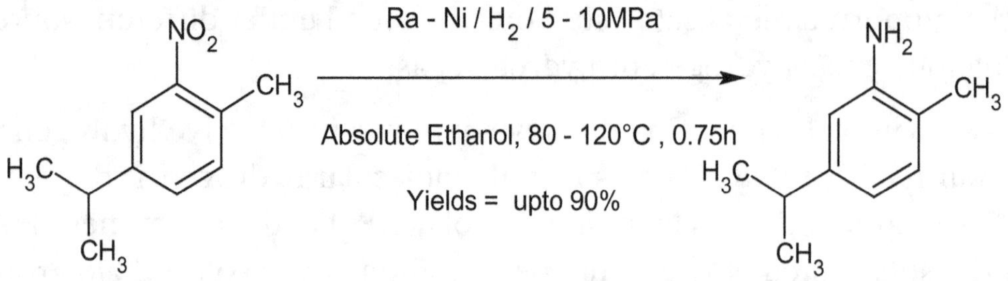

--Scheme-3

The nuclear substitutions with bulky groups on the benzene rings of NAs, as expected through steric hindrance, in some cases reduce the yields of the isolated amines (2) (Scheme-3).

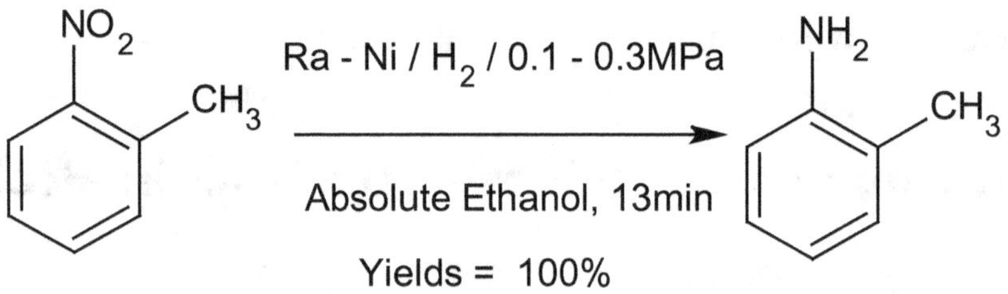

--Scheme-4

Adkins et al. have reported (3) Pt, Pd based noble metal catalysts and later chose to use Raney-Ni catalyst for catalytic hydrogenation of NAs to ANs in high to excellent (up to 100%) yields (Scheme-4).

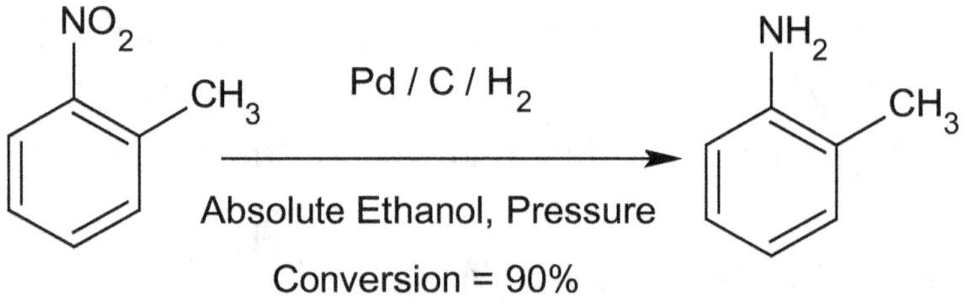

--Scheme-5

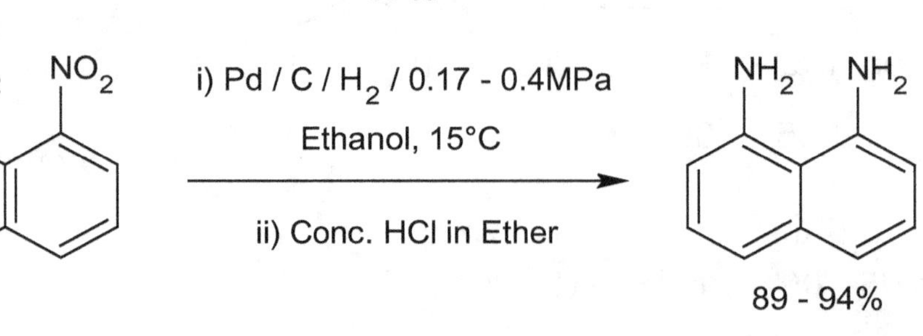

--Scheme-6

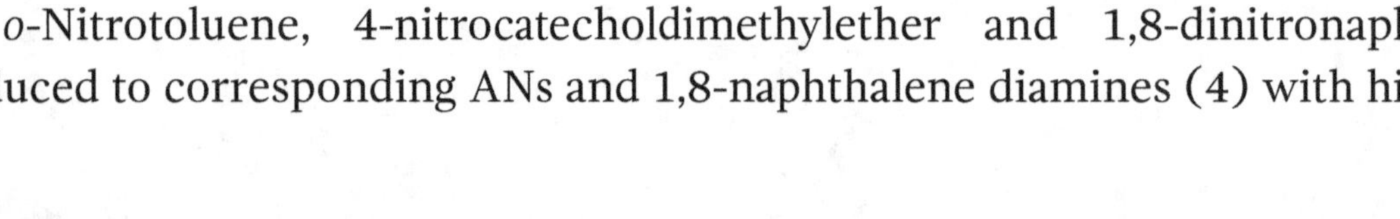

--Scheme-7

o-Nitrotoluene, 4-nitrocatecholdimethylether and 1,8-dinitronaphthalene were reduced to corresponding ANs and 1,8-naphthalene diamines (4) with high conversions and in high yields, but it required high catalyst loadings (30% Pd) (Scheme-5, 6 and 7). Although Pt is equally active catalyst as Pd, sometime one performs better than the other under particular set of reaction conditions. Freifelder et al have found that Pd is preferable over Pt for its higher activity and less tendency to saturate aromatic rings under optimized conditions (Scheme-8). The reaction with Pt took quite long time to complete. Generally, the reaction rates could be increased by the addition of a small amount of an additive as a promoter (5).

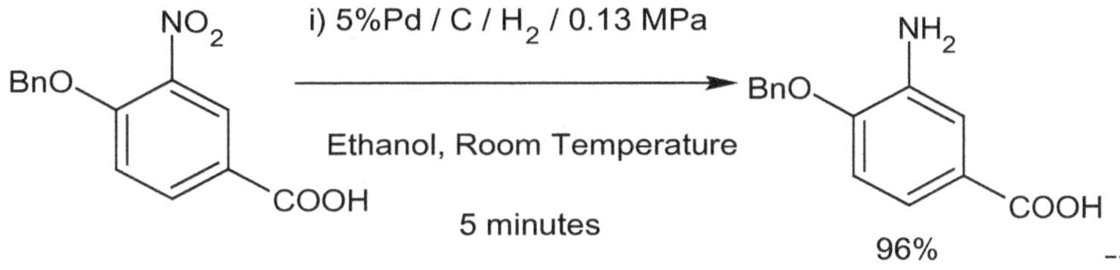

--Scheme-8

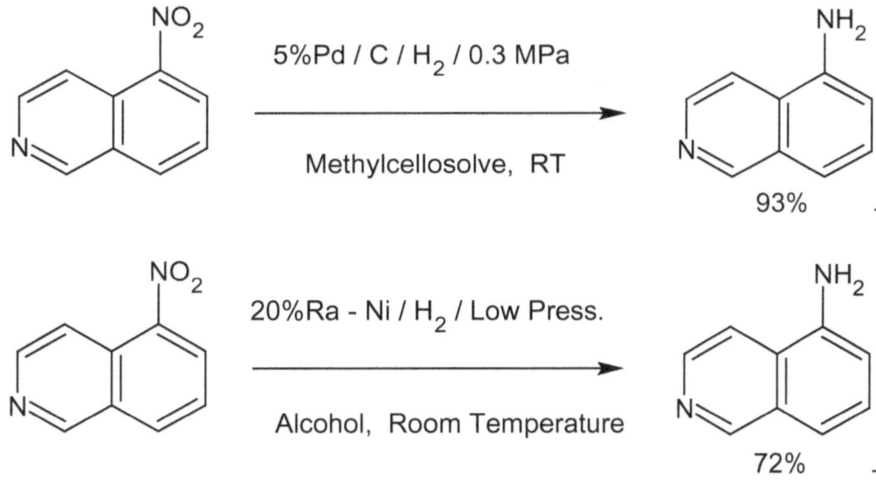

Scheme-9a, b

The hydrogenation of 5-nitroisoquiniline with Pd/C or on activated Raney-Ni (6) was completed in 18 hours with lower (73%) yield (Scheme-9b). In in some cases, the presence of acid additive enhances the reduction efficiency of the catalysts. For example, the reduction of 4-nitropyrogallol to 4-aminopyrogallol was accomplished faster using PtO_2 in the presence of one equivalent of HCl than without HCl in aqueous ethanol at room temperature and 0.35MPa H_2 (Scheme-10). Unlike Pd/C, Pt catalysts can be used under basic conditions, as 3 and 4-nitrophthalic acid is hydrogenated over PtO_2 under relatively milder conditions to the respective amino phthalic acids in quantitative yields (Scheme-11a, b).

Raney-Ni (8) also was equally effective, albeit at higher temperature (70°C) and pressure (10MPa).

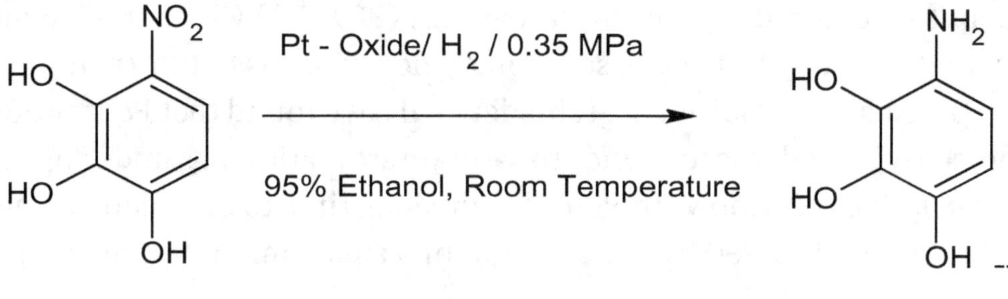

NO2
HO
HO
OH
Pt - Oxide/ H2 / 0.35 MPa
95% Ethanol, Room Temperature
NH2
HO
HO
OH

--Scheme-10

NO2
HOOC
HOOC
Pt - Oxide/ H2 / 0.30 MPa
Aq. NaOH / 45°C / 4h
100%
NH2
HOOC
HOOC

--(a)

HOOC
HOOC
NO2
Pt - Oxide/ H2 / 0.30 MPa
Aq. NaOH / 45°C / 4h
100%
HOOC
HOOC
NH2

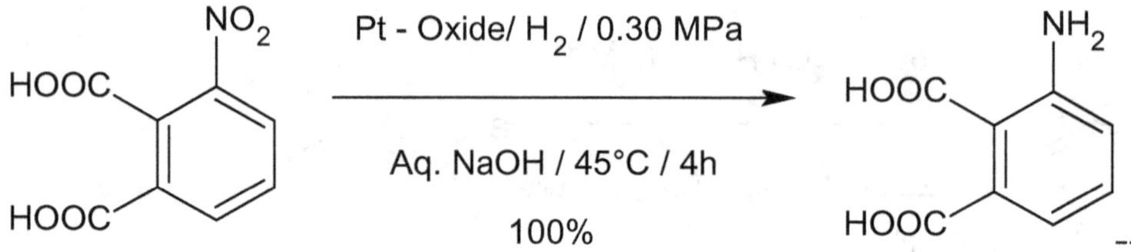

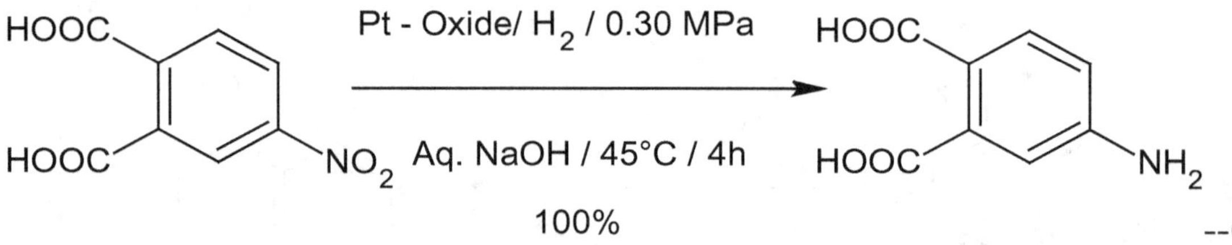

--(b)

Scheme-11a, 11b

Some substrates are quite insensitive towards some of the catalysts, while they are easily reduced with other catalysts.

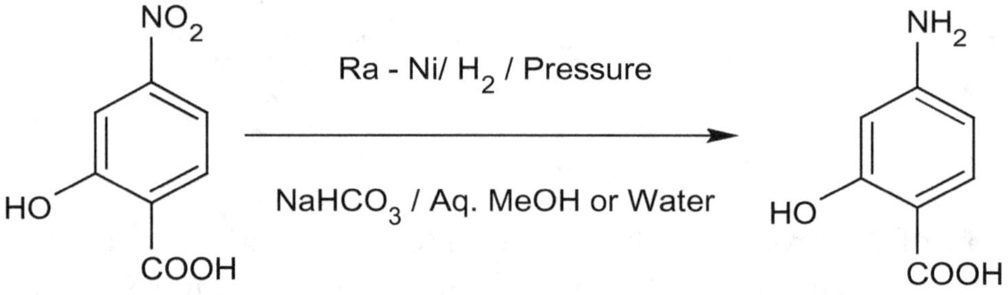

NO2
HO
COOH
Ra - Ni/ H2 / Pressure
NaHCO3 / Aq. MeOH or Water
NH2
HO
COOH

--Scheme-12

For example, 4-nitrosalicylic acid can be reduced to 4-amino salicylic acid (4-ASA), but quite low yields were obtained when the reduction was done using Pd/C while the reaction rate was very slow when platinum oxide was used. However, 4-ASA was obtained in good yield over Raney-Ni catalyst in methanol in the presence of a weaker base as $NaHCO_3$ (Scheme-12). Reduction of 4-diazo derivative by sodium dithionate is also reported. http://www.prepchem.com/synthesis-of-4-aminosalicylic-acid/.

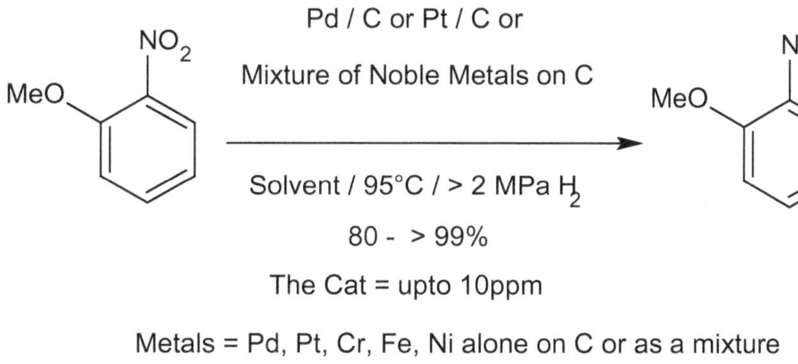
NO2
MeO
Pd / C or Pt / C or
Mixture of Noble Metals on C
Solvent / 95°C / > 2 MPa H2
80 - > 99%
The Cat = upto 10ppm
NH2
MeO
Metals = Pd, Pt, Cr, Fe, Ni alone on C or as a mixture

--Scheme-13

The hydrogenation of *o*-nitroanisole using transition metal catalyst (9) on carbon is reduced to *o*-aminoanisole in more than 99% yields (Scheme-13). The hydrogenation of *m*-nitrotrifluorotoluene over Pd-Fe/TiO_2 catalyst was studied under atmospheric pressure. The influence of reaction temperature and catalyst mass concentration on the hydrogenation reaction kinetics is investigated (Scheme-14).

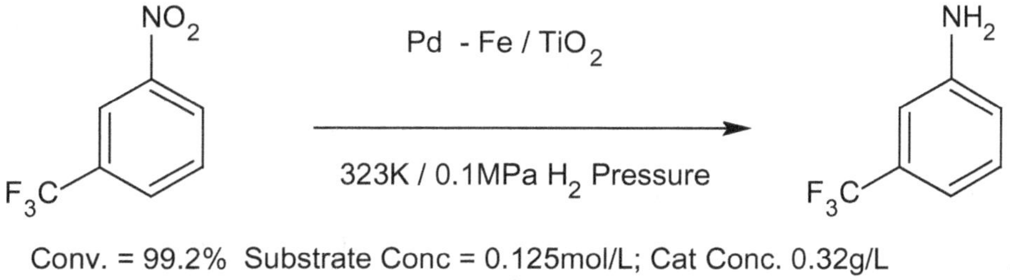
NO2
F3C
Pd - Fe / TiO2
323K / 0.1MPa H2 Pressure
NH2
F3C
Conv. = 99.2% Substrate Conc = 0.125mol/L; Cat Conc. 0.32g/L

--Scheme-14

The results show that *m*-nitrotrifluorotoluene can be converted to 99.2% under the condition of initial *m*-nitrotrifluorotoluene under optimized conditions. The calculated values by the kinetics equation are in good agreement with the experimental ones (10). In an interesting case, selective and sequential hydrogenation of NAs is accomplished using a homogeneous catalytic system (Scheme-15) (11).

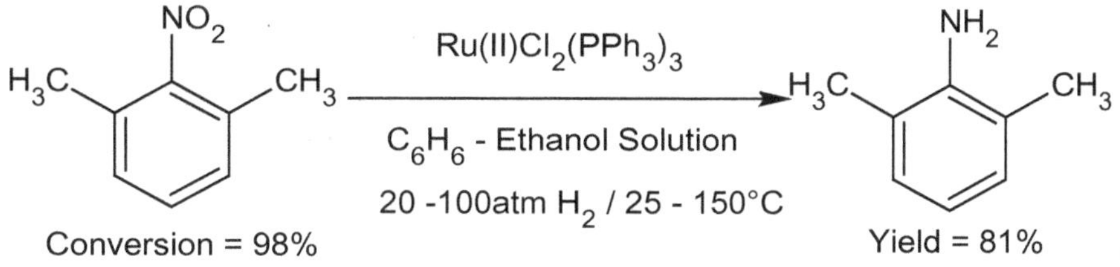
NO2
H3C
CH3
Ru(II)Cl2(PPh3)3
C6H6 - Ethanol Solution
20 -100atm H2 / 25 - 150°C
NH2
H3C
CH3
Conversion = 98%
Yield = 81%

--Scheme-15

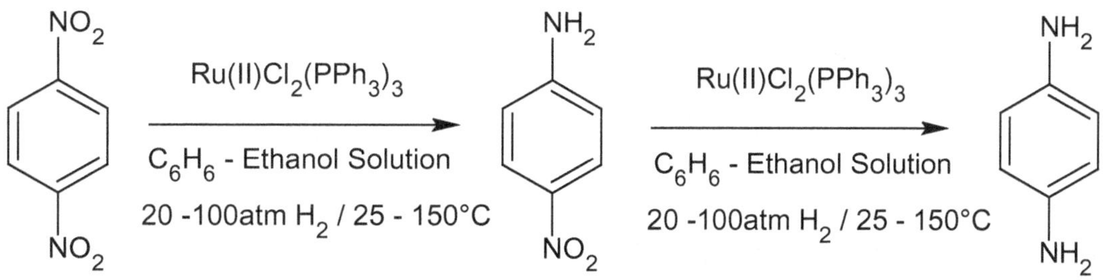
NO2
NO2
Ru(II)Cl2(PPh3)3
C6H6 - Ethanol Solution
20 -100atm H2 / 25 - 150°C
NH2
NO2
Ru(II)Cl2(PPh3)3
C6H6 - Ethanol Solution
20 -100atm H2 / 25 - 150°C
NH2
NH2

Scheme-16

Ru-complex can reduce 2, 6-dimethyl-NB (98% conversion) under catalytic hydrogenation conditions to 2,6-dimethylAN in 81% yield. The ruthenium or iron based catalysts are also highly selective for *p*-dinitroaromatic compounds and do form the inetermediate mononitroamine, which sequentially can be further reduced to diamine under the same experimental conditions (Scheme-16). The selective and sequential, first one nitro group then the other one, hydrogenation of NACs in quantitative yields at room temperature and atmospheric pressure is demonstrated in ethanol (Scheme-17).

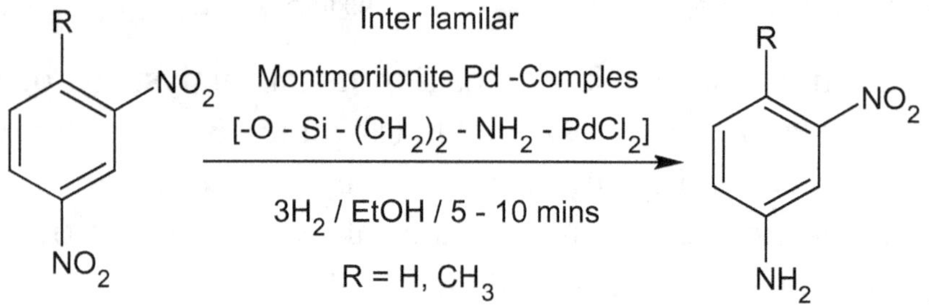

--Scheme-17

The wonderful results are obtained by an inter lamellar montmorillonite Pd(II) complex (12) (montmorillonitesilylamine-Pd(II)complex) as a heterogenized homogeneous catalyst.

Gold in association with nickel on alumina also gives preferential/sequential reduction products (13). The intermediate niroamine can be further reduced with hydrogenation over Ni/Al$_2$O$_3$, if needed. The catalytic gas phase hydrogenation of 1, 3-dinitrobenzene over Au/Al$_2$O$_3$ yielded exclusively reduction of a single –NO$_2$ group to generate 1,3-nitroaniline (E_a = 131 kJ mol^{-1}). Further hydrogenation over Ni/Al$_2$O$_3$ gave 1, 3-phenylenediamine (E_a = 38 kJ mol^{-1}), whereas both products were isolated over Au–Ni/Al$_2$O$_3$ under similar conditions (Scheme-18).

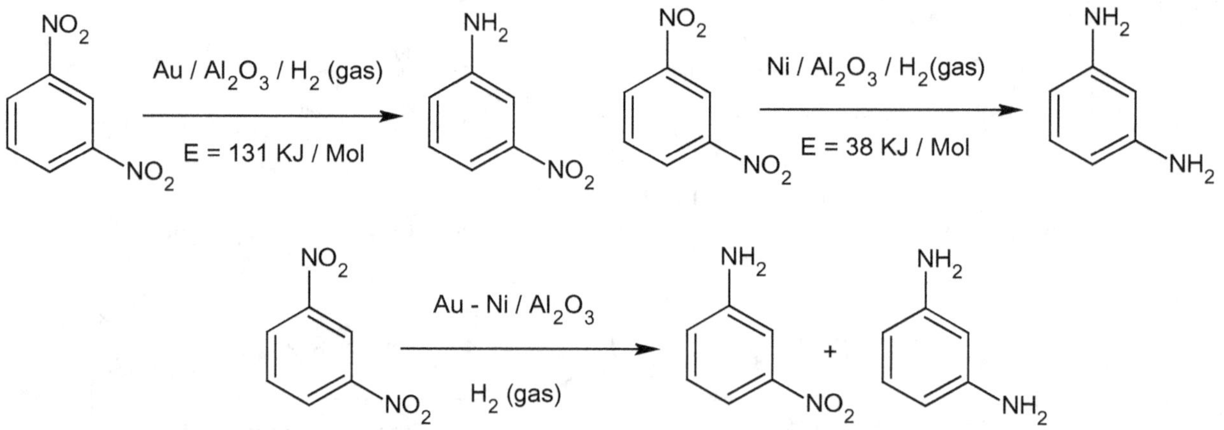

Scheme-18

Exactly similar results were obtained in case of hydrogenation of 1, 3, 5-trinitrobenzene over these catalysts. The analysis of the results demonstrated that the control of selectivity in poly nitroarene hydrogenation through the use of particular catalysts is governed by the surface composition of the catalyst. The idea of utilizing active surface compositions is reported by Cheng et al. with the selective hydrogenation of *m*-dinitrobenzene to *m*-nitroaniline over Ru-SnO$_x$/Al$_2$O$_3$ catalyst using different tools.

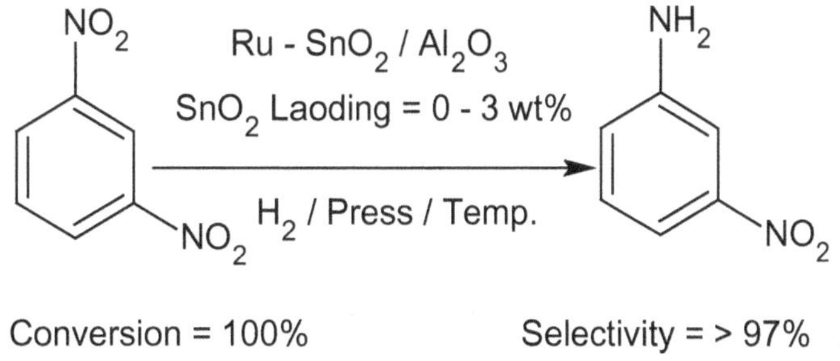

Conversion = 100%   Selectivity = > 97%   --Scheme-19

The *m*-DNB conversion was 100% while the selectivity for *m*-NAN was more than 97% (Scheme-19) (14).

Pd nanoclusters generated *in-situ* from Pd(acac)$_2$ was developed and used for the synthesis of various aromatic azo compounds by hydrogenation of the corresponding NAs using H$_2$ as the sole reductant under mild reaction conditions (15). If necessary, the azo compounds can be further reduced to aniline molecules (Scheme-20).

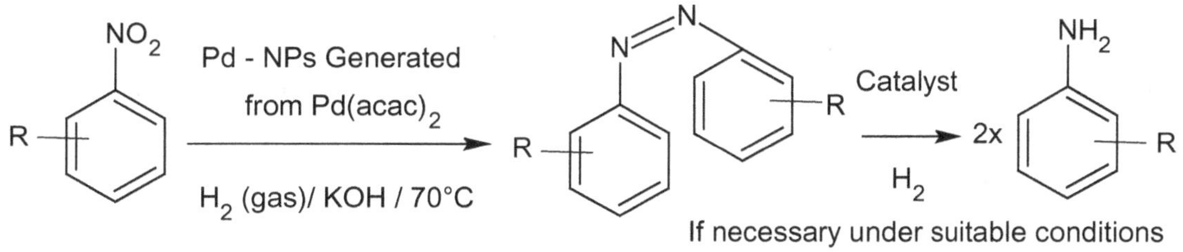

Scheme-20

We have looked in great detail about the mechanism of NB reduction through its intermediates in the first chapter, here the following example illustrates the mechanism of reduction of NB to AN with intermediates. Ordered hexagonal mesoporous Fe-SBA-15 with a large pore diameter was prepared and its catalytic activity was studied for the reduction of NACs into amino compounds. The regio-selectivity of the reduction of different nitro substituted compounds with Fe-SBA-15 is discussed (16).

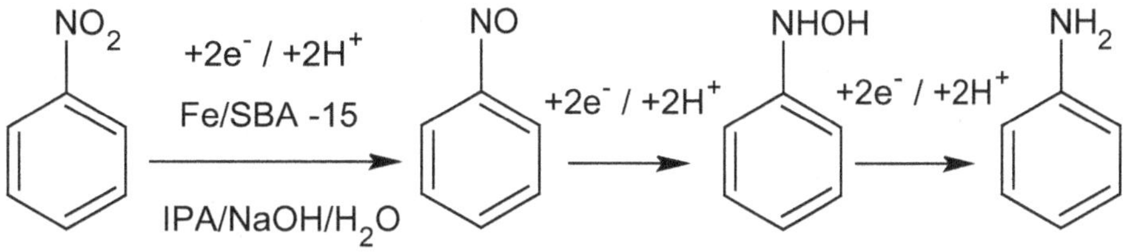

Scheme-21: Mechanism of NB reduction to AN

Mechanistically, the reduction pathway follows the standard NB reduction sequence as NB to PhNO to PhNHOH to AN in IPA/aq.NaOH (Scheme-21). Actually, there are some other intermediates as azo, azoxy, hydrazo etc, those can be selectively prepared and isolated or if needed can be further reduced to AN. Various azobenzenes and azoxybenzenes were selectively and conveniently reduced almost quantitatively to the corresponding hydrazobenzenes using sodium dithionite and dioctyl viologen as an electron-transfer catalyst. The reaction is performed in acetonitrile-water system under mild conditions without forming aniline derivatives. We know that the resulting hydrazobenzenes could be isolated and used for specific applications (17) or can be further reduced to the corresponding anilines (Scheme-22).

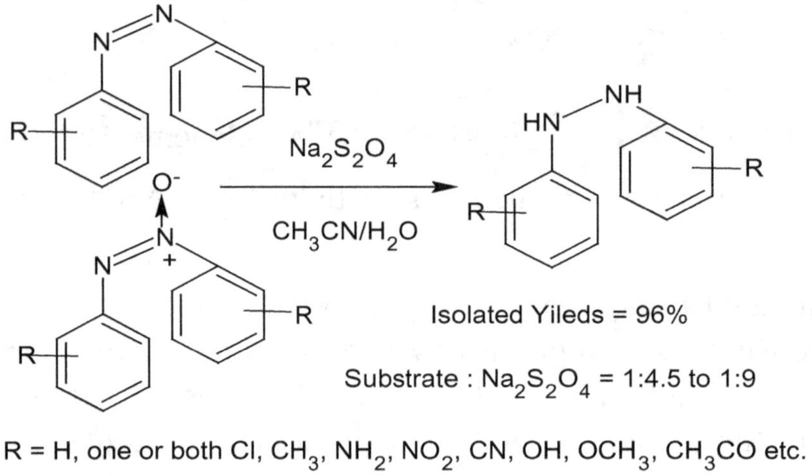

Na2S2O4
CH3CN/H2O
Isolated Yileds = 96%
Substrate : Na2S2O4 = 1:4.5 to 1:9
R = H, one or both Cl, CH3, NH2, NO2, CN, OH, OCH3, CH3CO etc.

--Scheme-22

NAs are selectively reduced to the corresponding *N*-arylhydroxylamines (PHA) with high selectivity using Zn dust in a CO_2/H_2O system under mild conditions. The yield of PHA from NB is 88% when the reaction is carried out at 25°C for 1.5 hours, 0.1MPa CO_2, with Zn: NB molar ratio equal to 3.

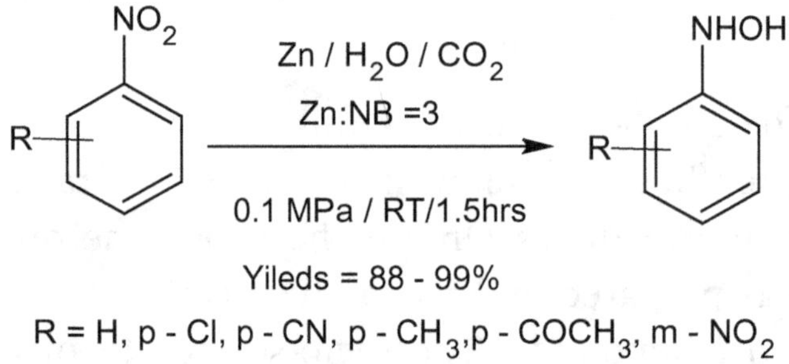

Zn / H2O / CO2
Zn:NB =3
0.1 MPa / RT/1.5hrs
Yileds = 88 - 99%
R = H, p - Cl, p - CN, p - CH3,p - COCH3, m - NO2

--Scheme-23

Other NAs which contain reducible functionality other than a nitro group are also reduced to the corresponding PHAs with yield from 88% to 99%, without affecting the other groups (18). The process fully removes the need to use NH_4Cl and is environmentally benign (Scheme-23). A supported Pt-NPs based catalyst was used in the chemoselective hydrogenation of NAs to PHA in 97% yield (Scheme-24).

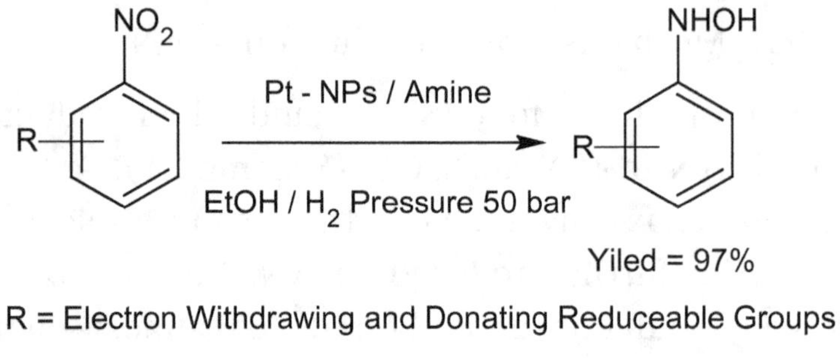

Pt - NPs / Amine
EtOH / H2 Pressure 50 bar
Yiled = 97%
R = Electron Withdrawing and Donating Reduceable Groups

--Scheme-24

The decreased PHA hydrogenation rate at high H_2 pressure and in the presence of TMEDA allow for selective transformation of a range of other NAs containing electron-withdrawing and -donating (reducible) functional groups to their corresponding PHAs in excellent (more than 90%) yields (Scheme-24) (19). Ultrathin Pt wires with ethylenediamine are used as a highly selective catalyst for the synthesis of PHAs (20).

Selective reduction of 5-nitro-2-chloro-2',4'-dimethylbenzenesulfonanilide to 5-amino-2-chloro-n-(2',4'-dimethyl) benzenesulfonamide (ACD) is achieved on Pd-Ru/γ-Al_2O_3 catalysts in ionic liquids(ILs) (Scheme-25).

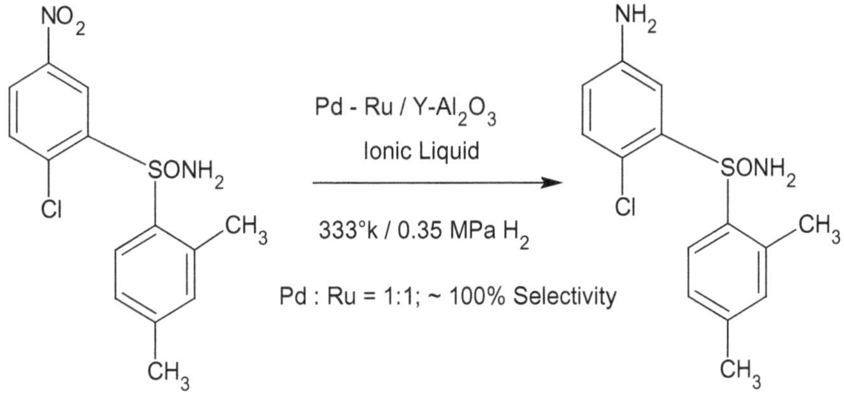

--Scheme-25

The ionic liquids and the molar ratio of Pd to Ru showed an influence on the selectivity of catalysts (21) and the highest-selectivity was achieved when Pd:Ru molar ratio was 1:1 for hydrogenation of NCD in [XPy]Br (Scheme-25). Aniline analogs are obtained by catalytic hydrogenation of NAs using metal catalysts over solid supports.

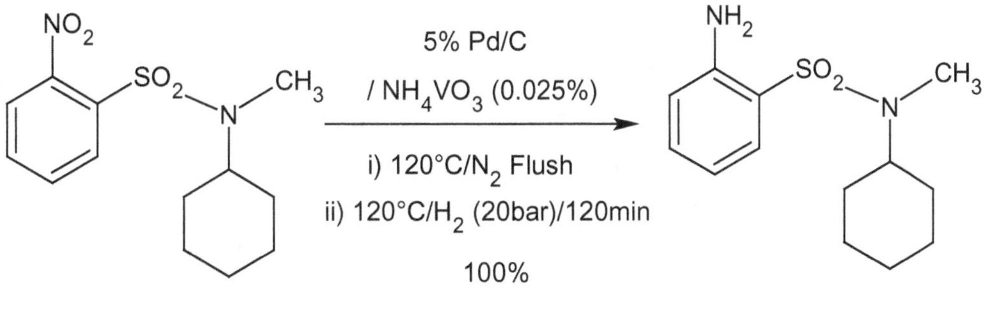

--Scheme-26

In some cases the aryl amines are obtained in 100% yields (Scheme-26). https://www.google.com/patents/EP0825979B1?cl=en.

4.02.00: Bi- and Multinuclear AN Analogs/Derivatives:

As pointed out in the beginning of this chapter, now we will briefly highlight the bi and multinuclear AN analogs. A brief account is given about the preparation/synthesis processes with relevant references.

4.02.10: Raw Materials or Feedstocks for Aniline:

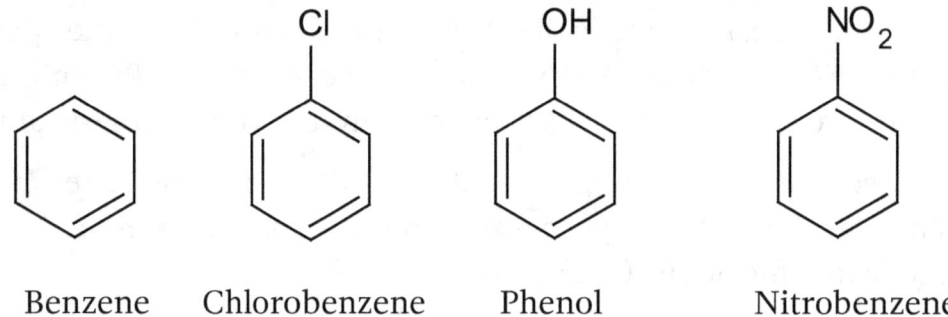

Benzene   Chlorobenzene   Phenol   Nitrobenzene

4.02.11: Primary Aniline Analogs:

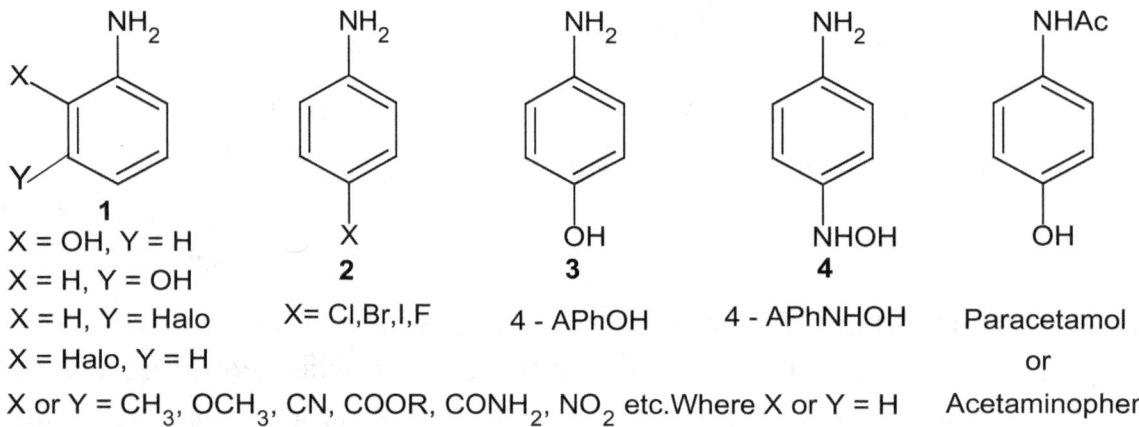

1
X = OH, Y = H
X = H, Y = OH
X = H, Y = Halo
X = Halo, Y = H

2
X= Cl,Br,I,F

3
4 - APhOH

4
4 - APhNHOH

Paracetamol
or
Acetaminophen

X or Y = CH$_3$, OCH$_3$, CN, COOR, CONH$_2$, NO$_2$ etc.Where X or Y = H

4.02.12: N-AlkylAnilines:

In general, selective mono-alkylation of primary amines is very important inorder to get single N-alkylated products. Under primary aniline derivatives, N-alkylANs are important and are made by highly selective N-monoalkylation of AN. For example, selective N-alkylation/methylation of aniline with methanol over a heteropolyacid on montmorillonite K-10 is achieved in 99% selectivity to N-methylAN and with 79% AN conversion (22). There are many reports on selective N-monoalkylation of both alkyl and aromatic primary amines (23). N-Mono and also N, N-dialkylANs can also be made under suitable conditions (24).

4.02.13: Representative Binuclear, Polycyclic Analogs:

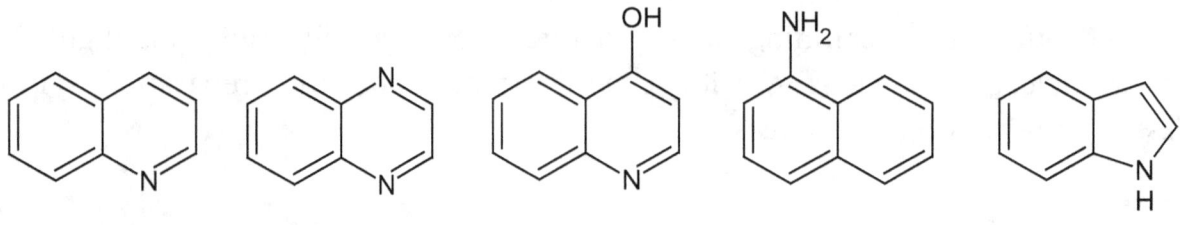

Quinoline   Quinoxaline   4-Hydroxyquinoline   Naphthylamine   Indole

http://www.organicchemistry.org/synthesis/heterocycles/benzofused/quinolines. shtm. https://en.wikipedia.org/wiki/Quinoline.

Catalytic transfer hydrogenations using hydrogen sources other than H_2 gas are also performed using different catalysts. Nickel boride/hydrazine hydrate (25) reduces NACs to 4-benzyloxy indole derivatives (Scheme-27).

--Scheme-27

A simple, two-step process for 4, 5 and 6-benzyloxy indole is reported in overall good to high yields. For example, 2-benzyloxy-6-nitro toluene is transformed into 4-benzyloxyindole in 80% yield. As a part of aniline and its analogs, the N-substituted derivatives (26) of aniline are made by using gold, platinum and other M-NPs or nanowires and H_2 and also using other hydrogen sources.

Chemoselective reduction of nitro groups in the presence of activated heteroaryl halides with a commercially available sulfided platinum catalyst (27) is accomplished in 99% selectivity (Scheme-28).

11 Examples. No Dehalogenation and Selectivity = > 99%   --Scheme-28

The optimized conditions employ low temperature, low pressure, and low catalyst loading (<0.1 mol % Pt) to afford heteroaromatic amines in almost quantitative yields

and with minimal or no hydrodehalogenation byproducts.

For Quninoxalines from O-(Alkoxycarbonylmethylamino)-nitrobenzene see Ref: *The Chemistry of Heterocyclic Compounds* (2004), D. J. Brown et al pp5 and 6.

[1-Benzyl-7-Cl-2,3(1H,4H) Quinoxalinedione]   [6-CF$_3$,3,4-Dihydro-2(1H)-Quinoxalinone]   [1-Hydroxy-4-Me-2, 3(1H,4H)-Quinoxalinedione]

Although quinoxaline itself is mainly of academic interest, quinoxaline derivatives are used as antibiotics (28) such as olaquindox, carbadox, echinomycin, levomycin and actinoleutin. Substituted quinoxalines (29) are prepared by employing a new method which involves a tandem oxidative azidation/cyclization reaction of *N*-arylenamines (Scheme-29).

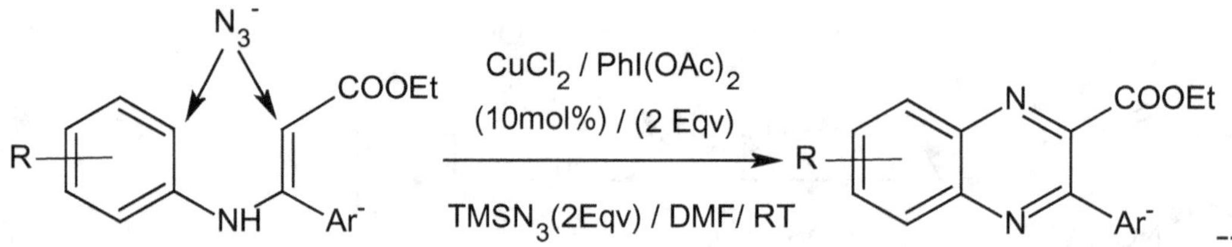

--Scheme-29

Some other quinoxalines are easily obtained on a gram scale and converted to various useful scaffolds, as the compound LASSBio-1022 was prepared in 83% yield in two steps (30). Similarly, benzimidazoles (a), phenathradines (b), carbazoles (c), quinolones (d) have been synthesized by different groups (31). Substituted quinoxalines can be formed by condensing *ortho*-diamines with1,2-diketones(Scheme-30). https://en.wikipedia.org/wiki/Quinoxaline.

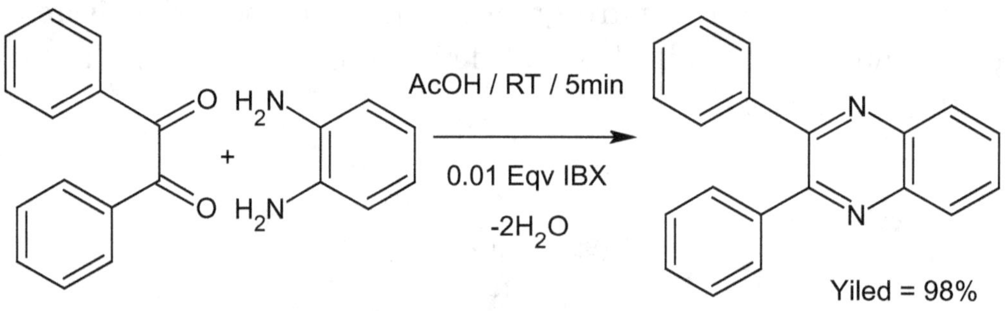

Scheme-30

Quinoxaline and its analogues (32) may also be formed by reduction of amino acids -substituted 1, 5-difluoro-2,4-dinitrobenzene (DFDNB). A novel iodide-catalyzed reduction of NAs with H_3PO_2 or H_3PO_3 is studied and its application in the synthesis of a potential anticancer agent (33) is discussed (Scheme-31).

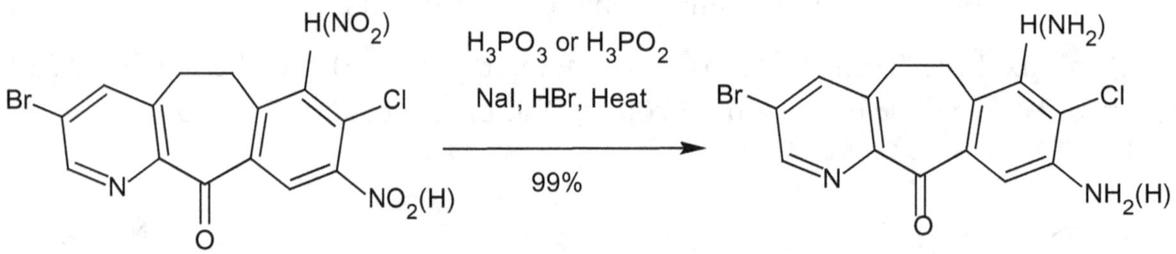

Scheme-31

A novel, high yielding iodide-catalyzed reduction method using hypophosphorous and/or phosphorus acids was developed to reduce both diaryl ketones and NAs. The reaction is highly chemoselective in the presence of chloro and bromo substituents and the corresponding ANs are obtained in high yields (99%). This efficient and practical method has been successfully applied to a large scale production of a potential anticancer agent.

4.03.00: Fluoroquinolones:

There are many fluoroquinolones which are available in the market as antibiotics, as Ciprofloxacin, Levofloxacin, Oflofloxacin, Moxifloxacin, Norfloxaxin, Gatfloxacin, etc.

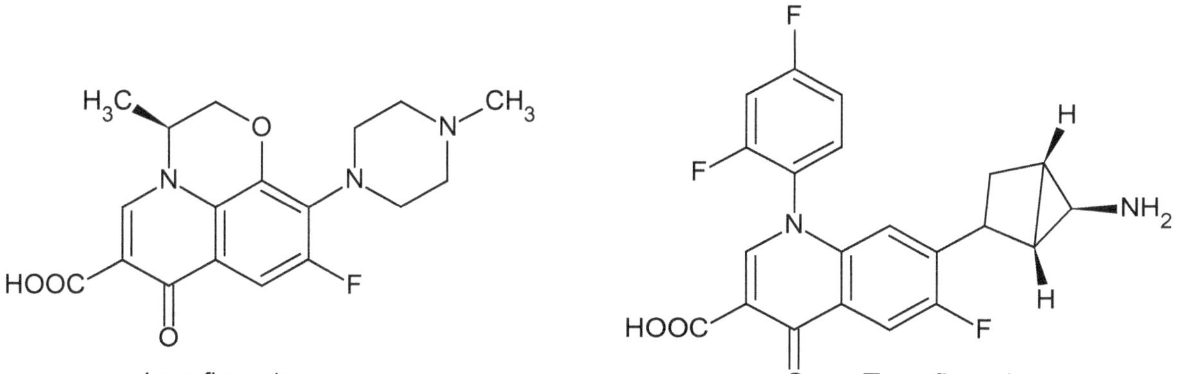

Levofloxacin   Trovafloxacin

Levofloxacin and Trovafloxacin are presented here with their structures, as examples of fluoroquinolone antibiotics under multinuclear pharmaceurical AN analogs, which are manufactured at commercial scale. Bicalutamide is an AN based anticancer drug available in the market since 1995 and is primarily used in the treatment of early and advanced prostate cancer.

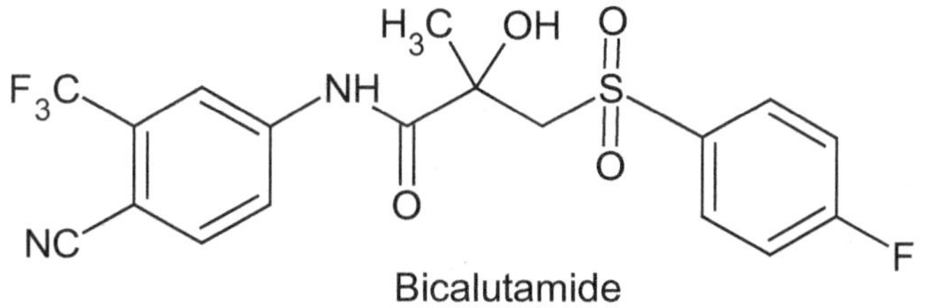

Bicalutamide

Bicalutamide is mainly used in and is only approved for the following indications: i). Metastatic prostate cancer (MPC) in men in combination with a gonadotropin-releasing hormone (GnRH) analogue and ii) Locally advanced prostate cancer (LAPC) in men as a monotherapy in high doses. Other related drugs are Flutamide, Topilutamide, Acetothiolutamide, Enobosarm, etc. Lamture et al. have developed a commercial process for Bicalutamide. The technology was transferred to a multinational pharma company (34). An improved convergent and economical method has been developed for the synthesis of Erlotinib, a 4-anilinoquinazoline and an EGFR-tyrosine kinase inhibitor for the treatment of non-small-cell lung cancer.

Erlotinib

The final two steps for the formation of this 4-anilinoquinazoline from suitable 2-aminobenzonitrile intermediate and 3-ethynylaniline were modified and were performed in a simple one-pot reaction. The ring-closing mechanism is determined to proceed via the formation of phenyl benzamidine intermediate, which was isolated for the first time and characterized (35).

References:

1). L. W. Covert, et al. *J. American Chemical Society*, 54(04):1651–1663(1932).

2). C. F. Allen, et al. *J. Organic Synthesis*, Coll. Vol. 03, P63 (1995).

3). H. Adkins, H. R. Billica, *J. American Chemical Society*, 70(02): 695–698(1948).

4). L. H. Klemm, et al. *J. Organic Chemistry*, 22:161(1957). Gerald Booth "Naphthalene Derivatives" in Ullmann's Encyclopedia of Industrial Chemistry, 2005, Wiley-VCH, Weinheim. doi:10.1002/14356007.a17_009

5). M. Freifelder, *Catalytic Hydrogenation in Organic Syntheses: Proceeding and Commentary*, Wiley-Interscience, NY, 1978, p27-30.p27, refs 86-90

6a). M. G. Banell, et al. *Organic & Biomolecular Chemistry*, 12(38):7433-7444(2014). 6b). E. Lieber, G. B. L. Smith *J. American Chemical Society*, 58(08):1417–1419(1991).

7a). R. B. Moffete, et al. *J. Medicinal Chemistry*, 09,:475-478(1966). 7b). H. Jiang, et al. *Molecules*, 21:833-894 (2016), & cross reference no. 19.

8). S. H. Merril, et al. *J. Organic Chemistry*, 25:1882-1883(1960).

9). J. R. Kosakin, *Catalaysis of Organic Reactions*, Mercel Dekker, NY, 1981, PP461-471.

10a). D. Jin, C. Liu, www.cnki.com.cn: 10b), H. Yoshida, et al. *J. Physical Chemistry C*, 115:2257-2267(2011),

10c). S. Fujita, et al. *J. Supercritical Fluids*: 60:106-112(2011).

11). J. F. Knifton, *J. Organic Chemistry*, 41(07):1200–1206(1976).

12). K. Mukkanti, et al. *Tetrahedron Letters*, 30(02):251–252(1989).

http://www.organicchemistry.org/chemicals/reductions/iron.shtm

13). F. C-Lizana, et al. *J. Physical Chemistry. C*, 116 (20):11166–11180(2012).

14). H. Cheng, W. Lin, et al. *Catalysts*, 4(3), 276-288 (2014).

15). J. Wang, L. Hu et al. *RSC Advances*, 03:4899-4902(2013).

16). N. S. Sanjini & S. Velmathi, *RSC Advances*, 04:15381-15388 (2014).

17a). K. K. Park, S. Y. Han, *Tetrahedron Letters*, 37(37):6721–6724(1996), 17b). G. Q. Gao, et al. *New J. Chemistry*: 38:4661-4665(2014). 17c). J. M. Khurana, S. Singh, *J. Chemical Society - Perkin Transactions*-1, (13):1893-1895 (1999), 17d). K. Ohe, S. Uemura, et al. *Toru J. Organic Chemistry*, 54(17):4169-4174 (1989).

18a). S. Liu, Y. Wang, et al. *Green Chem*istry: 11:1397-1400(2009), 18b). M. J. Gibian , A. L. Baumstark, *J. Organic Chemistry*, 36(10):1389–1393(1971).

19). E. H. Boymans, et al. *Catalysis Science & Technology*, 05:176-183 (2015).

20). G. Chen, C. Xu, et al. *Nature Materials*, 15, 564–569 (2016).

21). Y. Zhao, et al. *Modern Applied Science*, 04(05):155-161(2010).

22). M. Nehate, V. V. Bokade, *Applied Clay Science*, 44(03-04):255-258(2009).

23a). A. C. Bayer, US-5030759 A**(1991)**. **23**b) A. Liang, Y. Lin, *Applied Catalysis A: General*, 134(01):53-66(1996). 23c) V. Pace, et al. *Organic Letters*, 9(14):2661–2664(2007). 23d) S. Naskar, M. Bhattacharjee, *Tetrahedron Letters*, 48(19):3367-3370(2007). 23e) R. Byun, et al. *J. Organic Chemistry*, 72(25):9815–9817(2007). 23f**)** R. Martinez, et al. *Organic & Biomolecular Chemistry*, 07(10):2176-2181(2009). 23g) *N. Shankarajan, et al. J. Organic Chemistry*, 76(17):7017–7026(2011). h) A. Bruneau-Voisine, et al. *Journal of Catalysis*, 347:57-62(2017).

24a). C. Siswanto, J. F. Rathman, *J Colloid Interface Sciences*, 196(01):99-102(1997). 24b). N. Iranpoor, et al. *Tetrahedron*, 65(19):3893-3899(2009).

25). D. H. Lloyd , D. E. Nichols, *J. Organic Chemistry*, 51 (22):4294–4295 (1986).

26a). J. Wang, et al. *Current Organic Chemistry*, 2015: Vol. 19, 14 pages

b) http://shodhganga.inflibnet.ac.in/bitstream/10603/144604/10/10_chapter%202.pdf.

27a). A. J. Kasparian, et al. *J. Organic* Chemistry, 76(23):9841–9844(2011). 27b). Synthesis of quinolines: C. S. Cho, et al. *Bulletin Korean Chemical Society*, 23(04):541-542(2002). 27c). Chemoselective reduction of NAs to benzotriazoles: V. Zimmermann, et al. *J. Combinatorial Chemistry*, 09(02):200-203.(2007).

28). The Chemistry of Heterocyclic Compounds, Quinoxalines: Supplement II, D. J. Brown, E. C. Taylor, Jonathan A. Ellman, John Wiley & Sons, 15-Feb-2004 - Science - 510 pages, PP 5 and 6. https://en.wikipedia.org/wiki/Quinoxaline.

29). H. Ma, D. Li, and W. Yu, *Organic Letters*, 18 (4), 868–871 (2016).

30). Y. Jiao, L. Wu, et al. *J. Organic Chemistry*, 82 (08):4407–4414 (2017).

31a). O. Ravi, A. Shaikh, et al. *J. Organic Chemistry*, 82(08):4422–4428 (2017). 31b) W. Song, P. Yan, et al. *J. Organic Chemistry*, 82(08):4444–4448(2017). 31c) S. Maiti, et al. *Organic Letters,* 19(08):2006–2009(2017). 31d). S. Maiti, et al. *Organic Letters,* 19(09):2454–2457(2017). 31e) G. S. Kumar, et al. *Organic Letters,* 19(10):2494–2497(2017).

32). X. Wu, G. Liu, et al. *Molecular Diversity*, 08(2):165–174. (2004).

33). G. G. Wu, F. X. Chen, et al. *Organic Letters*, 13(19):5220–5223(2011).

34a). Jagannath B. Lamture, et al. Unpublished work (2003-2004). 34b). https://en.wikipedia.org/wiki/Bicalutamide

35). D. Asgari, et al. *Bull. Korean Chemical So*ciety, 32(03): 909-914(2011).

CHAPTER-5

SUBSTITUTED ANILINE ANALOGS-II

5.00.00: Catalytic Transfer Hydrogenation (CTH)

Introduction: Two types of hydrogenation processes are in practice for the transformations of organic functional groups and they are based on the source of hydrogen used. The first and the foremost, which has been used for a long time now, is the catalytic hydrogenation using molecular hydrogen or hydrogen gas. The other one is the catalytic transfer hydrogenation (CTH) using hydrogen sources other than molecular hydrogen, and they are alcohols, boranes, formic acid (FA), hydrazine hydrtate (HH), etc. Hydrogenation is a chemical process in which a hydrogen molecule (dihydrogen or two hydrogen atoms) are added to a group in a molecule with multiple bonds as olefins, acetylenes, carbonyls, nitro groups etc. Generally, catalytic hydrogenation or transfer hydrogenation needs a catalyst, preferably a metal catalyst for an efficient addition of hydrogen to the recipient moiety till it reaches saturation. With few exceptions, both types of reactions are catalyzed by metal catalysts, which could be as metals, metal salts, bi-metals, mono- and bi-metallic-nanoparticles (M-NPs) etc. These metal catalysts could be either unsupported or supported on solid supports as SiO_2, TiO_2, Al_2O_3, Chitosan, Polymers, Resins, Foams, etc. The mono or bi- M-NPs are highly active and selective towards the end products and can be recycled several times without loss of their activity. Because of certain limitations with processes using molecular hydrogen, the second option is favoured as these sources are abundantly available, cause less pollution, available at lower costs and perform with high efficiency.

The recent developments in green catalytic transformations using biogenic products such as alcohols and FA as efficient reagents make it an interesting case for energy storage materials. Based on their origin from natural resources, they are already basic and popular chemicals both in bulk and fine chemical synthesis. With growing population and developments each and every field has created a huge demand for energy. This energy is derived from different natural resources as coal, petrol, food, electricity etc. In order to meet the ever-increasing energy demand, the development of effective, renewable, and environmental friendly sources of alternative energy are of paramount importance. Amongst the natural resources, hydrogen (H_2) is a renewable and one of the most promising alternative energy carriers for clean energy future. The hydrogen storage techniques based on liquid-phase chemical hydrogen storage materials have become an attractive choice. The currently known H_2 sources, other than hydrogen gas, can be used as readily available, efficient reducing agents in organic functional groups transformations inlcuidng NAs to ANs.

As we have seen under Haloanilines and also as covered in previous chapter, the catalytic hydrogenation involves molecular hydrogen for the reduction of substituted HNAs and other substituted NAs. In this section, we will focus on catalytic transfer hydrogenations (CTH) of substituted NAs including HNBs, using hydrogen sources other than molecular hydrogen. These hydrogen sorces or hydrogen donors are inorganic or organic molecules which undergo dehydrogenation and evolve hydrogen during hydrogenation process and they are: Alcohols, Formic Acid (FA), Formates as $HCOONH_4$ or HCOOM (M = K, Na, Zn etc.), Hydrazine Hydrate (HH), NH_4Cl, Boranes, Borohydrides, Glucose, $CO+H_2O$, CO_2+H_2O, $scCO_2$ etc.Generally these are metal catalyzed reactions of hydrogen generation. through dehydrogenation of these hydrogen sources.

Aromatic amines are widely used as important intermediates in the synthesis of a variety of chemicals and are generally produced from their corresponding NACs by catalytic hydrogenation or through other reduction processes. These processes mainly utilize a low-cost but potentially explosive H_2. On the other hand, the NA hydrogenations using the above listed other hydrogen donors are simple and are conducted under mild conditions. Moreover, they are abundantly available at lower costs, produce ANs in good yields and excellent selectivities and hence they have an added advantage over the conventional hydrogenation using molecular hydrogen. However, the use of noble metals limits their scope and hence continuous efforts are being made in finding solutions without using noble metal catalysts, especially using the earth abundant metals as Fe, Cu, Zn, Sn, Mg etc. These new developments have been recently summarized in detail (1). Selective reduction of NAs is an important synthetic tool for the preparation of aromatic amines as they are industrially important group of chemicals. The reduction of NACs is also of considerable interest because these chemicals are common groundwater contaminants and reduction reactions can play a central role in their environmental fate or cleanup. With this brief background now we will focus on the developments made by using different hydrogen sources with an emphasis on quantitative substrate conversions, excellent yields and selectivity to corresponding ANs.

Note: In this chapter, the hydrogen sources are arranged with the chapter number (5) and alpha-betical order of the source as Alcohol (5A-), Boranes (5B-), Carbon Monoxide and Water ($CO+H_2O$) (5C), etc. Another important point the reader must keep in mind that one aspect is discussed under different sub-headings. For example, "Water" is a separate sub-heading which covers reductions in water as a solvent and water as a hydrogen source. The water is also a component of a system under a particular noble or non-noble metal catalyst, and so is the case with Slianes, Sulfides, Hydrazine etc, although they are covered under individual sub-headings. But, if we are talking about metal catalyzed hydrogenations then all these hydrogen sources, besides the information covered under their individual sub-headings, may be seen in there as well with relevant references.

Chapter-5A: Alcohols as Hydrogen Sources or Donors:

5A.01.00: Alcohols as Hydrogen Sources:

The roles of alcohols as solvent in the catalytic hydrogenation of NAs using metal catalysts and molecular hydrogen or hydrogen gas are discussed in the previous chapter under "Ctalytic Hydrogenation of NAs". Alcohols such as methanol, ethanol, 2-propanol or isopropanol (IPA), ethylene glycol, glycerol etc. have been used as solvents as well as the sources of hydrogen in the transfer hydrogenations of functional groups. Chemoselective catalytic transfer hydrogenations of NAs in alcohols both as hydrogen donors and solvents are discussed below.

5A.01.10: Methanol: Methanol is the first member of the straight chain aliphatic alcohols and is used both as a solvent and as a hydrogen donor. In an attempt to employ it as a hydrogen donor, NB is used as a substrate to obtain AN- through CTH over heterogeneous Pd-based catalyst in a fixed-bed reactor. With the increase of the molar ratio of water to methanol from 0 to 1, the conversions of NB is increased from and 7.1% to 31.9%, and the selectivity to AN- is increased from 22.0 to 94.5mol% (2). A mild and efficient electron-transfer method for the selective reduction of NACs using Sm(0) metal in the presence of a catalytic amount of 1,1'- dioctyl-4,4'-bipyridinium dibromide gives aryl amines (Scheme-1) in excellent yields-(77-99%) without affecting other functional groups (3).

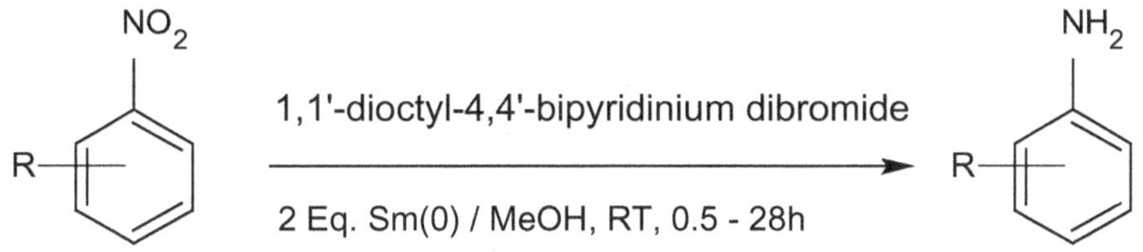

$$R\text{—}C_6H_4\text{—}NO_2 \xrightarrow[\text{2 Eq. Sm(0) / MeOH, RT, 0.5 - 28h}]{\text{1,1'-dioctyl-4,4'-bipyridinium dibromide}} R\text{—}C_6H_4\text{—}NH_2$$

Yields = 77 -99% --Scheme-1

Using supercritical methanol (MeOH-scCO$_2$) as a hydrogen source and alumina, NB and HNBs are initially reduced to AN, and HANs. Aniline is later converted to N, N-dimethylaniline with methanol as an alkylating agent. The reaction is performed at 545K and 20MPa pressure. It was shown that the use of scCO$_2$ as a co-solvent had a significant impact on the course of the reduction process including secondary and side reactions. The conditions for the reduction of HNBs into respective HANs with high conversion, nearly 100% selectivity and almost without dehalogenations have been accomplished in methanol as a hydrogen donor (4).

5A.01.11: Ethanol: A one-pot hydration and reduction of 1-ethynyl-2-NBs strategy is applied for the preparation of the corresponding 2'-aminoacetophenones using SnCl$_2$ in ethanol and water mixture (Scheme-2).

--Scheme-2

Besides Sn/EtOH/H$_2$O as a hydrogenation system, a range of common reducing agents as Fe/HCl, Zn/HCl, Ni$_2$B etc. is described. The authors have noted that the SnCl$_2$/EtOH/H$_2$O reduction must be happening in two steps as i) first the substrates get reduced to ethynylANs, ii) in the second step the ethynylANs undergo hydration to yield the aminoacetophenones in good yields (5).

6 -Aminochrysene

Reduction of aromatic and heteroaromatic nitro compounds to the corresponding amino compounds was achieved by indium/ammonium chloride induced reaction in aqueous ethanol. This method was extended for the preparation of large quantities of 6-aminochrysene in excellent yields (6).

5A.01.12: Isoprpanol: There are several reducing agents as boranes, borohydrides, hydrazine, formic acid and glucose etc. used in the CTH of NAs. Some of these systems are expensive. Thus, relatively a simple and inexpensive alternative reducing agent that offers more environmental friendly reaction conditions and higher efficiency is desirable. In this context, 2-Propanol or IPA is a very popular hydrogen donor because it is less expensive, non-toxic, possesses good solvent properties and is readily transformed into acetone. Acetone is environmental friendly, easy to remove from the reaction system and has commercial applications. With supercritical IPA, NB and HNBs have been reduced to AN, and corresponding HANs. The conditions for the reduction of HNBs into respective amines with high conversion and nearly 100% selectivity almost without dehalogenation were found. In these reactions, IPA acts both as a solvent and hydrogen donor and gets converted to acetone. 2-Methylpentanol also is used as a hydrogen donor in the reductions of NACs (4, 7a, 7b). In another case NB and other NAs are reduced with IPA and aq. KOH to AN in 100% yield, while other corresponding ANs are isolated in relatively lower yields (75-89%) (8).

Heterogenous CTH of NAs to corresponding ANs is reported using novel cobalt (II)-substituted hexagonal mesoporous aluminophosphate molecular sieves as a catalyst. The end products are obtained in good to high yields (Scheme-3) (9).

$R = H, CH_3, CN, CHO, NH_2, NO_2, OCH_3, Cl$ etc. --Scheme-3

The reduction of various NACs into their corresponding amines is achieved by zeolite-supported Cu-NPs with IPA as a sustainable reducing agent with good selectivity and excellent yields (Scheme-4).

R = 21 Different Substituents   --Scheme-4

Different reducing agents, metals, solid supports and their applications in selective reductions of NAs is discussed. Here, both the low-cost and readily available IPA and Cu-NPs over environmental friendly zeolites as solid support are used for chemoselelctive transfer hydrogenation of functionalized NAs (10).

An easily accessible in-situ catalyst composed of [{RuCl$_2$(p-cymene)}$_2$] and terpyridine has been developed for the selective transfer hydrogenation of NAs and aromatic azo compounds to corresponding anilines (Scheme-5a, 5b).

--Scheme-5a

$$[(RuCl_2)(pcyemene)_2] \ (2.5mol\%)$$

Terpyridine (5mol%)

R -Ar - N=N - Ar - R $\longrightarrow$ 2 x R - Ar - NH$_2$

KOH / iPrOH (15mol%)

100°C / 15h ——Scheme-5b

The procedure is general and the selectivity of the catalyst has been demonstrated by applying it to a series of structurally diverse NAs and azo compounds (11).

Fe_3O_4-NPs on bio-mass derived activated carbon are highly active and chemoselective catalysts for the CTH of NAs (RNO_2; R = H, OH, NH_2, CH_3, and COOH) with IPA as a hydrogen donor as well as solvent and KOH as a promoter. The reaction system not only exhibit excellent activity with high AN yields but also represents a green and durable catalytic process, which facilitates facile operation, easy separation and the catalyst recyclability (12).

Metal or metal salts have an important role in the catalytic reductions of organic functional groups. SmI_2 constitutes an excellent complement to these methods as it combines high reactivity with a high chemoselectivity (Scheme-6).

NO_2 SmI_2 / IPA / THF - H_2O NH_2

RT / Yield = 90% ——Scheme-6

The reduction of NB and other substituted NAs with a combination of SmI_2, isopropylamine and water also gave high yields of corresponding ANs (86-90%) (13).

5A.01.13: Glycerol: In general, Glycerol is a promising green solvent and reducing agent for metal-catalyzed transfer hydrogenation reactions. For example, CTH of NAs in glycerol using magnetic Fe_3O_4-Ni -NPs. gives corresponding ANs in high yields (Scheme-7).

NO_2 Fe_3O_4 - Ni MNPs(8.85mol% Ni) NH_2

KOH (2Eqv)

R ——   Glycerol / 80°C / 2.5 - 4h   —— R

Yields = 84 - 94%

R = H, 2CH$_3$, 3CH$_3$, 2OMe, 3OMe, 4OMe, 2Cl, 3Cl, 3Br, 4Br

4F, 2,5Cl$_2$, 4OH, 2NH$_2$, 3NO$_2$, 4CN, 4COOR, 4COCH$_3$ ——Scheme-7

There are many methods, but Raney-Ni and excess of NaOH (2.5 eqvs) delivered the desired ANs in moderate to good yields (44%–81%) after 24h at 100°C. The substrates containing electron withdrawing substituents were reduced more easily than those with electron-donating groups.

Much faster and chemoselective transformations (NO_2 and CO) were subsequently described under milder reaction conditions using Fe_3O_4-Ni MNPs/KOH in glycerol, which gave better results (84-94% yields) than Raney-Ni and NaOH. Particularly, it is also worthy of note that starting from 1, 3-dinitrobenzene, regioselective reduction of only one NO_2 group took place under the standard reaction conditions. Usually, the interactions of the nitro group to the catalyst surface compared to the carbonyl group are responsible for the chemoselectivity observed (14, 15). Garcia et al. have published a review on glycerol based solvents, their synthesis, properties and applications (16).

References:

1). M. Kumar et al. RSC Adv., 03:4894–4898(2013).

2). Y. Xiang, et al. *Applied Catalysis A: General,* 375(02):289-294(2010).

3). C. Yu , et al. *J. Organic Chemistry,* 66(03):919–924(2001).

4). V. P. Sivcev, et al. *Journal of Supercritical Fluids,* 86:137-144(2014).

5). E. Bosch & L. Jeffries, *Tetrahedron Letters,* 42(46): 8141-8142 (2001).

6). B. K. Banik, et al. *Synthetic Communications,* 30(20):3745-3754(2000).

7a).V. P. Sivcev, et al. *J. of Supercritical Fluids,* 61:115–118(2012).

7b). V. P. Sivcev, et al. *J. Fluorine Chemistry,* 04(03): 5 Pages (2014).

8). T. M. Jyothi, B. S. Rao, *Indian J. Chemistry,* 39A:1041-1043(2000).

9). S. K. Mohapatra, S. Sonavane, et al. *Tetrahedron Letters,* 43:8527-8529(2002).

10). T. Subramanian et al. *ChemCat Chem,* 4(12):1917-1921(2012).

11). R. V. Jagadeesh, *Chemistry A European J.,* 17(51):14375-14379 (2011).

12). P. V. Kumar et al. *ACS Sustainable Chem & Engg,* 04(12):6772-6782(2016).

13). T. Anker et al. *Tetrahedron Letters* 48:5707–5710(2007).

14). M. B. Gawande, et al. *Chemistry A European J..,* 18: 12628-12632(2012).

15). A. E. Díaz-Álvarez & V. Cadierno, *Applied. Science,* 3: 55-69 (2013).

16). J. I. García et al. *Green Chemistry,* 12: 426-434(2010).

Chapter-5B: Boranes and Borohydrides

5B.00.00: Introduction:

As with other hydrogen sources, the examples of CTH by boranes also will be seen under different headings throughout this text book. Here are some of the selected examples, which highlight the essence of boranes and borohydrides in the chemoselective reductions of NAs. Hydrogen has a great potential as energy carrier, particularly in fuel cell applications. However, it has serious storage problems. Therefore, among the various possibilities the boranes and metal borohydrides as KBH_4, $NaBH_4$ have high gravimetric/volumetric hydrogen storage capacities and hence they are considered as potential hydrogen storage materials (1, 2).

5B.01.00: Boron, Borane, Diborane:

Boron with symbol-"B" is an element occupying the fifth position in the periodic table (Atomic. No. = 5 and Atomic Wt. =10.81). It falls under the non-alkali elements as nitrogen, sulfur and halogens. Boron was discovered by Lussac and Thenard in 1808 and at the same time Sir Humphry Davy in London is also credited with Boron discovery. Boron has many uses in human life as it is an important and often underutilized trace mineral element. It is naturally present in certain foods and also within the environment. Boron uses include the ability to help keep the skeletal structure strong by adding to bone density, preventing osteoporosis, treating conditions like arthritis, and improving strength and muscle mass. It also helps in body building, improving brain functions and treating infections etc. https://draxe.com/boron-uses/.

5B.01.10: Borane:

Borane (trihydridoboron or borine) is an inorganic compound with the chemical formula BH_3 and is the first member of the borane family of 13 compounds with a general chemical formula as B_xH_y. Borane is only known to exist as a transient intermediate since it dimerises to form diborane, B_2H_6. The most important boranes are diborane B_2H_6 and two of its pyrolysis products, pentaborane B_5H_9 and decaborane $B_{10}H_{14}$. Borane chemistry has led to new experimental techniques and theoretical concepts. Boron hydrides have been studied as potential fuels for rockets and automobiles. https://en.wikipedia.org/wiki/Boranes.

Boron exists in the form of borax, boric acid, borates etc. It is a colorless gas and an explosive type of material (3). It is believed to be a reaction product during pyrolysis of diborane while making higher boranes as pentaborane and decaborane. Boron is routinely used in chiral reductions, Suziki-Miyaura cross coupling, and selective reduction of many functional groups in organic compounds. The borane (BH_3) is an electron deficient molecule in nature which makes it to attach to electron rich sites in a functional group thereby giving selective and specific reduction of that particular group. Borane forms stable boron-ligand complexes, therefore borane in THF as BH_3-THF complex is the choice of borane source at commercial scale. Recently borane chemistry has led to the development of number of pharmaceuticals and industrial chemicals with high selectivity and excellent yields of the desired products. About ten years ago Matos et al had reviewed such development of boron reagents in process chemistry as excellent tools for selectivity (4).

5B.01.11: Diborane: B_2H_6:

Diborane is made up of boron and hydrogen and is dimer of borane BH_3 with the formula B_2H_6. It is a colorless, highly unstable and pyrophoric gas at room temperature and forms explosive mixtures with air. Diborane is known since twenties in the last century and is characterized through electronic structure as diamagnetic in nature (5). In 1936, diborane, was a rare substance, prepared in less than gram quantities in only two laboratories, that of Alfred Stock at Karlsruhe, Germany, and of H. I. Schlesinger, at the University of Chicago. The existence of the simplest hydrogen compound of boron, not as BH_3, but as B_2H_6, was considered to constitute a serious problem for the electronic theory of G. N. Lewis. "FROM LITTLE ACORNS TO TALL OAKS, FROM BORANES THROUGH ORGANOBORANES", Nobel Lecture, 8 December, 1979, by Herbert C. Brown. http://www.nobelprize.org/nobel_prizes/chemistry/laureates/1979/brown-lecture.pdf.

5B.01.12: Diborane Applications in Chemistry:

Diborane is a key boron compound with a variety of applications, especially extensively used in the reductions of multiple functional groups in organic compounds. Diborane is manufactured by reacting boron halides (BF_3, BCl_3) with metal hydrides as LiH, NaH etc. But much safer and cheaper method is the reaction between BF_3 and $NaBH_4$ or $NaBH_4$ (SBH) and iodine.

$$3BF_3 + 4NaBH_4 = 2 B_2H_6 + 3NaBF_4; 2NaBH_4 + I_2 = 2NaI + B_2H_6 + H_2$$

Diborane reacts with alkenes, alkynes, water, methanol, HCl etc. With water it forms boric acid and hydrogen, while with methanol it forms trimethoxyborate ester and hydrogen. $B_2H_6 + 6H_2O = 2B(OH)_3 + 6H_2$; $B_2H_6 + 6MeOH = 2B(OMe)_3 + 6H_2$

One of the boron chemistry pioneer, H. C. Brown et al have used diborane, since late thirties in the last century, for reducing multiple functional groups in organic molecules including NAs (6). Reduction of organic compounds with many functional groups using diborane is known for a long time and is extensively reviewed (7).

5B.02.00: Amine-Boranes:

Amine boranes are widely used as selective reducing agents or hydrogen sources in organic chemistry. For example, Amine boranes are used in deracemization of dl- amino acids in high yield and quantitative selectivity (yields ~ 90% and ee. >99%) and are commonly preferred over $NaBH_4$ and $NaBH_3CN$ (8). Some of the most prominent members of this group are as discussed below.

5B.02.10: Ammonia Borane: NH_3BH_3 (AB):

Ammonia borane (NH_3BH_3, AB) has a hydrogen capacity of 19.6wt%, exceeding that of gasoline and making itself an attractive candidate for hydrogen storage applications in fuel-cell. Because of certain limitations for $LiBH_4$ and $LiAlBH_4$ as hydrogen sources, since the middle of last century AB emerged as an alternative to these borohydrides owing to its higher gravimetric hydrogen density (10.8wt% H).

$$H_3NBH_3 + H_2O = NH_4BO_2.2H_2O + 3H_2,$$

It showed slightly better net storage capacities in specific conditions (7.8wt% H) than $NaBH_4$ (7.3wt% H). Sutton et al. have demonstrated that after dehydrogenation, the polyborazylene (PB) can be converted back to AB nearly quantitatively (9). As per the DOE (2015) report, the on-board hydrogen storage system should have at least 9 wt% H. This situation makes AB with 19.6 wt% H as a potential candidate for hydrogen storage materials. Besides AB, many other amine-boranes have been prepared from methyl amine, dimethyl amine, trimethyl amine, ethylene diamine, pyridine and $LiBH_4$ and $NaBH_4$.

Borane tert-butylamine ($^tBuNH_2$-BH_3)

Borane trimethylamine (Me_3N-BH_3)

Borane isopropylamine ($^iPrNH_2$-BH_3)

Although, AB is recognized as hydrogen storage material, it usually requires an efficient catalyst for its dehydrogenation to release hydrogen needed for fuel and also in chemical reductions. Therefore, the development of just not only efficient but also economical catalysts to further improve the kinetic properties under moderate conditions is important for the applications of this commercially important system. In the past few years, a number of highly active M-NPs catalysts have been developed for AB dehydrogenation (10). Such M-NPs are easy to handle and separate. A very simple synthesis of bimetallic Pd–Pt–Fe_3O_4 nanoflake-shaped alloy NPs is used for dehydrogenation of AB followed by the reduction of NAs to ANs in methanol at ambient temperature with > 99% yields (11). Yan et al. have prepared and used Fe-NPs for hydrolytic dehydrogenation of AB (12). A variety of NAs were reduced selectively into ANs via G-$Ni_{30}Pd_{70}$ catalyzed tandem dehydrogenation of AB with conversion yields of up to 100% (13). Chemical storage materials based on their properties such as low molecular weight and high gravimetric hydrogen density are highly promising as hydrogen sources. Amongst the list of materials in this category, recently ammonia borane (AB, NH_3BH_3) and hydrazine borane (HB, $N_2H_4BH_3$) have attracted much attention. AB hydrolysis releases 3 moles of hydrogen per mole of AB at room temperature. $NH_3BH_3 + 2H_2O = NH_4BO_2 + 3H_2$

Copper has attracted much attention for being as reactive as and much more cost effective than noble metals (14). Yang et al. have reviewed the research progresses of the Cu based nanocatalysts in dehydrogenation of AB for hydrogen generation (15). Both AB and HB having hydrogen contents as high as 19.6wt% and 15.4wt%, from which 7.8wt% and 12.2wt% hydrogen of the starting materials respectively show their high potential for chemical hydrogen storage materials(16). Recently, a new approach has been developed using monodispersed Ni-Pd-NPs (3.4nm) on graphene and used for the tandem dehydrogenation of AB and hydrogenation of NAs (Scheme-1).

Conversion Yields = 100%   --Scheme-1

G-$Ni_{30}Pd_{70}$ catalyzed tandem reactions in 5–30 min with the conversion yields reaching up to 100% in methanol at room temperature are performed (13, 17). Au-NPs/TiO_2 even at a ppm loading level catalyze the quantitative reduction of NAs to ANs by AB complex, without dehalogenation of HANs (Cl, Br) while keeping the other groups intact.

--Scheme-2

The heterogeneous catalysis of NA reductions by supported Au-NPs is explored with the AB as a reductant for the first time (Scheme-2). The authors have used Au-NPs/TiO_2 system for reduction of a variety of NAs (H, OH, CN, -CH=CH_2, BnO, MeO, Cl, Br, etc.) using AB in ethanol at 25°C. The isolated yields of the corresponding anilines were between 84-95%. Except 3-nitrostyrene all other conversion yields were 100%. About 10% 3-ethylAN was obtained from over-reduction of 3-NS (18).

By now we have realized that many groups in the world have been working on hydrogen storage materials and there is a plethora of publications in the literature about amine-boranes (AB, MAB, DMAB etc.). The newly published reports have the following common points in their studies: i) the method of preparation of the catalyst, ii) characterization of the catalyst by X-ray, NMR, etc. iii) the solid surface structure, iv) the size of nanoparticles and v) the factors responsible for the activity of the catalyst, yields and the selectivity of the desired products. The kinetics and the reaction mechanisms using DFT calculations and other instrumental studies are looked at in detail with potential for hydrogen generation and its applications at industrial scale. Most of the metal catalysts are in the form of NPs as such or on solid supports as SiO_2, TiO_2, Carbon, Graphene, Charcoal, etc. So far, we have been citing the relevant references after highlighting the results in brief. However, keeping the above aspects in mind, henceforth we will just list the titles and the corresponding references instead of describing individual findings in detail, so that the readers can visit the references of their interests for further details.

5B.02.11: Metal-NPs on Solid Support:

Recently, some other metal catalysts as Ru or Ru/TiO_2, Ru-Pd, $Zn-Fe/TiO_2$ nanofibres/nanorods, grapheme (G) or reduced grapheme oxide (rGO) have been used for hydrolysis of AB under moderate conditions. The most recent developments along with their titles and relevant references are given below:

A-(i)). N. Kous *Synth & Reactivity in Inorg, Metal-Org, & Nano-Metal Chem*, 46(04):534-542(2016). A(ii). M. Rakap, et al. *i). Intl J. Green Energy*, 12(02):1288-1300(2015). ii). Barakat, *Intl J. Green Energy*, 13(07):642-649(2016).

B). Fe-Pd (3.5nm)-NPs on rGO for AB (ANs Yields > 99%). i). O. Metin, et al. *Catalysis Science & Technology*, 6(15):6137-6143(2016). ii). O. Metin, et al. *J Colloid Interface Science*, 498:378-386(2017).

C). Metal Ions (Co (2+), Ni (2+), Cu (2+)) assisted AB hydrolysis. S. B. Kalidindi, et al. *Inorganic Chemistry*, 47(16):7424-7429(2008).

D). Hollow Ru-NPs from Ni@Ru Core@shell for AB hydrolysis. G. Chen, et al. *Chemical Communications* (Camb), 48(64):8009-8011(2012).

E). Synthesis of triple-layered Ag@Co@Ni core-shell NPs for the dehydrogenation of AB. F Qiu et al. *Chemistry*, 20 (02):505-509(2013)

F). Size-controlled Ag@Co@Ni/Graphene core-shell NPs for hydrolysis of AB. F Qiu et al. *Chemistry An Asian J*, 09(02):487-493(2013).

G). G-supported metal/metaloxide nanohybrids in catalysis. Y. Cheng, et al., *Catalysis Science & Technology*, 05:3903-3916(2015).

H). Tetrametallic Ag@CoNiFe Core–Shell-NPs on Graphene for AB Hydrolysis. L. Yang, et al. *ChemCatChem,* 06(06):1617-1625(2014).

I). Graphene-Supported Trimetallic Core–Shell Cu@CoNi-NPs for AB Hydrolysis.: X. Meng, et al. *ChemPlusChem*, 79(02):325–332(2014).

J). MIL-101- Ni-Ru-Core Shell Alloy NPs for Dehydrogenation of AB. S. Roy, et al. *European J Inorganic Chemistry*, (27):4353–4357(2016).

K). Pt/CeO$_2$ Hybrids on rGO Catalyst in Reverse Micelles for Hydrolysis of AB. Q. Yao, Y. Shi, *Chemistry An Asian J.*, 11(22):3251–3257(2016).

L). Nanoporous Pt/Ru Alloys for Hydrolytic Dehydrogenation of AB. Q. Zhou, C. Xu, *Chemistry An Asian J.*, 11(05):705–712(2016).

M). Nanocatalysts for hydrogen generation from AB and HB: Z. Lu, et al. *J. Nanomaterials*, Vol.2014 ArticleID 729029, 11pages (2014).

5B.02.12: Carbon/Mono- and DiCarbenes as Solid Support:

N). AB dehydrogenation by Ni monocarbene and dicarbene catalysts. P. M. Zimmerman et al. *Inorganic Chemistry*, 48(12):5418-5433(2009).

O). Catalytic hydrolysis of AB via Co-Pd-NPs as catalyst. D. Sun et al. *ACS Nano*, 5(8):6458-6464(2011).

P). Ru-M/C (M =Co, Ni, Fe) Core–Shell NPs for Dehydrogenation of AB. N. Cao, et al. *Chemistry An Asian J*, 09(02):562–571(2014).

Q). Ru/SBA-15: Catalyst for dehydrogenation of AB and HB. Q Yao, et al. *Scientific Reports*: 05:15186(2015).

R). Synthesis of Pt- NPs (M = Fe, Co, Ni) for Dehydrogenation of AB. S. Wang et al. *ACS Applied Mater Interfaces*, 06(15):12429-12435(2014).

S). Bimetallic Au-Ni-NPs in SiO$_2$ Nanospheres for Dehydrogenation of AB. H. L. Jiang et al. *Chemistry*, 16 (10), 3132-3137(2010).

T). Synthesis of Water/Air-Stable Ni-NPs for Hydrolysis of AB. J. M. Yan et al. *Inorganic Chemistry*, 48(15):7389-7393(2009).

U). Co-Co$_2$B, Ni-Ni$_3$B and Co-Ni-B NCs for AB Methanolysis. S. B. Kalidindi, et al. *Physical Chemistry & Chemical Physics*, 11(05):770-775(2008).

V). PVP-PD-NCs for AB Methanolysis. S. K. Kim et al. *J. American Chemical Society*, 132(29):9954-9955(2010).

W). Ru(cod)(cot) Precatalyst Dehydrogenation of AB at Room Temperature. M. Zahmakiran, et al. *Langmuir*, 28(11):4908-4914(2012).

X). Au–Co@CN nanocatalyst: a 3x more active for photohydrolysis of aqueous AB. L. Guo, et al. *ACS Catalysis*, 05(01):388–392(2015).

Considering the growing importance of amine boranes as the energy storage materials, safer and efficient synthetic methods for their practical synthesis and productions are vital to chemists. The recent discovery by researchers at Purdue University could lead to a safer and cheaper production of amine-boranes. A non-dissociative open-flask with AB for ready synthesis of ammonia-trialkylboranes and aminodialkylboranes is reported by P. V. Ramachandran: chandran@purdue.edu.

https://www.purdue.edu/newsroom/releases/2016/Q3/purdue-researchers-discovery-could-lead-to-safer,-cheaper-production-of-amine-boranes.html.

5B.02.13: MOF Motif for Dehydrogenation of AB and MAB:

Metal Organic Framework (MOF) is a relatively new concpt of using metal catalysts with enhanced acivities. Bimetallic Core Shell NPs (as Pd@Co Core-Shell NPs or Au@Ag Core-Shell NPs) either confined inside MOF or immobilized on or stabilized by MOF have gained interest in the recent past as emerging MOF materials for their synergistic effects in catalysis. MOFs are porous crystalline materials constructed from metal ions or clusters and multidentate organic ligands and can be categorized by the type of active centers as follows: (i) open metal centers and functional organic linkers in the MOF structure, (ii) active guest sites in the pores and active sites in the MOF structure, and (iii) bimetallic NPs on MOF supports (19-23). AB Confined by a MOF for chemical hydrogen storage is synthesized for the first time at 78°C and used as a hydrogen storage and size-selective reduction material in organic synthesis (24). Highly dispersed bimetallic Rh–Ni-NPs (1.1 ± 0.2 nm) have been successfully anchored on MOFs (ZIF-8) by using a simple liquid impregnation method. Among all the tested catalysts with different compositions, $Rh_{15}Ni_{85}$@ZIF-8 exhibits the highest catalytic activity toward hydrolysis of AB, with high turnover frequency and 100% hydrogen selectivity at room temperature (25, 26). Hydrogen generation from the hydrolysis of AB using non-noble metals as Co-Ni-NPs, Cu-Co-NPs, Ni-Core Shell NPs and noble-metals as Pd-NPs, and Ag@CO for dehydrogenation of AB and MAB are listed in chapter-5 (Refs: 96-118) in the book, "Hydrogen Production Technologies", by M.Sankir, et al. John Wiley & Sons, 2017 - Science - 656 pages, PP 229-233.

5B.02.14: Methyl Amino Borane (MeNH$_2$BH$_3$: MAB):

There is a demand for a sufficient and sustainable energy supply. Lithium borohydride (LBH), sodium borohydride (SBH), ammonia borane (AB), hydrazine hydrate (HH), and formic acid (FA) have been extensively investigated as promising hydrogen storage materials based on their relatively high hydrogen contents. The area of dehydrogenation of hydrogen storage materials and the use of liberated hydrogen for the selective transfer hydrogenation of functional groups in organic chemistry is of great interest in the current times. The following examples highlight the new developments in this area, using MAB as a hydrogen source.

Ag@Co/G-NPs: Well-dispersed magnetically recyclable core-shell Ag@M (M = Co, Ni, Fe)-NPs supported on graphene (G) have been synthesized via a facile in-situ one-step procedure using MAB and AB as reducing agents under ambient conditions. Although the Ag@Fe/G-NPs are almost inactive towards the hydrolysis of AB, the Ag@Co/G-NPs are the most reactive catalysts, followed by Ag@Ni/G-NPs. Compared with AB and SBH, the as-synthesized Ag@Co/G and the catalyst on graphene support with MAB as a hydrogen source exert the highest catalytic activity. The following examples are about metal-nanoparticles (M-NPs) on solid supports.

a). MNPs on SS: L. Yang, et al. *ACS Appl. Mater Interfaces*, 05(16):8231-8240(2013).

b). Ru@Co/C-NPs and carbon black on SiO_2, Al_2O_3 etc. give the highest activities. N. Cao, et al., *Chemistry Asian J*, 09 (2):562-571(2013).

c). Ni@Ru Core-Shell-NPs (i) and Ru(0)/CeO_2, (ii) Ru(0)@MWCNTs, for MAB dehydrogenations.: i). G. Chen, et al., *Chemistry*, 18(25):7925-7930(2012). ii). S. Akbayrak, et al. *Dalton Transanctions*, 45(27):10969-10978(2016). iii). S. Akbayrak, et al., *ACS Applied Materials & Interfaces*, 04(11):6302-6310(2012).

d). Fe(0)-NPs Embedded on BN-Polymer: S. Duman, et al. *J Nanoscience & Nanotechnology*, 13(07):4954-4961(2013).

e). RGO/Pd for dehydrogenation: P. Xi, et al. *Nanoscale*, 04(18):5597-5601(2012).

f). Catalytic H_2 generation under mild conditions. i). H. L. Jiang, et al. *ChemSusChem*, 03 (05):541-549(2010). ii). U. Sanyal, et al. *ChemSusChem*, 4(12):1731-1739(2011).

5B.02.15: Dimethyl Amine – Borane (Me$_2$NHBH$_3$: DMAB):

DMAB is made by reacting dimethylamine hydrochloride salt with $LiBH_4$ or $NaBH_4$:
$$Me_2NH.HCl + NaBH_4 = Me_2NH\text{-}BH_3 + NaCl + H_2$$

However, it seems DMAB is unstable and forms a dimer under experimental conditions. The use of photo-irradiation of a solution of DMAB ($R_2NH\text{-}BH_3$) containing a catalytic amount of metal carbonyl complex, $[M(CO)_6]$ (M = Cr, Mo, W), produces aminoborane dimers $[R_2N\text{-}BH_2]_2$ in high yields. Under similar conditions bulkier aminoboranes gave monomeric products, believed to proceed via an intra-molecular pathway and the active catalyst is $[Cr(CO_4)]$. The reaction follows a stepwise mechanism involving NH- and

BH- activation. Dehydrocoupling of borane-primary amine adducts RNH_2-BH_3. (R = Me, Et, t-Bu) gave borazine derivatives (27). Simple introduction of an *i*Pr group to the Cp ligand in titanocene alkyne complexes serves as a highly active catalyst for DMAB dehydrogenation (28). Mechanism of DMAB dehydrogenation catalyzed by an Ir(III) PCP-pincer complex is reported (29). Transfer hydrogenation of NAs to ANs by Pd-NPs *via* dehydrogenation of DMAB complex is investigated using a simple and highly efficient protocol (Scheme-3) (30).

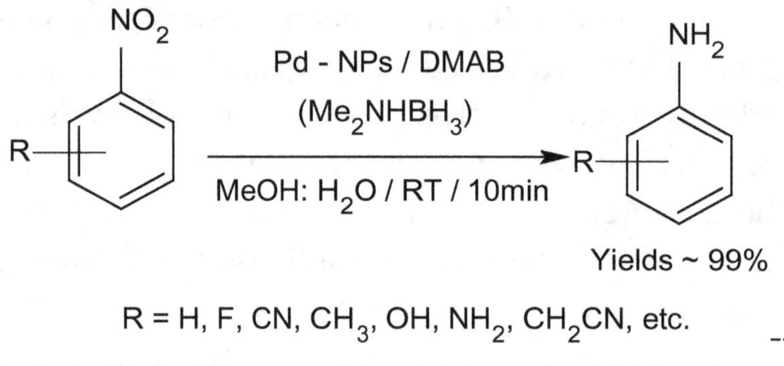

--Scheme-3

Novel Ru(0)-nanocatalyst and Rh(0)-nanocatalyst for the hydrolytic dehydrogenation of DMAB in water at room temperature have been developed (31).

5B.02.16: Trimethylamine-Borane (Me₃N-BH₃: TMAB):

Hydrolytically stable aminoboranes can be activated in-situ through Pd catalysis. Hence, Pd/TMAB is an efficient hydrogen-transfer reagent (Scheme-4) for reductions of NAs to ANs in high yields (90—>99%)

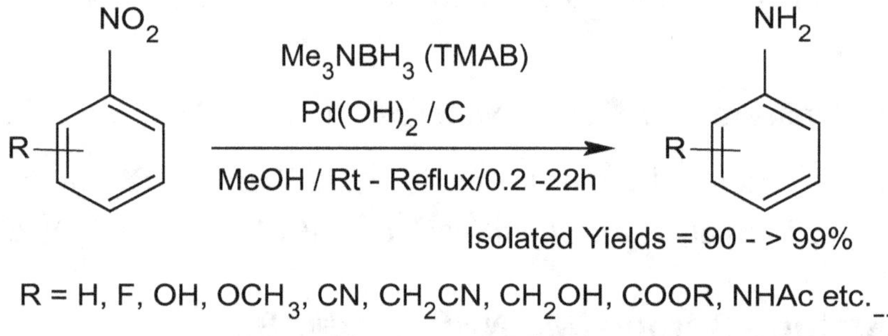

--Scheme-4

Likewise, the Pd catalyzed methanolysis/decomplexation of stable borane–amine adducts is accelerated by the action of NB (32).

5B.02.17: Ethylene Diamine Bisborane; (H₃BNH₂(CH₂)₂NH₂BH₃: EDAB):

Ethylene diamine (EDA) itself has found applications in selectively reducing NAs to the corresponding symmetrical azo compounds (33). However, there are no other reports in public domain about the reductions of NAs to ANs using EDA as a hydrogen source. EDAB is reported as a suitable hydrogen source for selective transfer hydrogenation of C=C and carbonyl compounds by M-NPs. http://shodhganga.inflibnet.ac.in/bitstream/10603/176/2/11_chapter4.pdf.

Salher et al. have investigated the liberation of H_2 from EDAB in ionic liquids (ILs) media as a potential hydrogen storage material. The correlation between polarity of the ILs and hydrogen yield was investigated and the suitability for hydrogen storage systems is evaluated and discussed (34). Recently, EDAB is synthesized and characterized as a hypergolic hybrid rocket fuel additive (35).

5B.02.18: Hydrazine Borane (N_2H_4-BH_3: HB) and Hydrazine Bisborane (N_2H_4-$(BH_3)_2$: HBB):

Hydrazine borane (HB, 15.4wt% H_2)) was first reported by Goubeau and Ricker in 1961. HB is one of the most recent boron- and nitrogen-based materials in the field of hydrogen storage materials. HB releases hydrogen as per the following equation:

$$N_2H_4BH_3 + 2H_2O = N_2H_5BO_2 + 3H_2$$

AB (NH_3BH_3) and HB ($N_2H_4BH_3$), having hydrogen content as high as 16-19.6wt% and 15.4wt%, which can ideally release 7.8wt% and 12.2wt% hydrogen of the starting materials respectively. Since 2009 HB and its alkali derivatives ($MN_2H_3BH_3$ with M = Li, Na and K), have positioned themselves as being potential candidates for chemical hydrogen storage. The thermolytic dehydrogenation of HB should be avoided for certain risks, but its alkali metal derivatives can be used with relative safety. Moury and Lu et al. have independently reviewed these aspects in detail. Further, Moury et al have performed HB hydrolysis in water under mild conditions in the presence of a Ni-Pt nanocatalyst (36). Although HB is an important source of hydrogen storage material it needs a promoter to enhance its activity. By adding an equimolar amount of lithium hydride (LiH) as a promoter, it showed excellent hydrogen release rates at a reasonable temperature range of 100–150°C. The resulting mixture contains 14.8wt% H_2. It is observed that between HB and the HB+LiH mixture, the latter is a better hydrogen releasing material (37). A bimetallic catalyst Ni-Ag has given the best (100%) performance in HB dehydrogenation (38).

Hydrazine bisborane (HBB:$N_2H_4(BH_3)_2$ is the latest hydrogen-rich B–N–H compound (16.8wt% H_2) as a potential chemical hydrogen storage material. In contrast to early reports stating the hazardous properties, the present work is focused on fundamentals to provide a complete data sheet on the material and shed light on the real potential of HBB both for solid and liquid-state hydrogen storage systems (39, 40). A long list of publications about this topic is given at the web site: http://pubs.rsc.org/-/content/forwardlinks?doi=10.1039%2Fc2ee21508j

The high pressure behavior of another hydrogen storage material (41), "Guanidinium Borohydride, (GBH) $[C(NH_2)_3]^+[BH_4]$" is also reported. Another hydrogen storage material, aminoimidazole borohydride (Im-NH$_2$BH$_4$) is synthesized *via* the reaction between 2-aminoimidazole hemisulfate ((Im-NH$_2$)$_2$SO$_4$) and NaBH$_4$ by a simple ball milling method. This compound holds theoretical hydrogen capacity of 8.1wt%H. Thermal dehydrogenation analyses demonstrate that it can release about 3.2wt% hydrogen below 320°C, giving C$_3$N$_3$H$_3$BH as a product (42).

5B.02.19: Nano Catalysts in H$_2$ Generation from HB:

Considering the importance of hydrogen storage materials as energy sources in the future, many groups have been focusing on enhancing the dehydrogenation processes of such potential hydrogen storage materials. The following examples highlight their results of using metal based catalysts for this purpose, especially using PVP-stabilized Cu–NPs (43) and Ni-NPs for HB (44a) and aq. HB (44b). Hydrogen Generation from AB, HB, HBB, is reported by different groups as given in references (19-23) below.

5B.02.20: Polyaminoboranes:

The development of effective catalysts and processes for dehydrocoupling and dehydrogenation of AB under mild conditions remains a challenge in the field of hydrogen economy and material science. Over the past decade, metal catalyzed dehydrogenation/ dehydrocoupling reactions for releasing H$_2$ from AB have gained significant momentum (45). Ru-Nanocatalyst (Ru(cod)(cot) precatalyst (cod = 1,5-η(2)-cyclooctadiene; cot = 1,3,5-η(3)-cyclooctatriene) is generated in-situ and used in dehydrogenation of MAB and AB in THF at 25°C in the absence of any stabilizing agent. With the help of previous reports (46) it is revealed that in situ formed Ru(n) clusters (not Ru(0)(n)-NPs) are kinetically dominant catalytically active species in this catalytic system. The resulting Ru catalyst generates more than 1.0 equivalent of H$_2$ at the complete conversion of AB to polyaminoborane (PAB; [NH$_2$BH$_2$]$_n$ and polyborazylene (PB; [NHBH]$_n$ units at room temperature. These high molecular weight polyaminoboranes have been characterized and used in the dehydrogenation and hydrogen generation processes (47). In another case, the catalytic dehydrocoupling/dehydrogenation of MAB to yield the soluble, high molecular weight poly(N-methylaminoborane)[MeNH-BH$_2$]$_n$ (MW >20,000), at 20°C using Brookhart's Ir(III) pincer complex as a catalyst is reported. The analogous reaction with AB gave an insoluble product, [NH$_2$-BH$_2$]$_n$, but copolymerization with MAB gave soluble random copolymers, [MeNH-BH$_2$]$_n$-r-[NH$_2$-BH$_2$]$_m$ The structures of polyaminoborane and copolymers were analyzed and characterized by analytical tools. Basically, it is a two stage polymerization mechanism involving: i). the Ir-catalyst dehydrogenates the starting material to afford the monomer MeNH=BH$_2$ and, ii) the same catalyst effects the subsequent polymerization of these species. A wide range of other catalysts based on Ru, Rh, and Pd were also found to be effective in the preparation of polyaminoborane. For example, polyaminoborane was obtained from the initial stage of the dehydrocoupling/

dehydrogenation of the starting materials with $[Rh(\mu\text{-}Cl)(1,5\text{-}cod)]_2$ as the catalyst at 20°C. The catalyst is reported to give the N,N,N-trimethyl borazine, $[Me_3N\text{-}BH_3)$, under different conditions (dimethoxyethane, 45°C). The ability to use a variety of catalysts to prepare polyaminoboranes suggests that the synthetic strategy should be applicable to a broad range of amine-borane precursors and is a promising development for this new class of inorganic polymers as polyaminoboranes and borazines (48). Amine–borane dehydrogenation chemistry: Here, the metal-free hydrogen transfer using new catalysts, the mechanisms and the synthesis of polyaminoboranes is reported (49). Bhunya et al. have unfolded the crucial role of a nucleophile in Ziegler–Natta type Ir catalyzed polyaminoborane formation. DFT investigations reveal that a transition metal bound amino-borane unit is responsible for chain initiation and chain growth in polyaminoborane formation catalyzed by Brookhart's iridium pincer catalyst. This mechanistic manifold reported here is also implicated in other transition metal catalyzed polyaminoborane formation processes (50). Amine-boranes undergo efficient catalytic dehydrocoupling/ dehydrogenation using transition metal or main group precatalysts to produce various classes of products including aminoboranes, borazanes, borazines, and most recently the first examples of soluble high molecular weight polyaminoboranes. Metal-free hydrogen exchanges are also possible between amine-boranes and aminoboranes under mild conditions. Stubb et al. have recently reviewed this matter in detail (51). The crucial role of Metal-Free Catalysis in Borazine and Polyborazylene formation in transition-metal-catalyzed AB dehydrogenation is unravelled (52).

References:

1a). A. Züttel, et al. *Nature*, 414(6861):353–358(2001).

1b). R. Moury, et al. *International J. Hydrogen Energy:* 37(21)15938-15991(2012).

2). Hydrogen, Fuel Cell, Electrochemical and Experimental Technologies: 2013 Edition, Chapter-3: 2013 - Tech & Engg - 1205 pages, PP489-490.

3). W. H. Bauer et al. *BORAX TO BORANES, Advances in Chemistry,* Vol. 32, Chapter-13, 115–126 (1961).

4). K. Matos, et al. *Chemical Reviews,* 106(7):2617–2650(2006).

5a). H. I. Schlesinger, et al. *J. Am Chemical Society:* 53:4321(1931).

5b). R. S. Mulliken, *Journal of Chemical Physics:* 3:635 (1935).

6a). H. C. Brown, et al. *J. Am. Chemical Society:* 61:673-680(1939).

6b). H. C. Brown, et al. *J. Am Chemical Society,* 81(24):6428–6434(1959).

6c). H.C. Brown, et al. *J. Am Chemical Society,* 92(6):1637–1646(1970).

6d). H. C. Brown, et al. *J. Am Chemical Soc.,* 92 (24):7161–7167(1970).

6e). H. C. Brown, et al. *J. Organic Chemistry*, 37(19):2942–2950(1972).

7a). C. F. Lane, *Chemical Reviews*, 76 (06):773-799(1976).

7b). S. P. Cummings, et al. *J. Am. Chem. Society* 138:6107-6110(2016).

8). F. R. Alexandre, et al. *Tetrahedron Letters*, 43:707–710(2002).

9). A. D. Sutto, A. K. Burrel, et al. *Science*, 331: 1426-1429(2011).

10a). W. W. Zhan, et al. *ACS Catalysis*, 06(10):6892–6905(2016). 10b). P. Liu et al. *ACS Applied Materials & Interfaces*, 09(12):10759–10767(2017).

11). B. M. Kim, et al. *ACS Applied Material & Interfaces*, 08 (23):14637-14647(2016).

12). J. Yan, et al. *Angewandte Chemie Intl Edition*. 47:2287-2289(2008).

13). S. Sun et al. US 20160279619 A1 (2016).

14a). I. Nakamula, et al. *Tetrahedron Letters*: 42:2285-2288(2001).

14b). H. Nishihara, et al. *Chemical Communications*: 5716-5718(2008).

15a). Y. W. Yang, et al. *Materials Technology*, 30(02): Volume 30, 2015: Part-A2: Nanoscience, A89-A93 (2015). 15b). Q. Li et al. *Integrated Ferroelectrics, An Intl Journal*, 170(01):73-82(2016).

16). Z. Lu, Q. Yao, et al. A Review by Hindawi Publishing Corporation, *Journal of Nanomaterials*, Volume 2014, Article ID 729029, 11 pages, (2014).

17). H. Goksu, S. F. Ho, et al. *ACS Catalysis*, 04(6):1777-1782(2014).

18). C. Gryparis, et al. *Advanced Synthesis & Catalysis*, 355:907-911(2013).

19). Y. B. Huang, et al. *Chemical Society Reviews*, 46(01):126-157(2017).

20a). Y. Chen, Q. Xu, *Nano Micro-SMALL*, 11(01):71-76(2015). 20b).H. L. Jiang et al. *J American Chemical Society*, 133(05):1304-1306(2011).

21). Y. Z. Chen, et al. *Small*, 11(01):71-76(2014).

22). B. Li et al. *Advanced Materials*, 28(40):8819-8860(2016).

23a). M. P. Suh, et al. *Chemical Reviews*, 2012, 112(02):782–835(2012).

23b). M. Zahmakiran, et al. *Langmuir*, 28(11):4908-4914(2012). 23c). N. Blaquiere, et al. *J. American Chemical Society*, 130(43):14034–14035(2008). 23d). X. Zhang, L. Kam, et al. *Accounts of Chemical Research.*, 50(1): 86–95(2017). 23e). *R. T. Baker, et al. J. American Chemical Society*, 134(12):5598–5609(2012).

24a). Z. Li, et al., *J. American Chemical Society*, 132 (05):1490–1491(2010).

24b). X. Wang, et al., *Chemical Communications*: 51:7610-7613(2015).

25). B. Xia, C. Lu, et al. *Intl. J. Hydrogen Energy*, 40(46):16391-16397(2015).

26a). i) W. Zhan, et al. *ACS Catalysis*, 06(10):6892–6905(2016).

26b). A. Rossin & M. Peruzzini, *Chemical Reviews*, 116(15):8848–8872(2016).

27). C. A. Jaska, et al. *Chemical Communications*, 962-963(2001).

28). Y Kawano et al. *J American Chemical Society*, 131 (41):14946-14957(2009).

29). T. Breweries, et al. *ChemCatChem*, 03(12):1865-1868(2011).

30). E. M. Titova, *ACS Catalysis*, 07(04):2325–2333(2017).

31). N. M. Patil, et al. *RSC Advances*, 05:86529-86535(2015).

32). M. Zahmakiran et al. *Dalton Transactions*, 41(16):4976-4984(2012).

33). M. Couturier, et al. *Tetrahedron Letters*, 42:2285–2288(2001).

34). T. F. Chung, et al. *J. Organic Chemistry*, 49 (7): 1215-1217(1984).

35a). S. Sahler, et al., *Intl. J. Hydrogen Energy*: 38:3283-3290(2013). 35b). S. Sahler, et al. *Molecules*: 20:17058-17069(2015).

36a). M. A. Pfeil, P. V. Ramachandran, et al. *J. Propulsion and Power*, 31(01): 365-372(2015). 36b). P. V. Ramachandran, P. D. Gagare, et al. i) WO2012006347-A2(2011), ii) US 9534002(2017).

37a). R. Moury et al. *Energies*: 8:3118-3141(2015). 11b). Ibid, *Intl. J. Hydrogen Energy*, 37(21):15983-15991(2012). 37c). Z. Lu, et al. *Scientific Reports*, No. 5, Article: 15186(2015). Doi:10:1038/srep15186

38). T. Hügle, et al. J. Am. Chem. Soc., 131(21):7444-7446(2009).

39a). J. F. Petit, et al: The 2013 World Congress: *Adv. In Nano, Biomechanics, Robotics & Energy Res*: (ANBRER-13), Seoul, South Korea, August 25-28 (2013). 39b). J. Hannauer, et al. *Energy & Environmental Sci*ence: 04:3355-3358(2011).

40a). S. Pylypko, et al. *Inorganic Chemistry*, 54 (09):4574–4583(2015). 40b). G. Qi, K. Wang, et al. *J. Physical Chemistry C*, 120(38):21293–21298 (2016).

41). G. Qi, et al. *J. Physical Chemistry. C*, 120(38):21293-21298(2016).

42). G. Qi, et al. *J. Physical Chemistry C*, 120 (25):13414-13420(2016).

43). Y. Wu, Y. Qi, et al. *RSC Advances*, 6(105):103299-103303(2016).

44). M. Yurderi, et al. *Applied Catalysis B: Environ*, 165:169-175(2015).

45a). D. Ozhava, et al. *Applied Catalysis B: Environ*, 162:573-582(2015).

45b). W. B. Aziza, et al. *Intl. J. Hydrogen Energy*, 39(30):16919-16926(2014).

46a). V. S. Nguyen et al., *J Physical Chemistry A*, 111(36): 8844-8856(2007).

46b). V. S. Nguyen, et al. *Physical Chem & Chem Physics,* 13(14):6649-6656(2011).

47). D. C. Zhong et al. *Chemical Communications,* 48(98):11945-11947(2012).

48). Z. Lu, Q. Yao et al. *J. Nanomaterials,* Vol.2014, ID 729029, 11 pages (2014),

49). S. Bhunya, et al. *ACS Catalysis,* 06(11):7907–7934(2016).

50). S. Bhunya, et al. *Chemical Communications,* 50(44): 5919-5922(2014).

51). N. E. Stubbs, et al. *J. Organometallic Chemistry,* 730: 84-89(2013).

52a). S. Bhunya et al. *ACS Catalysis,* 05(06):3478–3493(2015).

52b). A. Staubitz, et al. *J American Chemical Society,* 132 (38):13332-13345(2010).

52c). E. M. Titova, et al. *ACS Catalysis,* 07(04):2325–2333(2017).

52d). M. A. Esteruelas, et al. *Organometallics,* 36(12):2298–2307(2017).

5B.03.00: BOROHYDRIDES:

For over about 60 years metal borohydrides, including $Ca(BH_4)_2$, KBH_4, $NaBH_4$, have been used in the reduction of functional groups in organic molecules. Benzaldehyde to benzylalcohol (90%) (1a) and ethyl-*p*-nitrobenzoate to *p*-nitrobenzyl alcohol (87%) (1b) was performed using sodium borohydride (SBH, $NaBH_4$). Over the past two decades, the utility of SBH directly or in modified form has been greatly expanded. The combination systems of metal borohydrides (M=Na, K, Li) and borohydride exchange resin (BER) in the presence of metals or metal salts are highly effective for the reduction of NAs (2). Like some other boranes, SBH also is inactive or less active as a reducing agent, but its activity can be enhanced by the addition of promoters. For example, addition of iodine to SBH in THF provides H_3B–THF that is useful for hydroborations and reductions of various functional groups with enhanced selectivities. Metals as Sb, Bi, Se, etc. also act as a co-catalyst to the system enhancing the SBH activity and selectivity. Periasamy et al have summarized the various methods used for the enhancement of reactivity and selectivity of SBH in organic synthesis (3). NACs are chemoselectively reduced to the corresponding aryl amines effectively using the SBH-$CuSO_4$ system without affecting other functional groups. When NB treated with SBH/$CuSO_4$ in aqueous ethanol aniline was obtained in 94% yield.

5B.03.10: Boron & Boric Acid:

The hydrolysis of boron produces boron oxide/boric acid and hydrogen gas. It means boron itself can act as a hydrogen source needed in the transfer hydrogenation of many functional groups. The range of hydrogen production efficiency of 95–100% was obtained for all experiments as well as a full conversion of the boron particles to boron oxide, and the boron oxide to orthoboric acid (4). Boron is a promising element for hydrogen storage material with its chemical hydrides and nanostructural forms. It is also used as an additive in nickel metal hydride battery systems to enhance hydrogen compatibility and performance. A brief summary of hydrogen storage systems and technology in general and focus on possible uses of boron and its compounds is presented as a summary or a review (5).

5B.03.11: Sodium Borohydride (NaBH$_4$) as a Reductant:

Boron based materials, especially borohydrides and boranes have gained importance as chemical hydrogen storage materials. This is attributed to their low molecular weight and high gravimetric hydrogen density, typically up to 10^{e20}wt% H. Some of these most intensely investigated materials are LiBH$_4$ (18.4wt% H), NaBH$_4$ (10.8wt% H), and AB (19.5wt% H). The first two have certain limitations in hydrolytic and thermolytic dehydrogenations respectively. Actually, SBH was the first member of the group to be investigated for its hydrolysis (6) behavior (7.3wt% H).

$$NaBH_4 + 2H_2O = NaBO_2 \cdot H_2O + 4H_2.$$

Since its discovery in 1940, SBH as a hydrogen source and hydrogen storage material is reviewed in great detail (7). Hydrogen generation from SBH in boric acid-water mixture is reported. https://www.researchgate.net/publication/263610956

http://research.sabanciuniv.edu/1522/1/review_of_hydrogen.pdf

5B.03.12: NaBH$_4$ and KBH$_4$ with Metal Catalysts:

The reduction of NAs with hydrides (NaH, LiAlH$_4$) is known for a long time (8).

Borohydrides as LiBH$_4$, Ca(BH$_4$)$_2$, NaBH$_4$ etc. are also used as reducing agents. However, SBH being a weak reducing agent it requires a promoter, so that its activity and efficiency is enhanced by the addition of a catalytic amount of metal salts (9).

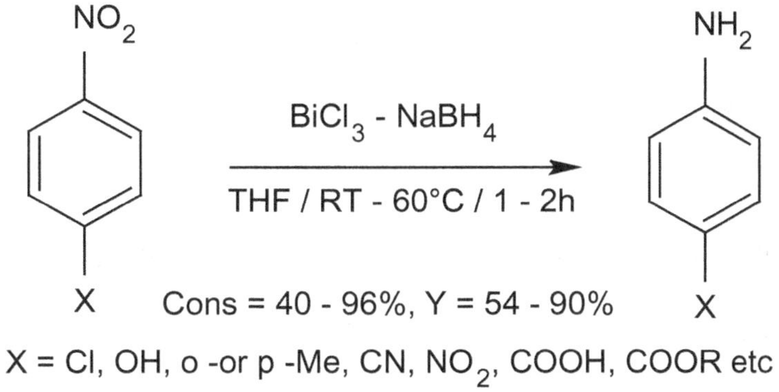

NO$_2$ → NH$_2$

BiCl$_3$ - NaBH$_4$

THF / RT - 60°C / 1 - 2h

X   Cons = 40 - 96%, Y = 54 - 90%   X

X = Cl, OH, o -or p -Me, CN, NO$_2$, COOH, COOR etc   --Scheme-5

Novel reduction systems for conversion of NAs to corresponding amines are developed in good to excellent conversions and yields by SBH in the presence of bismuth chloride (NaBH$_4$-BiCl$_3$) or antimony chloride (NaBH$_4$-SbCl$_3$) as promoters (Scheme-5) (10). Recyclable aluminium oxy-hydroxide supported Pd-NPs (0.5wt% Pd), Pd/AlO(OH)) as catalysts have been used for the selective hydrogenation of NACs via SBH hydrolysis in > 99% yields in short times (<1-13min) (Scheme-6).

$$NO_2 \xrightarrow[\text{NaBH}_4 / H_2O / RT / \sim 1 - 13\text{min}]{\text{Pd -NPs / AlO(OH)}} NH_2$$

High Selectivity & Yields = > 99%   --Scheme-6

This process can be assessed as an eco-friendly method involving both reusable catalysts (Pd/AlO(OH)-NPs) and a hydrogen source (SBH) (11).

Some other metal-catalysts, as listed below, have been used in the dehydrogenation of SBH (12) SBH hydrolysis catalyzed by carbon aerogels (CA) supported Co-NPs, Co_3O_4 macrocubes, Co-Cu, (13a-13e), the effect of Ag doping on Co-B-NPs in SBH hydrolysis (14), and the cobalt-based catalysts for the hydrolysis of SBH and AB (15) have been studied. Porphyrinatoiron/SBH and a novel phthalocyanatoiron/SBH catalyst systems were investigated for the reductions of NAs and found that the latter being superior to the former in yields and chemoselectivities of ANs. As a part of the enhancement of SBH activity, the rate of reduction by the PcFe/SBH system increased upon the addition of 2-bromoethanol (16). Aryl amines are synthesized by novel, practical & efficient reagents as LiCl/SBH in aqueous medium by the reduction of aryl azides (17). NAs are efficiently reduced to corresponding ANs using $CuSO_4$ as a catalyst and SBH as a hydrogen source (18). Novel Cu-NPs were synthesized and used for the reduction of NAs to corresponding arylamines with SBH in high yields (up to 95-100%)(Scheme-7).

$$NO_2 \xrightarrow[\text{EtOH / } H_2O / RT/10\text{min}]{\text{Cu -NPs / NaBH}_4} NH_2$$

When R = CH$_3$CO = : 0°C - 5°C / 5min

Yields = R = H (AN: 94%), 4 -CH$_3$CO (4 -CH$_3$CHOH =AN; 100%)   --Scheme-7

Additionally, the catalyst is stable and recyclable (19). A new efficient, mild and practical method for reduction of NACs employing Raney-Ni/SBH system is reported.

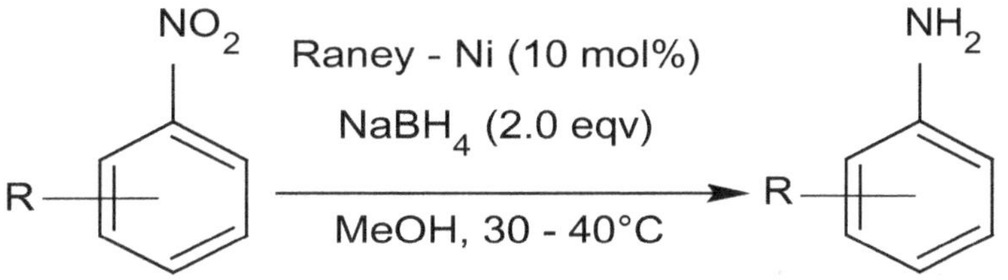

R = 13 Different Substituents   Yields = 85 - 97% --Scheme-8

The method is simple, inexpensive, can be easily scaled-up and applicable for large scale preparation of different substituted anilines (Scheme-8)(20).

A selenium and activated carbon (AC) catalyst has been applied for the selective reduction of NAs to their corresponding amines using SBH as a reducing source under mild conditions (Scheme-9).

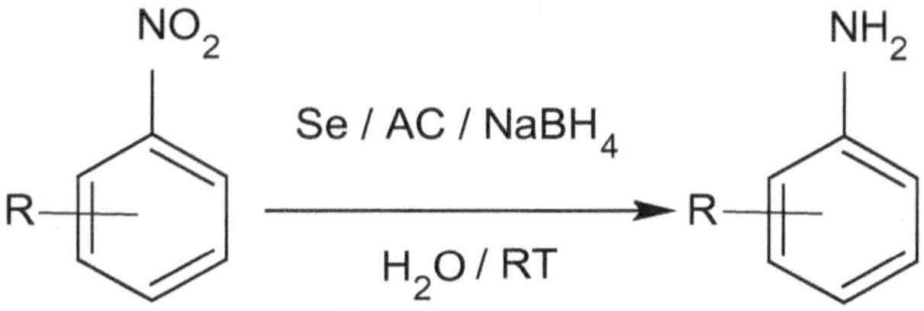

Yields = 84 - > 99%

R = H, o or m or p -Me, , o or ,m, or p -Cl, p -OCH$_3$ etc. --Scheme-9

The catalyst is stable, recyclable and cost-effective and the ANs are obtained in high to excellent yields (84- > 99%) (21). In another case, some aryl amines are obtained in 86–98% yields in the presence of other functional groups (X= 4-CN, 4-CO 2 Et, H, 4-Cl, 2-Me, 3-Me, 4-Me, 4-OMe) catalyzed by Se activated molybdenum (IV) oxide (22). Bimetallic Pt–Ni-NPs/SBH (bi-MNPs) is found efficient in the reductions of 2-NP, 4-NP and 4-nitroaniline (4-NA) using growing microelectrode (GME) and full grown microelectrode (FGME) as catalyst (23). Palladium at ppm level on Fe-NPs on TPGS-750-M surfactant selectively reduces NAs using KBH$_4$ in THF-water at room temperature (Scheme-10) (24, 25).

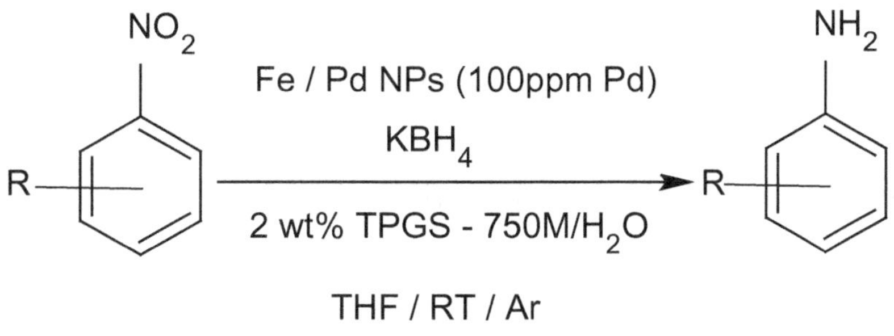

--Scheme-10

5B.03.13: BI/TRIMETALLICS –NaBH$_4$:

Reduction of 4-NP to 4-AP:

The cooperative effects of bi-metallic hydrogenations are always better than single metal catalysts. Here is an example where NAs have been hydrogenated with inexpensive and highly efficient catalyst systems as CuSO$_4$/CoCl$_2$, Pt–Ni-NPs, Fe-Ni/Fe$_3$O$_4$ etc. which are prepared *in-situ* using SBH as the reducing agent (26, 27). Some additives a Sb, Bi, I$_2$ etc. have been found effective in this context. Recently, Saka et al have found out that a Cu-Co- based bimetallic alloy catalyst is effective in generation of hydrogen by hydrolysis of SBH under aqueous medium (28).

Dopamine-directed in-situ and one-step synthesis of Au@Ag core–shell NPs (i) immobilized on a MOF and Ag/Fe$_3$O$_4$@SiO$_2$-APTES nanostructures are used as a synergistic catalyst in the reduction of 4-NP to-4-AP using SBH as a hydrogen source (29). Bimetallic catalysts such as Pt/Cu, Pd/Cu, Pd/Au, Pt/Au, and Au/Cu DENs were synthesized and evaluated as catalysts for 4-NP to 4-AP reduction with SBH in methanol. Pt/Cu was found the most effective, while Au/Cu catalyzed it much slower than expected. The authors have demonstrated that their catalytic properties are dependent on the adsorbate's binding energy (30). Bao et al. have extensively reviewed the applications of M-NPs in different solvents for catalytic reduction of NAs and other different functional groups (31).

--Scheme-11

The catalytic activity of trinuclear (Cu$_x$Ni$_{100-x}$–CeO$_2$; x = 20, 40, 60 etc.) nanocomposites (NCs) was investigated in 4-NP reduction to 4-AP (Scheme-11). Among the synthesized nano composites (NCs), Cu$_{60}$Ni$_{40}$–CeO$_2$ exhibited the best catalytic activity. The kinetic studies are performed (32).

5B.03.14: Metal-Nano Particles-SBH (M-NPs/SBH):

M-NPs have found applications in many fields as nano medicines, site specific drug delivery to tumors, in anti-bacterial treatments, nano catalysts in organic synthesis and manufacturing, in energy and electronics environment and others. Some of the noble metals as Rh, Ru, Pd, Ag, Au, have got preferential choices, especially in the catalytic hydrogenations of functional groups in organic molecules. The following examples using

Ag-NPs and Au-NPs are illustrative examples in the metal catalyzed reduction of NAs with SBH as a hydrogen source, especially for the reduction of 4-NP and -4-nitroaniline. The metal surface of the catalysts remains active and the activity remains unaltered throughout the course of the reduction (33). M-NPs mediated electron transfer from BH_4^- ion to the nitro groups and the concentrations of SBH were important factors when Ag-NPs on GME and FGME surfaces were applied as catalysts in the reductions of 2-nitrophenol (2-NP), 4-nitrophenol (4-NP) and 4-nitroaniline (4-NA) with SBH (34).

5B.03.14a: M-NPs on Solid Supports:

The solid supports used are of wide range such as chitosan, polymeric materials, carbon, grapheme, and inorganic materials as SiO_2, Al_2O_3, Iron oxides etc. Some these products are used during preparations of M-NPs which help stabilize the NPs and some provide active solid surface area. The mesoporous assemblies of silver on TiO_2-NPs (Ag/MTA) are fabricated and their catalytic efficiency for the selective reduction of NAs with SBH and ammonia-borane (AB, NH_3BH_3) is demonstrated with almost quantitative selectivity and yields.

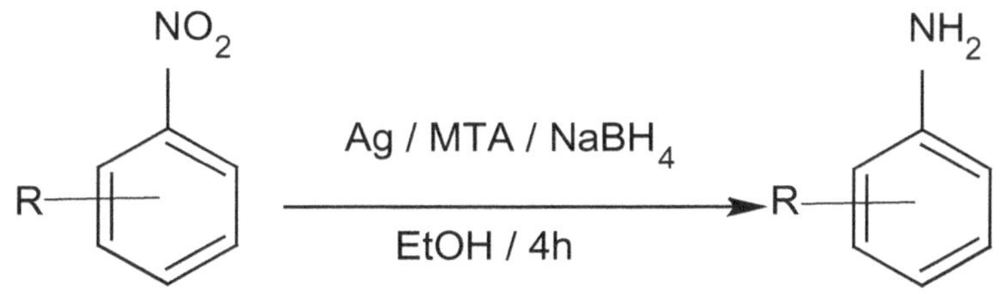

R = 4 -Me: 4 -Toluidine Yield > 98% ; 4 - Toluidine Sel = >99% --Scheme-12

Because of the high observed chemoselectivities and the clean reaction processes, the present catalytic systems, i.e., Ag/MTA-NaBH$_4$ and Ag/MTA-NH$_3$BH$_3$, show promise for the efficient synthesis of aryl amines and N-aryl hydroxylamines at industrial levels (Scheme-12) (35). Catalytic effects of Au@Fe$_3$O$_4$ yolk–shell nanostructures with various Au sizes (8-15nm) were found efficient for the reduction of NAs as 2-NP, 4-NP, 4-NT and, 4-CNB, in the presence of NaBH$_4$ in water. The catalytic performance is highly dependent on the Au-NPs size and core sizes and also on the nature of the substituent groups (36a). Photocatalytic enhancements to the reduction of 4-NP by resonantly excited triangular Gold–Copper Nanostructures is reported with 32 fold enhancement in the reaction rate (36b). The recent developments in noble metal based nano catalysts (Ag, Au, Pd, Rh, Ru etc.) in the reduction of NAs have been reviewed (37).

CoB, CoB/SiO$_2$ and Co–Co$_2$B- NCs are used for the catalytic and chemoselective reduction of NAs using two different hydrogen sources as SBH and hydrazine hydrate (HH). Both the systems display an interesting chemoselective reduction switch (38). Unsupported CoB/SiO$_2$ catalyst demonstrated high activity for the catalytic hydrolysis of SBH solution. The structure effect caused by the SiO$_2$ support helped to prevent

Co-B-nano clusters from aggregation and therefore the activity increased significantly on hydrolysis of alkaline SBH solution (39).

5B.03.14b: Polymers and Microgel as Supports:

The Au-NPs with supports as: a) chitosan-coated iron oxide magnetic nanocarriers, b) on commercially available poly(methyl methacrylate) (PMMA) beads, and c) on oxidized mesoporous carbon have been tested. Au/PMMA with an excess amount of SBH in water gave the best results in the reduction of 4-NP to 4-AP with highest catalytic activity among polymer supported Au-NPs previously reported. Ag-NPs on hybrid polymer microgels are found effective in 4-NP reduction to 4-AP using aq. SBH at 27°C. Ag-NPs were fabricated inside the microgels by in-situ reduction of silver ions (Scheme-13).

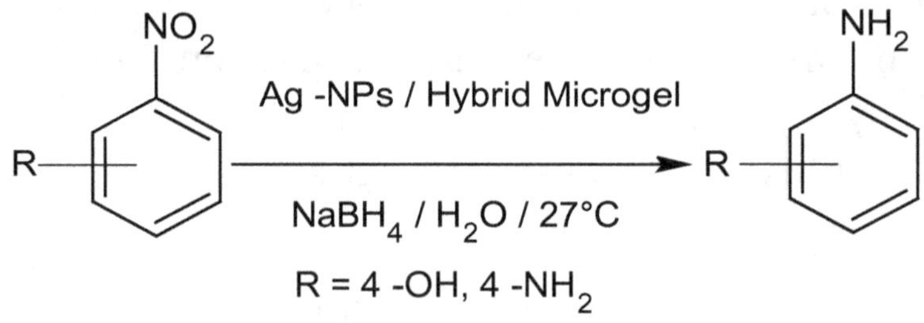

--Scheme-13

Ag-NPs were found to be a more effective and efficient catalyst for reduction of 4-NP and 4-nitroaniline as compared to NB (40--44).

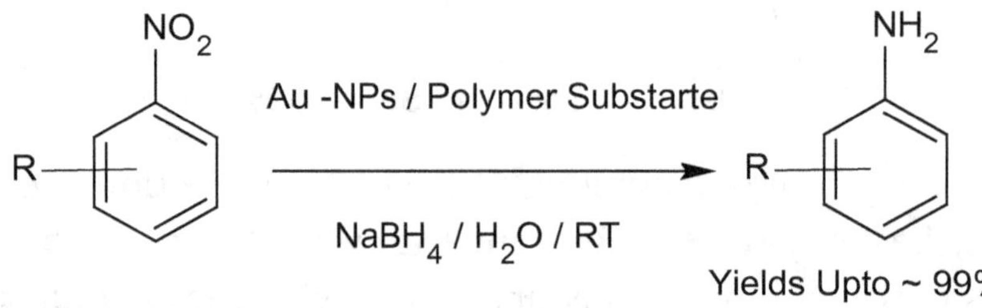

Yields Upto ~ 99% --Scheme-14

Water-soluble Au-NPs have been synthesized using a nitrogen-rich polyoxyethylenated substrate and were found suitable for their applications in catalytic selective reduction of NACs using SBH in water. The new nano catalyst could be easily recycled taking advantage of its solubility properties (Scheme-14) (45). Pd-NPs clusters supported on a spherical polymer network for the reduction of 4-NP to 4-AP are successfully used (46-47). Ag-NPs on polymeric cationic support were found more active than other catalytic systems in the reduction of 4-NP to 4-AP with SBH as a reducing agent (48).

5B.03.14c: Aryl Amines with Simple Reduction Systems:

A general, operationally simple and highly applicable protocol for the CTH of NAs has been developed under environmentally benign reaction conditions using a simple iron salt $Fe(OTf)_3$ (10 mol%) as a catalyst and SBH as the stoichiometric reductant (Scheme-15).

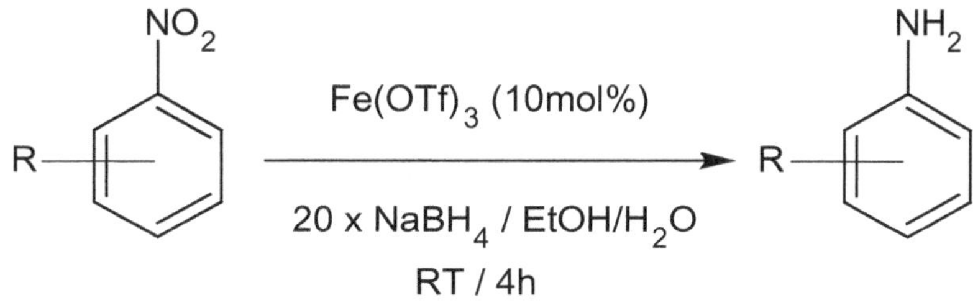

NO$_2$ → NH$_2$

Fe(OTf)$_3$ (10mol%)

20 x NaBH$_4$ / EtOH/H$_2$O

RT / 4h

NB Conversion > 99%   Yields = Up to 80%   --Scheme-15

Quantitative conversions and poor to good AN yields are obtained (49). Pd-PCs have been established as recyclable heterogeneous catalysts for the chemoselective reduction of NACs to arylamines using SBH in ethanol (Scheme-16) (50).

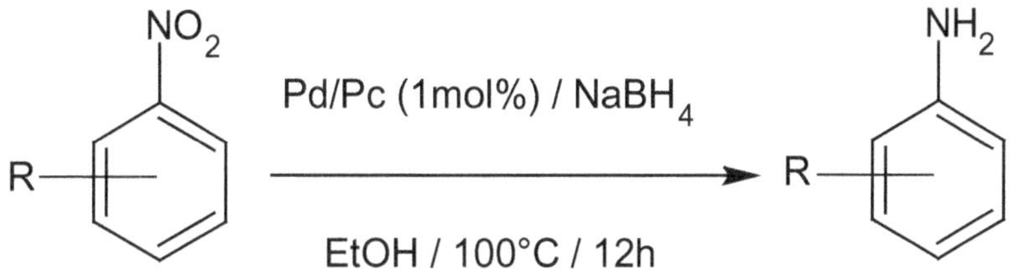

NO$_2$ → NH$_2$

Pd/Pc (1mol%) / NaBH$_4$

EtOH / 100°C / 12h

20 Examples & Isolated Yields up to 94%   --Scheme-16

Mechanistic studies showed that the reduction of nitro group proceeds through nitroso pathway and possibly Pd-PCs activate the nitro group for effective reduction. PdPCs also facilitated the hydrogen generation from the combination of SBH-EtOH. Kiasat et al. have synthesized Ag-NPs on SiO$_2$ and are used in the chemoselective reduction of NAs to the corresponding arylamines (100% yields) in aqueous SBH. This process is compared with methods by different groups using NaBH$_4$ and different metal/solvent combinations as: 1). DMSO (74%), 2).Sulfolane (50%), 3).Se/NaOAc/DMF (100%), 4). Sm/MeOH/ Bipyridinium Salt (100%): 5). NH4Cl, LiCl, Diglyme (100%), 6). ClP-(OEt)$_2$, (iPr)$_2$NEt, CHCl$_3$, N$_2$, (100%), 7). ZrCl$_4$/THF-(100%). The % AN yields in each system are given in the parentheses. These reactions are performed at different temperatures and time. The current method by Kiasat et al. with Ag-NPs/SiO$_2$, SBH in water at reflux temperature for 1h gives 100% ANs (51).

A series of half-sandwich Ru(II) p-cymene complexes containing Schiff-base ligands have been synthesized and used in the reduction of NAs to ANs in the presence of SBH in water as a H$_2$ source. Some of the catalysts systems were found compatible with NAs containing various functional groups (52).

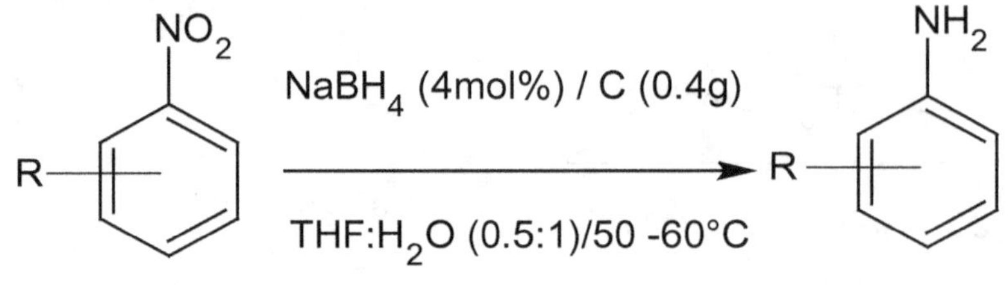

NO$_2$ → NH$_2$

NaBH$_4$ (4mol%) / C (0.4g)

THF:H$_2$O (0.5:1)/50 -60°C

R = H : Yield = 100%; No H$_2$O: Yield= 0%   --Scheme-17

A new synthetic method for a mild and convenient reduction of NAs to ANs using SBH on charcoal in a mixture of H$_2$O-THF (1:0.5 mL) at 50–60°C with high to excellent yields of products is developed (Scheme-17). In the absence of water AN was not formed at all (53). Pyridine zinc tetrahydroborate [(Py)Zn(BH$_4$)$_2$], as a new stable ligand-metal-borohydride is prepared quantitatively by complexation of 1:1 zinc borohydride and pyridine at room temperature (Scheme-18).

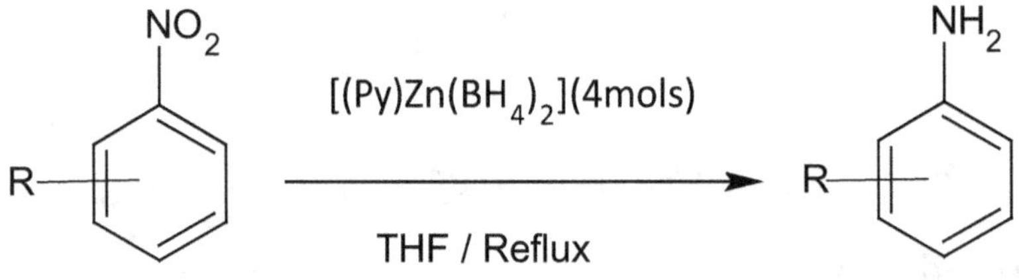

NO$_2$ → NH$_2$

[(Py)Zn(BH$_4$)$_2$](4mols)

THF / Reflux

Yields = 80 - 98% --Scheme-18

Zeynizadeh et al. have used as-prepared Zinc tetrahydroborate [Zn(BH$_4$)$_2$] for efficient reduction of NACs to their corresponding primary amines in refluxing THF(54). A chemoselective and green reduction of nitro arenes to aromatic amines with FeSO$_4$, NaBH$_4$, H$_3$PW$_{12}$O$_{40}$ in water at room temperature is reported. AN and o-toluidine are obtained in 100% yields while others are also obtained between 92 and 99% yields. The method is simple, inexpensive, easily scalable and applicable for the large scale preparation of substituted anilines (55).

5B.03.15: Other Borohydrides:

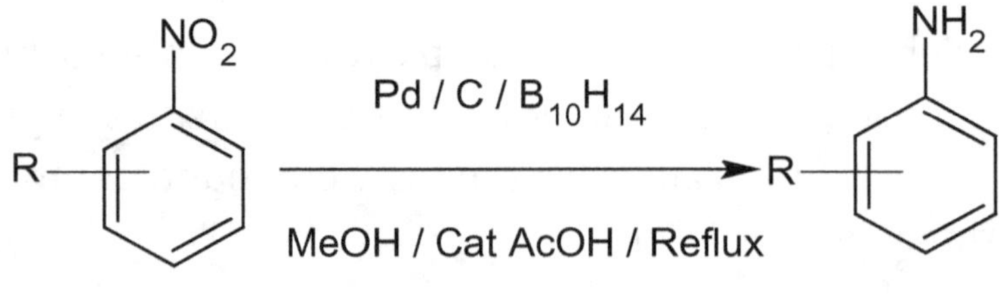

NO$_2$ → NH$_2$

Pd / C / B$_{10}$H$_{14}$

MeOH / Cat AcOH / Reflux

Scheme - 19

81 - 97%

R =4-COOH, 4-COOMe, 4-COOBn, 4-CH$_2$CN, 4-CH$_3$, 3 -CH$_2$OH, 4-Br-3-Me

5B.03.15a: Decaborane:

Chemoselective reduction of NACs to corresponding ANs using decaborane in methanol is investigated (56).

NAs were chemoselectively reduced to the corresponding ANs using decaborane ($B_{10}H_{14}$) in the presence of Pd/C and two drops of acetic acid under reflux conditions in methanol (Scheme-19).

5B.03.15b: Sulfurated Calcium Borohydride:

An efficient and practically modified new reagent, Sulfurated Calcium Borohydride $Ca(BH_2S_3)_2$ is easily prepared by metathetical reaction between $NaBH_2S_3$ and $CaCl_2$ in THF. It is used in the reduction of aryl azides and ANCs to their corresponding amines in high yields with SBH in aq. THF solution (Scheme-20)(57).

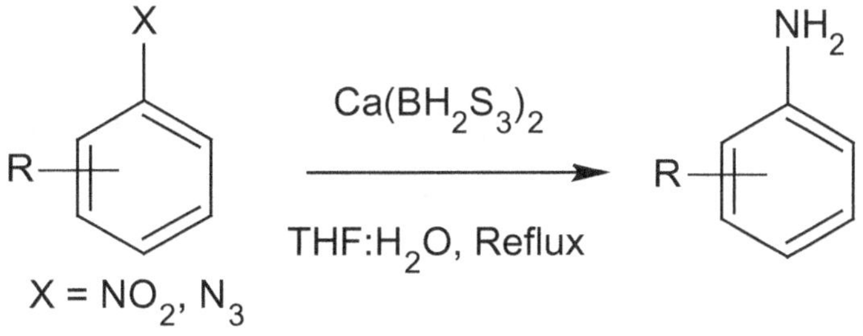

--Scheme-20

Reduction of nitro groups is accompanied with regioselectivity, while very high chemoselectivity is also observed for the reduction of an azido functional group in the presence of a nitro group. Unsubstituted, substituted NAs reductions are also accomplished by sulfurated SBH (58).

5B.04.00: Additive Effect on NaBH$_4$ Activity:

In scientific field the scientists keep unfolding new things through their visionary efforts. In chemical reactions quantitative substrate conversions and 100% yield and selectivity are the most desired objectives. However, unless the reaction conditions are optimized for a particular case it doesn't deliver the expected results and hence it needs further optimizations. We have seen earlier that some of the additives act as promoters of catalytic activities and hence the yields and selectivities of end products. Besides the substrates and other main chemicals, solvent and additives as promoter do help in enhancing the catalytic efficiency. Because of its lower efficiency, SBH requires promoters in reducing organic functional groups efficiently. For example, we have seen earlier that other reagents as $SbCl_3$, $BiCl_3$, I_2, $CuSO_4$ etc. do enhance SBH activity and selectivity, especially in the reductions of NAs. Addition of $CuSO_4$ to SBH enhances the reduction of NAs under most mild conditions (Scheme-21) (59). However, this system also reduces ketones, aliphatic esters, olefins and nitriles.

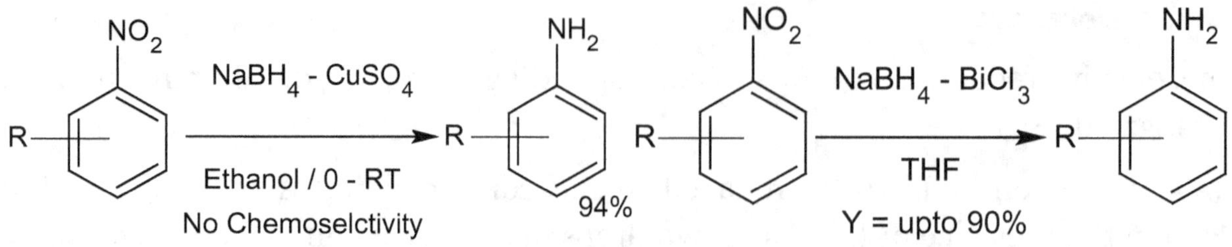

Scheme-21

On the other hand, Sn, Bi, Sb salts are quite effective in chemoselective reduction of substituted NAs (60). The rate of reduction by the PC-Fe/SBH system increased upon the addition of 2-bromoethanol (61). Arylamines are obtained in quantitative yields with NAs reductions in aq. SBH catalyzed by M-NPs on solid supports (62). Copper acetylacetonate/SBH in aq. ethanol and Cu nanorods and nanospheres of desired size are found efficient catalysts in the chemoslective reductions of NAs, 4-NP and 5-nitroisophthalic acid using aqueous SBH (63).

Brack et al have extensively reviewed the heterogeneous and homogenous catalysts for hydrogen generation by hydrolysis of aqueous SBH solutions (64). Demirci et al have covered SBH hydrolysis as hydrogen generator: issues, the state of the art and applicability upstream from a fuel cell viewpoint are presented (65). Yadav and Periasamy have independently reviewed dehydrogenation of SBH (66).

References:

1). H. C. Brown, et al. i). US2683721-A(1954). ii).US 2856428-A(1958).

2a). H. Firouzabadi, et al. *Iranian J. Science & Technology Transactions. A,* 19:103(1995), 2b) H. Firouzabadi, et al. *Bulletin Chemical Society Japan,* 70:155(1997),

3). M. Periasamy, et al. *J. Organometallic Chemistry,* 609:137-151(2000).

4). B. Wahbeh, et al. *Intl. J. Hydrogen Energy,* 38(14):6210-6214(2013).

5). E. Fakioglu, et al. *Renewable Energy,* 48:10-15(2012).

6). E. Petit, et al., *Intl J. Hydrogen Energy,* 37(21):15983-15991(2012).

7). U. B. Demirci, et al. SBH Hydrolysis as Hydrogen Generator: Issues, State of the Art, Applicability Upstream for Mobile. Fuel Cells, Wiley-VCH Verlag, 10(03), (2010), 60 pages pp.335.

8). H. C. Brown & S. Krishnamurthy, *Tetrahedron,* 35:567-607(1979).

9). B. Ganem, et al., *J. American Chemical Society,* 104:6801(1982).

10). P. Ren, S. Pan, et al. *Synthetic Communications,* 25(23):3799-3803(1995). 11). H. N. Borah et al. *J Chemical Research* (S): 228-229(1994). 10c). A Review: R. Retnamma, et al. *Intl J. Hydrogen Energy,* 36(16):9772-9790(2011).

12). H. Goksu, *New J. Chemistry*, 39:8498-8504 (2015).

13a). J. Zhu, et al. *Intl J Hydrogen Energy*, 38(25):10864-10870(2013). 13b).H. Kim, et al. *Energy*, 121:238-245(2017). 13c). H. Kahri, et al. *RSC Advances*, 06:102498-102503(2016). 13d). Review: P. Brack, et al. *Energy Science & Engineering*, 03(03):174-188(2015). 13e). $NaBH_4$ as a fuel for the future: D M F Santos et al. *Renewable & Sustainable Energy Review*, 15:3980–4001(2011).

14). L. Wei et al. *AIP Conference Proceedings* 1794:020004 (2017). doi:10.1063/1.4971886,

15). U. B. Demirci, et al. *Phys. Chemistry & Chem. Physics*, 16:6872-6885 (2014).

16). H. S. Wilkinson, et al. *Tetrahedron Letters*, 42(02):167-170(2001).

17). S. Ram, et al. *Synthetic Communications*, 30(24):4495-4500(2000).

18). S. Yoo, S. Lee, *Synlett*, 07:419-420(1990).

19). Z. Duan et al. *Bull. Korean Chemical Society,* 33(12):4003-4006(2012).

20). I. Pogorelic et al. *J. Molecular Catalysis A: Chemical*, 274(1-2):202–207(2007).

21). K. Cai, etal. *BullChemReactionEngg & Catalysis,*10(03):275-280(2015).

22). K. Yanada, et al. *Tetrahedron Letters,* 33(11):1463-1464(1992).

23). N. Pradhan, et al. *Colloids & Surfaces A: Physicochemical & Engineering Aspects,* 196(02-03):247-257(2002).

24). C. M. Gabriel et al. *Organic Process Research & Dev., 21* (02): 247–252(2017).

25a). Y. Liu, et al. *Molecules*, 16(05):3563-3568(2011). 25b). P. Ren et al. *Synthetic Communications*, 27(20):3497-3503(1997).

26). M. Ficker, et al. *Synthetic Communications*, 46(2):176-182(2014).

27a). D. Wu, Y. Zhang et al. *Inorganic Chemistry*, 56(9):5152–5157(2017). 27b). S. K. Ghosh, M. Mandal, et al. *Applied Catalysis A: General*, 268(01-02):61-66(2014).

28). C. Saka, et al., *Energy Sources, Part A: Recovery, Utilization, and Environ Effects,* 37(09):956-964(2015).

29a). P. Huang, et al. *Chem An Asian J*, 11(19):2705–2709(2016). 29b). R. Ahmadi, et al. *Monatshefte für Chemie-Chemical Monthly,* 148(08):1423–1431(2017).

30). Z. D. Pozun, et al. *J. Physical Chemistry C:* 117:7598-7604(2013).

31). T. D. Bao, et al. *Catalysts:* 07:207-239(2017).

32). M. Kohantorabi, et al. *Industrial Engg & Chemical Res.,* 56(05):1159-1167(2017).

33a). P. Liu, et al. *Applied Surface Science*, 255(07):3989-3993(2009). 33b).F. A. Al-Marhaby, et al. *World J. Nano Science & Engineering*, 06:29-37(2016).

34). D. Androeau et al. *Nanomaterials*, 06:54-66(2016).

35a). F. Lin, et al. *J. Physical Chemistry C*, 121 (14):7844–7853(2017).

35b). K. Layek, et al. *Green Chemistry:* 14:3164-3174(2012).

36a). R. Sedghi, et al. *Current Organic Chemistry*, 20(6):696-734(2016).

36b). M. Hajfathalian, et al. *J. Physical Chemistry: C*, 119(30):17308–17315(2015).

37). S. Patil, et al. *RSC Advances*, 3(32):13243-13250 (2013).

38). C. Yang et al. *Intl J. Hydrogen Energy*, 36(2):1418-1423(2011).

39a). Y. Chang, & D. Chen, *J. Hazardous Materials*, 165(1-3):664-669(2009). 39b). K. Kuroda, et al. *J. Molecular. Catalysis .A: Chemical*, 298(01-02):07-11(2009). 39c). P. Guo, et al. *J. Colloid and Interface Sci*ence: 469:78–85(2016).

40). M. Ajmal, et al. *J. Chemical Engineering*, 30(11):2030-2036(2013).

41). Z. H. Farooqi, et al. *Turkish J. Chemistry:* 39: 576-588(2015).

42). Z. H. Farooqi, et al. *Materials Science-Poland*, 33(3):627-634 (2015).

43). Z. H. Farooqi, et al. *J. Polymer Engineering*, 36(01): p82 (2016).

44). J. Najeeb, et al. *Imperial J. Interdisciplinary Res.*, (IJIR), 02(09):266-273(2016).

45). W. Guo, et al. *Chemistry An Asian J*, 10(11):2437-2443(2015).

46). Science & Technology of Separation Membranes, 2 Vols, Tadashi Uragami, John Wiley & Sons, 2017, Technol & Engg - 848 pages

47). E. D. Sultanova, et al. *Chemical Communications*, 51(68):13317-13320(2015).

48). B. Baruah & G. J. Gabriel, *Langmuir*, 29(13):4225–4234(2013).

49). A. J. MacNair, et al. *Organic& Biomolecular Chemistry*, 12:5082–5088(2014).

50). P. K. Bala, et al. *Catalysis Letters*, 144(07):1258–1267(2014).

51). A. R. Kiasat, et al. *Chinese Chemical Letters*, 21:1015–1019(2010)

52). W. Jia, H. Zhang, *Organometallics*, 35 (4):503–512(2016).

53). B. Zeynizadeh et al. *Synthetic Communications*, 36:2699-2704(2006).

54a). S. Narasimhan et al. *Aldrichimica Acta*, 31(01):19-26 (1998).

54b). B. Zeynizadeh, et al. *J. Chinese Chemical Society*, 50:267-271(2003).

55). H. Aliyan, et al. *Nanostructures:* 01:21-26(2012).

56a). J. W. Bae, et al. *Chemical Communications:* 1857–1858(2000). 56b) J. W. Bae, Y. J. Cho, et al. *Tetrahedron Letters*, 41:175–177(2000). 56c). S. H. Lee et al. *Synthetic Communications*, 36(17): 2469-2474(2006).

57). A. R. Kiasat, et al. *Synthetic Communications*, 30(4):587-589(2000).

58). L. Brindle, et al. *Canadian J Chemistry*: 49:2990 (1971).

59). S. Yoo, S. Lee, *Synlett*, 419-420(1990).

60a). H. N. Borah, et al. *J. Chemical Research* (S), 228-229(1994). 60b).P. Ren, et al. *Synthetic Communications*: 25(23):3799-3803(1995).

61). H. S. Wilkinson, et al. *Tetrahedron Letters*, 42(02):167-170(2001).

62). D. M. Dotzauer, Y. Wen et al. *Langmuir*, 25(03):1865-1871(2009).

63a). K. Hanaya, et al. *J. Chemical Society Perkin Trans-1:* 00:2409-2410 (1979).

63b). A. Dutta, et al. *Catalysis Communications:* 11:651-655(2010).

64). P. Brack, et al. *Energy Science & Engineering,* 03(03):174–188(2015).

65). U. B. Demirci, et al. *Fuel Cells,* 10(03): 335 (2010).

66a). M. Yadav, et al. *Energy & Environmental Science,* 05(12): 9698-9725 (2012).

66b). M. Periasamy, et al. *J. Organometallic Chemistry:* 609:137–151(2000).

5B.05.00: Synthesis of 4-Amino Phenol (4-AP) Using SBH (NaBH$_4$):

Para-Aminophenol (PAP) or 4-Aminophenol (4-AP) is an important AN derivative, as it has many applications, including manufacturing of Paracetamol (4-acetaminophenol), an anti-pyretic, anti-inflammatory drug produced in thousands of tons per year in the world. Since we are looking at the applications of borohydrides in the reduction of NAs to ANs, we will focus our attention on highlighting the recent developments in the synthesis of 4-AP from 4-NP using SBH as a reducing agent. Although, other borohydrides as LiBH$_4$, KBH$_4$, Ca(BH$_4$)$_2$ etc. are used for the reduction 4-NP, NaBH$_4$ is the most commonly used reagent for this purpose because of its availability in the market at commercial scale and its low price for making the process economical. Since the general procedures and the parameters of study during the development of new methodologies for NA reductions are fairly similar, we have given only the publication titles and the corresponding references. Wherever it is necessary, we will highlight the salient features or important observations made by a particular research group or publication authors.

5B.05.10: Metal-Nanoparticles (M-NPs):

A general M-NPs catalyzed reduction of 4-NP to 4-AP using SBH as a reducing agent is as shown in the following scheme (Scheme-22). The metal catalyst could be either from the noble metal group (Pd, Pt, Ag, Au, Ru, Rh etc.) or non-noble metals (Fe, Co, Cu, Zn, etc.) and they could be single M-NPs or Bi-M-NPs. The reactions are usually carried out in aqueous phase in the presence or absence of a base (NaOH or KOH) and $NaBH_4$ (BH_4^-) as a reducing agent and MeOH as a solvent..

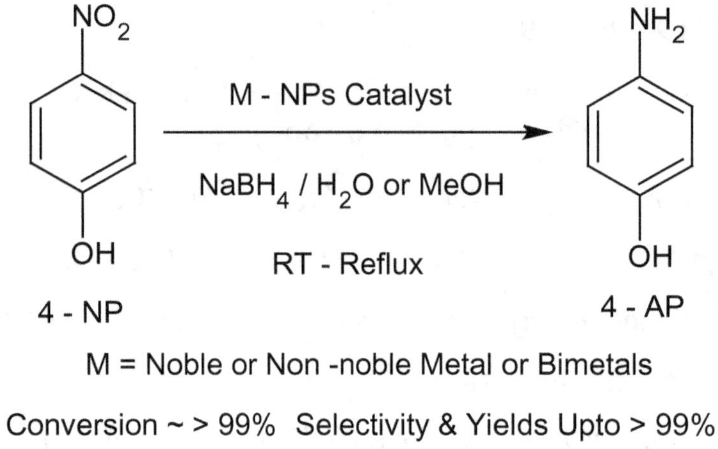

M = Noble or Non -noble Metal or Bimetals

Conversion ~ > 99%   Selectivity & Yields Upto > 99% --Scheme-22

Some of the work cited below is highly innovative leading to a proposition or postulation of developing better strategies/new catalytic systems for better performances. For example, the published work is an example to explore the catalytic behavior of amorphous Ni-B/C catalyst which may shed light on developing other advanced amorphous catalysts for 4-NP reduction using an economical source of hydrogen generation (1).

5B.05.11: Noble Metal Catalysts:

Generally, the noble metals as Au, Ag, Ru, Rh, Pd, have been used in the selective reduction of NAs. The reduction of NB and HNBs by chemically bound hydrogen in the presence of immobilized Rh-complexes highlights the catalytic activity of metal complexes attached to a solid support (2). Rh complexes as [(RhCl(COD)]$_2$, RhCl(PPh$_3$)$_3$, and RhCl$_3$ immobilized on silica gel modified by amine and aminophosphine groups catalyze selective reduction of NB to AN in aqueous medium at 25–82°C by transfer of hydrogen from 2-propanol or IPA/KOH and SBH. Under experimental conditions, it was observed that the rate of reduction of NB by SBH is one to two orders of magnitude higher than with IPA. More recent work using mono- and bimetallic noble metal catalysts is highlighted below.

5B.05.12: Noble Mono-Metal-NPs:

Hallo porous Au-NPs (i) and Ag-NPs (ii) have been used in the reduction of 4-NP to 4-AP with SBH, and also can be used in water pollutant removal and environmental remediation (3). A novel CTA$^+$ AuCl$_4^-$ complex on the silica support and their synergism in the catalytic reduction of 4-NP to 4-AP with SBH shows higher catalytic activity compared to Au(0) supported on silica. Such synergism is presumed to be due to the surfactant bilayer with high concentration of 4-nitrophenolate ions near the Au(0)-NPs embedded on the SMS leading to highly efficient contact between them (4).

In this context, Au, Ag, Pd, NPs in the catalytic reduction of 4-NP to 4-AP using SBH in aqueous solution with a quantitative assessment of the role of dissolved oxygen in determining the induction time using spectroscopy is investigated (5). A highly ordered silver and gold nanotubes arrays were embedded in 100 nm pores of PET TeMs *via* electroless deposition technique at 4°C during 1hour and used for reduction of 4-NP to 4-AP by SBH and found that Au/PET composites exhibit not only more powerful catalytic activity but also exhibited non-linear dependence of rate constant from temperature (6). Synthesis and catalytic application of Pd-NPs supported on kaolinite-based nanohybrid materials and their use for the reduction of 4-NP to 4-AP is performed (7).

5B.05.13: Noble Bimetal-NPs:

Pt-Au-PDA/RGO dendrimer-like NPs supported on the surface of polydopamine-functionalized/rGO have been synthesized and their activity tested on 4-NP to 4-AP reduction with SBH. The catalyst activity is based on two types of electron transfers: (i) from the PDA coating to both Au and Pt atoms; (ii) from Au to Pt atoms, and the Pt:Au(3:1)-gives 100% conversions and decolorizes colored water in 8s (8). Similarly, a well-defined bimetallic yolk-shell of Ni@Pt-Ni nanocrystals with porous shells were uniformly deposited on reduced graphene oxide (Ni@Pt-Ni NCs-rGO) under hydrothermal conditions. The catalyst exhibited better catalytic performance toward the reduction of 4-NP in comparison with i) commercial Pt/C (50 wt%), ii) monometallic Pt -NPs/rGO and iii) Ni-NPs/rGO catalysts (9). Monodispersed bimetallic Rh-Ag/rGO in the presence of the tris(triazolyl)-polyethylene glycol (tristrz-PEG) ligand as a weak stabilizing agent are found efficient in 4-NP to 4-AP reduction with SBH (10). Au/Pd (75:25) nanoalloys were synthesized using spherical polyelectrolyte brushes (SPB) as carrier system and used to reduce 4-NP to 4-AP with SBH through hydroxylamine intermediate (11). Two types of M-NPs onto block copolymer nanospheres (Pd–Au@ PS-P2VP nanotubes) show efficient catalytic activity for the reduction of 4-NP by SBH (12). Au/Ag-NPs bilayer and Au/Ag/Au-NPs trilayer have potential applications as cost-effective catalysts in 4-NP reduction with SBH (13). Co-NPs-(i) and ligand free Au-NPs-(ii) are used as efficient catalyst in the reduction 4-NP to 4-AP using SBH under mild conditions (14). Au-Ag alloy NPs on GO as both a reducing agent and a support with an Au-rich shell prepared in two steps. The catalyst exhibited excellent catalytic activity for the reduction of 4-NP to 4-AP in water at room temperature. The catalyst activity was superior to that of GO decorated with Au–Ag alloy-NPs that was prepared by a one-step method and GO decorated with either Au or Ag-NPs alone (15).

5B.05.14: Non-Noble Metal-NPs:

One-step catalytic reduction of 4-NP through the direct injection of metal salts into oxygen-depleted reactants such as 4-NP and metal borohydride is presented. This single-step procedure when carried out in the absence of dissolved oxygen leads to extraordinary catalytic activity (16) with turnover frequencies as high as 65000 h^{-1}. A simple, economic and one pot synthetic protocol was followed to synthesize SiO$_2$ capped/stabilized Fe$_3$O$_4$ nanostructures and used in the heterogeneous catalytic reduction of 4-NP to 4-AP using SBH under microwave irradiations (Scheme-23).

SiO$_2$ / Fe$_3$O$_4$ - Nanostructures

NaBH$_4$ / H$_2$O or MeOH

MW Irradiations / Few Minutes

4 - NP   4 -NP Conversion = 99.5%   4 - AP   --Scheme-23

The effect of different parameters such as concentration of reducing agent (NaBH$_4$), quantity of catalyst applied and the effect of microwave irradiation time was evaluated. The 99.5% 4-NP reduction was achieved by using 100 µg of Fe$_3$O$_4$/SiO$_2$ nanocatalyst in a short reaction time. The catalyst and the procedure are simple, economic, environmental friendly and bio-compatible and hence it can be used in environmentally benign and industrially important reactions (17). Magnetic Fe$_2$O$_3$-Cu$_2$O-TiO$_2$ NCs were synthesized by using sol gel method and used for the catalytic reduction of 4-NP to 4-AP by using SBH as the hydrogen donor. The as-prepared catalyst showed better catalytic ability than pure TiO$_2$ (18). Cu-NPs loaded on rGO (Cu/rGO) acted as the co-catalyst and the rGO acted as the substance of electron transfer. The catalytic mechanism is also suggested (19). From the graphene hydrogels embedded with three different metals, i.e. Au, Ag and Cu, stable and recyclable NPs were synthesized and tested as heterogeneous catalysts in a model reaction of 4-NP reduction with SBH in water. Cu-NPs gave the highest activity. The outstanding catalytic activity arises from the synergistic effects between graphene and M-NPs (20).

A ternary Cu$_2$O–Cu–CuO-nanocomposites are effective catalyst with intriguing activity for 4-NP reduction to 4-AP in aqueous solution in the presence of SBH at room temperature. The catalyst system is more active than other CuO-NP based systems and the methodology was successfully applied to diverse NAs (21). Reduction of a variety of NBs such as NB, 4-ethy-NB, 4-isopropyl-NB, 4-nitroacetophenone etc. with CuCl$_2$/ MeOH/SBH provided some of the corresponding ANs in good to quantitative yields, including 4-AP (22a).

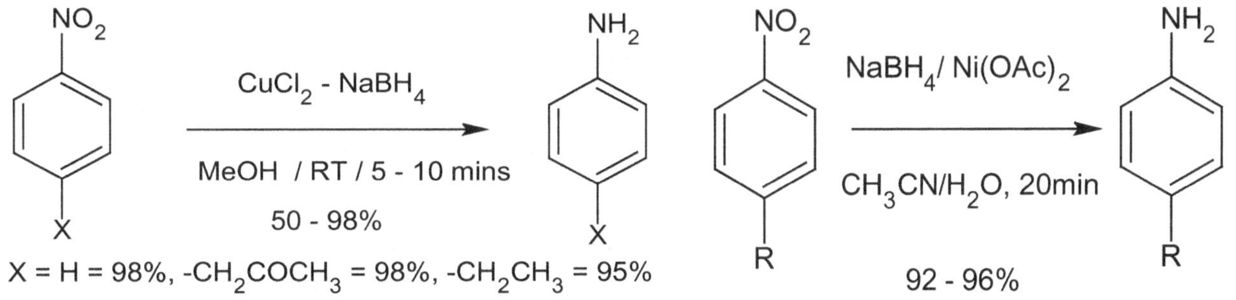

Scheme-24

Amongst Cu based catalysts $CuCl_2$ was super active system, although $CuSO_4$ and $Cu(NO_3)_2$ were also equally good and methanol being the best solvent evolved. Ni metal as $SBH/Ni(OAc)_2.4H_2O$ complex also was found effective in these transformations (Scheme-24). A bimetallic Cu_{60}-Ni_{40}-CeO_2 NPs catalyst exhibited the best catalytic activity towards the reduction of 4-NP to 4-AP with SBH.

Contrary to oxygen depleted systems as reported above, nanoscale zero-valent iron (NZVI)/$NaBH_4$ system in an oxygen environment reduced 4-NP to 4-AP with complete conversion of 4-NP and high conversion efficiency of 4-AP (>98%). This technology can be effectively used for wastewater reductive treatment in oxygen environments (23).The catalytic activity and the after-use stability of the NiS-NPs in the reduction of 4-NP to 4-AP was higher than those of the other nickel sulfide phases. Also, NiO/NiS composites were found as highly active and efficient catalysts in decolorization of organic wastewater pollutants (24, 25).

5B.05.15: Metal-NPs on Solid Supports with $NaBH_4$ (SBH):

Catalytic membranes have an extensive range of desirable applications in chemical fields. Here, an Au-coated carbon nanotube (Au/CNT) hollow fibre membrane was fabricated by depositing Au-NPs on a CNT hollow fibre membrane, which exhibited excellent catalytic activity for the reduction of 4-NP to 4-AP. The observed improvement in the catalytic activity can be ascribed to the synergistic effect between the Au-NPs and CNT membrane and it offers a new avenue for the exploration of membranes for practical applications (26). Urania-palladium-graphene nanohybrids were synthesized via a solvothermal process in ethylene glycol. This ternary hybrid Pd-nanocatalyst showed considerably higher catalytic activity than palladium-graphene hybrids towards the reduction of 4-NP to 4-AP by SBH (27). Au-NPs (~2.3nm) were prepared in the channels of mesoporous silica SBA-15(Au/SBA-15) and were used for the reduction of 4-NP to 4-AP with SBH and also in the treatment of different waste water pollutants (28).

In order to investigate structure-property relationships, the catalytic properties of gold nanospheres and gold nanostars were evaluated in the reduction of 4-NP to 4-AP by SBH. The metal particle size structures played a key role in determining the activity of the catalysts. The nanostars showed increased activity and the higher activity compared to spherical gold particles was mainly because of the abundance of flat/spiky features on these particles, which show high metal utilization (29). $Fe_3O_4@SiO_2@Au$ nanoparticles synthesized were found to be catalytically active for the reduction of 4-NP to 4-AP using SBH and giving very high pseudo-first-order rate constant (30). Ag-NPs/silica nanowires (Ag-NPs/SiO_2NWs; 35 ± 5 nm) were fabricated by two steps consisting of the preparation of the silica nanowires (SNWs) by chemical dispersing and acid leaching from the natural mineral chrysotile and an in-situ reduction approach (for Ag-NPs). The catalyst was used for the effective reduction of 4-NP to 4-AP in the presence of SBH as hydrogen source (31).

5B.05.16: M-NPs Supported on Natural Products and NaBH$_4$:

The development of new synthesis methods with cost-effectiveness, green routes, environment friendly, involving nontoxic chemicals, benign solvents, non-hazardous process of making pharmaceutical/agricultural products of human interest are challenges for the scientific community. The use of M-NPs has drawn a great deal of attention in the recent years. One of the challenges in their development is lack of methodologies with appropriate precaustionary measures about the concerns as raised above. In recent times the natural products have emerged as complementary sources in developing suitable methods for making M-NPs. Aromal et al. have reported the green synthesis of Au-NPs (15-25nm) using the aqueous extract of fenugreek (*Trigonella foenum-graecum*) as a reducing and protecting agent. The process involves the reduction of $AuCl_4^-$ with fenugreek seed extract. This method is simple, efficient, economic and nontoxic, with a stable and recyclable catalyst. The FTIR spectrum studies indicated the presence of different functional groups present in the biomolecule capping the nanoparticles. The synthesized Au-NPs show good catalytic activity for the reduction of 4-NP to 4-AP by excess SBH. The catalytic activity is found to be size-dependent, the smaller nanoparticles showing faster/higher activity (32). Ag-NPs supported by reductant-free cellulose microgels under vacuum filtration rapidly catalyze the reduction of 4-NP to 4-AP by SBH (33). Rice husks as a sustainable silica source for hierarchical flower-like metal silicate architectures assembled into ultrathin nanosheets for adsorption and catalysis. Ni-NPs/silica (Ni-NPs/SiO_2) exhibited high catalytic activity and good stability for 4-NP reduction to 4-AP with SBH within 160s, which can be attributed to the ultra-small particle size (~6.8 nm), good dispersion and high loading capacity of Ni-NPs. Considering the abundance and renewability of rice husks, metal silicate with complex architecture can be easily produced at a large scale and become a sustainable and reliable resource for multifunctional applications (34). Kim et al have similarly synthesized Au-NPs using $AuCl_4^-$ with furfuryl derivatives, which are expected to act as surface capping agents during AU-NPs formation. Small size Au-NPs (AuFFA-2 and AuFFA-4) were found highly effective for the catalytic hydrogenation of 4-NP to 4-AP in the presence of SBH at room temperature (35). Development of eco-friendly and novel method for the synthesis of M-NPs is one of the most popular and emerging aspect of nano-biotechnology. Gnidia glauca leaf and stem extract are used to synthesize Au-NPs. As compared to other biological methods, the syntheses were extremely rapid leading to an efficient and complete reduction of 4-NP to 4-AP within 20 min in the presence of as-prepared Au-NPs from Gnidia and using SBH as hydrogen source(36).

Ag-NPs and Au-NPs were synthesized on calcium alginate gel beds using a green photochemical approach and used for the efficient reduction of 4-NP to 4-AP (37). Cyclodextrins as effective additives in Au-NPs-catalyzed reduction of NAs in a ball-mill in aqueous system using SBH is presented. It is observed that the catalytic performance strongly depends upon the nature of the saccharide additive, the nature and location of the substituent on the benzene, water and the ball-milling conditions. Interestingly, the amount of reductive agent ($NaBH_4$) was drastically reduced compared to reductions performed in solution. Additionally, the catalytic system could be recycled over three consecutive runs without significant loss in activity, thus highlighting the efficacy of the combination of mechanochemistry, supramolecular chemistry, and catalysis (38).

Au-NPs of various shapes including sphere, triangle and hexagon were biogenetically synthesized using the yeast cells of *Magnusiomyces ingens* LH-F1. The Au-NPs exhibited excellent catalytic activities for the reduction of NPs (i.e. 4-NP, 3-NP and 2-NP) to APs in the presence of SBH. This is the first report on Au-NPs biosynthesis by *Magnusiomyces ingens*, which may serve as an efficient candidate for green synthesis of M-NPs (39). Cucurbit [7]uril-stabilized Au-NPs (5.7nm) as catalysts for the reduction of the pollutant 4-NP to 4-AP compound is reported. When compared with recently reported work, these NPs are found as efficient catalyst in 4-NP reduction with SBH (40). Application of ferrocene-resorcinarene in Ag-NPs (20-30nm) synthesis and applications in the reduction of 4-NP to 4-AP is studied. The Ag-NPs at 40 nanomoles quantity is sufficient to complete the reduction of 4-NP *over* ten minutes (41). The spherical Ag-NPs using Date Palm fruit extract as stabilizing agent were prepared, characterized and used in the reduction of toxic nitro compounds into less toxic corresponding amines by using SBH (42).

5B.05.17: Photocatalytic Reduction of NAs with NaBH$_4$ (SBH):

Clinopti-lolate-based silver bromide and silver bromide-titanium dioxide photocatalysts were prepared, characterized and used for the photocatalytic reduction of nitro compounds using SBH in aqueous medium. With this new approach the best efficiency is obtained over the Ag/AgBr/TiO$_2$/NZ in natural pH under UV light. It was observed that the photoreduction efficiency was affected by changing the SBH concentration and temperature (43). Ag-NPs were synthesized in various sizes. The UV-VIS results show that the positions of the surface plasmon resonance bands depend on the size of the Ag-NPs. In the prepared samples, Ag-NPs as a catalyst reductively transform 4-NP to 4-AP. The smallest NPs had the highest catalytic activity and the rate constant of chemical reduction decreased with increasing size of the Ag-NPs (44). Ag-NPs/TiO$_2$ photocatalysts were synthesized and tested for 4-NP reduction in the presence of aqueous SBH under UV irradiation (45). Direct evidence of plasmonic enhancement on catalytic reduction of 4-NP to 4-AP over Ag-NPs supported on flexible fibrous networks gave very good activity and recyclability of the catalyst. Ag-NPs supported on polyacrylonitrile (Ag-NPs/PAN) microfibrous network were synthesized and used as a direct evidence of plasmon-enhanced reduction of 4-NP to 4-AP through photo-assisted catalytic process with SBH as a hydrogen donor. The catalyst was about three times more active than normal catalytic system (46).

5B.06.00: Metal Borohydrides as Energy Storage Materials:

Considering the importance of a sustainable and green environment the use of suitable hydrogen storage source is critical. In recent times many natural sources, chemicals with ability to regenerate hydrogen are looked at as energy storage materials. In this context, besides boranes, alcohols, formic acid etc. metal borohydrides (M = Li, K, Na, Ca) have drawn attention as hydrogen storage materials. An ideal on-board hydrogen storage material will have a low molar weight, be inexpensive, have rapid kinetics for absorbing and desorbing H$_2$ in the 25–120°C temperature range, and store large quantities of hydrogen reversibly. With these aspects in mind, amongst metal borohydrides SBH has got least attention because of its high thermodynamic stability and slow hydrogen exchange kinetics. On the other hand with recent developments in promoting H$_2$ release and tuning the thermodynamics of the thermal decomposition of solid SBH and owing to its low cost and high hydrogen density (10.6wt%), SBH has received extensive attention as a promising hydrogen storage medium. Recently, these aspects of SBH, including the fundamental dehydrogenation and rehydrogenation pathways and many more such as catalytic doping, nanoengineering, additive destabilization and chemical modifications have been reviewed at length (47).

Different metal borohydrides, MBH$_4$, have been discovered, characterized and used as reducing agents and energy storage materials during the past decade revealing an extremely rich chemistry including fascinating structural flexibility and a wide range of compositions and physical properties. The possibility of lanthanide-bearing borohydrides related to solid state phosphorus and magnetic refrigeration is explored. Two major classes of metal borohydride derivatives have also been discovered: i) anion-substituted compounds where the complex borohydride anion, BH$_4^-$, is replaced by another anion, *i.e.* a halide or amide ion and ii) metal borohydrides modified with neutral molecules, such as NH$_3$, NH$_3$BH$_3$, N$_2$H$_4$, *etc.* Here, a review with new synthetic strategies along with structural, physical and chemical properties for metal borohydrides, revealing a number of new trends correlating composition, structure, bonding and thermal properties is presented. All these issues related to metal borohydrides and derivatives, their synthesis, structure and properties towards the rational design of novel functional materials are reviewed in detail (48).

References:

1). W. Liu, R. Chen, *RSC Advances*, 06(97):94451-94458(2016).

2). V. Z. Sharf, et al. *Bulletin of Academy Sciences, USSR*, 38(03):468 472(1998).

3a). M. Guo, et al. *J. Hazardous Materials*: 310:89-97(2016).

3b). C. Kastner, et al. *Langmuir*, 32(29): 7383–7391(2016).

4). A. Pal, et al. *RSC Advances*, 05(95):78006-78016 (2015).

5). E. Menumerov, et al. *Nano Letters,* 16(12):7791–7797(2016).

6). A. Mashentseva, et al. Nuclear Instruments and Methods in Physics Research Section B: Beam Interactions with Materials and Atoms, 365: Part-A, 70–77(2015).

7). G. Ngnie, et al. *Dalton Transactions*, 45(22):9065-9072(2016).

8). W. Ye, et al. *Applied Catalysis B: Environment*, 181:371-378(2016).

9). L. Mei, et al. *New J. Chemistry*, 40(03):2315-2320(2016).

10). C. Wang, et al. *J. Material Science* 52(16):9465–9476(2017).

11a). S. Gu, et al. *J. Physical Chemistry C*, 118(32):18618-18625(2014).

11b). S. Gu, et al. *Physical Chem.&Chemical.Physics.*, 17(42): 28137-28143(2015).

12). S. Mei, et al. *J. Materials Chemistry A,* 03(07):3382-3389(2015).

13). F. Zhao, et al. *Chin J. Polymer Science*: 33:1421–1430(2015).

14). A. Mondal, et al. *Bulletin Material Science*: 40:321–328(2017).

15). T. Wu, et al. *J. Material Chemistry A,* 01(25):7384-7390(2013).

16). E. Menumerov, et al. *Catalysis Science &Technology,* 07(07): 1460-1464(2017).

17). M. T. Shah, et al. *Microsystalline Technology*: 01-14 (2017).

18). P. Babji & V. L. Rao, *Intl. J. Chemical Studies*, 4(5):123-127(2016).

19). L. Qin, H. Xu, et al. *Catalysis Letters*: 147:1315-1321(2017).

20). K. Zelechowsk, et al. *Polish J. Chemical Technology*, 18(04):47-55(2016).

21). A. K. Sasmal, et al. *Dalton Transactions*, 45(07):3139-3150(2016).

22a). A. Rahman, et al. *Intl. J. Advances in Engineering & Tech*nology, 01(04):278-282(2011). 22b) Setamdideh et al. *Orient. J. Chemistry*, 27(3):991-996 (2011).

22c). Benington et al., *J. Organic Chemistry*: 18:1508 (1953); 22d). Finger et al., *J. Am. Chemical Society:* 81:98 (1959).

23). S. Bae, et al., *Applied Catalysis B: Environ,* 182:541-549(2016).

24). M. Gholami, et al. *Ind. Engineering & Chemical Res.,* 56(05):1159-1167(2017).

25). D. Kuo, et al. *RSC Advances:* 07:4353–4362(2017).

26). Q. Zhang, X. Fan, et al., *RSC Advances,* 06(47):41114-41121(2016).

27). D. Li, Z. Li, *Science China Mater*ials: 60:399–406(2017).

28). A. T. Mah, et al., *Powder Technology:* 315:147-150(2017).

29). T. Ma, W. Yang, et al. *Catalysts:* 07:38-47(2017).

30). P. Alonso-Cristobal, et al. *RSC Advances,* 06:100614-1006229(2016).

31). T. Duan, et al., *J. Physical Chemistry. C,* 119(37):21465-21472(2015).

32). S. Aromal, *Spectrochimica Acta-A: Molecular & Bio-molecular Spectroscopy,* 97:01-05(2012).

33). Y. Han, et al., *ACS Sustainable Chemistry & Engg,* 04(12): 6322–6331(2016).

34). S. Zhang, et al. *J. Hazardous Materials:* 321: 92–102(2017).

35). K. J. Kim, et al. *Applied Surface Science,* 414(31):325-334(2017).

36). S. Patil, et al. *J. Nanomedicine Nanotechnology,* 07(02):358-367(2016).

37). S. Saha, A. Pal, et al., *Langmuir,* 26(04):2885–2893(2010).

38). S. Menuel, et al. *Green Chemistry,* 18(20):5500-5509(2016).

39). X. Zhang, *Colloids, SurfacesA: Physicochem Engg Aspects,* 497:280-285(2016).

40). E. Blanco, I. E. Adell, *RSC Advances,* 06(89): 86309-86315(2016).

41). T. Sergeeva, et al. *RSC Advances,* 06(90):87128-87133(2016).

42a). A. R. Kiasat, et al. *Chinese Chemistry Letters,* 21:1015–1019 (2010).

2b). S. Farhadi, et al. *Acta Chimica Slovikia:* 64:129–143(2017).

42c). Z. Li, X. Xu, et al. *RSC Advances,* 05(38):30062-30066(2015).

43). H. Salari, et al. *Intl. J. Materials & Chemistry,* 01(01):49-52(2011).

44). R. Seoudi, et al. *World J. Nano Science & Engineering:* 06:29-37(2016).

45). M. M. Mohamed, *Applied Catalysis B: Environmental,* 142-143:432–441(2013).

46). S. Gao, et al. *Applied Catalysis B: Environmental,* 188:245-252(2016).

47). J. Mao & D. H. Gregory, *Energies:* 08:430-453(2015).

48). L. H. Jepsen, et al. *Chemical Society Reviews,* 46(5):1565-1634(2017).

Chapter-5C:
5C.00: CO + H$_2$O; CO$_2$ + H$_2$O and scCO$_2$ As H$_2$ Sources

5C.00.00: Introduction: Development of efficient methods with renewable sources, compact, flexible processes with high selectivity towards valuable end products of multiple applications are of great importance in the modern synthetic chemistry. We have seen that metal catalysts in the form nanoparticles (NPs)/nanoclusters or nanostructures or nanocomposites (NCs) have been used for efficient reductions of nitroarenes (NAs). Amongst different hydrogen sources, other than hydrogen gas, CO+H$_2$O is also a unique hydrogen source used for the hydrogenation of NAs and other substituted NACs. Aniline (AN) and its analogs are important chemical intermediates for many commercial products and hence they need economical, greener and safer synthetic and manufacturing processes. The major commercial synthetic pathways for manufacturing AN, is through the reduction of NB, which in turn is obtained by the nitration of benzene with a nitrating mixture (Conc.HNO$_3$+Conc.H$_2$SO$_4$). However, the nitration is an exothermic reaction and needs temperature control, and produces lots of waste through acid neutralizations and in most cases the use of noble metal catalysts for reduction of NB to AN. These factors contribute to higher process costs and environmental loads.

Chemoselctive hydrogenation of NB to AN, and its substituted or functional analogs can be made by utilizing different synthetic approaches. One of them is the use of transition metal catalyzed hydrogenation in the presence of CO+H$_2$O, CO$_2$+H$_2$O and super critical CO$_2$ (scCO$_2$). The reduction process can be performed with or without other solvents. Sometime alcohols and alkanes as hexane or heptane are used. Other hydrogen sources as alcohols, HCOOH or HCOONH$_4$, boranes, silanes etc. as hydrogen donors are covered under their individual names in this chapter only. These are unique hydrogen sources, other than hydrogen gas, used for the hydrogenation of NAs. These are just hydrogen sources and if used alone will give mixed results with low yields of the desired products. Strategies as adding promoters to catalytic systems help enhance the catalyst's activity and the product selectivity. The transition metals as Cu, Ni, Fe, Pd, Pt, Ru, Rh, Au, Ag, etc. have proved exceptionally great in achieving quantitative conversion and selectivity of the end products. For example, Au-NPs on TiO$_2$ or SiO$_2$ or Pt/Co(OH)$_2$ combinations are used in certain cases. Especially, in the latter case the Pt acts as a hydrogenation catalyst, while Co helps prohibit or prevent hydrodehalogenation of HANs obtained in-situ by reduction of HNBs. This is the best example of a co-operative, synergistic effect of two metals in a reaction. Let us look at individual cases as covered under different headings based on the metals used as catalysts and CO+H$_2$O, or CO$_2$+H$_2$O or scCO$_2$ or dense phase CO$_2$ as hydrogen sources under given experimental conditions. So far the catalysts are concerned, they are as metals (especially non-noble or noble metals as listed above), bi-metals, metal salts, metal nanoparticles (M-NPs) as such or on solid supports (as TiO$_2$, SiO$_2$, Al$_2$O$_3$ ZrO$_2$, Fe(OH)$_2$, Activated Carbon, Carbon Filaments, Graphene and Reduced

Grapheme Oxide (rGO), resins, foams, natural products etc). We will begin with Noble Metals in alfa-betical orders as gold, palladium, etc. as a metal catalysts and the above listed hydrogen sources.

5C.01.00: Noble-Metal Catalysts:

1). Gold (Au): The discovery of new catalysts as Au-NPs/CO/H_2O-mediated systems for chemoselective synthesis of multi-functional organic molecules has drawn attention in the recent past. So far, Platinum group metal catalysts (PGM: Pt, Ir, Rh and Pd over PtO_2) for CO/H_2O-mediated reduction at ambient temperatures is not reported. Li et al have demonstrated that by modulating the PGM catalyst particle size and electronic activities over the TiO_2 surface can be used as selective catalysts for CO/H_2O-mediated reduction of NAs under mild conditions. This might help to establish a more sustainable and industrially-relevant processes (1). A novel process for hydrogenation of NAs over ferric hydroxide supported Au-NCs in the presence of CO+H_2O is developed (Scheme-1).

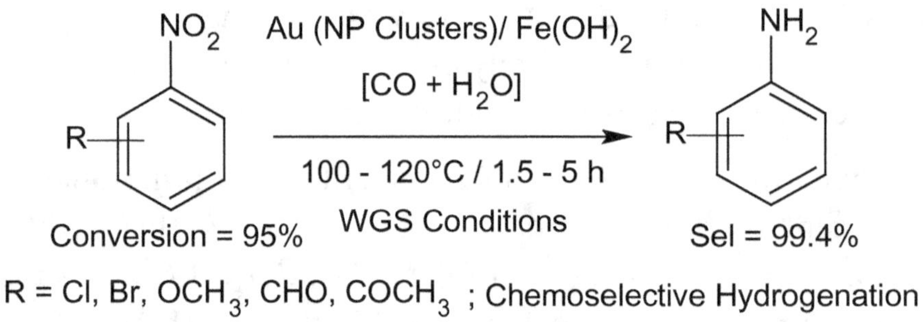

R = Cl, Br, OCH_3, CHO, $COCH_3$; Chemoselective Hydrogenation --Scheme-1

Different substrates depending on the substituent groups (R= Cl, Br, OCH_3, CHO, $COCH_3$ etc.) gave varying selectivities to ANs. Both the nitro substrate conversions and the AN selectivity were excellent (2). NAs were selectively reduced using Au/TiO_2-VS (VS=very small) and CO+H_2O as hydrogen source in ethanol-water and DMF-water as solvents.

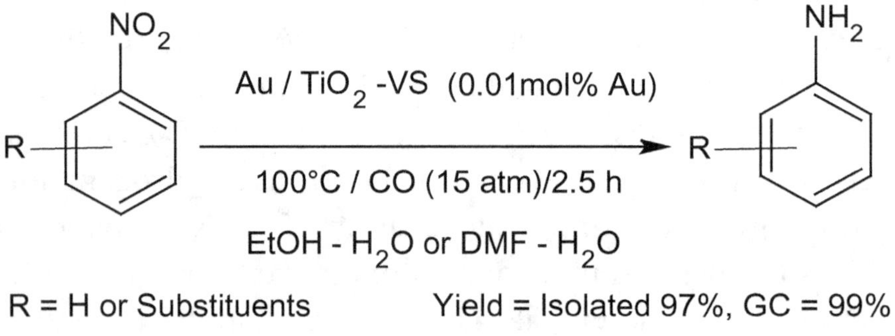

R = H or Substituents   Yield = Isolated 97%, GC = 99% --Scheme-2

The reaction was performed at room temperature and 5atm CO pressure.

Compared to other metals on support catalyst Au/TiO_2-VS were found most efficient. NB was reduced to AN with 99% conversion and 99% AN selectivity at 100°C and 15atm CO pressure for 2.5h in a solvent system (Scheme-2). Using this system many substituted NAs with highly sensitive groups such as CH=CH, Cl, CN, CHO etc. have been chemoselectively reduced to the corresponding ANs in almost quantitative yields (up to > 99%) and 96-99% NAs conversions (3).

Other groups also have used titania (TiO_2) supported Au-NPs and HCOOH or $HCOONH_4$ as hydrogen sources for chemo- and regioselective reductions of NAs to the corresponding ANs in ethanol at room temperature. This is because, under experimental condition HCOOH decomposes to $CO+H_2O$ (4), ideally giving the same system as Au-NPs/$CO+H_2O$. $HCOONH_4$ produces $CO+H_2O$ plus NH_3 gas (5). Blaser and Tafesh independently have reviewed the selective reduction of ANCs to aromatic amines using $CO+H_2O$ and metal complexes (6).

2). Palladium (Pd): As we have seen earlier that there are constant efforts to minimize the cost of manufacturing of aryl amines obviously for their regular demands in the market. One of the aspects of these studies is the use of alternate hydrogen donors in place of molecular hydrogen as a reducing agent. Dense phase CO_2 (DPCD) is usually safer than gas phase CO_2 (7), especially during transport and handling. In a study, AN selectivity was higher in DPCD than in ethanol for all the catalysts used and was almost 100% for Pd based catalysts. The $scCO_2$ is a suitable alternative as a solvent for good yields and 100% AN selectivity. Pd catalyzed hydrogenation of NB in $scCO_2$ (20MPa) at 90°C and H_2(5MPa) and with and without IPA as a traditional solvent is studied. The process of NB hydrogenation in the $scCO_2$ is seen as accelerated one. The rate of formation of AN in the $scCO_2$ medium is 3.5–5-fold higher than the reaction in IPA or without the solvent (8). The process selectivity towards aniline is 92–95%.

In-situ synthesized Pd-NPs supported on B-MCM-41 is found as an efficient catalyst for hydrogenation of NAs in $scCO_2$. Highly ordered materials of Pd-NPs supported on B-MCM-41 were obtained depending on the Si/B ratio through in-situ synthesis. The particle size of Pd, as well as its dispersion on solid support makes Pd/B-MCM-41 as an efficient catalyst for the reduction of NB in $scCO_2$ with exceptionally faster reaction rate (5min) and high yield of AN (100%).

--Scheme-3

The Pd particle size related to Si/B ratio and physical properties of CO_2 such as pressure- and temperature-dependent solvent power influenced the reaction rate. The mechanism is studied. The catalyst was found superior to other similar systems with different metals as Al and Ga. (9,10a, 10b). The concept of using $scCO_2$, $CO+H_2O$ and CO_2+H_2O has been successfully applied in the metal catalyzed selective reductions of NAs to ANs for economical manufacturing of aryl amines. The rate of NB hydrogenation is increased, and the selectivity to AN (>90%) is altered in the CO_2-expanded cyclohexane phase in the absence of dense phase CO_2. However, it increases up to >95% on CO_2 pressurization of the liquid reaction phase (11). NB reduction catalyzed by Pd and co-powered by $FeCl_3$, Fe_2O_3 or Fe powder in the presence of I_2 and pyridine and $CO+H_2O$ as a reducing agent is reported (Scheme-4). The reduction is carried out at 180°C, 2.5-4MPa pressure for 2hrs. NB conversion is at 98-100% and a 100% selectivity to AN is achieved (12).

$$NO_2 \xrightarrow[\text{CO (2.5 -4MPa) / 180°C/2h}]{\begin{array}{c}PdCl_2/Fe/I_2 \\ CO + H_2O\end{array}} NH_2$$

Conversions = 98 - 100%   Selectivity = 100% --Scheme-4

Mechanistic studies of Pd-Ligand catalyzed reduction of NB to AN with CO in methanol is performed. A comparative study between diphosphane and 1, 10-phenanthroline complexes is the focus of the study (13-14).

3). Platinum (Pt): As we know that two brains are better than one, this proverb is very much relevant in synthetic organic chemistry especially when it comes to enhancing the activity of one component by another. Scientists have realized that a combination of solvents, hydrogen sources, bases, metals etc. do give better results than going with a single component. These days there are countless examples following this protocol for achieving results which were not possible otherwise. Here, one of them is the study of understanding the molecular mechanism for the chemoselective hydrogenation of substituted NAs as nitrostyrene or HNBs with Au-NPs on TiO_2 support (15). Here, the TiO_2 is not acting just as a solid support to the Au-NPs, but also it shows a cooperative effect between gold and the support. By combining kinetics and in-situ IR spectroscopy, it has been found that the nitro groups adsorb weakly on the Au (111) and Au(001) surfaces. Although a stronger adsorption occurs on low-coordinated atoms in Au-NPs this adsorption is not selective. On the other hand, an energetically and geometrically favored adsorption through the nitro group occurs on the TiO_2 support and in the interface between the Au-NPs and the TiO_2 support. Such preferential adsorption is not observed with Au-NPs on silica and is not selective for the reduction of NACs, which is contrary to the Au/TiO_2 catalyst. Therefore, the high chemoselectiviy of the Au/TiO_2 catalyst can be attributed to the cooperation between the Au-NPs and the TiO_2 support (16) that preferentially activates the nitro groups.

Chemoselective hydrogenation of 3-nitrostyrene (3-NS) over a Pt/FeO$_x$ pseudo-single-atom-catalyst in CO$_2$-expanded liquids gives 3-aminostyrene (3-AS) with >95% selectivity. Chemoselective hydrogenation of NAs containing two reducible groups in one molecule is a highly desired approach to the synthesis of functionalized ANs. To make this process environmentally benign, a pseudo-single-atom-catalyst Pt/FeO$_x$ was used in scCO$_2$ and CO$_2$-expanded toluene. The results showed that scCO$_2$ afforded excellent selectivity but low reactivity due to the limited substrate solubility in the reaction medium. By contrast, when the reaction proceeded in CO$_2$ expanded toluene, both the conversion of 3-NS and the selectivity of 3-AS reached above 95% under optimum conditions, while the organic toluene amount could be reduced by 90% compared to that without CO$_2$. The thermodynamic calculations revealed that the solubility of H$_2$ increased while the viscosity of the system decreased with the CO$_2$ pressure, which facilitated the mass transfer and therefore increased the reaction rate while keeping the selectivity high.

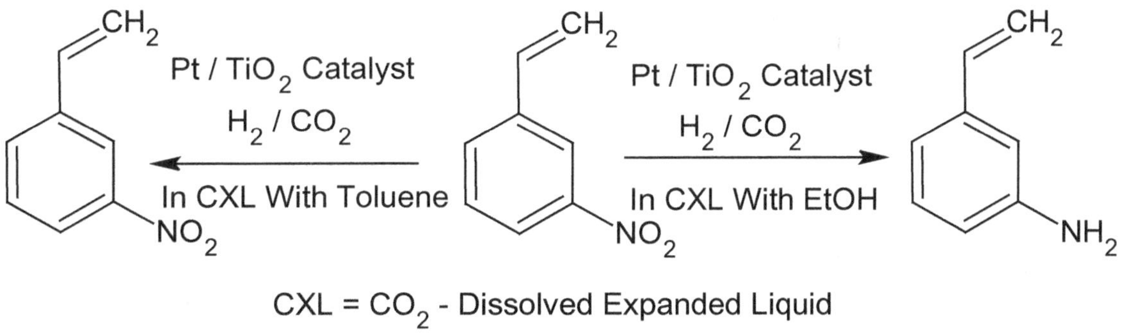

CXL = CO$_2$ - Dissolved Expanded Liquid

Scheme-5

Yoshida et al have studied the hydrogenation of 3-NS with a Pt/TiO$_2$ catalyst in CO$_2$-dissolved expanded polar and nonpolar organic liquids (CXL) (Scheme-5) (17). Hydrophobic solvent like toluene will have undesired effect on the local composition. This is because of the difference in the CO$_2$ pressure in these two different solvents. Therefore under identical experimental conditions 3-NS in ethanol was transformed into 3-AS but not in toluene, and accordingly the product selectivity in these two solvents is different (18). The same group has observed that during the selective hydrogenation of 3-NS to 3-AS over Pt/TiO$_2$ catalysts and CO$_2$-dissolved expanded liquid (CXL) phase at 12 MPa the phase behavior and pressurized CO$_2$ are important factors. This was ascribed to: i) retardation effects of scCO$_2$ on the hydrogenation of 3-NS to 3-ethylAN, ii) CXL phase gave better AS selectivity than in scCO$_2$ homogenous phase and iii) lower Pt loadings and temperatures were favorable for higher 3-AS selectivity.

3a). Pd/C and/or Pt/C as Catalysts: Water as a clean solvent and promoter in the organic synthesis has attracted more attention in recent times. Cheng et al. have studied the effect of water for the hydrogenation of o-CNB over Pt/C and Pd/C catalysts in ethanol, n-heptane and compressed CO$_2$. Interesting enough the reaction rate decreased in ethanol but increased in n-heptane and compressed CO$_2$ with the addition of water. For catalyst activity it is necessary to activate the functional group (NO_2) through the interactions via a hydrogen bonding and the other thing is to affect the solubility of hydrogen in ethanol and n-heptane, which is a key factor in the outcome of the reduction. The combination of H_2O and CO_2 is more efficient than the pure H_2O, CO_2 and H_2O–n-heptane systems. The o-CNB phase was expanded in the compressed CO_2 (9MPa) and so was the concentration of H_2 in o-CNB phase increased due to the miscibility of CO_2 and H_2 resulting in the enhancement of reaction rate and the maximum conversion. Between the two catalysts, it was found that the H_2O–CO_2 media and Pt/C catalyst is one of the most effective systems for the hydrogenation of o-CNB (19). The preparation of p-Aminophenol (PAP) from NB by one-pot catalytic hydrogenation and in situ generated acid-catalyzed Bamberger rearrangement was first realized in a pressurized CO_2/H_2O as a green system (Scheme-6).

Pd or Pt Catalyst

CO_2 + H_2O → H^+

80 - 160°C

CO_2 (5.5MPa) / H_2 (0.2 MPa)   Selectivity = ~ 85% --Scheme-6

Amongst Pd and Pt as catalysts, by employing Pt-Sn/Al_2O_3 as a catalyst complex NB could be converted to PAP with selectivity as high as 85% when the reaction was carried out at 140°C under 5.5MPa CO_2 and 0.2 MPa H_2 (20). Various surface species originating from the reaction between CO_2 and H_2 over Al_2O_3-supported on Pt, Pd, Rh, and Ru model catalysts (the Pt metal group, PGM) were investigated by attenuated total reflection infrared (ATR-IR) spectroscopy under high-pressure conditions. The end product distributions are dependent on the catalyst surface (21).

Pt, Et_3N, PPh_3
$SnCl_4$

CO/H_2O; 30Kg/cm^2

Y = 96%   --Scheme-7

NAs are readily and selectively transformed to aryl amines in excellent yields under mild conditions with CO+H_2O in the presence of a Pt catalyst. The presence of Et_3N,

$SnCl_4$ and PPh_3 are essential for the high catalytic activity (Scheme-7). NAs are reduced selectively by this procedure under even 30 kg/cm^2 of CO pressure producing AN in 90-96% yields (22). The choice of a solvent in organic synthesis is one of the key factors in achieving best results under given conditions. For example, hydrogenation of NB with Pt/C catalysts in scCO_2 and ethanol is performed with AN selectivity $\geqslant$80% at

10MPa, which is two times higher than in ethanol. The Pt particle size does interfere in the product selectivity in scCO$_2$ vs ethanol (23).

Ichikawa et al have found that chemoselective hydrogenation of HNBs over Pt/C catalysts proceeds effectively in scCO$_2$ to produce HANs with excellent selectivity. The rate of the hydrogenation of nitro groups is markedly enhanced in scCO$_2$ compared to the neat reaction, and the dehalogenation reaction is significantly suppressed (24). Ichikawa et al have identified right conditions to achieve complete conversion of substituted NAs (SNAs) and almost quantitative selectivity to ANs. In a particular case, p-CNB is reduced to p-CAN over Pt/C in scCO$_2$ with 99.7% selectivity which was attributed to the suppression of undesired dehalogenation. An interesting observation is revealed that CO generated in-situ by the reverse WGSR (CO$_2$+H$_2$→CO+H$_2$O) poisoned selectively platinum active sites. This feature of the catalyst and scCO$_2$ system inhibited undesired side reactions including dehalogenation and hydrogenation of aromatic rings resulting in the excellent product selectivity (99.7%) in scCO$_2$ (Scheme-8).

Conversions = 100%   Sel ; 99.7%   --Scheme-8

This process is applicable to other NAs as well with 100% NA conversions and >99% AN selectivity (25). Selective hydrogenation of HNBs without HAN's dehalogenation is a big challenge. We have seen earlier that in Pt/Co(OH)$_2$ bi-metallic catalyzed hydrogenation of HNBs the Co(OH)$_2$ is known to prohibit the dehalogenation effectively. Even in sensitive cases of BNB, INB the BAN and IAN, the selectivity reached >99.6% at 100% conversion of both the NBs (26a). Similar results are obtained when Pt/Co(OH)$_2$ is used as a catalyst in the presence of CO+H$_2$O, where the high selectivity to p-BAN (99.7%) was also obtained in the hydrogenation of p-BNB at 100% conversion (26b).

3b). Metal Nano Particles (M-NPs): M-NPs have emerged as a good alternative to molecular complexes, and currently much attention is being paid to their synthesis in order to achieve efficient and selective reductions of NAs with CO+H$_2$O mixture as a hydrogen source (Scheme-9) (27a).

--Scheme-9

Non-supported and supported NPs have been used for this catalytic reaction, and among them the Pt-NPs appear as promising catalysts due to the fact that they combine a good activity in nitro reduction while keeping a high degree of selectivity (27b).

High chemoselelctivity without delalogenations were obtained using PEG stabilized Pt-NPs (PS/Pt-NPs) immobilized on solid supports such as γ-Al$_2$O$_3$, SBA-15, TiO$_2$ and active carbon, etc. In the hydrogenation of p-CNB in scCO$_2$ the catalyst showed high selectivity to p-CAN >99.3% in the whole range of nitro conversions. Such high selectivity (>99.1%) was also obtained in the hydrogenation of other HNBs or HNTs (as o-CNB, m-CNB p-BNB, m-INB, and 2-chloro-6-nitrotoluene (2-CNT). The dehalogenation and the accumulation of intermediates were simultaneously and fully suppressed in scCO$_2$ (28).

3c). Selective Synthesis of PAP or 4-AP: An efficient green process was developed for production of PAP or 4-AP from hydrogenation of NB through the promoting effects of CO$_2$ and H$_2$O. The reaction is catalyzed by Pt/Al$_2$O$_3$ catalyst in scCO$_2$ and H$_2$O without any additional acid catalyst involved. The acidity aroused by the *in-situ* formed carbonic acid increased with CO$_2$ pressure, which acted as an acid catalyst during the reaction and it can easily self-neutralize by depressurization of CO$_2$. The process is a green route for PAP as it removed the final neutralization and salts disposal of the traditional process using liquid acids. Under the optimal reaction conditions PAP was obtained in high yield (68.9%) at 98.5% conversion of NB. The authors have observed through FTIR study that CO was generated by the reverse WGSR (CO$_2$ + H$_2$ $\rightarrow$ CO + H$_2$O). The in-situ generated CO gets adsorbed on the surface of Pt/Al$_2$O$_3$ to suppress the undesired deep hydrogenation of PHA to AN, which leads to high selectivity to PAP through Bamberger rearrangement (29). Supported bimetallic catalyst Pt-Pb/SiO$_2$ hydrogenation selectively reduces NB to PAP in an environmentally benign CO$_2$/H$_2$O pressurized system (Scheme-10).

--Scheme-10

Among the various bimetallic catalysts prepared, Pt-Pb/SiO$_2$ is the best with 82% PAP selectivity from NB at 110°C under 5MPa CO$_2$ and 0.2 MPa H$_2$ (30).

4). Rhodium (Rh): The reduction of NB by transition metal catalysts especially from group VIII (Rh, Ru) in association with ligands as phospines or phenathroline and in the presence of CO+H$_2$O (31a-c) is relatively a new approach. Ragaini et al. have done quite a work on water gas systems, especially utilizing CO+H$_2$O as a reducing agent and different catalysts for different functional groups as nitro, olefins etc. For example,

[Rh(CO)$_4$]$^-$M (M = K$^+$ or Cs$^+$ or PPN$^+$) was used in the reduction of NB to AN in water as a solvent. The process is selective for the reduction of nitro group only (31a). Activation of Rh$_6$(CO)$_{16}$ with 1,10-phenanthroline and substituted derivatives in the catalytic reduction of NB to AN with CO+H$_2$O has led to complete conversion of NB and quantitative yields of AN. The most active system is obtained using the ligand 3, 4, 5, 6, 7, 8-Me$_6$phen in the form of a chelate: Rh ratio of 3. The yields of 100% (substrate/Rh=1000) are easily achieved at 165°C without any basic co-catalyst (31b). Bartloni et al have reported NB reductions using carbonyl-Rh-pyridine complexes as [Rh(CO)$_2$(pyridine)$_2$](PF$_6$) in methanol-water under an atmosphere of CO (Scheme11).

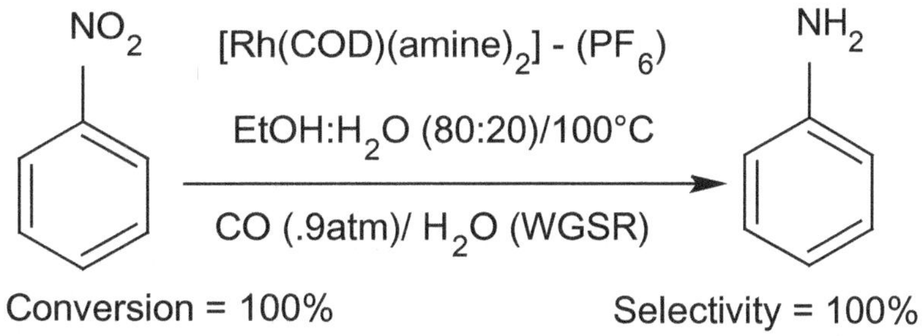

NO$_2$ [Rh(COD)(amine)$_2$] - (PF$_6$)   NH$_2$

EtOH:H$_2$O (80:20)/100°C

CO (.9atm)/ H$_2$O (WGSR)

Conversion = 100%   Selectivity = 100% --Scheme-11

NB conversions with 3-methyl and 4-methyl pyridine complexes with Rh were 100% and aniline as a sole product is obtained with 100% selectivity. The amount of product formed depends upon the nature of the amine coordinated to the Rh center (31c). The Rh cluster complex, Rh$_6$(CO)$_{16}$ has been found to catalyze the homogenous reduction of NB to AN at >80°C in the presence of N,N,dimethylbenzylamine and any of the following reducing gases: H$_2$/CO, H$_2$, or CO/H$_2$O. The reductions are highly selective and AN was the only product detected (32). Rh–amine complexes also have been applied successfully for NB reduction with CO+H$_2$O system under WGSR conditions. The emphasis is on quantitative NB conversions and 100% AN selectivities with full protection of other functional groups, high yields and environmental aspects (Scheme-12) (33a-d).

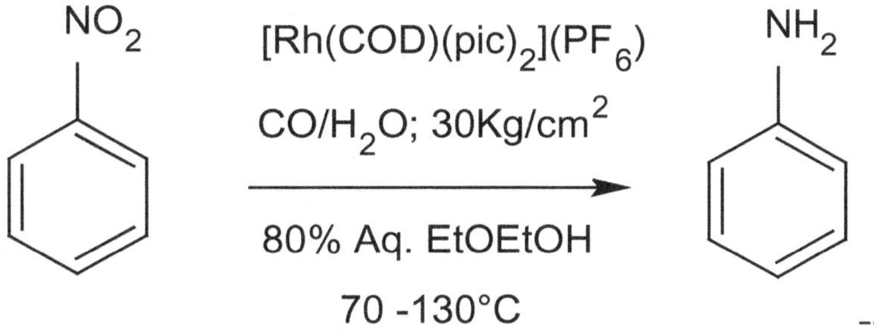

NO$_2$ [Rh(COD)(pic)$_2$](PF$_6$)   NH$_2$

CO/H$_2$O; 30Kg/cm^2

80% Aq. EtOEtOH

70 -130°C   --Scheme-12

The influence of the reaction conditions using CO pressure (0-2atm) and metal catalyst as Pt, Rh contents (~1-12 wt %) and temperature (70-130°C) on the catalytic reduction of NB to AN by [Rh(COD)(2-pic)$_2$](PF$_6$) is presented (Scheme-12). Reaction is performed in 80% aqueous 2-ethoxyethanol under WGSR conditions (34). Chemoselective reduction of NB to AN by CO/H$_2$O catalyzed by [Rh(CO)$_4$] in a homogeneous catalysis in water is performed. The [Rh(CO)$_4^-$ complexes are found as very active catalyst for the reduction of NB to AN by CO/H_2O in water as a solvent in the absence of a ligand and or a base (31a, 35).

5). Ruthenium (Ru): Reduction of NB to AN with $CO+H_2O$ catalyzed by $Ru_3(CO)_{12}$ with an alfa-diamine ligand at Ru: Ligand ratio of 1.5 at 150°C-180°C is studied. The NB conversion was low at 25-35% while the AN selectivity was at 91- 99% (36). Ragaini et al. have also reported a selective method of NB reduction using $Ru_3(CO)_{12}$ catalyst with P,N-ligands in the presence of $CO+H_2O$ (Scheme-13).

NB Conversion upto 51% Sel to Aniline = upto 99% --Scheme-13

$Ru(CO)_{12}$ complexes with Ru-Bis(arylimino)acenapthene (Ar-BIAN) and bis(phenylimino)phenanthrene (Ph-BIP) are found as most effective catalysts in NA reductions using $CO+H_2O$ system. The ligands are found quite superior to earlier ones and the reduction is highly chemoselective to nitro group in the presence of olefins and keto groups (37a-c).

The $Ru_3(CO)_{12}$/PEDPA (P,N-Containing Bidentate Phosphine Ligand, PEDPA)) complex has been found to be effective in the CO selective reduction of 4-propylthio-2-nitroaniline to 4-propylthiobenzene-1,2-diamine (Scheme-14).

Conversions upto 77%   Selectivity = 97 - 98% --Scheme-14

Under the optimum conditions, as 140°C, CO (5.0MPa) and substrate/catalyst molar ratio=300, the substrate conversion and the amine selectivity were 77% and 98% respectively. The functional chemoselectivity and the recyclability of the catalysts remained unaffected till the sixth run (38). Considering the importance of novel processes many groups have published their work using Rh, Ru and $CO+H_2O$ systems. These catalytic reactions are useful from both synthetic and industrial viewpoints, because the after-treatment of by-products can be simplified in comparison to the conventional methods as the reactions proceed with high selectivity. Nomura et al have published a review which highlights recent progresses in the selective reduction of NAs to aryl amines catalyzed by homogeneous Ru or Rh catalysts under CO/H_2O conditions (37).

Catalytic reductions of ANCs to amines using transition metal (Rh, Ru, Os, Ir & Pt etc.) carbonyl complexes as catalysts and carbon $CO+H_2O$ in place of hydrogen is reported (39). Under experimental conditions the carbonyls act as catalyst to convert $CO+H_2O$ to CO_2+H_2. The metal carbonyls which are more active than $Fe(CO)_5$ and together with it in the presence of H_2O, CO, and a weak base such as Me_3N, serve as catalysts for the conversion of NB, DNB, and 2,4- and 2,6-DNTs to the corresponding ANs.

A new method for continuous hydrogenation in supercritical fluids (CO_2 or propane) using heterogeneous noble metal catalysts on Deloxan aminopolysiloxane supports is investigated, which has potential both for laboratory-scale hydrogenation and also for the industrial production of fine chemicals (40). In a unique case the catalytic hydrogenation of NB over Pd/C, Pt/C, Ru/C, and Rh/C catalysts exhibit 100% selectivity to AN in $scCO_2$, which is higher than in ethanol as a solvent (41).

6). Silver (Ag): Kaneda et al. have reviewed NA reductions using noble metals supported on hydrotalcite (Metal/HT) titled, "Development of Heterogeneous Olympic Medal Metal-Nanoparticles Catalysts for Environmentally Benign Molecular Transformations Based on the Surface Properties of Hydrotalcite (HT)". As an illustrative example of multi-functional NAs with sensitive groups, 3-nitrostyrene (3-NS) is reduced to 3-aminostyrene (3-AS) (Scheme-15) using HT-supported M-NPs as Ag/-NPsHT, Au-NPs/HT, Pd-NPs/HT, Rh-NPs/HT and Pt-NPs/HT etc.

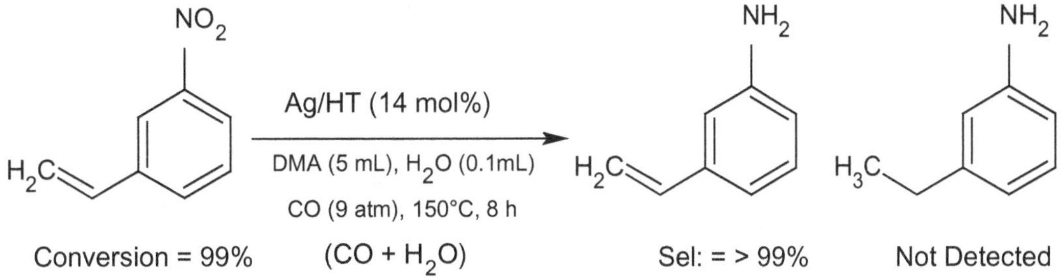

Scheme-15

Ag/HT with $CO+H_2O$ as a reducing agent gave >99% selectivity to corresponding ANs at 94-99% conversions of the substrates under 9atm of CO pressure at 150°C. 3-AS was formed in >99% yield in 3h without any side products such as 3-ethyl-AN or 3-ethyl-NB. The time course for the reduction of 3-NS using Ag/HT and other HT-supported M-NPs (Au/HT, Pd/HT, Pt/HT and Rh/HT) was studied. Although the activity of the Au/HT catalyst was higher than that of Ag/HT, an over reduction to 3-ethyl-AN was observed resulting in lower selectivity toward 3-AS. Pt/HT and Rh/HT functioned as good catalysts, giving 3-AS in moderate yields with high selectivity, while Pd/HT did not exhibit high activity or chemoselectivity in the intermolecular competitive reaction of NB and styrene. The table below shows the % conversions of NAs and respective selectivity towards corresponding ANs using this procedure. The authors believed that an integrated heterogeneous catalyst that rely on the concerted actions between Olympic medal M-NPs and the surface properties of inorganic crystallites will open new avenues for development of novel organic transformations and contribute to the realization of Green and Sustainable Chemistry

Table-1: Reduction of Substituted NBs (0.25mmol) with Ag/HT (14 mol %) in DMA-H_2O and CO (9atm) at 150°C (42).

No.	4-Substituent in NB	Time (h)	% Conversion	%Selectivity
1	CH_2=CH -	8	99	>99
2	CH_3 – CH=CH -	8	96	>99
3	Ph – CH=CH -	24	95	>99
4	3,4-Pyrrole -	24	94	>99

7). Noble-Metals on Carbon and scCO$_2$: Besides CO+H_2O as a reducing agent for the reduction of NAs, scCO$_2$ in water and some alcohols also gives excellent chemoselectivity to corresponding ANs. See the following illustrative examples.

Table 2: Results of hydrogenation of NB in scCO$_2$ at 35°C

Medium	Catalyst	Time(min)	%Conversion	%AN Sel
CO	5% Pd/C	10	52	100
14MPa	5%Pd/C	50	100	100
	5%Pt/C	10	71	79
	5%Pt/C	50	100	100
	5%Ru/C	10	4	100
	5%Rh/C	10	12	100

F. Zhao et al have done a very good work on hydrogenation of NAs over transition metals on solid supports as Pt/C or Pd/C, and scCO$_2$ and obtained excellent results with respect to the product selectivity (Table-2) (43).

Supercritical CO$_2$ (scCO$_2$) Reviews: Most of the solvents used in the traditional organic reactions are flammable, toxic and generate waste and pollution and hence must be replaced by other non-toxic solvent as scCO$_2$. The properties of scCO$_2$ as viscosity, density, coefficient of mass and heat transfer, dielectric constant can be adjusted by altering pressure or temperature slightly near the supercritical point. Therefore, the H_2 and organic substrates could dissolve into the scCO$_2$ to form a homogeneous phase or the CO_2 carrying H_2 dissolves into the organic substrates to form an expansive phase resulting in an increase of H_2 concentration in the reaction systems. This further helps in improving reaction rates, the interaction between CO_2, substrate and intermediates etc., which in turn could lead to improve the selectivity of the target product. The participation with the substrate molecules through hydrogen bonding could accelerate reaction rate. The molecular interactions between CO_2 or H_2O with the functional groups of reactants, higher H_2 concentration in the reaction system, and the acidity of the CO_2/H_2O system, etc. improve the reaction rate and product selectivity. The authors have reviewed the characteristics of the CO_2/H_2O as a green solvent system and their promoting functions played in the catalytic hydrogenation (44-45).

References:

1a). S. Li, F. Wang et al. *Chinese J. Chemistry,* 35(05): 591-595(2017).

1b). Z. Rappoport, et al.The Chemistry of Organogold Compounds: 2 Volume Set, John Wiley & Sons, 20-Apr-2015 - Science - 1232 pages.

2a). L. Liu, et al. *Chemical Communications*: 653-655(2009). 2b). L. Liu, et al. *Dalton Transactions,* (19):2542-2548(2008).

3). L. He, et al. *Angewandte Chemie Intl Edition,* 48:9538-9541(2009).

4a). L. Tao et al. *Advanced Synthesis & Catalysis,* 357(4):753–760(2015). 4b). L. Yu, Q. Zhang, *ChemSusChe*m, 08(18):3029–3035(2015).

5). X. Lou, L. He, *Advanced Synthesis & Catalysis,* 353(2-3): 281-286(2011).

6a). A. M. Tafesh, et al. *Chemical Reviews,* 96:2035-2052(1996). 6b).H. U. Blaser, et al. *ChemCatChem,* 1:210-221(2009), 6c). Ichikawa et al. *Synfacts* 9(01):107-118(2013). 6d).X. Liu, L. He, et al. *Accounts Chemical Res,* 47(3):793–804(2014). 6e). Gold Catalysis: Preparation, Characterization, and Applications: Laura Prati, Alberto Villa, CRC Press, 05-Jan-2016 - Science - 526 pages, PP 459-462.

7). F. Eldevik, *Pipeline and Gas Journa*l, Vol 235(11):(2008). https://pgjonline. com/2009/03/20/safe-pipeline-transmission-of-co2/

8). M. Y. Rakitin, et al. *Catalysis in Industry,* 07:01-05(2015).

9a). T. Ishizaka, et al. *Green Chemistry,* 14(12):3415-3422 (2012). 9b).M. Chatterjee, et al. *Advanced Synthesis & Catalysi*s, 354(10):2009-2018(2012).

10a). T. Dang-Bao, et al. *Catalysts,* 07:207-240(2017). 10b). M. Chatterjee, et al. *Green Chemistry,* 14(12):3415-3422(2012).

11). J. Hao, et al. *Industrial Engineering.& Chemical Res,* 47(17):6796–6800(2008).

12a). J. Skupinspa, et al. *Reaction Kinetics & Catalysis Letters*, 72(01):21-27(2001). 12b).A. Krogul, et al. *J. Molecular Catalysis A: Chemical*, 337(1-2): 9-16(2011). 12c). A. Krogul et al. *Organic Process Research & Development*, 19(12):2017–2021(2015).

13). T. J. Mooibroek, et al. *Organometallics*, 31(11):4142–4156(2012).

14). J. M. Thomas "Catalysis Making the World a Better Place", Catalysis, CO_2 as a green feedstock and using single-site catalysts is summarized., Published on 11 January 2016. http://rsta.royalsocietypublishing.org/content/374/2061/20150226.

15a). A. Corma et al. *J. American Chemical Society,* 130(27):8748–8753 (2008). 15b). K. Shimizu, et al. *J. Physical Chemistry C*, 113 (41):17803–17810 (2009).

16a). M. Boronatet al. *J. American Chemical Society,* 129(51):16230-16237(2007). 16b). L. Wang, et al. *ACS Catalysis, 06* (07):4110–4116(2016).

17a). G. Xu, H. Wei, et al. *Green Chemistry*, 18(05):1332-1338 (2016). 17b). A. Corma, et al. *Faraday Discussions*: 188:09-20(2016).

18a). H. Yoshida, et al. *J. Physical Chemistry C,* 115 (5):2257–2267(2011).

18b). S. Fujita, et al. *J. Supercritical Fluids:* 60:106-112(2011).

19). H. Cheng et al. *Applied Catalysis A: General*, 455:08-15(2013).

20). T. Zhang, et al. *Org. Proc Research &Development,* 19(12):2050–2054(2015).

21). D. Ferri, et al. *J. Physical Chemistry B.* 109(35):16794-16800 (2005).

22). Y. Watanabe, et al. *Tetrahedron Letters*, 24(38):4121-4122(1983).

23). F. Zhao, et al. *J. Catalysis,* 224(02):479-483(2004).

24). S Ichikawa et al. *Chemical Communications (Camb),* (07):924-926(2004).

25a). S. Ichikawa et al. *Chemical Communications:* 07:924-926 (2005). 25b). S. Ichikawa, et al. *Advanced Synthesis & Catalysis,* 356(11-12):2643-2652(2014).

26a). H. Cheng, et al. *J Colloid Interface Sci.* 377(01):322-327(2012).

26b). X. Chao et al. *Current Organic Chemistry,* 16(02):280-296(2012).

27a). J. Li, L. Liu et al. *Energy & Environmental Science,* 07(01):393-398(2014).

27b). P. Lara & K. Philippot, *Catalysis Science & Technology,* 04:2445-2465(2014).

28). H. Cheng, X. Meng, et al. *J Colloid Interface Science,* 415:01-06(2014).

29). L. Zhao, H. Cheng, et al. *J. CO_2 Utilization,* 18:229-236(2017).

30a). T. Zhang, et al *Chinese Chemical Letters,* 28(2):307-311(2017). 30b).S. K. Tanielyan, et al. *Org. Process Research & Development, 11* (04): 681–688(2007).

31a). F. Ragaini & S. Cenini, *J. of Molecular Catalysis A: Chemical*, 105(03): 145-148(1996).

31b). E. Alessio, et al. *J. of Molecular Catalysis* 22(3): 327-339(1984). 31c).M. Bartolini et al. *J. Chilean Chemical Society*, 52(03):1254-1256(2007).

32). Ryan et al. *J. Molecular Catalysis*, 05:319-330(1979).

33a). A.J. Pardey, et al. *J. of Molecular Catalysis A: Chemical*, 164: 225–234 (2000)), 33b). A. J. Pardey et al *Inorganica Chimica Acta* 329:22–30(2002), 33c). M. M. Mdleleni, et al. *J. of Molecular Catalysis A: Chemical*, 204–205: 125–131(2003). 33d). C. Longo, et. al. *Polyhedron*, 19:487-493(2000).

34a). J. Mayora, et al. *Bulletin Society of Chilean Quím.*, 46 (2):121-129 (2001). 34b). C. Linares, et al, *Catalysis Letters*, 50(03-04):183-185(1998). 34c). K. Kaneda et al. *J. Molecular Catalysis:* 12:385-387(1981).

35). F. Ragaini et al. *J. of Molecular Catalysis A: Chemical*, 174(01-02): 51-57(2001).

36). F. Ragaini, et al. *J. Molecular Catalysis*. 85: Ll-L5 (1993).

37a). M. Vigaino, F. Ragaini et al. *ChemCatChem*, 02(09):1150-1164(2010). 37b). K. Nomura, *Catalysis Surveys from Asia*, 02(01):59-69(1998). 37c).F. Ragaini, et al. *Organometallics*, 14(01):387-400(1995).

38). J. Y. Jiang, et al. *Chinese Chemical Letters*, 15(4): 394-396 (2004).

39a). K. Cann , T. Cole , et al., *J. American Chemical Society*, 100(12):3969-3971(1978). 39b).G. Mestroni et al. EP0097592 A2(1984). 39c). R. Ugo, Aspects of Homogeneous Catalysis: A Series of Advances, Springer Science & Business Media, 06-Dec-2012 - Science - 220 pages.

40). M. G. Hitzler, et al. *Organic Process Res. & Dev.*, 02(03):137–146(1998). 41). F. Zhao, et al. *Preprints Papers-Am. Chem. Soc., Div. Fuel Chem.* 49(01):13-14(2004).

42). K. Kaneda et al. *Molecules* 15:8988-9007(2010).

43). F. Zhao, et al. *Advanced Synthesis & Catalysis*, 346(06):661-668(2004). 43c). F. Zhao, S. Fujita, et al. *Catalysis Today*, 98(04):523-528(2004). 43d).F. Zhao et al. *J. Catalysis*, 224:479-483(2004). 43e). F. Zhao et al. *J. Catalysis*, 264(01):01-10 (2009).

44a). H. Cheng, & F. Zhao, *Chinese Science Bulletin*, 60(26): 2482-2489(2015). 44b). H. Cheng & F. Zhao, *Scientia Sinica Chimica*, 45(03):308-326(2015).

45a). Catalytic reduction of nitrobenzene in scCO$_2$ by Wu & Man, PP 320, in a Book: "Innovations in Green Chemistry and Green Engineering: P. T. Anastas, & J. B. Zimmerman, Springer Science & Business Media, - Science - (2012), 334 pages.

45b). M. Chatterjee , et al. Advances in CO$_2$ Capture, Sequestration, and Conversion: *ACS Symposium Series*, Vol. 1194, Chapter 9, pp 191–250 (2015).

5C.02.00: Non-Noble Metal Catalysts:

1). Cobalt (Co): Earth abundant metal catalysts, alternate hydrogen donors in place of molecular hydrogen and cheaper solvents are the way forward for economical processes in regularly needed chemicals at commercial scales.

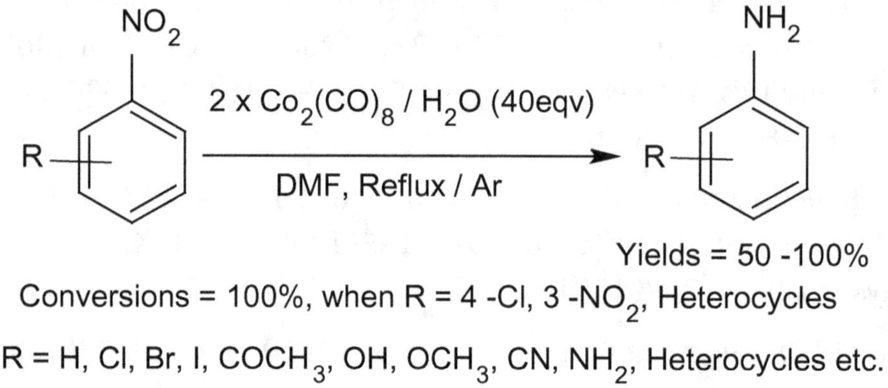

$$2 \times Co_2(CO)_8 / H_2O \text{ (40eqv)}$$

DMF, Reflux / Ar

Yields = 50 -100%

Conversions = 100%, when R = 4 -Cl, 3 -NO_2, Heterocycles

R = H, Cl, Br, I, $COCH_3$, OH, OCH_3, CN, NH_2, Heterocycles etc. --Scheme-16

A wide variety of substituted NBs were reduced in high to quantitative yields (some at 100%) using $Co_2(CO)_8$ (2eqv) and water (40eqv) in DME under reflux conditions (Scheme-16). The reaction is completed within 30-60 minutes with a high level of chemoselectivity, making it as a general procedure for large scale preparation of aryl amines under selective, rapid and mild conditions (46).

Table-3: Selective Reduction of Nitroaromatics Using $Co_2(CO)_8/H_2O$:

No	Substrate	Product	% Yield(Isolated)
1	4-CNB	4-CAN	99%(100%Conversion)
2	4-BNB	4-BAN	80%
3	2-BNB	2-BAN	58%
4	3-AminoNB	3-AminoAN	93%(100%Conversion)
5	4-Nitrotoluene	4-Toluidine	100%
6	2,4-DN-CB	2,4-DA-CB	100%

CNB =ChloroNB, BNB=BromoNB, DN-CB=Dinitro-Chlorobenzene

Tafesh et al have reviewed the selective reduction of ANCs to aryl amines using CO and metal complexes (47). Hydrogen-free cobalt–rhodium hetero-bimetallic-NPs catalyst is found useful in synthesizing AN-analogs by reductive amination of carbonyl compounds with NAs in the presence of CO+H_2O. The water added and generated in-situ produces hydrogen gas via WGSR system. The catalyst is stable and can be recycled and has a potential of using at commercial scale (48). Co catalyzed (Co_3O_4-NGr/C) hydrogenation of NAs leads to corresponding ANs with 85->99% substrate conversions and the corresponding ANs were obtained in yields >87-99% (49a). Here, HCOOH is used as a hydrogen donor, which in the presence of Et_3N undergoes in-situ decomposition on the metal/solid surface to produce CO_2 and H_2 (Scheme-17) (49b). HCOOH = CO_2 + H_2

> 50 Examples, Yields = 89 - 97% --Scheme-17

Other Co-Fe-Graphene based catalyst systems (Co-Co_3O_4/NGR@C or Fe_2O_3/NGR@C) have been used for selective hydrogenations of NAs, but involve the use of hydrogen gas and hence they are covered under bi-metallic catalytic hydrogenation of NAs. Amongst the other ways or mechanisms of generating H_2 from CO_2 and H_2O, the photocatalytic reduction of CO_2 and H_2O with Et_3N occurred efficiently using a cobalt (II) chloride complex adsorbed on multi-walled CNTs as a CO_2 reduction catalyst. $[Ru^{II}(Me_2phen)_3]^{2+}$ (Me_2phen = 4,7-dimethyl-1,10-phenanthroline) functions as a photocatalyst to yield CO and H_2 (2.4:1) and a high turnover number of 710 (50).

2). Copper (Cu): Aniline is prepared from NB and $CO+H_2O$ at 75--100°C with copper carbonate-ammonia complex as a catalyst (NH_3: Cu = 2:1) at pH 10. The substrate-product list is given as: *p*-Nitroanisole to *p*-Anisidine, *o*-CNB to *o*-CAN, *p*-Nitrobenzoic acid to *p*-Aminobenzoic acid, 1-Nitronaphthalene to 1-Aminonaphthalene (51). Yanez et al. have reported the reduction of NB using Cu (II) immobilized on poly(vinylpyridine) and Rh-pyridine complexes under $CO+H_2O$ in ethoxyethanol to AN in 98% yield, while azobenzene was present at 2% (52, 53).

3). Nickel (Ni): As a general case, Ni catalyzed chemoselective hydrogenation of NB with 99.9% conversion to AN with 99.5% selectivity is reported. Similar transformations are seen in the liquid-phase hydrogenation of *o*-CNB to *o*-CAN over supported nickel catalysts on solid supports like SiO_2, ZrO_2, TiO_2 and Al_2O_3 (Scheme-18) (54, 55).

Conversion = 99.9%   Selectivity = 99.5% --Scheme-18

Similarly, Ni on solid support is used as an efficient catalyst for hydrogenation of NB with 100% AN selectivity in the presence of dense phase CO_2 as a hydrogen source and other organic solvents. The selective hydrogenation of NB over Ni/γ-Al_2O_3 catalysts in ethanol, and in n-hexane was found better than using other solvents under similar reaction conditions. This catalytic system shows significant molecular interactions which lead to almost 100% selectivity to AN over the whole NB conversion range of 0-100%. Dense phase CO_2 strongly interacts with NB, NSB (nitrosobenzene), and PHA, which reflects in the reaction rate and the selectivity to AN (56). The salient features of this catalyst system are an effective medium of H_2O and low-pressure CO_2 (1-18MPa). The metal catalyst and the type of solid support play an important role in the outcomes of the reductions of NAs (Scheme-19). Meng et al. have realized 100% selectivity to aniline at any conversion range of NB using Ni on alumina ($Ni/y-Al_2O_3$) and $scCO_2$ in ethanol as a solvent system. The reduction proceeds through PHA as an intermediate (Scheme-19).

NO_2 NH_2

$Ni/Y -Al_2O_3$ / EtOH

$scCO_2$(dense phase) / Pressure

Conversion = 0 -100%   Selectivity = 100%

at any conversion --Scheme-19

The CO_2+H_2O system was a better solvent than CO_2 alone, Ethanol+CO_2 and Hexane+CO_2. At about 80% nitro conversions more than 99% corresponding ANs are obtained (Scheme-20) (56, 57-58).

NO_2 NH_2

Ni / TiO_2

R   R

H_2O + CO_2 / 35 - 50°C / 1 - 18MPa

Conversiosn Upto 80%   Yields > 99% --Scheme-20

The significance of molecular interactions and hydrogenation mechanisms has been studied during selective hydrogenation of CNBs to CANs over Ni/TiO_2 at 35°C in $scCO_2$, ethanol, and n-hexane. The reaction rate was solvent dependent and followed the order of $scCO_2$ > n-hexane > ethanol. In $scCO_2$, the selectivity to CANs and AN over Ni/TiO_2 were 97–99.5% and <1% AN respectively, with the conversion of NAs in the range of 0–100% with no dehalogenation in cases of HANs (56, 59).

4). Ce, Cu, Ni, as Catalysts: Organic Carboxylic Acids as H_2 Source: The CO, CO_2 and $scCO_2$ in water are highly efficient hydrogen sources and alternatives for molecular hydrogen in the catalytic hydrogenations of organic functional groups including NAs. There are some examples where organic carboxylic acids such as formic acid and oxalic acid serve as a source CO or CO_2. The following examples illustrate these observations. Metal catalyzed reductions of NAs with HCOOH as a hydrogen source is one of the important reactions in hydrogenation processes. HCOOH decomposes under experimental conditions on metal surfaces to give $CO+H_2O$ or CO_2 and H_2. For example, on reduced ceria (CeO_2) only CO_2 and H_2 were produced at 450-600K (60). Formic acid, HCOOH, decomposes to CO and H_2O in gas phase but to CO_2 and H_2 in aqueous phase (61). Cu (110) oriented surface and Ni surface also decompose HCOOH into CO_2 and H_2 (62, 63). Supercritical CO_2-($scCO_2$) is already being used in large scale production of CeO_2 nanowires, which find applications in solvent-free selective hydrogenation of NAs and HNBs (64). A process for manufacturing ANs is developed by the reduction of NACs in the presence of a catalyst and a reducing agent as hydrogen or carbon monoxide and water or an aliphatic alcohol at 100-200°C. The catalysts used are a mixture of two or three heavy metals (binary or ternary metal mixtures) as Mn, Fe, Co, Ni, Cu, Ag, Ce, etc. as their oxides, hydroxides or carbonates (65).

6). Selenium (Se): Selective reduction of substituted NAs to corresponding aryl amines is accomplished by selenium catalyzed hydrogenations with $CO+H_2O$ in the presence of Et_3N. H_2Se is proposed as the reduction species in this reaction.

Conv = 100%   Selectivity = 100% --Scheme-21

$$ArylNO_2 + 3CO + H_2O = ArylNH_2 + 3CO_2$$

In a control experiment, NB was reduced to AN with 100% selectivity with H_2Se and quantitative conversion of NB under N_2 atmosphere (Scheme-21) (66). Always, the hunt of new and improved methodologies leads to better technologies. In the presence of metal-ligand catalysts, $CO+H_2O$ is one of the best hydrogen source for reducing NAs to ANs under mild conditions as atmospheric pressure and room temperature. As an illustrative example, Se catalyzed reduction of NB to AN at moderate temperatures by CO/H_2O system at 4MPs CO pressure is reported.

Se/Nb = 0.02; H_2O/NB =10

Conversion = 98.2%   Selectivity = 100% --Scheme-22

Under specific conditions the process was found at its best with NB conversion >98% and the AN selectivity at 100% (Scheme-22) (67). Besides noble metals as Pd and Pt, Se/CO/H_2O has shown a consistent performance in heterogenous catalytic hydrogenations of NAs. The progress of the metal-catalyzed reduction of NBs including Se/CO/H_2O system to prepare ANs is reviewed in detail (68). In the presence of selenium as a catalyst the ANCs are efficiently and quantitatively reduced by CO/H_2O to form the corresponding amines under atmospheric pressure.

R = 12 Substrtates Conv = 100%, Sel.= 100%, Exceot o -& p -CNB --Scheme-23

The reduction occurs in high selectivity regardless of other reducible functionalities present on the aromatic ring, making it one of the highly chemoselective hydrogenation process for NAs (Scheme-23) (69).

7). Sulfur & Vandium (S & V): In an effort to directly synthesize N-formyl derivatives of AN, catalytic reductive carbonylation of NB to methyl N-phenylcarbamate is possible. A new three-component catalytic system (S or its Compounds +CH_3COONa+V_2O_5) can be used for reductive carbonylation of NB with CO and CH_3OH which gives methyl N-phenylcarbamate and AN. The initial CO pressure is equal to 14MPa at 298K and the reaction takes place at 360-470°C (70). It has been proved that the hydrogen from water molecule is taking part only indirectly in the reaction with NB to give AN (Scheme-24).

--Scheme-24

The reaction mechanism involves the formation of COS from S and CO and in the presence of CH_3COONa a complex with NB is formed. Thus, the NO_2 groups are reduced to give a nitrene group, which in turn reacts with CO and the intermediary phenyl isocyanate quickly reacts either with $H_2$0 to give AN or with CH_3OH to give methyl-

N-phenyl carbamate (PhNHCOOCH$_3$). This idea is supported by the favorable effect of oxygen addition to the reaction system as well as by the positive effect of vanadium oxide (71). High conversion of NAs to ANs is achieved through a multi-component system involving S, Amines, and V as a catalyst (COS, or H$_2$S, + Et$_3$N + NH$_4$VO$_3$) at 12MPa pressure of CO and at 25-165°C (Scheme-25).

Conv ~ 99% H$_2$O: NB= >5   Sel = > 96%   --Scheme-25

The feedstock conversion was up to 99.8%. The CO pressure, H$_2$O: NB ratio and the presence of solvents like methanol and dioxane did help in achieving high aminophenol selectivity. Sulfur based catalytic system was found to be very effective and its low price and non-sensitivity to the common catalytic poisons makes it more advantageous than the other catalytic systems (72).

8). Titanium (TiO$_2$): Titanium(IV)oxide in the presence of photolight and oxalic acid selectively reduced NB to AN. Oxalic acid acts as a hole scavenger and enhances the AN yield in the presence of molecular oxygen or dioxygen (Scheme-26).

--Scheme-26

The oxalic acid decomposes to give CO$_2$ and H$_2$O. The in-situ formed CO$_2$+H$_2$O system acts as a reducing agent. NB and PHA species were not detected (73).

9. Zinc (Zn): An eco-friendly Zn-H$_2$O-CO$_2$ system is presented for chemoselective reduction of NAs to ANs with high yields (80% –97%) in scCO$_2$ (Scheme-27).

scCO$_2$ = Super Critical CO$_2$ --Scheme-27

This process brings together efficient and important green chemistry technologies as it involves the use of CO_2 as a solvent and the use of water as a H_2 donor (74).

$$CO_2 + H_2O \longrightarrow H_2CO_3$$

$$2 \times H_2CO_3 \xrightarrow{M} 2HCO_3^- + H_2$$

$$2HCO_3^- \xrightarrow{M} 2CO_3^- + H_2$$

M = Metal Catalyst Surface   --Scheme-28

Mechanistically, CO_2 and H_2O react to form H_2CO_3, which in turn decomposes to give HCO_3^- and H_2 gas, which then reduces NB to AN (Scheme-28). Electrochemical reduction of CO on various metal electrodes at low temperature in aqueous $KHCO_3$ medium is reported (75).

PHA Synthesis with Zn: NAs are chemoselectively reduced to the corresponding N-PHAs with high selectivity (88%) using Zn dust in a CO_2/H_2O system under mild conditions (Scheme-29) (76).

R = H, p -Cl, p -CN, p -CH$_3$, p -COCH$_3$, m- NO$_2$ --Scheme-29

Other NAs which contain reducible functionality other than a nitro group are also reduced to the corresponding N-PHAs with yield from 88% to 99%. The use of NH_4Cl and Ultrasound (US) are advantageous for the benign process with Zn and $CO+H_2O$. The yield of N-PHA reaches 95% when the reaction is carried out with a Zn: NB molar ratio of 2.2 under US (40 KHz) at 25°C and normal pressure of CO_2 in shorter reaction time and consumption of less Zn (77).

References:

46). H. Lee, M. An, *Bulletin of Korean Chemical Society*, 25(11):1717-1719(2004).

47). A. M. Tafesh, et al. *Chemical Reviews*, 96 (6):2035–2052 (1996).

48). J. W. Park & Y. K. Chung, *ACS Catalysis*, 5(8):4846–4850(2015).

49a). R. V. Jagadeesh, M. Beller et al. *Green Chemistry*, 17(2): 898-902(2015).

49b). H. S. David et al. *J. Catalysis*, 61(1):48-56(1980).

50). S. Aoi, et al. *Catalysis Science & Technology*, 06(12):4077-4080(2016).

51). H. R. Apell, US3290377: (1966).

52). J. E. Yanez, et al. *J. of Coordination Chemistry*, 59(15): 1719–1728 (2006).

53a). A. J. Pardey, et al. *J. Molecular Catalysis A: Chemical*: 164:225–234 (2000),

53b). Ibid, *Inorganica Chimica Acta* 329:22–30(2002),

54a). J. Xiong, et al. *Catalysis Communications*, 8(3):345-350(2007).

54b). N. Mahata et al. *Catalysis Communications*,10(8)1203-1206(2009).

55). X. Liu, X. Ma, et al. *RSC Advances*, 05(46):36423-36427 (2015).

56). X. Meng et al. *J. Catalysis*, 264(01):01-10 (2009).

57). X. Meng, H. Cheng, et al. *Green Chemistry*, 13(03):570-572 (2011).

The applications of CO_2 in organic synthesis are covered in the following two books.

58a). New and Future Developments in Catalysis: Activation of Carbon Dioxide. Steven L Suib, Newnes, 11-Jul-2013 - Technology & Engineering - 658 pages.

58b). Transformation and Utilization of Carbon Dioxide: B. M. Bhanage, M. Arai, Springer Science & Business Media, 27-Jan-2014 - Science - 388 pages.

59). X. Meng, S. I. Fujita et al. *J. Catalysis*, 269(01):131-139(2010).

60). S. D. Senanayake, et al. *J. Physical Chemistry C*, 112:9744–9752(2008).

61). N. Akiya, et al. *AIChE Journal*, 44(02): 405-414(1998).

62). H. S. David et al. *Journal of Catalysis*, 61(01):48-56(1980).

63). M. Rozenberg, et al. *J. Physical Chemistry: A*, 119 (31):8497–8502(2015).

64). Z. Sun, H. Zhang, *J. Materials & Chemistry*, 20:1947-1952(2010).

65). Dodman et al. US3637820 A(1972).

66). X. Liu, S. Lu, *J. Molecular Catalysis A: Chemical*, 212(01-02):127-130(2004).

67a).A. D. Peng, et al. *Chinese J. Catalysis*, 23(05):457-459(2002).

67b). X. Z. Liu, et al. *Progress in Chemistry -Beijing-* 18(05):550-555(2006).

68). H. U. Blaser, et al. *Chem.Cat.Chem*, 01:210-221(2009).

69a). X. Liu & S. Lu, *J Molecular Catalysis A: Chemical*, 212(01):127-130(2004).

69b). J. Li, et al. *Chem. J. Chinese Universities*, 27(04):708-710(2009).

70). K. V. Macho, et al. *Chemical Papers* 45 (3) 363-368 (1991).

71). V. Macho, et al. *J. Molecular Catalysis A: Chemical*, 209:69-73(2004).

72). V. Macho, et al. *Collect. Czech Chemical Communications*: 60:514-520(1995).

73). K. Hiroshi, et al, *Chemistry Letters*, 38(5): 410-411(2009).

74). H. F.Jiang, et al. *Chinese J. Chemistry*, 26(08):1407–1410(2008).

75a). A. Wieckowski, et al. *Electrochimica Acta*: 28:1619-1626(1983).

75b). A. Wieckowski, et al. *Electrochimica Acta*: 28:1627-1633(1983).

75c). M. Azuma, et al. *J. Electrochemical Society*: 137:1772-1778(1990).

76). S. Liu, Y. Wang et al. *Green Chemistry*, 11:1397-1400(2009).

77). S. Liu et al. *Research on Chemical Intermediates*, 38(09):2471-2478 (2012).

5F.00.00: Formic Acid and Formates:

Formic acid (FA) is a sustainable, green and affordable product based on renewable natural resources with a great potential as a hydrogen storage materials. Formic acid (FA) is a simple C-1 organic carboxylic acid with a molecular formula of CH_2O_2 and a chemical structure as HCOOH. FA and its salts as formates can be easily dehydrogenated using different metal catalysts. FA under special conditions can undergo decomposition to give carbon dioxide and hydrogen molecule.

$$HCOOH = CO_2 + H_2$$

Chemically, FA is a non-toxic, biodegradable, cheap and readily available from bio-refineries. More importantly, FA with a high volumetric capacity of $53gH_2/L$ or containing 4.4 wt% of hydrogen is a non-toxic liquid at ambient temperature and therefore is an ideal candidate for a potential hydrogen storage material. FA can be generated via catalytic hydrogenation of CO_2 or bicarbonate in the presence of an amine with suitable ruthenium catalysts. With an appropriate and suitable catalyst of desired activity, FA can be used as a hydrogen source to prepare ANs with great efficiency and selectivity. Koh et al. (1) have demonstrated that Pd-NPs (~2-5nm) have potentials to meet the current requirements of dehydrogenation of FA. Some metal catalysts are required for the generation of free hydrogen from FA or its ammonium or hydrazinium salts or even metal salts as sodium, potassium magnesium or zinc formates by the process of their dehydrogenation (2-4). The following examples illustrate applications of these techniques to the synthesis of ANs by the CTH of NAs with FA or its salts as hydrogen donors.

5F.01.00: Formic Acid Decomposition/Dehydrogenation:

5F.01.10: Non-Noble Metals

As an alternative to expensive noble metals as Au, Pt, Pd, different research groups have used Ni, Cu and Zn metals for chemoslective reduction of NAs to corresponding ANs in good to high yields using formic acid and ammonium formates as hydrogen donors. Gowda et al have developed highly chemoselective methods for the preparation of AN by NB reduction based on economical and ecological parameters. A novel method for the chemoselective reduction of NACs to corresponding ANs using Raney-Ni-catalyzed formic acid dehydrogenation process is employed (Scheme-1).

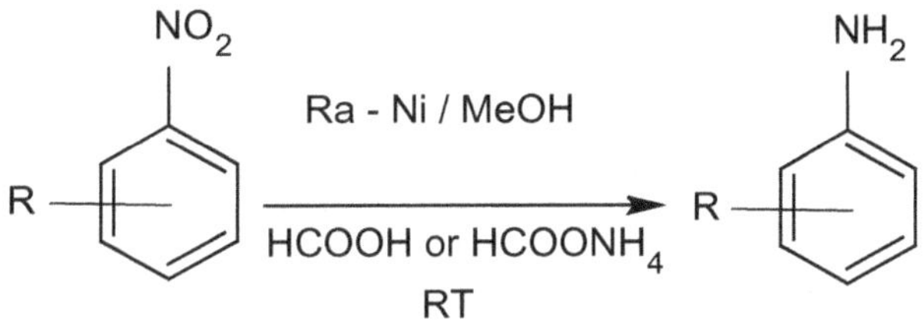

R = -OH, -OCH$_3$, -CHO, -COCH$_3$, -COC$_6$H$_5$, -COOH,

-COOC$_2$H$_5$, -CONH$_2$, -CN, -CH=CH-COOH, -NHCOCH$_3$ --Scheme-1

The reduction can be carried out not only with HCOOH (FA) but also with HCOONH$_4$ (5). Saha et al. have used Cu-NPs and HCOONH$_4$ for a highly chemoselective reduction of NACs to the aryl amines in ethylene glycol at 120°C (Scheme-2).

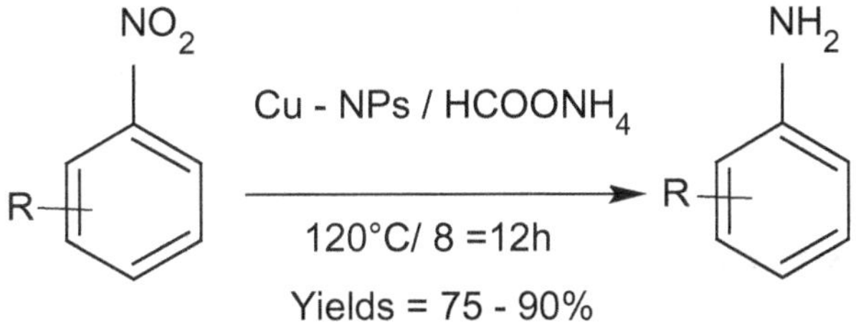

--Scheme-2

The reductions are very clean and are successfully carried out in the presence of a wide variety of other reducible functional groups to produce ANs in high yields (6). The efficiency of a process, including the catalysts, the hydrogen donor, the solvent, pressure and temperature, is realized when it is at its best in delivering quantitative conversions, yields and the selectivity of the end products. For example, in the reduction of NB with HCOONH$_4$ (4Eqv) and HCOOH (1Eqv) in the presence of the catalyst (Mg-Il-Pd) at 90°C in 24h AN was obtained in 96% yield and 99% selectivity, without any trace of N-formylated product. In some cases the nitro conversions were 100% with 100% selectivity to ANs (7). Rapid and facile reduction of nitro compounds with multiple functional groups in the same molecule is of paramount importance. This is achieved by CTH of NAs using zinc and HCOONH$_4$ or zinc and FA systems under mild conditions in methanol as a solvent (Scheme-3).

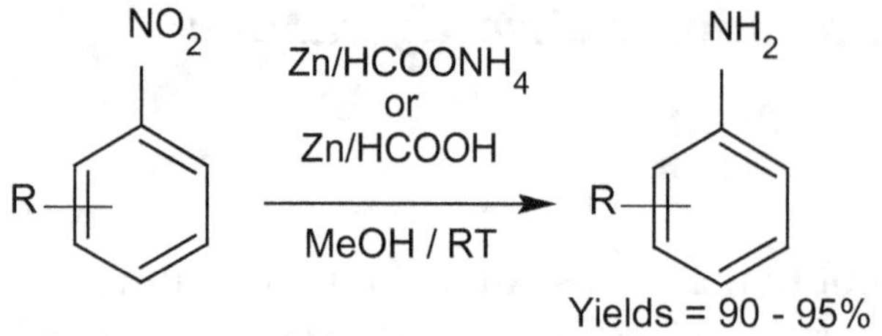

Yields = 90 - 95%

R = H, CH₃, OCH₃, OH, Cl, Br, CHO, CN, COOH, COOR etc. --Scheme-3

The importance of the method is based on the fact that it utilizes cheap or low-cost and commercially available raw materials and produces ANs in high to excellent yields. Moreover, it is a highly chemoselective process for the reduction of nitro group without affecting other functional groups (8). Under same conditions as stirring the reaction mixture at room temperature in MeOH exactly the same results with chemoselectivity are obtained (AN yields 90-95%) using economical and commercially available zinc dust and hydrazinium monoformate, which is found more effective than hydrazine or FA (9). On similar lines, Gowda et al have used Raney-Ni and hydrazinium monoformate as hydrogen source for chemoselective reduction of number of functionalized NAs to corresponding ANs in high yields (70-95%). Actually, the M-Hydrazinium monoformate (M = Zn, Ra-Ni) catalyst system is more effective than Pd/C-HCOOEt$_3$, Pt/C-HCOONH$_4$, Ra-Ni-HCOONH$_4$ etc. Unlike ammonium formate, hydrazinium monoformate is freely soluble in methanol, which helps in achieving better results (5, 10). Magnesium-hydrazinium monoformate is found as an equally efficient catalyst in CTH OF NAs to ANs, including sensitive HANs with yields between 80% and 95% (11, 12). All the three systems (Zn-HCOONHNH$_2$, Mg-HCOONHNH$_2$ and Ra-Ni-HCOONHNH$_2$) are found highly compatible with several sensitive functionalities such as halogens, CHO, -COR -COOH, -COOC$_2$H$_5$ -CN, -CH=CH-COOH, etc.

5F.01.11: Metal-Ligand Catalysts:

Iron and FA pair is used for a chemoselective CTH of NAs in ethanol, which not only has tolerance towards other functional groups but also delivers corresponding ANs in good to excellent yields under mild conditions (Scheme-4).

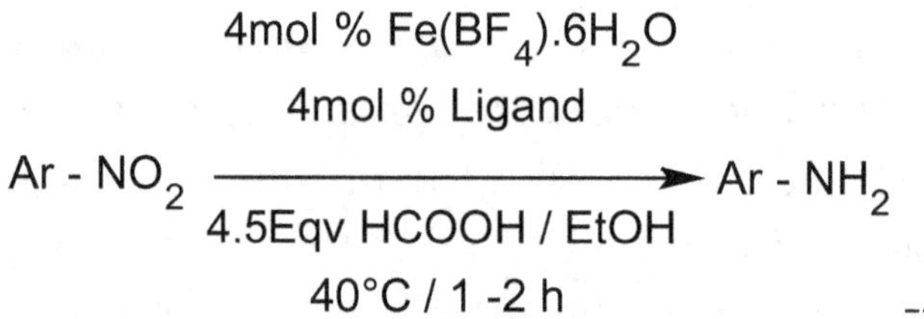

$$Ar - NO_2 \xrightarrow[\substack{4.5Eqv\ HCOOH\ /\ EtOH \\ 40°C\ /\ 1\ -2\ h}]{\substack{4mol\ \%\ Fe(BF_4).6H_2O \\ 4mol\ \%\ Ligand}} Ar - NH_2$$

--Scheme-4

Notably, the process constitutes a rare example of base-free transfer hydrogenations (13). A highly chemoselective and efficient process is developed for the transfer hydrogenation of NAs with sensitive functional groups using flower-shaped micro-mesoporous iron oxide (MMIO) with FA as the reducing agent and tris[(2-diphenylphosphino)-ethyl] phosphine as the ligand (14).

Chemoselective Cubes: Cubane-type $[Mo_3S_4X_3(dmpe)(3)](+)$ clusters (dmpe=1,2-(bis) dimethylphosphinoethane), in combination with an azeotropic 5:2 mixture of HCOOH and triethyl amine (Et_3N) as the reducing agent act as a selective cluster catalysts (X=H) or precatalysts (X=Cl). The catalyst is effective in transfer hydrogenation of functionalized NAs, without the formation of hazardous hydroxylamines (15).

5F.01.12: Anilines from Azo, Azides and Formanilides:

NACs with other functional groups are reduced to corresponding azo compounds using lead (Pb) as a catalyst and triethylammonium formate as a hydrogen donor. The conversion is reasonably fast, clean, high yielding and occurs at room temperature in methanol (Scheme-5 (16).

$$2 \text{ x X - ArNO}_2 \xrightarrow[\text{MeOH / RT}]{\text{Pb / HCOONEt}_3} \text{ArN=NAr} \xrightarrow{\text{Pb/NH}_4\text{Br}} 2 \text{ x Ar NH}_2$$

Scheme-5: X = Cl, Br, CN, CH_3, OCH_3, COOH, $COCH_3$, OH

If necessary, one can reduce the azo intermediates to ANs with Pb/NH_4Br (Scheme-12). A variety of functionalized aryl azides and aryl sulfonyl azides are also reduced to the corresponding amines with excellent chemoselectivity in high yields (70-90%) by Cu-NPs and $HCOONH_4$ in water (Scheme-6).

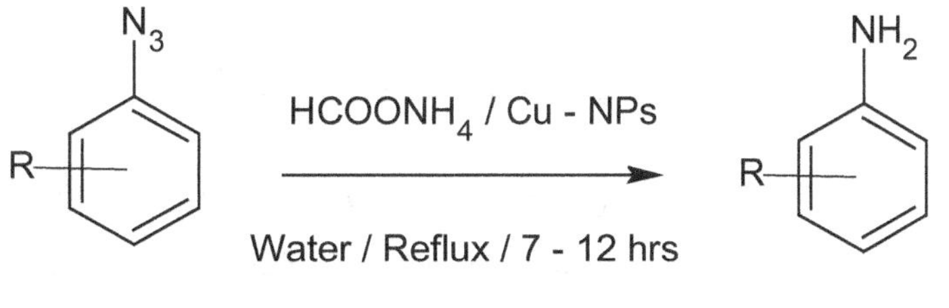

High Yields and High Chemoselctivity   Yields = 70 - 90% --Scheme-6

It is observed that the surface hydrogen on Cu-NPs is considered to be the active reducing species (17).

5F.01.13: Metals and Ammonium Salts:

Under metal and metal salts, Zn and aq. ammonium salts (as NH_4Cl, NH_4Br) are used for the chemoselective reductions of NAs to the corresponding amines in ionic liquids (ILs) as a safe and recyclable reaction medium. The effect of ammonium salts in the process were seen and the combination of Zn/NH_4Cl in $[bmim][PF_6]$ or Zn/HCO_2NH_4 in [bmim] $[BF_4]$ were the suitable conditions for the reduction of NAs (Scheme-14a). Azobenzenes were also smoothly reduced to hydrazobenzenes (Scheme-14b) with Zn/HCO_2NH_4 (aq.) in a recyclable [bmim][BF_4] without any over reduction to the corresponding ANs (Scheme-7a, b).

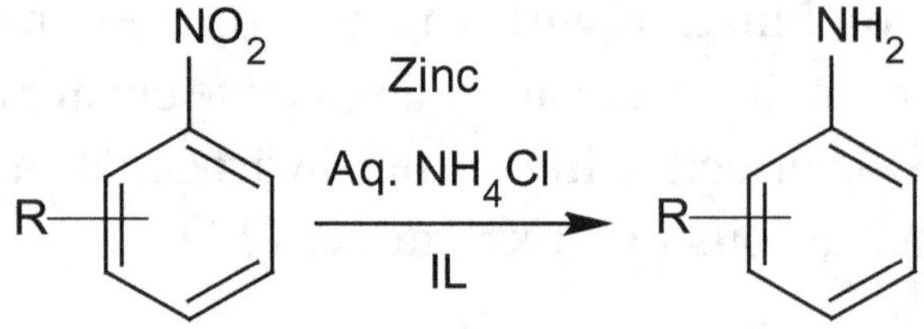

--Scheme-7a

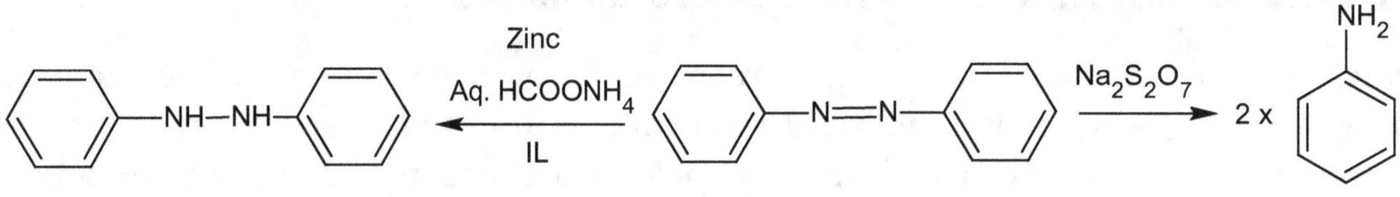

Scheme-7b

However, the azo compounds can be reduced to ANs using other efficient reducing agents as $Na_2S_2O_7$, Pb/NH_4Br, Pd/C/HCOONH$_4$ etc. as highlighted earlier in this section. In effect, an efficient reduction of NAs and azo compounds in ILs as safe and recyclable reaction medium using zinc and ammonium salts is presented (18). In a reverse case, the azobenzenes (Scheme-8) are synthesized through room temperature activation of oxygen by monodispersed M-NPs (19). It is highly challenging but desirable to develop efficient catalysts for the activation of oxygen under mild conditions.

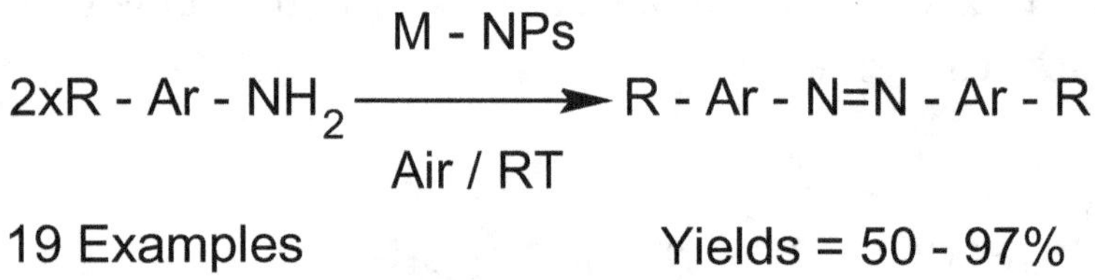

$$2x \text{ R - Ar - NH}_2 \xrightarrow[\text{Air / RT}]{\text{M - NPs}} \text{R - Ar - N=N - Ar - R}$$

19 Examples   Yields = 50 - 97%   --Scheme-8

Here, the various monodispersed M-NPs (M=Ag, Pt, Co, Cu, Ni, Pd, and Au) efficiently activated molecular oxygen under mild conditions, illustrated by the aerobic oxidation of ANs to form symmetric or asymmetric aromatic azo compounds through oxidative dehydrogenative coupling of ANs (19, 20). This discovery indicates that exploiting the catalytic power of M-NPs could enable sustainable chemistry suitable for important oxidation reactions (Scheme-8).

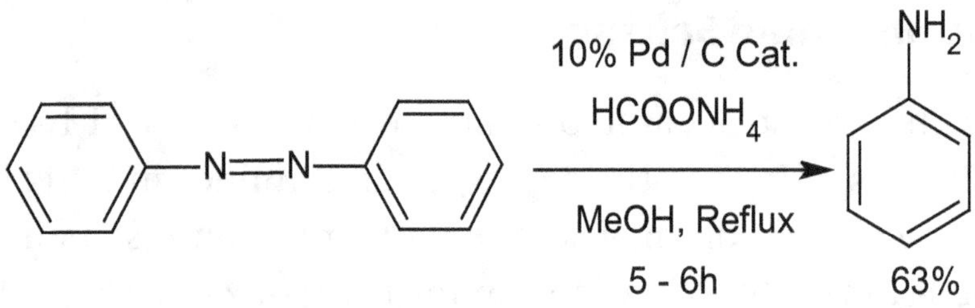

--Scheme-9

Jnaneshwara et al. have reported the reductive cleavage of azobenzenes through hydrazobenzenes (Scheme-9) to their corresponding amines with Pd(0) catalyst and HCOONH$_4$ (21, 22). Rollas et al have reviewed the reduction of aromatic and heteroaromatic azo compounds to corresponding ANs (23).

5F.02.00: Noble Metal Catalyzed Dehydrogenation of Formic Acid:

5F.02.10: Gold (Au): Gold-catalyzed reductive transformation of nitro compounds using FA as a mild, efficient, and versatile system is developed (24). In one case, 2-nitrotoluene is reduced to 2-aminotoluene or *o*-toluedine in 90% yield (25). Wener et al have developed a process for the reduction of organic compounds using alkali formate salts in the absence of a PTC, where AN from NB is obtained in 95% yield, while 4-aminophenetole is isolated in 94% yield (26). An efficient subnanometric Au-catalyzed hydrogen generation process via FA decomposition under ambient conditions is developed using different metal/support catalysts (27), as Au/C, Au/TiO$_2$, Au/TiO$_2$-NCs, Pd-Au/ZrO$_2$. Au/ZrO$_2$, Ag/ZrO$_2$NCs etc.

Bimetallic Au-Pd-NPs were successfully immobilized in the metal-organic frameworks (MOFs) (ED-MIL-101) and the resulting composites, Au-Pd/MIL-101 and Au-Pd/ED-MIL-101 have been used in the dehydrogenation of FA for chemical hydrogen storage. Au-Pd NPs with strong bimetallic synergistic effects have a much higher catalytic activity and a higher tolerance with respect to CO poisoning than monometallic Au and Pd counterparts (28). Chemo- and regioselective reduction of a wide diversity of NACs to the corresponding amines has been achieved by a combination of Au-NPs supported on titania and ammonium formate (HCOONH$_4$) in ethanol at room temperature (29, 30). 5-chloro-2(2,4-dichlorophenoxy)-AN is synthesized by CTH of corresponding nitro ether (NE) substrate using 5%Pt/C and ammonium formate in butanol. The conversion of NE and percent selectivity towards AN obtained was found to be 99.76% and 99.67%, respectively (31).

5F.02.11: Palladium (Pd): The hydrogenation of NAs with formate salts (HCOONH$_4$ or HCOONEt$_3$) as hydrogen donor and Pd/C as a catalyst under liquid-liquid/solid system was found exceptionally good for CTH process-(32). Mechanistically, NA, formate anion and water are attached to one and the same active side of the catalyst. It is also performed using a soluble triarylphosphine-Pd-acetate catalyst (33, 34) (Scheme-10).

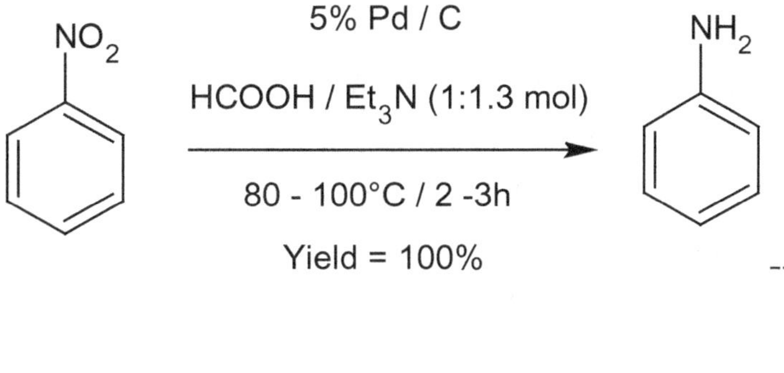

5% Pd / C

HCOOH / Et$_3$N (1:1.3 mol)

80 - 100°C / 2 -3h

Yield = 100%

--Scheme-10

Wiener et al. have reported the results on the mechanism of heterogenous transfer hydrogenation of NAs by formate salts catalyzed by palladium/carbon (35).

The hydrodehalogenation is a serious problem during the reduction of HNBs to HANs. However, if someone is interested in reductive dehalogenation of HANs, Pd/MCM-41 catalyst with $HCOONH_4$ in methanol at 356K is used. In some of the functionalized NBs both the conversion and the AN selectivity are 100%. In case of 4-CNB and 4-FNB the hydrodehalogenated ANs were obtained in 85% and 90% respectively. NB with substituents as 4-$CONH_2$, 4-COOMe, 3-NO_2, 4-NO_2, the conversions in all the cases were 100% and the AN selectivities were 100% for first three derivatives, while 4-NO_2 derivative gave 99% AN selectivity.

--Scheme-11

However, during the chemoselective reduction of 3-chloro, 4-fluoro nitrobenzene using 10% Pd/C (instead of Pd-MCM-41) and $HCOONH_4$ in refluxing methanol the 4-fluoroaniline was obtained only in 56% yield (36, 37) (Scheme-11).

Although it is not about FA, but as a reference we have noted that 5% Pd/C and hydrogen gas in methanol also can reduce 2,4-difluor-5-chloronitrobenzene to 2,4-difluoroaniline in 70% yield when the mixture under hydrogen pressure of 22lbs was stirred for 3hrs at 60°C (38). A supported Pd based heterogenous catalyst Zr-P (Pd/Zr-P) and FA proved to be the best catalyst for CTH of NB and also was found to be highly chemoselective when it was explored for base-free hydrogenation of various substituted NAs (39). Highly dispersed Pd-Cu alloy NPs have been successfully prepared within a macroreticular basic resin bearing -$N(CH_3)_2$ functional groups, which is proven to be responsible for the efficient production of high-purity H_2 from FA dehydrogenation for chemical hydrogen storage. By the addition of Cu, the electronically promoted Pd sites show significantly higher catalytic activity as well as a better tolerance towards CO poisoning as compared to their monometallic Pd counterparts. The present catalytic system is particularly desirable for an ideal hydrogen vector in terms of its potential industrial application for fuel cells (40). The liquid phase reduction of substituted NAs with the $HCOONEt_3$ system has been carried out by the use of a supported palladium (0,6%) on $AlPO_4/SIO_2$ (20:80) catalyst. The reduction rate of NAs containing electron-acceptor substituents is much higher than that of electron donor-containing substrates (41). Simple and rapid hydrogenation of 4-NP with aqueous FA in catalytic flow reactors with a mixed distributed bimetallic (Pd-Ag) layer is accomplished with excellent results. Preferential acid leaching of Ag from the Pd–Ag layer produced a porous Pd surface. Hydrogenation of 4-NP was examined in the presence of FA simply by passing the reaction solution through the catalytic tubular reactors (Scheme-12).

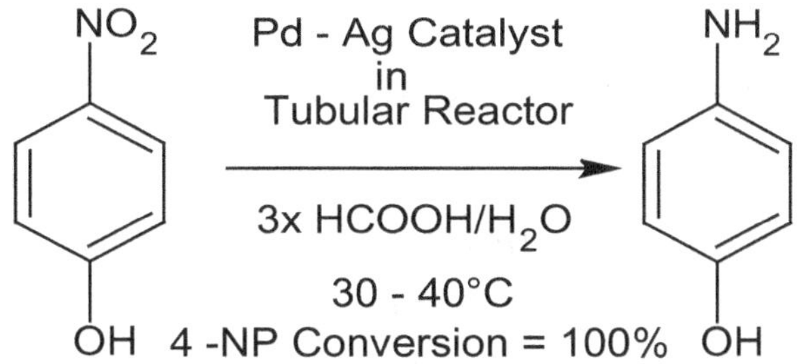

--Scheme-12

4-AP or PAP was the sole product (i.e. 100% selectivity) of hydrogenation with complete conversion of 4-NP within 15s of residence time in the porous PdO reactor at 40°C. No side product was detected. Since the dehydrogenation of FA did not occur to any practical degree, the nitro group was reduced via hydrogen transfer from FA to 4-NP (42). In an interesting case, many nitronapthalene derivatives have been reduced to corresponding amines using Pd and aqueous HCOONa in high to excellent yields. For example, 1-aminonaphthalene-3, 6, 8-trisulphonic acid (yield 98%) is obtained by reducing the disodium salt of a 1-nitronaphthalene-3,6-8-trisulphonic acid with HCOONa in water containing 3% Pd/C catalyst. The reduction is carried out in refluxing conditions for 24hrs (43). The CTH of o-nitroanisole to o-anisidine was studied in the temperature range 35–85°C with $HCOONH_4$ as H-donor and IPA as solvent using Pd/C as catalyst (Scheme-13).

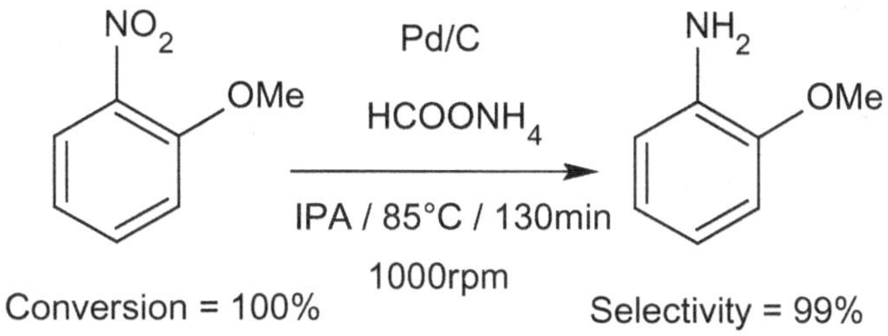

--Scheme-13

The reaction mixture was stirred with an agitation speed of 1000 rpm. In 130 min at 83°C all o-nitroanisole was converted (100% conversion) with 99% selectivity towards o-anisidine. The possibility of Fenton Chemistry to treat aqueous waste stream was explored and found suitable (44).

5F.03.00: Noble Metals on Polymer Supported Formates:

5F.03.10: Palladium: NAs can be reduced in high yields (70-90%) to the corresponding ANs by transfer hydrogenation using a stable H-donor, polymer-supported formate (PSF), in combination with palladium acetate catalyst. The reactions occurs at 100–120°C in DMF and the PSF can be recycled at least for three more runs (Scheme-14).

R = Halide, Ketone, Aldehyde, Ester remain unaffected. --Scheme-14

The procedure is chemoselective for nitro group in the presence of other sensitive groups which remain unaffected (45). CTH of NACs using recyclable PSF as a hydrogen donor and Pd/C as a catalyst produces corresponding amines (46, 47) in excellent yields (90–98%). Ram et al have first reported results of their studies on chemoselective reductions by CTH of NAs using Pd/C and $HCOONH_4$, and then they have published a brief review "Ammonium Formate in Organic Synthesis", which summarizes the developments in the reductions of NAs to ANs using noble metals and ammonium formates in methanol (48, 49).

5F.03.11: Platinum (Pt): Gowda and Aken independently have reduced the ANCs to respective amines in high yields by CTH using 5% Pt/C with $HCOONH_4$ or FA as hydrogen donor in methanol at ambient temperature (Scheme-15).

R = Halide, NO_2, $CONH_2$, CN etc. --Scheme-15

It was observed that the former was more efficient donor than the later. The reduction of nitro groups occurs without hydrogenolysis of halogens and the reducible substituents remain unaffected under the reaction conditions (8, 46, 50).

Pd/C and $HCOONEt_3$ reduce HNBs to ANs but with complete hydrodehalogenation of the halogen group (Scheme-16a).

Scheme-16a, 16b

However, under the same reaction conditions Pt/C and HCOONEt$_3$ reduces 2-BrNB to 2-BrAN without any debromination (Scheme-16b) (51).

5F.03.12: Rhodium (Rh): In the presence of iodide ions, an efficient and selective rhodium-catalyzed transfer hydrogenation of NAs with FA as the hydrogen source takes place to give amines or formanilides (Scheme-17).

$$Ar - NO_2 \xrightarrow[\text{DMSO /100°C/ 20h}]{\begin{array}{c}\text{KI / [CP*RhCl}_2]_2 \\ \text{HCOOH / Et}_3\text{N}\end{array}} Ar - NH_2$$

--Scheme-17

The formanilides formed can be either isolated or further hydrolyzed to corresponding ANs (62). Entwistle et al have reduced NAs to ANs in high yields by using palladium, platinum, or rhodium metal catalyst with formic, phosphinic, or phosphorous acid as reported above.

5F.03.14: Ruthenium (Ru)-Complxes: Watanabe et al. have successfully employed RuCl$_2$(PPh$_3$)$_3$ catalyst with HCOOH./Et$_3$N for the highly chemoselective hydrogenation of NAs to ANs, as NB to AN (NB Conv 97%, AN Sel 97%), 2-CNB (99%conv) to 2-CAN(94% sel.), 4-CNB (Conv. 99%) to 4-CAN (98% sel) at 125°C in 5h (52). According to Taleb, et al NB can be selectively reduced to AN by aqueous HCOOMe in the presence of a catalytic system comprising [Ru$_3$(CO)$_{12}$], Pd(OAc)$_2$, tricyclohexylphosphine, and 1,10-phenanthroline (Scheme-18).

--Scheme-18

The process was found to be sequential involving deoxygenation of NB to azoxy- and azo-benzene,and then hydrogenation to aniline, formamidation, and eventually the hydrolysis of formanilide to AN can be accomplished (53, 54). Ru(III)(acac)$_3$) was synthesized and tested for the reduction NB to AN. Under optimal reaction parameters as 80°C, 4.0h, sodium formate, Ru(acac)$_3$, the conversion of NB is 100.0 %, and the selectivity to AN was 96.65 % with the same yield (96.65%). The catalyst system exhibited excellent catalytic properties in the hydrogen transfer hydrogenation of NB to AN (55). An unsymmetrically protonated PN(3)-pincer complex in which ruthenium is coordinated by one nitrogen and two phosphorus atoms was employed for the selective generation of hydrogen from FA. Mechanistic studies suggest that the imine arm participates in the FA activation/deprotonation step. A long life time of 150h with a turnover number over 1 million was achieved (56). Some of the Ru based catalytic systems are particularly interesting for the generation of H_2 for new applications in portable electric devices. Hydrogen generation at ambient conditions with application in Fuel Cells is presented (57). Ru catalyst and also with a Ru catalyst promoted by functionalized ionic liquids are investigated for FA decomposition and reported here (58, 59, 60). Some other reports on Ru catalyzed dehydrogenation of FA are covered under phtocatalysis in this section only. Mellone et al have dehydrogenated FA to a mixture of CO_2 and H_2 using homogeneous Ru catalysts bearing the polydentate tripodal ligands (61).

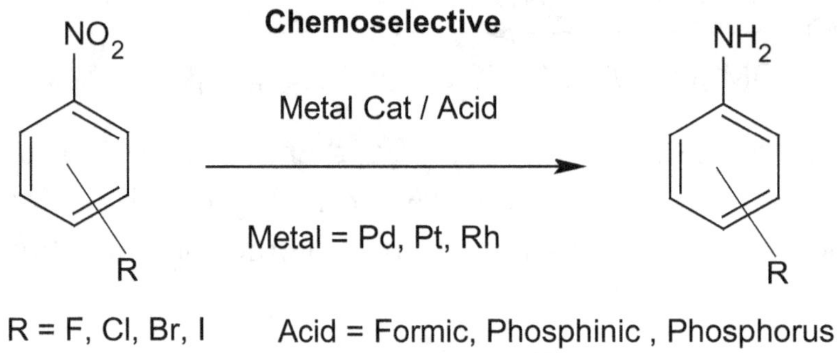

--Scheme-19

With formic acid, nitro-compounds containing fluorine were reduced but not those containing chlorine, bromine, or iodine. With the other acids, nitro-compounds containing any of the halogens were reduced with retention of the halogen groups (Schemes-19) (63).

5F.04.00: Formic Acid & Microwaves:

Highly selective reduction of NACs to their corresponding ANs in good yields using FA and CeY zeolite under monomode or single mode reactor is reported. This system is found to be compatible with several sensitive functionalities and the catalyst is quite stable (64a). Pd/C catalyzed reduction of substituted NAs with $HCOONH_4$ in ethanol under MWI gives corresponding ANs in 79-91% yields (64b).

5F.05.00: Formic Acid & Photocatalysis:

A highly efficient dehydrogenation of FA over a Pd-NPs-based Mott–Schottky gives an exceptional catalytic performance of a photocatalyst composed of Pd-NPs and mesoporous carbon nitride for the dehydrogenation of FA in water at room temperature to produce H_2 gas. This is due to the enhanced electron enrichment of the Pd-NPs through charge transfer at the interface of the Mott–Schottky contact (65a, 65b). A mild photochemical approach was applied to construct highly coupled metal–semiconductor dyads, which were found to facilitate the hydrogenation of NB to AN and a wide range of NAs in excellent yield (>99%, TOF: 1183) using FA as hydrogen source and water as solvent at room temperature (66). According to Mondal et al the DFT calculations showed that hydrogenation of NB over the Ru (0001) catalyst surface through a direct reaction pathway is more favorable than that through an indirect reaction pathway (67a). Reduction of *m*-nitrobenzene sulfonic acid (*m*-NBSA) to *m*-aminobenzene sulfonic acid over photoactivated TiO_2 is achieved in high yields with quantitative conversion of *m*-NBSA (67b, c). Pd-NPs supported on defective TiO_2 with oxygen vacancies have been proved very efficient in the reduction of NAs in water and $HCOONH_4$ as a hydrogen source under mild conditions. The process can be performed with solar light coupled with thermal energy for best results (68a, b), (69a, b). A highly efficient and synergistic catalysis of plasmonic Au@Pd nanoparticles supported on titanium-doped zirconium-based amine-functionalized MOFs (UiO-66($Zr_{100-x}Ti_x$)) for boosting room-temperature hydrogen production from FA under visible light irradiation is investigated (70).

A Ru-based biomimetic hydrogen cluster [Ru_2 $(CO)_6$ (μ-SCH_2CH CH_2S)] has been synthesized, which in the presence of the P ligand tri(o-tolyl)phosphine is efficient for FA decomposition. With less than 50 ppm of the catalyst the highest TOFs for Ru-complexes as well as the best efficiency for photocatalytic hydrogen production from FA is achieved (71). Also, with the [$RuCl_2$(benzene)]$_2$/dppe catalyst system a remarkable TON of 260,000 at room temperature was obtained. Moreover applying $Fe_3(CO)_{12}$ together with tribenzylphosphine and 2,2':6',2"-terpyridine under visible light irradiation a TON of 1266 was obtained, which is the highest activity known to date for selective dehydrogenation of FA applying non-precious metal catalysts (72).

5F.06.00: Novel New Ctalayst for Dehydrogenation of HCOOH/HCOONH$_4$:

Metal catalyzed reductions of NAs with HCOOH as a hydrogen source is one of the most important reactions in hydrogenation processes. HCOOH decomposes under experimental conditions on metal surfaces to give CO_2 and H_2. Some other highly efficient catalytic systems for the dehydrogenation of FA are recently reported. For example on reduced ceria (CeO_2) only CO_2 and H_2 were produced at 450-600K (73). Cu (110) oriented surface and Ni surface also decompose HCOOH into CO_2 and H_2 (74). Thus HCOOH is as good as $CO+H_2O$, CO_2+H_2O or $scCO_2$ in effecting reductions of organic functional groups including nitro group in NACs. HCOOH decomposes to give CO_2 and H_2 when exposed to a drying agent (75a-d).

5F.07.00: Bi and Multinuclear Aniline Derivatives:

In context of AN- and its analogs, indoles are biclyclic analogs and have fairly good applications in synthetic organic chemistry and are synthesized using metal catalyst complexes and FA/Et_3N as a hydrogen source (76, 77).

Gefitinib   R = (CH$_2$)$_3$ Norfloxacin

3-Chloro-4-fluoro aniline (3,4-CFAN) is an important intermediate in the synthesis of an anti-cancer drug Gefitinib (78a) and antibiotics like Norfloxacin. Norfloxacin ethyl ester was prepared from 3-chloro-4-fluoroaniline in ionic liquid in a one-pot procedure by condensation with EMME [ethoxymethylenemalonic diethyl ester], cyclization, ethylation and condensation with anhydrous piperazine. After its hydrolysis, Norfloxacin was obtained with an overall yield of 72.7% (78b). 3, 4-CFAN is synthesized by catalytic hydrogenation of 3-chloro-4-fluoro-NB in > 99% yield (79). A process for producing 3-chloro-4-fluoronitrobenzene is characterized by chlorinating 4-FNB in the presence of iodine and a metal as Fe, Sn etc. (80)

Combretastatins, including cis-Stilbene as the basic skeleton are provided with intensive mitosis inhibiting activity and intensive cytotoxicity. Therefore, investigations are on-going for the development of anticancer drugs using a derivative thereof as the active (effective) component. A process for producing an aminostilbene derivative with an enhanced efficiency and at industrial convenience as compared to the conventional techniques is reported. Aminostilbene, an intermediate of an anticancer drug, is synthesized by the catalytic transfer hydrogenation of the corresponding nitro derivative using Pt/C (0.5mmol Pt) catalyst and ammonium formate in acetonitrile (81).

References:

1). K. Koh, et al. *J. Materials Chemistry A*: 05:16150-16161(2017).

2). C. Guan, et al. *Inorganic Chemistry*, 56(01):438–445(2017). 2b). A. Matsunami, *et al. ACS Catalysis*, 07(07):4479–4484(2017). 2c). K. Sordakis, et al. *Chemical Reviews*, Article ASAP. **DOI:** 10.1021/acs.chemrev.7b00182.

3). S. Zhang et al. *ACS Applied Materials & Interfaces*, Article ASAP, June 28, 2017.

4a). Z. Li & Q. Xu, *Accounts Chemical Research*, 50(06):1449–1458(2017).

4b). X. Gu, et al., *J. American Chemical Society*, 133(31):11822–11825(2011), *4c)* M. We, *ACS Energy Letters*, 02(01):01–07(2017).

5). D. C. Gowda et al. *Synthetic Communications*, 30(16):2889-2895(2000).

6). A. Saha, B. C. Ranu, *J. Organic Chemistry*, 73(17):6867–6870(2008).

7). B. Karimi, et al. *ChemPlusChem*, 80(12):1750–1759(2015).

8). D. C. Gowda, et al. *Indian J. Chemistry*, 40B:75-77(2001).

9). S. Gowda et al. *Synthetic Communications*, 33(02):281-289(2003).

10). S. Gowda & D. C. Gowda, *Tetrahedron*, 58(11):2111-2113(2002).

11). K. Abiraj, et al. *Synthesis. & Reactivity in Inorganic & Metal-Organic Chemistry*, 32(08):1409-1416(2002).

12). B. Raju, et al. *Indian J. Chemistry*, 48B:1315-1318(2009).

13). M. Beller, et al. *J. American Chemical Society*, 133:12875-12879(2011).

14). K. J. Datta, A. K. Rathi, et al. *ChemCatChem*, 08(14):2351–2355(2016).

15). I. Sorribes, et al. *Angewandte Chemie Intl Edition*, 51(31):7794-7798(2012).

16). G. R. Srinivasa, et al. *Tetrahedron Letters*, 44(31):5835–5837(2003).

17). S. Ahammed, et al. *J Organic Chemistry*, 76 (17):7235-7239(2011).

18). F. A. Khan, et al. *Tetrahedron Letters*, 44:7783–7787(2003).

19). S. Cai, et al. *ACS Catalysis*, 03:478–486(2013).

20a). A. Corma, et al. *Nature Protocols*, 05(03):429-438(2010). 20b). Y. Zhu, Y. Shi, *Organic Letters*, 15(08):1942–1945(2013).

21). G. K. Jnaneshwara, et al. *J. Chemical Research (S)*, 160-161(1998).

22). K. Abiraj, D. C. Gowda, *Synthetic Communications*, 34(04):599-605(2004).

23). S. Rollas, *Marmara Pharmaeutical J.*, 14: 41-46(2010).

24). L. Yu & Q. Zhang, *ChemSusChem*, 08(18):3029–3035(2015).

25). Lou, et al. *Advanced Synthesis & Catalysis*, 353(02-03):281–286(2011).

26). H. Wener, et al. US 4792625 A (1988).

27). Q. Bi, et al. *J. American Chemical Society*, 134:8926–8933 (2012).

28). X. Gu, et al. *J. American Chemical Society*, 133(31):11822-11825(2011).

29). X. Lou, et al. *Advanced Synthesis & Catalysis*, 353(02):281–286(2011).

30). L. Yu, et al. *ChemSusChem*, 08(18):3029–3035(2015).

31). M. Subramanyam, a).WO2009090669A2 (2009): b).WO2009090669 A3 (2009).

32). G. C. Torres, *Applied Catalysis A: General*, 161(01–02):213–226(1997).

33). N. Cortese, et al. *J. Organic Chemistry*, 42(22):3491-3494(1977).

34). R. A. W. Johnstone, et al. *Chemical Reviews*, 85:129-170; pp155-157(1985).

35). H. Wiener, et al. *J. Organic Chemistry*, 56(14):4481–4486(1991).

36). M. K. Anwer, et al. *J. Organic Chemistry,* 54(06):1284–1289(1989).

37). P. Selvama, et al. *Applied Catalysis. B: Environmental,* 49:251–255(2004).

38). R. J. Tull, et al. EP 0001825 A1 (1979).

39). J. Tuteja, et al. *RSC Advances,* 04(72):38241-38249(2014).

40). K. Mori, et al. *Chemistry A European J.,* 21(34):12085–12092(2015).

41). M. S. Climent, et al. *Reaction Kinetics & Catal Letters,* 14(04):489-493(1980).

42). R. Jawaid, et al. *Beilstein J. Organic Chemistry,.* 09:1156–1163(2013).

43). P. Balmfield, et al. US 4093646 A (1978).

44). P. Haldar, V. V. Mahajani, *Chemistry.lEngineering J.,* 104:27–33(2004).

http://chemistry.mdma.ch/hiveboard/rhodium/pdf/cth.nitroarene2aniline.proc-dev.pdf

45). B. Basu, et al. *Molecular Diversity,* 09(04):259–262(2005).

46). D. C. Gowda, B. Mahesh, *Synthetic Communications,* 30(20):3639-3644(2000).

47). K. Abiraj, et al. *Synthetic Communications,* 35(02):223-230(2005).

48). S. Ram, et al. *Tetrahedron Letters,* 25(32):3415-3418(1984).

49). S. Ram, et al. *Synthesis,* 91-95(1988).

50). K. V. Aken, et al. *Beilstein J. Organic Chemistry,* 02(03)01-07(2006).

51). N. A. Cortese, R. F. Heck, *J. Organic Chemistry,* 42(22):3491–3494(1977).

52). Y. Watanabe, et al. *Bulletin Chemical Society,* Japan, 57:2440-2444(1984).

53). A. B. Taleb & G. Jenner, *J. Organometallic Chemistry,* 456(02):263–269(1993).

54). A. B. Taleb & G. Jenner, *J. Molecular Catalysis,* 91(02):L149–L153 (1994).

55). W. Wang, et al. *Research Chemical Intermediates,* 40(08):3109–3118(2014).

56). Y Pan et al. *Chemistry An Asian J,* 11(09):1357-1360(2016).

57). A. Boddien, et al. *ChemSusChem,* 01(08-09):751-758(2008).

58). C. Fellay, et al. *Angewandate Chemie Intl Edition,* 47(21):3966-3968(2008).

59). X. Li, et al. *ChemSusChem,* 03 (01):71-74(2010).

60). M. Czaun, et al. *ChemSusChem,* 04(09):1241-1248(2011).

61). I. Mellone, et al. *Dalton Transactions,* 42(07):2495-2501(2012).

62). Y. Wei, et al. *Synlett,* 25:1295-1298(2014).

63). I. D. Entwistle, et al. *J. Chem. Soc., Perkin Trans. 1,* issue 4, 443-444(1977).

64a). K. Arya, A. Dandia, *J. Korean Chemical Society,* 54(01): 55-58(2010).

64b). E. Cini, et al. *Catalysts,* 07:89-115(2017).

65a). Y. Cai, et al. *Angewandate Chmie Intl Edition*,52(45): 11822–11825(2013).

65b). Y. Cai, et al. *Angewandate Chemie*, 125(45): 12038–12041(2013).

66). X. Li & Y. Cai, *Chemistry A European J.*, 20(50):16732–16737(2014).

67a). J. Mondal, et al. *Chemistry A European J..*, 21(52):19016–19027(2015). A Book: Heterogeneous Photocatalysis: From Fundamentals to Green Applications. Juan Carlos Colmenares Quintero, Yi-Jun Xu, Springer, 24-Dec-2015 - Science - 416 pages. PP315. 67c**)**. The Golden Age of Transfer Hydrogenation, D. Wang & D. Astruc, *Chemical Reviews*, 115(13):6621–6686(2015).

68a). X. Pan & Y. Xu, *ACS Applied Materials & Interfaces*, 06(03):1879–1886(2014). 68b). Engineering vacancies for solar photocatalytic applications: A Minireview: M. Long & L. Zheng, *Chinese J. Catalysis*, 38(04):617-624(2017).

69a). K. Naseem, et al. *Environmental Science and Pollution Research*, 24(07):6446–6460(2017). 69b**)**. A Review: N. P. Radhika, et al. *Arabian Journal of Chemistry*, Available online 2 August 2016.

70). M. Wen & K. Mori, *ACS Energy Letters*, 02(01):01-07(2017).

71). C. H. Chang et al. *Chemistry*, 21(17):6617-6622(2015).

72). A. Boddien, et al. *Chimia* (Aarau), 65 (04):214-218(2011).

73). S. D. Senanayake et al. *J. Physical Chemistry C*, 112:9744-9752(2008).

74). H. S. David et al. *J. Catalysis*, 61(1):48-56(1980).

75a). M. Rozenberg, et al. *J. Physical Chemistry A*, 119 (31):8497–8502(2015). 75b) Z, Dobrovolna, et al. *Research on Chemical Intermediates*, 26(05):489–497(2000). 75c). K. Mori, *Chemistry Select*, 01(09):1879–1886(2016). 75d). Y. Wu, et al. *Chemistry An Asian J.*, 12(08):860–867(2017).

76a). E. Vickerstaffe, et al. *Organic & Biomolecular Chemistry*: 01:2419(2003). 76b). C. W. Holzapfel, C. Dwyer, *Heterocycles*: 48:1513(1998); 76c). G. W. Gribble *J. Organic Chemistry*: 38: 4074(1973). 76d). G. W. Gribble, A Review, *J. Chemical Society, Perkin Transactions.*-1, **00**(07): 1045-1075(2000).

77). S. Cacchi & G. Fabrizi, *Chemical Reviews*, 105: 2873–2920 (2005).

78a). D. R. Rao, et al. US 8350029 B2 (2013): 78b). S. Zhu, et al. *Letters in Organic Chemistry*, 05(01):01-05(2008).

79). G. Fan, L. Zhang, et al. *Catalysis Communications*, 11(05):451-455(2010).

80a). M. Tordeux, C. Wakselman, *J. Fluorine Chemistry*, 74(02):251-254(1995).

80b). T. Ishikura, US 4898996 A (1990): EP 0307481 B1 (1993).

81). T. Yamamoto, et al. US 6930205 B2 (2005).

Chapter-5G: Glucose and Other Sugars as H_2 Source:

5G.00.00: Generally sugars are source of energy. Here, they work as a hydrogen source in the reduction of NAs to cortresponding ANs.

5G.01.00: Glucose: Long time ago, Glucose-NaOH combination is reported as a hydrogen source for the reduction of NAs (1). Sugars, carbohydrates (glucose, fructose etc.) have been used in the catalytic transfer hydrogenation (CTH) (2). Catalyst-free water mediated reduction of NAs using glucose as a hydrogen source is investigated. The highly chemo- and regioselctive reduction of NAs to their corresponding amines has been carried out using D-glucose as a hydrogen source in H_2O: DMSO (1:1).

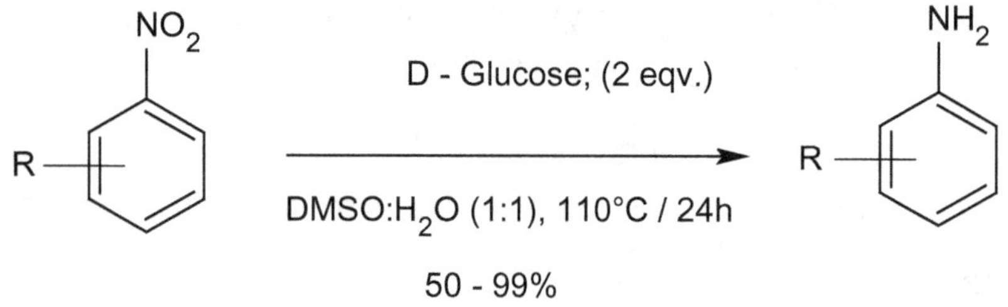

$$R = F, Cl, Br, I, OCH_3, CH_3, CN, CHO, COCH_3, CHCH_2, CH_2CH=CH\ COCH_3$$ --Scheme-1

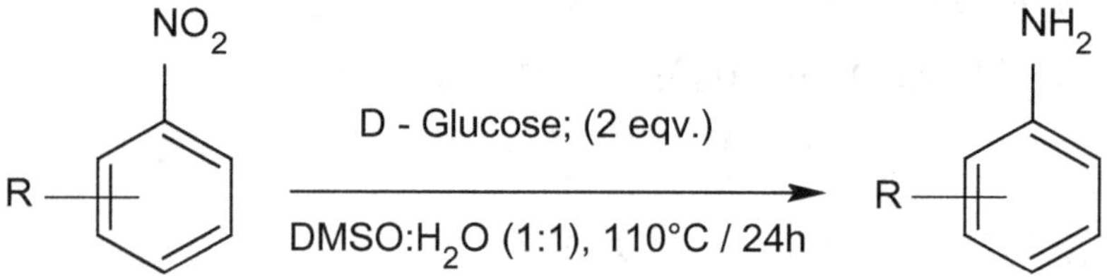

$$R = 2\text{-}NO_2; 3\text{-}NO_2; 4\text{-}NO_2 \qquad \text{Yield and Regiosel.} = > 99\%$$ --Scheme-2

The method is simple, inexpensive, easily scaled-up and applicable to the preparation of different substituted ANs (Schemes-1 & 2). As an example, 4-BrNB was reacted with Glucose in water and DMSO (1:1) and 4xKOH at 100-110°C and it was found that 4-BrNB conversion was > 99% and both the selectivity and the yield of 4-BrAN were also > 99%. Other substituents as 3-CH=CH_2, CH_3O-, 4-CH=CH-R, 4-Pyridine also gave the conversions, selectivity and yields > 99%. The present method does not require any catalyst, elaborated experimental setup or the use of high-pressure equipment compared with conventional hydrogenation methods (3).

Glucose is also used in the selective preparation of AN- intermediates. Both the selective conversion of glucose into arabinose and the reduction of nitrosobenzene to azoxybenzene are very important but challenging processes. For the first time, the simultaneous transformation of glucose to arabinose and of nitrosobenzene to azoxybenzene by photocatalysis is achieved by using Pd-NPs supported on mesoporous CdS as very active and selective catalyst for the reactions in water under visible light irradiation. Nitrosobenzene could be reduced to azoxybenzene with a selectivity of

92% by hydrogen from water and glucose. More importantly, glucose was oxidized to arabinose with high selectivity (70%) by photoexcited holes. The reaction mechanism and the reason for the high selectivity were studied using control experiments (4).

5G.01.10: D-Mannose: D-Mannose is a type of sugar that is related to glucose, but it does not behave like sugar in the body. It is used to help prevent urinary tract infections because it inhibits bacteria from adhering to the walls of the urinary tract. If bacteria can't latch onto your bladder it is less likely to cause an infection. Pd/mannose promoted tandem cross coupling-nitro reduction is employed in the expedient synthesis of aminobiphenyls and aminostilbenes in yields up to 99%. The dual role of D-mannose as a ligand for Pd catalyzed cross-coupling, and as a hydrogen source for nitro reduction is demonstrated in a modular cross coupling-nitro reduction sequence. The synthetic utility and generality of this green protocol has been illustrated by the synthesis of 20 aminobiphenyl and 10 aminostilbene derivatives in high yields through a one-pot Suzuki coupling-nitro reduction and a Heck coupling-nitro reduction, respectively starting from halonitroarenes as substrates (5).

5G.01.11: Cyclodextrins (CD): Cyclodextrins as effective additives in Au-NPs-catalyzed reduction of NAs in a ball-mill are studied in water and a small amount of $NaBH_4$. At the boundary between mechanochemistry, supramolecular chemistry and catalysis, the present study explores the role of CDs and other saccharide additives in the mechanosynthesis of Au-NPs and their use as catalysts in the reduction of NAs into their corresponding ANs. Water also plays a key role in both the reduction mechanism of the nitro groups and the supramolecular interactions with the substrate. Very interestingly, the amount of reductive agent ($NaBH_4$) was drastically reduced compared to reductions performed in solution (6).

References:

1). H. W. Galbrait, et al. *J. American Chemical Society*, 73:1323–1324(1951).

2). X. Du, L. He, et al. *Angewandte Chemie Intl. Edition*, 50:7815-7819(2011).

3). M. Kumar, et al. *RSC Advances*, 03:4894-4898(2013).

4). B. Zhou, J. Song, et al. *Green Chemistry*, 18(13): 3852-3857 (2016).

5). S. Rohilla, *RSC Advances*, 05(40):31311-31317(2015).

6). S. Menuel, *Green Chemistry,* 18(20): 5500-5509(2016).

Chapter-5H: Hydrazine Hydrate (HH):

5H.00.00: Introduction:

The reductions of NACs with hydrazine hydrate (HH) as a hydrogen source is reported since the second half of the ninetieth century. As alcohols or formic acid, hydrazine is also one of the hydrogen storage materials and deserves recognition for its applications in future fuel cells and as an energy source. Metals have been used as catalysts from the beginning of chemical technologies. However, for more than one reason their activity is compromised so that the desired results are not achieved. As a common saying goes, two heads are better than one, here too the co-operative effect from another partner or additive helps enhance the metal catalytic activity. For example, a bimetallic catalyst, another metal as an additive or promoter, solid supports, ionic liquids, combination of solvents etc. help enhance the activity of the main catalyst and also its selectivity towards the end products. For example, metals in association with phthalocyanines (PCs) give a much better performance than alone as a catalyst. We will see some of the examples of this co-operative effect in the selective reduction of NAs using HH as a hydrogen source. As usual, in most of the cases the authors give highlights of the catalyst with salient features as a stable, recyclable, high yields, economical, and green system, etc. So we will not repeat it under each and every summary of the work that we will be presenting here. Any unique properties of the catalytic system and the results obtained will be certainly highlighted. As stated earlier, the other functional groups are halo (cl, Br, I, F), CHO, COOH, COCH$_3$, CH$_2$OH, OBn, COOR, CN, NO$_2$, NH$_2$, aryl or heteroaryl etc.

5H.01.00: Metals on Solid Supports & HH:

Solid Supports as Metal Oxides, Silica, Clay, Zeolites, Polymers, etc are used.

5H.01.10: Non-Noble Metal Catalysts and HH:

Noble metals perform as very good catalysts in organic transformations. However, their use is prohibitive due to their high cost. Although noble metals are known in the catalysis for a long time, the Earth-abundant metal-based catalysts in the same time have remained underdeveloped. These non-noble metal catalysts were used to a lesser extent from the beginning of the 20th century. The discovery and applications of noble metals nanocatalysts have revolutionized the world of catalysis as they are extraordinarily efficient. Therefore, the development of Earth-abundant M-NPs is more recent, but it has appeared necessary because they are i) abundantly available, ii) economical and iii) lead to green technologies. The recent developments of efficient Earth-abundant transition-metal nanocatalysts are summarized in a review. This review highlights catalysis by NPs of Mn, Fe, Co, Ni, Cu, early transition metals as Ti, V, Cr, Zr, Nb and W and their nanocomposites with emphasis on basic principles and literature reported during the last 5 years. The last section highlights the catalytic activities of bi- and trimetallic NPs, which are very promising and simultaneously extend benefits from increased stability, efficiency and selectivity compared to monometallic NPs due to synergistic substrate activation (2). Generally, non-noble metals used as catalysts are iron, copper, cobalt, nickel, zinc, aluminium, magnesium, molybdenum, chromium, tin, lead, mercury etc. Lead and mercury are rarely used as catalysts alone or in combination with other metals as bimetallic catalysts.

5H.01.11: Metal-Phthalocyanines (PCs):

As PCs are quite stable compared to porphyrins, metal phthalocyanines (MPCs) are of great interest as catalysts for selective organic transformations including oxidation-reduction processes and electron transfer reactions. Iron-phthalocyanine with iron sulfate has been successfully applied for high chemo- and regioselective reduction of NACs to give the corresponding ANs in a green solvent system without using any toxic ligands. The catalytic systems were also compatible with a large range of other reducible functional groups. There are some instances where the dinitro compounds have been regioselectively reduced to the corresponding primary mono-amines in high yields (Scheme-1).

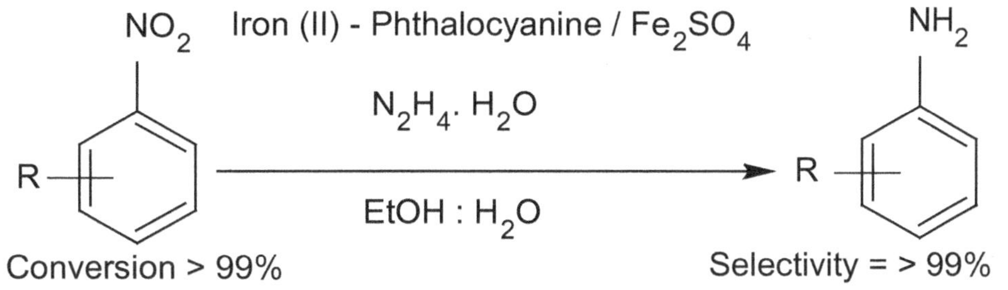

NO_2 Iron (II) - Phthalocyanine / Fe_2SO_4 NH_2

$N_2H_4 \cdot H_2O$

$EtOH : H_2O$

Conversion > 99%   Selectivity = > 99%

R = CO, CHO, CN, Halo, $CONH_2$, Halo, NO_2, OBn, OH, CH_2Cl etc. --Scheme-1

In most of the cases the conversion and the selectivity were greater than 99% as determined by GC-MS analysis (3). Zinc phthalocyanine with PEG-400 was established as a catalytic system for chemo and regioselective reduction of NACs to corresponding amines (Scheme-2).

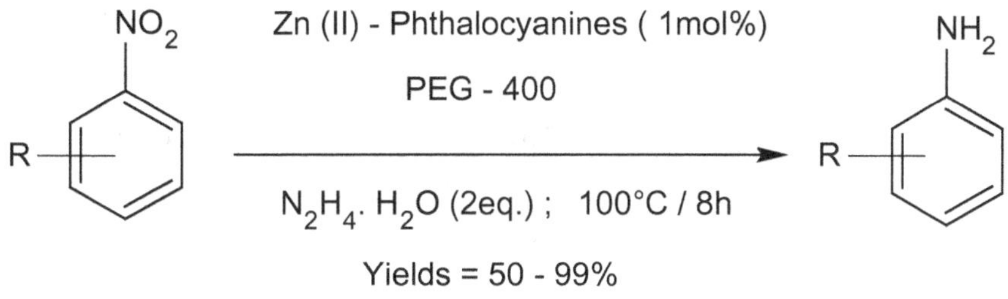

NO_2 Zn (II) - Phthalocyanines (1mol%) NH_2

PEG - 400

$N_2H_4 \cdot H_2O$ (2eq.) ;   100°C / 8h

Yields = 50 - 99%

R = CN, COOR, CHO, Lactone, OBn, NBn, Halo, COOH, OH, $CONH_2$ etc. --Scheme-2

In this (Zn(II)-PC) system a large range of reducible functional groups were well tolerated. The catalyst is required at low loading of metal, avoidance of toxic ligands and high isolated yields (4). Sharma, et al. have also used Cu(II) and Co(II) phthalocyanines as catalyst for highly chemo- and regioselective reductions of NAs. Copper/cobalt phthalocyanines were established for the first time as catalysts for this purpose in the presence of other functional groups (Scheme-3).

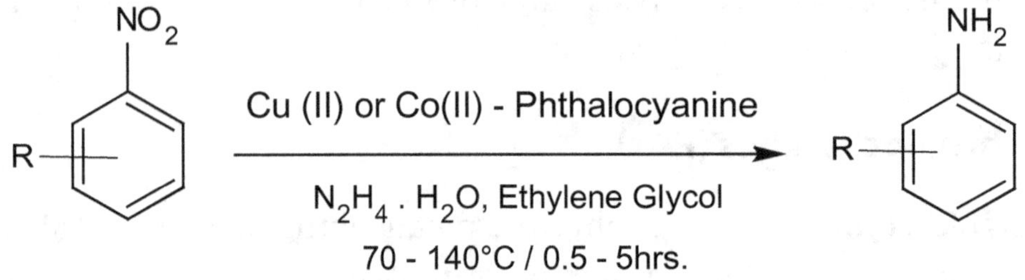

$$R = CONH_2, COOR, NO_2, CHO, Halo, Lactone, CN,$$
$$NBn, OBn, CH_2=CH_2, OH, OCH_3, COCH_3 \ etc.$$

In most Cases Conversions and Selectivty = > 99%   --Scheme-3

This method was also found to be highly regioselective towards the reduction of ANCs in a short time with high yields (>99%). In most of the cases the conversion and selectivity were >99% (5). N-Substituted isoindolinones are synthesized in one pot using Co-phthalocyanine (Co-PC) catalysts (6). Kumar et al. have briefly highlighted the recent developments and promising future of metal phthalocyanines catalyzed selective organic transformations (7).

5H.01.12: Iron (Fe):

Most of the work using iron based catalysts and hydrogen sources as silanes, water, hydrazine are covered under i) Water, ii) Solid Supports, iii) Metals and Metal Salts etc. Here a brief highlights or iron catalysed reductions are given with relevant references. Sulphur-containing NAs were rapidly reduced to the corresponding amines in high yields (up to 97-99%) by employing HH as a hydrogen donor in the presence of iron oxide hydroxide catalyst (8a) in IPA within 20-50 min. Iron(III) oxide-MgO catalyst also gives excellent yields of corresponding ANs (8b). Commercially available Fe_3O_4-NPs were utilized for efficient NA reductions, and could be recycled up to 10 times using magnetic separation, whilst retaining the activity as 99% AN yield in each case without any side-products. High chemoselectivity for reduction of the nitro versus other functional groups such as halogen, ester, O-benzyl, and N-Cbz groups was observed (9). An excellent chemoselectivity for the nitro group reduction with 100% conversions are demonstrated by Fe/Phenothroline/C with AN yields > 99% (10).

Sun et al. have developed a highly chemoselective method for the reduction of p-CNB to p-CAN in about 100% conversion and > 99.9% selectivity without dehalogenation product. The reaction is catalyzed by a bi-metallic Fe-Al catalyst complex with HH as a hydrogen source. H. Sun, H. Den, et al. Dept. of Chem., Central South Forestry University, Zhuzhou, Hunan 412006. Cai et al. have reported a simple but efficient method for preparing ANs. Under optimized conditions ANs were prepared with 96.3%-99.5% yields using PEG supported FeO(OH) and HH (11). A simple, easy to handle, economical and a practical method is developed for the chemoselective reduction of NAs to ANs using activated iron generated by Fe/HCl or Zn/FeSO$_4$ (12). Lauwnier et al have done a very good work in chemoselective reduction of diversely substituted NAs in high yields using specially prepared iron oxide hydroxide (Fe(O)OH; Ferrihydrite), Magnetite (Fe$_3$O$_4$) and Maghemite (Y-Fe$_2$O$_3$) and HH as a hydrogen donor (13).

As Professors Buchwald, Hartwig, Kappe, etc., Prof Beller also has made very good contributions to organic chemistry through valuable developments of simple, efficient, and economical green technologies. Here is an example from Prof. Beller's laboratory using pyrrolized iron–phenanthroline complexes supported on carbon which leads to the formation of highly selective catalysts for the reduction of structurally diverse NAs to ANs in 90–99% yields using HH as a hydrogen donor.

$$N_2H_4.H_2O + Catalyst = N_2 + H_2O + 2H_2$$

$$RArNO_2 + N_2H_4.H_2O + Fe-N/C = RARNH_2 + N_2 + H_2O$$

Excellent chemoselectivity for the nitro group reduction is demonstrated over 47 compounds (14). Octa-(aminophenyl)-silsesquioxane (OAPS) was prepared by reduction of Octa(nitrophenyl)silsesquioxane (ONPS) using HH as the reducing agent in the presence of Iron(III)Chloride catalyst with a yield of 87%. Hydrazine is a two-electron reducing agent whereas the nitro group is a four-electron reduction process. The activated carbon serves as an adsorbent and electrical conductor enabling the supply of the extra two electrons from many hydrazines making it possible the initial four-electron process successfully completed (15).

5H.01.13: Nickel (Ni):

Highly dispersed supported Ni-NPs on SiO$_2$ (Ni-NPs/SiO$_2$) were prepared by atomic layer deposition and tested for the chemoselective transfer hydrogenation of NAs with HH (16). Marko et al. have reported Ni-catalyzed reductions of NAs with HH and (Bu$_4$N) [Ni(tdt)$_2$] (tdt=toluene-3,4-dithiolate) and analogous Ni^{III}complexes in refluxing THF. Hydroxylamine derivatives are formed as intermediates (17). A Highly chemoselective heterogeneous nickel-molybdenum oxide catalysts supported on ordered mesoporous silica SBA-15 (Ni-MoO$_3$/CN@SBA-15) were prepared for the first time by treating SBA-15-supported nickel-molybdenum oxide materials with 1,10-phenanthroline. The catalyst exhibited unprecedented catalytic activity and high chemoselectivity for the reduction of various substituted NAs to the corresponding ANs in ethanol under mild conditions (Scheme-4). Owing to the yields of synergistic effect of metal Ni and MoO_3 species afforded excellent corresponding ANs in >99% within very short reaction periods (≤60 min).

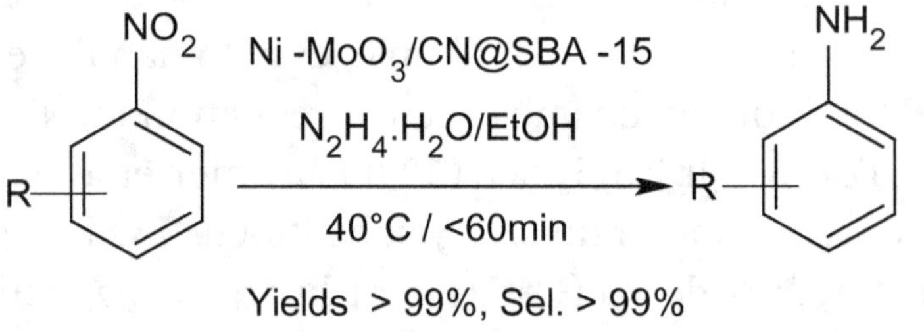

Yields > 99%, Sel. > 99%

R = Halogens, OH, NH_2, CHO, CN etc. 28 examples --Scheme-4

With 28 examples of substituted ANs were obtained in yields > 99% and selectivity > 99% (18). NAs are reduced efficiently in a mild, simple, clean, reusable and chemoselective manner by HH in the presence of trivalent cation-exchanged faujasite zeolites (19). In another case, the reduction of a series of NAs with HH catalyzed with colloid iron, cobalt, nickel, and copper particles has been studied.

M = Co, Ni, Fe, Cu in Colloidal Forms --Scheme-5

With Ni-NPs as a catalyst the reaction was performed in refluxing IPA and it was completed immediately at 60-80°C (Scheme-5). There was no N-alkylated product detected. High catalytic activities of cobalt, nickel and iron NPs have been demonstrated (20). Hee et al. have reported the synthesis of 6-amino quinolone in high yield (95%) by reducing 6-nitroquinoline with HH on montmorillonite clay K-10.

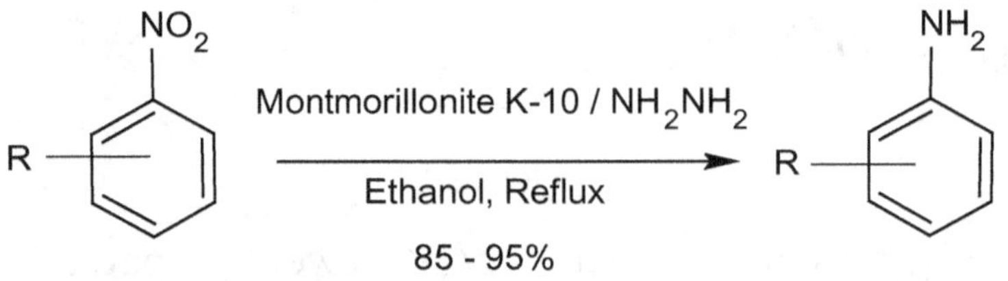

85 - 95%

R = o,m,p - CH_3, Cl, o - OH ; R = p - H, Br, OCH_3, CH_2OH --Scheme-6

NACs were readily reduced to the corresponding amino compounds in good yields with montmorillonite (K-10) and HH (Scheme-6) (21).

The reduction of NAs with $CO+H_2O$, and HH as reducing agents in the presence of metal catalysts is reviewed (22). A CeO_2 (10%)-SnO_2 catalyst prepared by a co-precipitation method efficiently catalyzes the transfer hydrogen reduction of a number of NACs with HH under mild conditions (23, 24).

5H.01.14: Raney-Ni (Ra-Ni):

Just as hydrazine has been used since a long time ago as a hydrogenation agent, Ra-Ni is also used for more than 60 years. HH is a very powerful reducing agent. Therefore, it cannot be used alone in the reduction of NAs as all the functional groups will be reduced. However, if Ra-Ni is used as a catalyst the selective reduction of nitro groups can be accomplished with HH in the presence of other functional groups. The selectivity is usually in the range of 80-90% (25). Gowda et al. have reported the application of hydrazinium monoformate as a new hydrogen donor with Ra-Ni for the facile reduction of NAs (Scheme-7) (26).

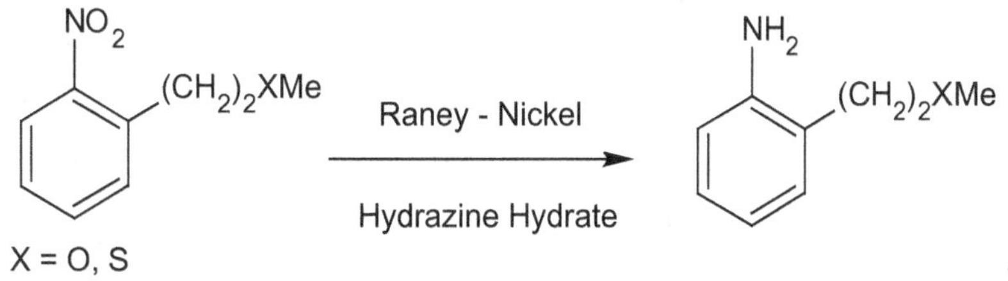

--Scheme-7

Synthesis of 2-(2-methoxyethyl)- and 2-(2-thiomethoxyethyl)-aniline and related compounds is investigated. The 2-(2-methoxyethyl)-1-NB was reduced with HH in the presence of Ra-Ni to 2-(2-methoxyethyl)-AN (27). HANs are very important organic compounds for their wide range of applications in many industries as agrochemicals, pharmaceuticals, etc. These HANs (o-, m-, p- (Cl, Br, F, I)) can be prepared by simple, convenient reduction of the respective HNBs in good yields using Raney-Ni (Ra-Ni) and HH in refluxing methanol (28). Ni and Co-NPs are used in the chemoselective reductions of NAs in aqueous phase or in water using HH as a reducing agent at room temperature providing industrially important fine chemicals or biologically important compounds (29, 30). NB is reduced to AN in high yields (~90%) with HH in the presence of Ra-Ni catalyst in methanol or in aqueous ethanol at room temperature (31, 32)

5H.01.15: Zinc (Zn) and Magnesium (Mg):

Chemoselective reduction of NAs using magnesium powder or zinc dust and hydrazine monoformate is reported (33, 34). NACs are reduced with in-house prepared and activated zinc-copper couple in the presence of HH as a hydrogen source (Scheme-8).

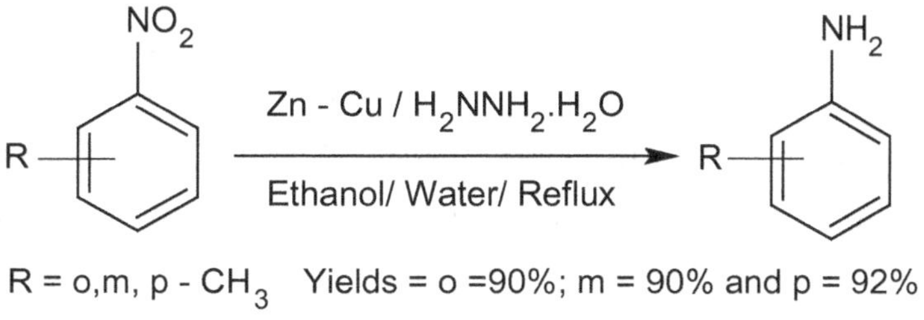

R = o,m, p - CH$_3$ Yields = o =90%; m = 90% and p = 92% --Scheme-8

The bi-metallic catalyst so obtained was found very active in the reduction of NACs (35). Selective reduction of nitro group in NAs (as -COOH, -CH$_3$, -OH etc.) is achieved using hydrazinium glyoxylate and Zn or Mg powder (36).

5H.02.00: HH in Continuous Flow Reactors:

We have covered this topic under Metals. Continuous-flow technology as a tool for the safe manufacturing of active pharmaceutical ingredients (APIs) is an emerging and highly promising methodology. Prof. Kappe's group is working on continuous-flow reactors. Iron-oxide NPs prepared in a homogenous system under microwave can be prepared and used in the highly chemoselelctive reduction of NAs to ANs nearly in quantitative yields. Some of the ANs with commercial applications have been synthesized using this continuous flow system (37-41). The stability of a catalyst is always a concern, therefore a novel egg-like nanosphere catalyst (Fe$_3$O$_4$@nSiO$_2$-NH$_2$-Fe$_2$O$_3$•xBi$_2$O$_3$@mSiO$_2$) is prepared and tested in the catalytic reduction of NAs to corresponding ANs in a fixed-bed continuous-flow reactor at 80°C with HH as reducing agent (42).

5H.03.00: Noble Metal Catalysts & HH:

Generally, the most commonly used noble metals as catalysts are gold, silver, palladium, iridium, platinum, ruthenium, rhodium etc. They are used as mono metals or bimetallic catalysts and also as M-NPs with or without supports.

5H.03.10: Gold (Au): Supported Au-NPs (<1mol %) are used first time for the activation of HH as a transfer hydrogenation agent in ethanol.

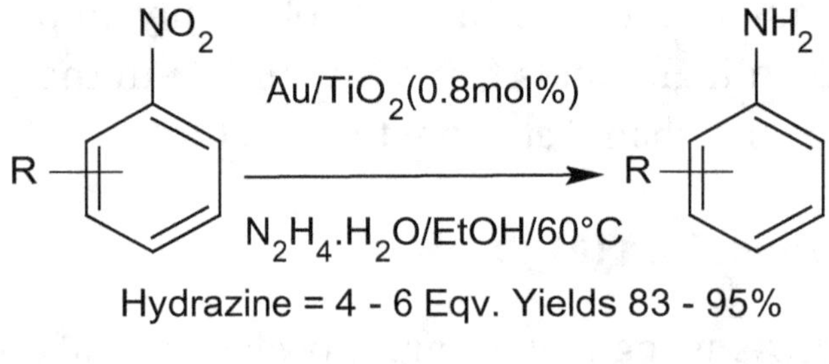

Hydrazine = 4 - 6 Eqv. Yields 83 - 95%   --Scheme-9

It takes 4 equivalents of hydrous hydrazine (Scheme-9) (43-46). The direct liquid N$_2$H$_4$/H$_2$O$_2$ fuel cell (DHHPFC) is known to be a unique power source for air-independent applications under extreme conditions such as outer space and underwater environments. Comparing to fuel cells using oxygen as the oxidizer, the replacement of oxygen by H$_2$O$_2$ allows much improved reaction kinetics on the cathode as well as higher power density and theoretical open circuit voltage of 1·2. The use of liquid fuel is also beneficial to the construction of more compact and portable power sources. A variety of materials have been designed and investigated as the hydrazine fuel cell catalysts Dealloyed nanoporous gold leaves (NPGLs) are found to exhibit high electrocatalytic properties toward both hydrazine (N$_2$H$_4$) oxidation and hydrogen peroxide (H$_2$O$_2$)

reduction, which allows the implementation of a direct hydrazine-hydrogen peroxide fuel cell (DHHPFC) based on these novel porous membrane catalysts. The activity of NPGLs complex under certain conditions is 22 times higher than that of commercial Pt/C electrocatalyst at the same noble metal loading. NPGLs thus hold great potential as effective, stable electrocatalysts for DHHPFCs (47).

The importance of an interfacial reaction to obtain mesoporous leafy nanostructures of Au and Pd have been synthesized and the as-synthesized porous architectures of palladium are efficient in the room-temperature hydrogenation of 4-NP by HH ($N_2H_4 \cdot H_2O$) (48). An environmentally safe approach for the synthesis of Au-NPs (23nm) using an aqueous leaf extract of *P. minus* was explored in this study, which is a simple, time saving (20min) and eco-friendly process. The biosynthesized Au-NPs are efficient in the reduction of 4-NP to 4-AP using HH as a reducing agent (49).

5H.03.11: Palladium: (Pd): Reduction of nitro compounds with Pd and hydrazine in refluxing ethanol is reported (Scheme-10) (50). CNBs are chemoselectively reduced to CANs using a noble metal (Pd or Pt) catalyst over carbon and HH as a hydrogen source (51).

--Scheme-10

Surfactants and ILs are used for reduction of NAs to ANs using Pd and HH (52, 53). Zhang et al have reported that 4-N-MethylaminoNB was reduced to corresponding ANs in 99% yield using a CTH with HH as hydrogen source (Scheme-11).

--Scheme-11

Zhang, et al. University of Technology, Dalian-116012 www.cnki.com.cn

5H.03.12: Platinum (Pt): *P*-Nitrobenzaldehyde is reduced to *p*-aminobenzaldehyde with platinum catalyst and hydrazine hydrate as a hydrogen donor (54). A stable, recyclable, economical solid supported Pt (SS-Pt) catalyst follows the principles of green chemistry (55).

5H.03.13: Rhodium (Rh): Reduction of simple NAs as substrates in the presence of Rh/C catalyst and HH produces the corresponding PHAs (56). A highly efficient and chemoselective reduction of NAs to ANs in quantitative yields is also accomplished with Rh nanocatalyst and HH under mild conditions, without affecting other sensitive functional groups (57).

5H.03.14: Ruthenium (Ru): Ru-NPs were assembled on CNTs and the resulting nanohybrid was used in the hydrazine-mediated catalytic selective hydrogenation of various NAs to PHA or AN at room temperature (58). Ru/C and hydrazine are used for reducing 2, 4-dinitroaniline to 4-nitro-o-phenylene diamine. See J. L. Miesel, et al. in Catalysis in Org Synth, Rylander & Greenfield (Eds), AP, NY, (1976), PP 273. Ru-NPs entrapped in ordered mesoporous carbons (OMCs) CMK-3 show superior efficiency in the green reduction process of NAs with HH as a hydrogen donor and using water as solvent. The higher activity than Ru/C is attributed to the interaction of Ru-NPs with CMK-3 and the mesoporous structure of CMK-3 (59). The ANCs are reduced to corresponding ANs using HH and photoinduced Ru-bipyridine complex.

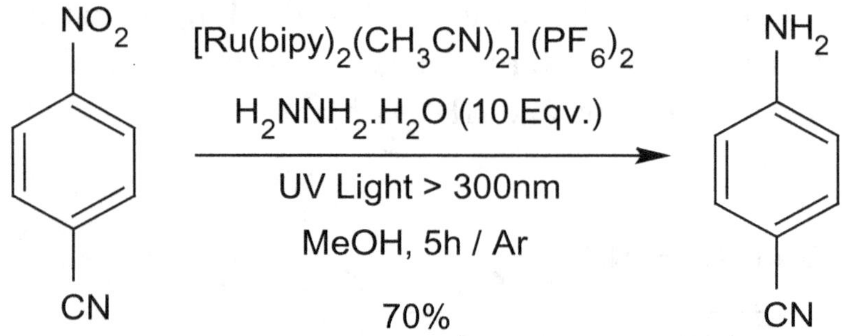

$[Ru(bipy)_2(CH_3CN)_2]\,(PF_6)_2$

$H_2NNH_2.H_2O$ (10 Eqv.)

UV Light > 300nm

MeOH, 5h / Ar

70%   --Scheme-12

Oxidative quenching pathway is proposed for photoexcited ruthenium catalyst in the reduction of NAs with HH (Scheme-12) (60, 61). In another case, among several catalysts prepared with different components the $[Ru(bpy)_2(MeCN)_2](PF6)_2$ was the most efficient catalyst and it gave the best yields(99%) of ANs in methanol (61).

5H.04.00: Selenium (Se):

The selective reduction of *m*-dinitrobenzene (*m*-DNB) to *m*-nitroaniline (*m*-NA) with Se as a catalyst and HH as hydrogen source is achieved under mild conditions. The effects of ratio of solvents, the molar ratios of other components on reduction of *m*-DNB to *m*-NA are investigated (Scheme-13).

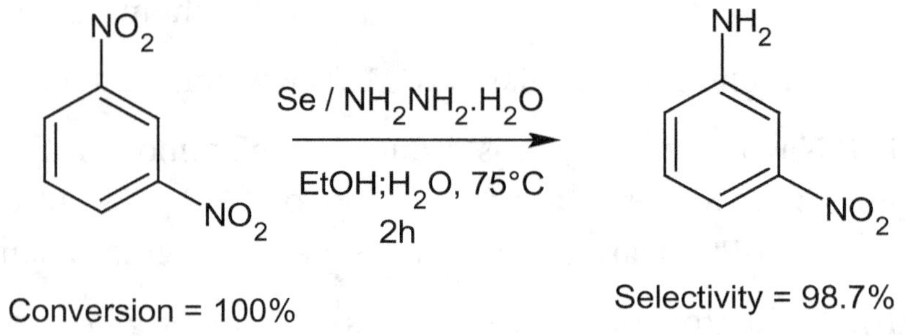

$Se / NH_2NH_2.H_2O$

$EtOH;H_2O$, 75°C

2h

Conversion = 100%   Selectivity = 98.7%   --Scheme-13

The results show that selenium has high catalytic activity and selectivity for the reduction under mild conditions. Under optimized conditions the *m*-DNB conversion and the *m*-NA selectivity had reached 100% and 98.7% respectively (62). It is quite interesting to note that a non-metal element (Se) exhibits such an excellent activity and regioselectivity towards *m*-NA.

5H.05.10: Bimetallic Catalysts: Mixed iron oxides [i) Iron(III) oxide-MgO, ii) Nickel-iron mixed oxide] are found efficient catalyst for NA reduction using hydrazine as hydrogen source (63, 64). In-house prepared bi-metallic catalyst (Ni-Pd-NPs) are found highly efficient and chemoselective towards the reduction of NAs with hydrazine in aqueous alkaline solution. Among the various compositions $Ni_{70}Pd_{30}$ shows maximum efficiency towards the catalytic reduction of 4-NP or PNP to 4-AP or PAP in the presence of HH at room temperature (65, 66). Some other examples of bimetallic catalysts will be found under other sub-sections as well, especially under continuous flow reactors and hydrazine as an energy storage material.

5H.05.11: Trinuclear Catalyst: As we have seen in case of other hydrogen sources all mono-, bi-, and trinuclear catalysts have been used for the chemoselective reduction of substituted NAs. Here is an example of synthesis of trinuclear $\{Co^{2+}-Co^{3+}-Co^{2+}\}$ and $\{Co^{2+}-Fe^{3+}-Co^{2+}\}$ catalyst complexes with accessible peripheral Co(II) ions.

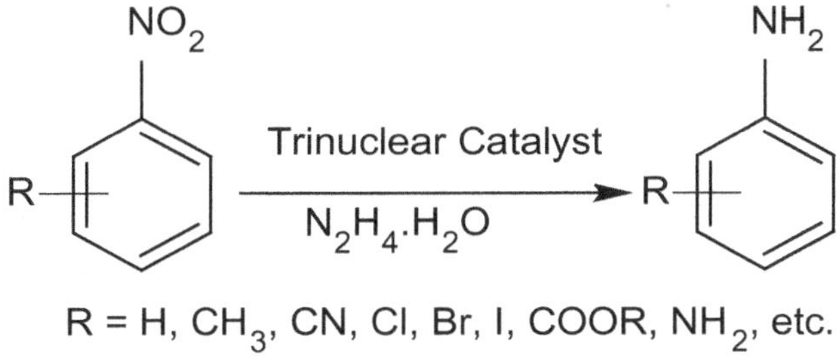

R = H, CH₃, CN, Cl, Br, I, COOR, NH₂, etc. --Scheme-14

Both trinuclear complexes function as efficient reusable heterogeneous catalysts for the selective reduction of assorted NAs to the corresponding ANs, using hydrazine as hydrogen donor (Scheme-14). The mechanistic investigations suggest the involvement of a Co(II)–Co(I) cycle in the catalysis process (67).

5H.06.00: NA Reduction over Polymer Supports:

Simple, eco-friendly, and rapid reduction of NAs to corresponding ANs is accomplished in excellent yields with ferrihydrite (FeO(OH) as a catalyst.

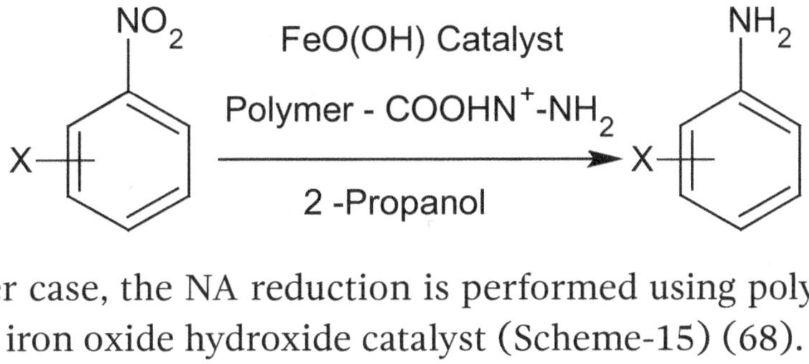

--Scheme-15

In another case, the NA reduction is performed using polymer-supported hydrazine hydrate over iron oxide hydroxide catalyst (Scheme-15) (68).

--Scheme-16

Polymer-supported nano-amorphous Ni–B particles (Scheme-16) prepared by ion-exchange-chemical reduction method exhibited good activity in the catalytic transfer hydrogenation of NACs with hydrazine hydrate as a hydrogen donor (69). Magnetically separable $SiO_2@Fe_3O_4$ core–shell supported nano-structured Cu (II) composites as a versatile catalyst for the reduction of NAs in aqueous medium at room temperature is reported with chemoselectivity and high yields (70). The deposition of Pd as a catalyst on polymer beads such as Merrifield resin or TentaGel allows the reduction of polymer-bound 4-nitrobenzoate to corresponding ANs (Scheme-17) with dilute HH solution in DMF at room temperature (71).

--Scheme-17

Also, polymeric PEG35K-Pd-NPs are insensitive to air and moisture and are found as an efficient and recyclable catalyst for reduction of NAs (72, 73). Several cation resins were prepared by ion-exchanging and used for catalytic reduction of NACs to ANs with HH in ethanol (Scheme-18).

Cation Resin - Hydrogen Transfer Catalyst

Metal Cations = Fe(III), NI (II), and Bi(III).   --Scheme-18

Among them, iron (III) resin has the best catalytic activity giving 85%-99% AN yields. K. Cai, et al. (School of Chem. and Chemical Enginnering, Xuzhou Institute of Technology, Xuzhou 221008, Jiangsu, China).

5H.06.10: HH on Activated Carbon with Oxygen Functional Groups:

It is worth noting that the CTH on activated carbon surface are performed with any metal catalyst. The following examples are representative of these interesting cases. Activated carbons modified by chemical treatment in aqueous solutions of HNO_3, HCl and H_2SO_4 or in gaseous N_2 and H_2 have been tested as catalysts in the reduction of NB using HH as hydrogen donor. Through analytical studies it was found that the chemical treatment leads to the formation of various oxygen functional groups on the surface of the activated carbon, which cause, i) the decomposition of HH and ii) adsorption of NB on the surface of activated carbon (Scheme-19).

$$ArNO_2 \xrightarrow[2H^+]{2e^-} ArNO \xrightarrow[2H^+]{2e^-} ArNHOH \xrightarrow[2H^+]{2e^-} ArNH_2$$

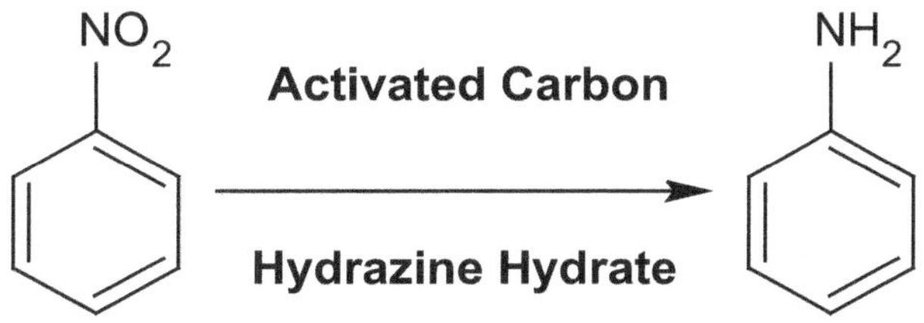

--Scheme-19

The catalytic activity is correlated to that of the oxygen functional groups, which suggests that an abundance of surface oxygen functional groups is favorable for this reaction. The activated carbon treated by hydrochloric acid had the largest amount of oxygen functional groups and exhibited the highest activity for NB reduction to AN. It is found that that hydrazine reaction at the carbon surface is the rate determining step (74).

$$3N_2H_4 = 4 NH_3 + N_2$$

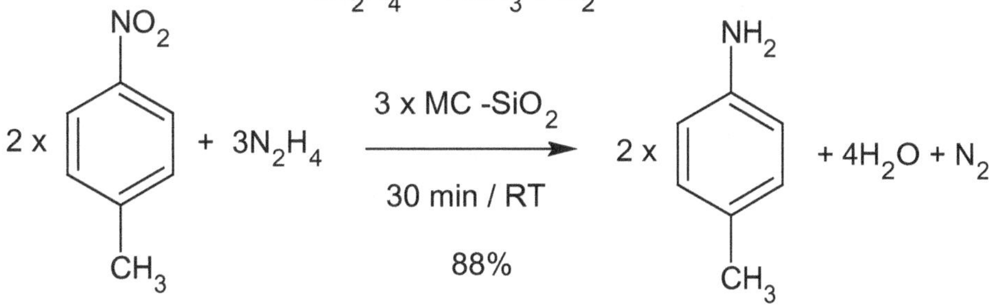

Catalyst Recycles 6 times : Y = 72 - 88%.

--Scheme-20

Wang et al have reported their study on the mesoporous carbons with tailored pore size, which were prepared by using sucrose as the carbon source and silica as the templates and using HH as the reducing agent (Scheme-20).

Compared with other carbon materials such as active carbon, these carbon materials exhibited much higher activity (75). Larsen & Figueras have independently reported the mechanism of the carbon catalyzed reduction of NB by hydrazine (76, 77). The γ-Fe_2O_3 NPs well dispersed in porous carbon through MOF process exhibits excellent catalytic activity, chemoselectivity and magnetic recyclability for the hydrogenation of diverse NAs under mild conditions with up to 100% conversions and 100% selectivity towards corresponding ANs (78).

5H.06.11: Hydrazine Hydrate & Microwave Irradiations (MWI):

HH ($N_2H_4 \cdot H_2O$) is a good reagent because it is less hazardous than other hydrogen donors and it generates only N_2 as a side product. Amongst many other methods of chemical reactions, MWI is one of the most utilized techniques. A process is developed for the synthesis of ANs by reduction of NAs with HH using Fe_3O_4 nanocrystals generated in-situ as a catalyst under MWI (Scheme-21).

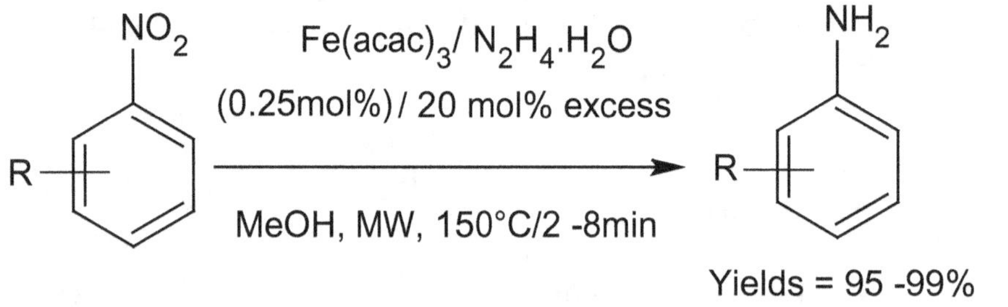

R = H, Cl, Br, I, $COCH_3$, OH, OCH_3, CN, NH_2, Heterocycles etc. --Scheme-21

The process is completed in few minutes generating pure ANs in 95-99% yields in all 20 cases (79). Solvent-free reduction of NACs to corresponding ANs in good to high yields with $FeCl_3 \cdot 6H_2O$, Fe(III) oxide hydroxide or Fe(III) oxides etc. on Al_2O_3 and HH as a reducing agent under MWIs is accomplished (Scheme-22) (80).

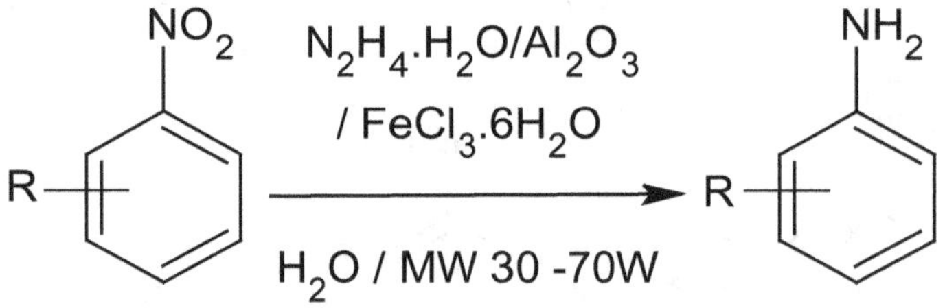

R = H, Cl, I, OH, CH_3, OCH_3, NH_2, etc. --Scheme-22

HNBs are chemoselectively reduced to HANs in high yields and selectivity using Pd/C and HH as a hydrogen source. However, using microwave irradiations at elevated temperature and pressure the dehalogenated products are formed (81).

5H.06.12: Hydrazine Hydrate in Synthesis of Multinuclear Analogs:

Majid et al. have reported the reduction of an intermediate nitroso compound to amine by HH without a catalyst. Convenient synthesis of 4-amino-3-disubstituted pyrazoles in one step from the corresponding diketo oximes is accomplished (82). Synthesis of 2-methyl benzimidazole from ethanol and o-nitroaniline catalyzed by a multifunctional

Cu-Pd-NPs/γ-Al$_2$O$_3$ is investigated. Bimetallic Cu-Pd catalyzed reduction of *o*-nitroaniline to *o*-phenylene diamine in ethanol and its use in the synthesis of benzimidazoles can smoothly be preformed (83, 84).

5H.07.00: Hydrazine Hydrate as Energy Storage Material:

The design and development of alloy catalysts has drawn a considerable attention in the recent past. This is because of the fact that the bimetallic alloy catalysts give better performance than the individual metals. Singh et al have done a very good study about this aspect using Ni-Pt, Ni-Ir and non-noble metals as Ni-Co alloys, which are prepared by special techniques. All these bimetallic alloys nano catalysts exhibit excellent catalytic activity towards the decomposition of HH producing hydrogen with 100% selectivity at room temperature, whereas the corresponding single-component counterparts are inactive for this reaction (Scheme-23). Hydrazine as an alternative for ammonia in hydrogen release in the presence of boranes (BH$_3$) and alanes (AlH$_3$) is reported (85).

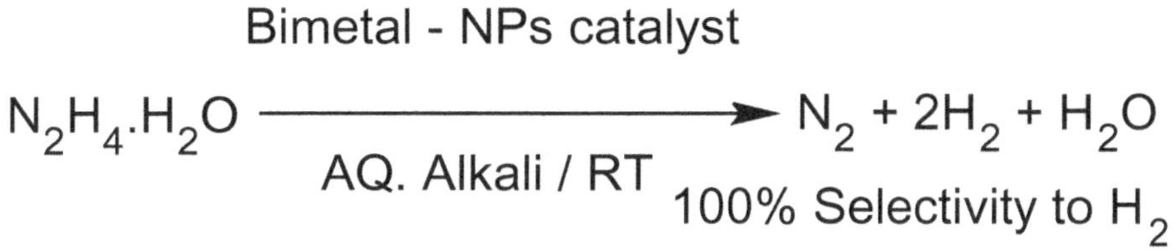

$$N_2H_4 \cdot H_2O \xrightarrow[\text{AQ. Alkali / RT}]{\text{Bimetal - NPs catalyst}} N_2 + 2H_2 + H_2O$$

Bimetal - NPs catalyst

100% Selectivity to H$_2$ --Scheme-23

These results confirm the fact that hydrazine has the potential to be used as hydrogen storage material (86, 87). In case of Ni-Ir, the use of surfactants enhances the activity by suppressing the agglomeration of NPs, but does not affect the bimetallic compositions of the NPs (88, 89). Noble-metal-free Ni-Fe alloy NPs (Ni-Fe nanocatalyst: 1:1) exhibits 100% hydrogen selectivity in basic solution (0.5 M NaOH) at 343K. These results encourage the effective application of HH as a promising hydrogen storage material (90). Cao et al. have developed a novel method wherein the immobilization of ultrafine bimetallic Ni-Pt-NPs inside the pores of MOF of MIL-101 and the NiPT@MIL-1-1 exhibited high catalytic activity, selectivity and durability toward hydrogen generation by dehydrogenation of alkaline HH (91). Well-dispersed bimetallic Ni-Pt-NPs with different compositions have been successfully grown on the MIL-96 (NiPt/MIL-96) by a simple liquid impregnation method using NaBH$_4$ as the reducing agent. Ni$_{64}$Pt$_{36}$/MIL-96 exhibits the highest catalytic activity among all the catalysts tested with a turnover frequency value of 114.3 h^1 and 100% hydrogen selectivity. This excellent catalytic performance might be due to the synergistic effect of the MIL-96 support and Ni-Pt-NPs. The Ni-Pt-NPs supported on other conventional supports, such as SiO$_2$, carbon black, γ-Al$_2$O$_3$, PVP, and even the physical mixture of Ni-Pt and MIL-96, exhibit inferior catalytic activity compared to that of Ni-Pt/MIL-96 (92). Ultrafine monodispersed bimetallic Ni-Pt-NPs on grapheme support with different compositions have been successfully synthesized by co-reduction method. Among all the catalysts tested, Ni$_{84}$Pt$_{16}$/graphene exhibited 100% hydrogen selectivity and marked high catalytic activity at 25°C for hydrogen generation from alkaline solution of hydrazine (93).

References:

1a). A. Furst, et al. *J. American Chemical Society*, 1957, 79 (20):5492-5493(1957),

1b). A. Furst , R. C. Berlo, et al. *Chemical Reviews*, 65(01):51-68(1965).

2). D. Wang & D. Astruc, *Chemical Society Reviews*, 46:816-854(2017).

3). U. Sharma, et al. *Chemistry.-A European J.*, 17(21):5903-5907(2011).

4). U. Sharma, N. Kumar, *Green Chemistry*, 14(8):2289-2293(2012).

5). U. Sharma, et al. *Advanced Synthesis & Catalysis*, 352(11-12):1834-1840 (2010).

6a). V. Kumar, et al. *Australian J. Chemistry*, 65:1594–1598(2012),

6b). V. Kumar et al. *Green Chemistry*, 14(12):3410-3414(2012).

7a). N. Kumar, et al. *Organic Chemistry, Current Research*, 01(03):1000e109(2012).

7b). A. B. Sorokin, *Chemical Reviews*, 113(10):8152–8191(2013).

8a). Q. Shi, et al. *Chinese Chemical Letters*, 17(08):1045-1047(2006).

8b). Q. Shi, et al. *Chinese Chemical Letters*, 17(04): 441-443(2006).

9). S. Kim, E. Kim, et al. *Chemistry – An Asian J*, 06(08):1921–1925(2011).

10). M. Belller, et al. *Chemical Communications*, 47:10972-10974(2011).

11). K. Y. Cai, et al. *Bull Chem Reaction Engg & Catalysis*, 11(03):363-368(2016).

12). Y. Liu, et al. *Advanced Synthesis & Catalysis*, 347(02-03)217-219(2005).

13a). M. Lauwiner, et al. *Applied Catalysis A: General*, 172(01):141-148(1998).

13b). M. Benza, et al. *Applied Catalysis A: General*, 172(01):149-157(1998).

13c). M. Lauwiner, et al. *Applied Catalysis A: General*, 177(01):09-14(1999).

14a). A. Felix, et al. *Chemical Communications*, 47(39):10972-10974(2011).

14b). *R. V. Jagadeesh, et al. Chemical Communications*, 47(39):10972-10974(2011).

15). Q. Shao, et al. *Advances Chemical Engineering & Science*, (03):93-97(2013).

16). C. Jiang, et al. *ACS Catalysis*, 05(08):4814–4818(2015).

17). A. Vizi-Orosz, L. Marko, *Transition Metal Chemistry*, 13(03):221-223(1988).

18). H. Huang, X. Wang, *Green Chemistry*, 19(03):809-815(2017).

19). M. Kumarraja, et al. *Applied Catalysis A: General*, 265(02):135–139(2004).

20). V. N. MOkhov, et al. *Russian J. General Chemistry,* 84(08):1515–1518(2014).

21). H. B. Hee & J. D. Gyu, *Tetrahedron Letters,* 31(08):1181-1182(1990).

22). R. Singh, R. Kaur, et al. *International Research Advances,* 03(02):30-39(2016).

23). R. W. Lu, et al. *Dyestuff and Coloration,* 40:160-162(2003).

24). T. M. Jyothi, R. Rajagopal, et al. *ChemInform;* 31(08): 1 page (2010).

25a). D. Balcom, A. Furst, *J. American Chemical. Society,* 75 (17):4334–4334(1953).

25b). A. Furst, et al. *J. American Chemical Society,* 79 (20):5492–5493(1957).

26). D. C. Gowda, et al. *Tetrahedron,* 58: 2211-2213(2002).

27). T. Türk, et al. *Monatshefte für Chemie,ChemMonthly,*126(06-07):725-732(1995).

28). B. E. Leggetter, et al. *Canadian J. Chemistry,* 38(12):2363-2366(1960).

29). R. K. Rai, A. Mahata, *Inorganic Chemistry,* 53(06):2904–2909(2014).

30). H. Li, R. Zhang, et al. *Synthetic Communications.* 27(17):3047-3052(1997).

31). L. P. Kuhn, US 2768209 A (1956),

32). N. Goswami, et al. *J. Chemical Technology & Biotechnology,* 34:195(1984).

33). K. Abiraj et al. *Synthesis and Reactivity in Inorganic and Metal-Organic Chemistry,* 32(8), 1409-1417(2002).

34). S. Gowda, et al. *Synthetic Communications,* 33(02):282-289(2003).

35). H. B. Hee, et al. *Bulletin Korean Chemical Soc*iety, 10(03):315-316(1989).

36). B. Raju, R. Ragul, et al. *Industrial J. Chemistry,* 14B:1315-1318(2009).

37). *D. Cantillo, et al. J. Organic Chemistry,* 78(09):4530–4542(2013).

38). B. Guttman, et al. *Angewandte Chmie,* 54(23):6688-6728(2015).

39). B. Guttman, et al. *European J. Organic Chemistry,* 2017(04):914-927(2017).

40). M. M. Moghaddam, et al. *ChemSusChem,* 07(11):3122-3131(2014).

41). P. L. Reddy, M. Tripathi et al. *Chemistry An Asian J.* 12(07):785-791(2017).

42). Y. Ai, Z. Hu, *Nano Research,* 01–13 (2017).

43). P. L. Gkizis, et al. *Catalysis Communications,* 36:48-51(2013).

44). A. Corma, P. Serna, *Catalysis Science,* 313: 332-334 (2006).

45). A. Corma, et al. *Angewandte Chemie Intl Edition,* 46:7266-7269(2007).

46). A. Corma et al. *Green Chemistry,* 09:849-851(2007).

47). X. Yan, F. Meng, et al. *Scientific Reports*-2, Article number:941 (2012).

48). S. Dutta, et al. *ACS Applied Materials & Interfaces.* 06(12):9134-9143(2014).

49). S. Borhamdin, et al. *J. Experimental Sciences*, 11(07):518-530(2016).

50). P. Silvo, *Annale Chime,* (Rome), 45:850-853(1955).

51). F. Li, et al. *Synlett*, 25(10):1403–1408(2014).

52). Z. Xu, G. Lu, et al. *Catalysis Communications*, 99:57-60(2017).

53). S. Rollas, *Marmara Pharmaceutical Journal*, 14: 41-46(2010).

54). L. P. Kuhn, US 2768209 A (1956).

55). A. K. Shil & P. Das, *Green Chemistry*, 15(12):3421-3428(2013).

56). A. Korte, et al. US 2013/0338373 A1 (2013).

57). P. Luo, K. Xu, et al. *Catalysis Science & Technology*, 02(02):301-304(2012).

58). D. V. Jawale, et al. *Chemical Communications*, 51(09):1739-1742(2015).

59). J. Hu, Y. Ding, et al. *RSC Advances*, 06(4):3235-3242(2016).

60). H. J. Shiori, et al. *Bulletin Chemical Society, Japan,* 77(09:1763-1764(2004).

61). T. F. Teply, *Collect. Czech. Chemical Communications*, 76(07):859–917(2011).

62). K. Cai, X. Zhou, et al. *J. Petrochemical Universities*, 2008-02.

63). P.S. Kumbhar, F. Figueras, et al. *Tetrahedron Letters*, 39(17):2573-2574(1998).

64) Q. Shi, et al. *Advanced Synthesis & Catalysis*, 349(11-12):1877–1881(2007).

65). D. Bhattacharjee, K. Mandal, et al. *RSC Advances*, 06(69)**:**64364-64373(2016).

66). A. B. Madavi, et al. *Particulate Science & Technology*, PP01-10(2016).

67). S. Srivastava, M. S. Dagur, et al. *Dalton Transactions*, 44:17453-17461(2015).

68). Q. Shi, R. Lu, *Green Chemistry,* 08:868-870(2006).

69). H. Wen, K. Yao, *Catalysis Communications*, 10(08):1207-1211(2009).

70). R. K. Sharma, et al. *J. Molecular Catalysis A: Chemical*, 293:84-95(2014).

71). M. Rodel, F. Thieme, et al. *Synthetic Communications*, 32(8):1181-1187(2002).

72). R. Kumar, V. Yadav et al. *Synthesis, Reactivity in Inorganic Metal-Organic and Nano-Metal Chemistry*, 41(01):114-119(2011).

73). V. Yadav, S. Gupta et al. *Synthetic Communications*, 42(2):213-223(2012).

74). L. Shi, et al. *Chinese J. Catalysis*, 33(09-10):1463-1469(2012).

75). H. Wang, et al. *Bulletin Korean Chemical Society,* 33(09):2961-2965(2012).

76). J. W. Larsen, M. Freund, et al. *Carbon*, 38:655-661(2000),

77). F. Figueras, et al. *J. Molecular Catalysis A: Chemical*, 173:223–230(2001).

78). L. Yang, X. Yu, et al. *Chemical Communications*, 52:4199-4202(2016).

79). D. Cantillo, et al. *Angewandte Chemie Intl. Edition*. 51: 10190-10193(2012).

80). A. Vass, J. Dudas, et al. *Tetrahedron Letters*, 42(32):5347-5349(2001).

81). F. Li, et al. *Synlett*, 25:1403-1408(2014).

82). T. Majid, C. R. Hopkins, et al. *Tetrahedron Letters*, 45: 2137-2139(2004).

83). F. Feng, J. Ye, et al. *RSC Advances*, 06:72750–72755(2016).

84). T. Bao, D. Pla, et al. *Catalysts*, 07:207(2017). doi:10.3390/catal7070207

85). N. Vinh-Son, et al. *Physical Chem & Chem Physics*, 11(30):6339-6344(2009).

86). S. K. Singh et al. *J American Chemical Society*, 131(29):9894-9895(2009).

87). S. K. Singh et al. *J American Chemical Society*, 131(50):18032-18033(2009).

88). S. K. Singh, et al. *Inorganic Chemistry*, 49(13):6148-6152(2010).

89). S. K. Singh, et al. *Chemical Communications* (Camb), 46 (35):6545-6547(2010).

90). S. K. Singh et al. *J American Chemical Society*, 133(49):19638-19641(2011).

91). N. Cao, et al. *Inorganic Chemistry*, 53(19):10122-10128(2014).

92). L. Wen, et al. *Dalton Transactions*, 44(13):6212-6218(2015).

93). Y. Du, et al. *ACS Applied Material & Interfaces*, 07(02):1031-1034(2015).

Chapter-5M: Metals, Metal Salts and Metal-NPs:

5M.00.00: Introduction:

Metal, Metal-Acid or Metal Salts, M-NPs and different hydrogen sources as alcohols, boranes, hydrazine etc are used for accomplishing efficient hydrogenations of NAs to ANs under a set of conditions. The developments using these hydrogen donors are covered under individual headings and also they appear under solid supports, metals and metal salts, mono- and bimetallic-NPs etc. Although we are focusing on CTH of NAs to ANs by hydrogen sources other than molecular hydrogen one must make a note that under some cases of metal catalysts even molecular hydrogen is used as a hydrogen source. Details on catalytic hydrogenations of NAs to ANs are covered in the previous chapter.

5M-01.00: Metals in Chemo- and Regioselective Synthesis of Anilines:

Since metals, metal nanoparticles (M-NPs) and metal salts are used as catalysts everywhere, some of the work presented here is also reported briefly under other sections as it is relevant there too. Care is exercised to keep the information at minimum in either of the two places, but not miss it as it is vital in both the contexts. Particularly, in this section the metals used as catalysts are in different forms as i) just in their original metallic form as Fe, Zn, Mg etc., ii) metals salts as $FeCl_3$, CuCl, $SnCl_2$ etc. iii) oxides as Fe_2O_3 or Fe_3O_4 Co_3O_4 etc iv) metals in powder forms as iron powder and v) in the form of M-NPs or BM-NPs. We have compiled the information on the basis of non-noble metals and noble metals and arranged in alpha-betical order of their names. Besides molecular hydrogen as reducing agent, the other hydrogen sources are: water, alcohols (methanol, ethanol, 2-propanol or isopropanol (IPA), glycerol, etc.), boranes, borohydrides, $CO+H_2O$, CO_2+H_2O, supercritical CO_2 ($scCO_2$), formic acid and formates (as metal formates from Na, K, Mg, Zn, ammonium, triethyl ammonium, and hydrazinium formate), inorganic salts as NH_4Cl, NH_4Br, silanes and sulfides, cyclohexene, HH etc. have been used. Of these, glycerol and IPA are covered under Ir and La metals, while others as $CO+H_2O$, HH, silanes, water as hydrogen sources are covered under their individual headings in this section.

5M.01.10: Non-Noble Metals:

5M.01.10a: Acids and Metals:

Besides the use of Fe/aq. HCl by Bechamp (1), other metals as Zn, Al, Sn or Fe and aqueous acid have been used to reduce NAs. These reactions produce nitroso and N-PHA intermediates, which are attributed to the electron transfer from the metal surface to the nitro group. Fe, Cu, Zn and Sn are abundantly available and in the presence of NH_4Cl or NH_4Br reduce NAs to ANs. The NACs are reduced to ANs by a variety of reagents which include metal-mediated reagents like $In/FeCl_3/H_2O$, Raney-Ni/NH_4Cl, $Ru_3(CO)_{12}$/chelating diimines, $FeS/NH_4Cl/CH_3OH/H_2O$, Sm/NH_4Cl, Sn/NH4Cl, Zn/NH_4Cl in ionic liquids and nanosized activated metallic iron powder in water. However, these methods have their own limitations while the noble metals' cost and tedious work ups are concerns. Gowda et al. have reported a chemoselective reduction of substituted NAs including HNBs using low cost commercial zinc dust and diammonium hydrogen phosphite at room temperature in methanol, without affecting other groups and especially there was no dehalogenation observed (2).

5M.01.10b: Iron (Fe):

Iron was the first metal (Fe/Aq. HCl) used in the reduction of NB to AN by Bechamp in 1854, and since then it is extensively used in its different forms as salts, oxides, powders, nanoparticles, plain or on solid supports for the chemoselective reduction of NAs to corresponding ANs in pure water, alcohol, water-alcohol mixture, aqueous organic solvents and with or without a hydrogen source. In such cases the alcohols or water act as a hydrogen source. There is a considerable wealth of information in the public domain using iron as a catalyst in the reduction of NAs. Hence we have taken it out of alfa-betical order and first highlighting the work based on iron metal catalyzed reduction of NAs here.

5M.01.10c: Iron and Different Hydrogen Sources:

AN and its analogs are important intermediates in pharmaceuticals, agro chemicals, polyurethanes, paint and dyes and all of them must be made at commercial scale. Bechamp reduction was first developed and used in the laboratory for the reduction of NB and nitronaphthalene and then it was used as a main reduction technology for commercial manufacturing of AN and other important aryl amines. AN required for pigment industry was still made till 2007 by Bechamp reduction of NB. Thomas Kahl et al. "Aniline" in Ullmann's Encyclopedia of Industrial Chemistry, 2007; John Wiley & Sons: New York. doi:10.1002/14356007.a02_303. However, under acidic conditions the intermediate PHA undergoes Bamberger rearrangement to give 4-AP or PAP (Scheme-1).

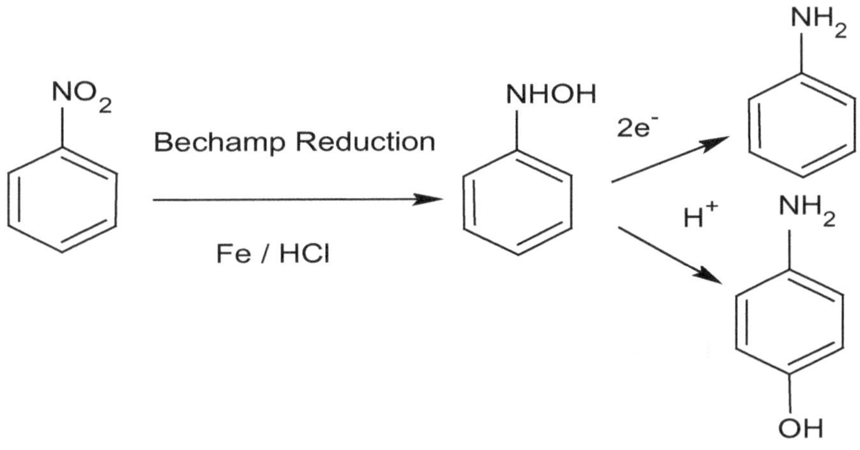

Scheme-1

PAP is prepared by following much more efficient and economical methods (2) from 4-NP to 4-AP. Recently, a novel reduced graphene oxide-zinc tungstate-iron oxide (rGO-ZnWO$_4$-Fe$_3$O$_4$) nanocomposites are prepared (3) and used for developing an economical and facile approach for the reduction of 4-NP to 4-AP using NaBH$_4$.

Reduction of NAs to ANs using activated iron (4) and a novel iodide-catalyzed reduction method using hypophosphorous and/or phosphorus acids are eported (5). NAs are reduced using a simple method to the corresponding ANs with iron or zinc catalyst and NH$_4$Cl under microwave heating. Mild acidity is provided by ammonium chloride in an aqueous medium in methanol. The conditions are tolerant to other functional groups (6), with the exception of bromoalkyl derivatives, which yield complex reaction mixtures, otherwise the yields are generally quite high (80–99%).

5M.01.10d: Alcohols:

Alcohols as glycerol and IPA are highlighted as hydrogen donors under La and Ir metals. More information on alcohols as hydrogen sources is available under "Alcohols" in other sections.

5M.01.10e: CO+H$_2$O:

When the hydrogenation is done using CO+H$_2$O in the presence of an alkoxide, AN carbamate is the sole product formed but only in 50% yield (7), which can be hydrolyzed to get ANs (Scheme-2). An intermediate species such as $[(ArN)Fe_3(CO)_9COOCH_3]$ is supposed to be formed that undergoes reductive elimination (and protonation) to afford the carbamate.

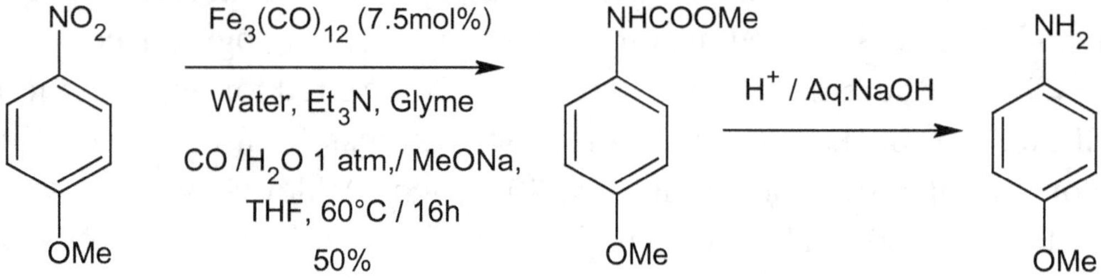

--Scheme-2

Further, the mechanistic studies on the reduction of NB to AN by iron carbonyl catalysts have shown that a radical anion $[Fe_3(CO)_{11}]^-$ was involved (8). Jiang et al have reported the environmentally benign and selective reduction of NAs with pressurized CO$_2$-H$_2$O medium (Scheme-3).

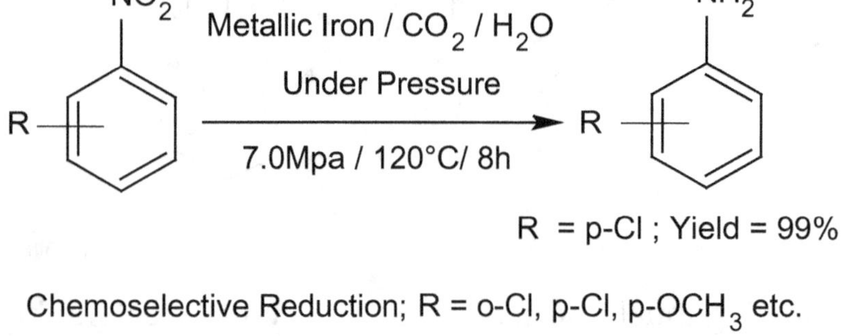

Chemoselective Reduction; R = o-Cl, p-Cl, p-OCH$_3$ etc. --Scheme-3

The tetraethylammonium undecacarbonylhydridotriferrate $[(C_2H_5)_4N^+HFe_3(CO)_{11}^-]$ (I) can be easily prepared in a two-step sequence from Fe(CO)$_5$, Et$_3$N and Et$_4$NCl. It selectively reduces NAs to ANs in good yields. The catalyst also reacts with some organic halides to give dehalogenated products (9a).

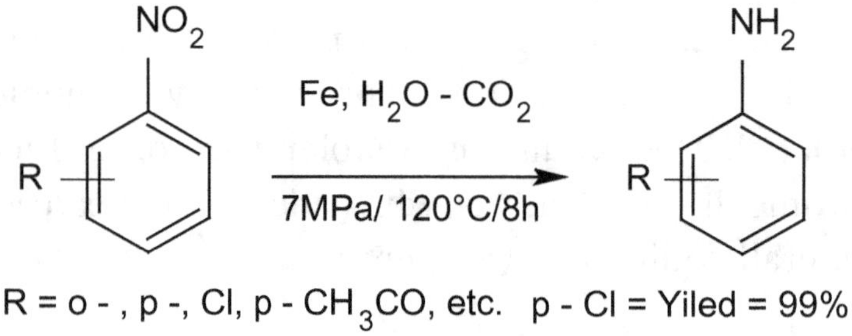

R = o -, p -, Cl, p - CH$_3$CO, etc. p - Cl = Yiled = 99% --Scheme-4

A new method for the reduction of NAs employing metallic iron in pressurized CO$_2$-H$_2$O medium has been developed. It is a general method and p-CNB can be reduced to p-CAN in 99% yield under optimized conditions. The method is applicable for the reduction of other substituted NAs as well (Scheme-4). After the reduction, the reaction mixture is self-neutralized on releasing of CO_2. The method is environmentally benign, easily scaled-up and applicable for the preparation of substituted ANs (9b).

5M.01.10f: Iron, Iron Oxides and HH:

Reduction of NACs with HH in the presence of the Iron(III)oxide-MgO catalyst (10a) can be smoothly accomplished (Scheme-5).

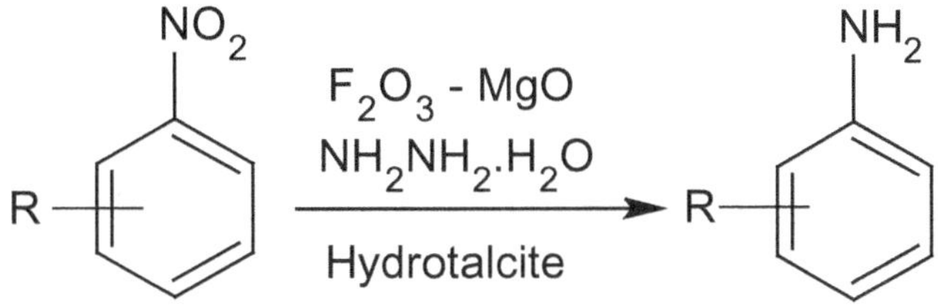

--Scheme-5

Iron (III) oxide-MgO, a stable and recyclable catalyst was prepared from a Mg-Fe hydrotalcite precursor which shows high activity and selectivity for reduction of NAs under mild reaction conditions. Natarajan et al (10b) have reinvestigated a process for the reduction of NAs containing other susceptible groups, which provides a viable economic route from the viewpoint of quantitative transformation, reduced reaction times and simple work up using iron powder and ammonium chloride solution in neutral medium. Desai et al. have used $FeS-NH_4Cl-CH_3OH-H_2O$ system to reduce NAs to ANs in good yields (11). A mild and efficient method for the preparation of ANs from NAs by treatment of a variety of NACs to corresponding ANs in high yields (>80-90%) with N, N-dimethyl-hydrazine (DMH) in the presence of catalytic amount of $FeCl_3·6H_2O$ in methanol is reported (Scheme-6).

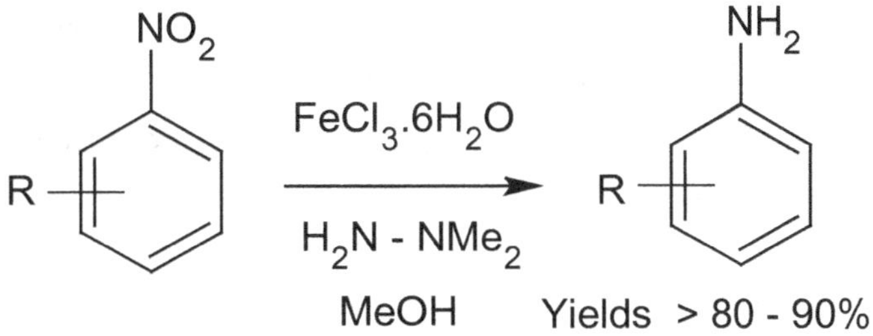

--Scheme-6

This reduction system is compatible with a wide range of functional groups (12). On similar lines, azides are reduced to amines using a mild method using DMH and catalytic amount of $FeCl_3$ (13). HH or DMH is used as one of the most efficient hydrogen source in the catalytic reduction of NAs.

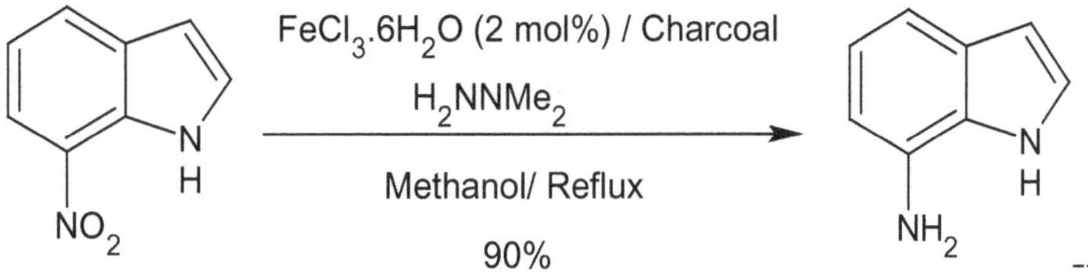

--Scheme-7

For example, 7-nitroindole is reduced to 7-aminoindole in 90% yield (14) using DMH in refluxing methanol (Scheme-7). Feng et al have reported selective reduction of NAs using iron catalyst and hydrazine as hydrogen donor (15). Sharma *et al.* have reported the selective nitro reduction with iron (II) phthalocyanine and iron sulfate as catalysts using hydrazine hydrate as the hydrogen source. The conversions and selectivity are up to 99% and the yields of ANs are also up to 99%. This protocol has attracted much attention as an eco-friendly method since it uses a green solvent system (H_2O-EtOH) (16). An efficient and highly selective iron-catalyzed reduction of NB is reported using 2-3 mole equivalent of HH as a hydrogen donor at 100°C in THF as a co-solvent. Pyrolysis of iron–phenanthroline complexes supported on carbon leads, especially when carbon concentration is 100%, to highly selective catalysts for the reduction of structurally diverse NAs to ANs in 90–99% yields (Scheme-8).

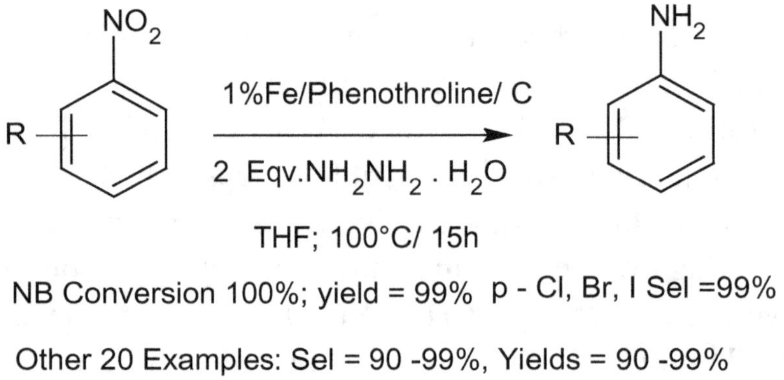

NB Conversion 100%; yield = 99% p - Cl, Br, I Sel =99%

Other 20 Examples: Sel = 90 -99%, Yields = 90 -99%   --Scheme-8

Excellent chemoselectivity for the nitro group reduction is demonstrated. Especially in case of HNBs both the selectivity and the yields of HANs are up to 99%, confirming that there was no dehalogenation at all. The catalyst system was effective till seven re-runs yielding 97-99% of ANs (17). The *o*-amino-*p*-methyl phenol was synthesized by using two reduction methods. 1) In the liquid-phase hydrogenation process with Raney-Ni as catalyst, the product was isolated in > 94% under optimal conditions (Scheme-9a). The activity of catalyst was stable over 7 runs.

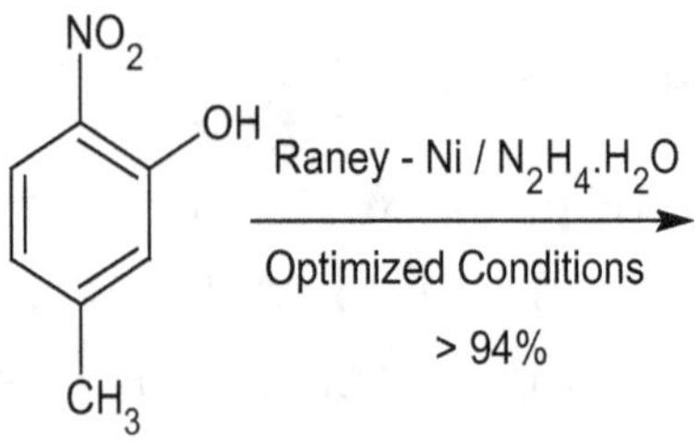

Scheme-9a 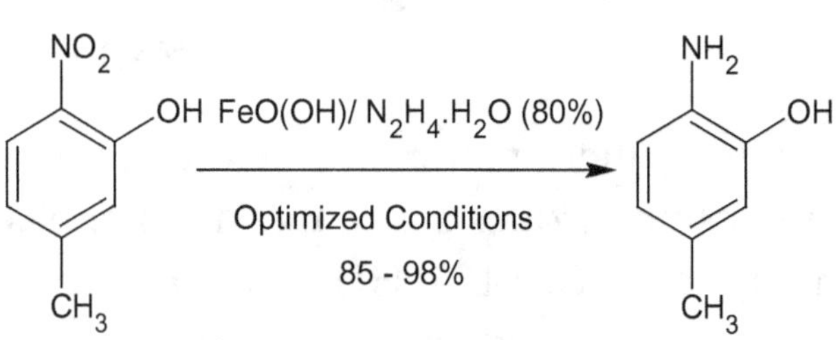   Scheme-9b

2). In the HH reduction method with iron oxide hydroxides (ferrihydrite) as catalyst and 80% HH as reducing agent, the yield of *o*-amino-*p*-methanephenol was 85%-98% (Scheme-9b). A comparative study of both the methods is done (18).

Simple and eco-friendly reduction of NAs to the corresponding ANs in high yields using polymer-supported HH over iron oxide hydroxide catalyst is reported. The polymer support and iron oxide paly a positive role in achieving high yields (19).

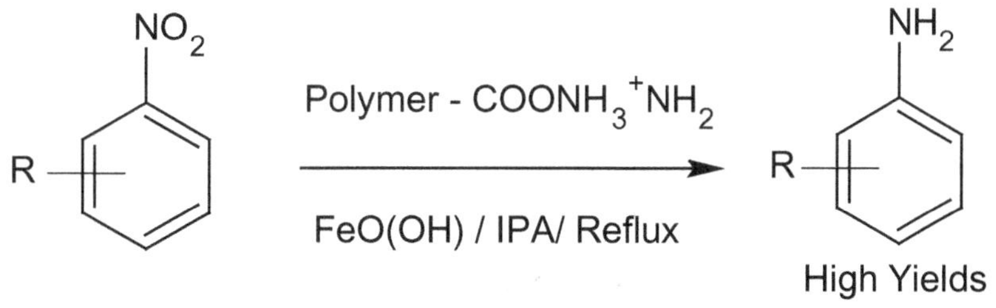

NO$_2$ → NH$_2$

Polymer - COONH$_3$$^+NH_2$

FeO(OH) / IPA/ Reflux

R —— R ——

High Yields

Chemoselective Reduction; R = Subtituents   --Scheme-10

The process is convenient, rapid, environmentally benign and chemoselective to NAs in IPA with polymer-supported HH in the presence of iron oxide hydroxide catalyst producing ANs in high yields (Scheme-10) (20). Fe$_3$O$_4$@graphene oxide nanocomposites as an efficient catalyst is prepared and used for the reduction of NAs with HH as hydrogen source in ethanol as a solvent (21).

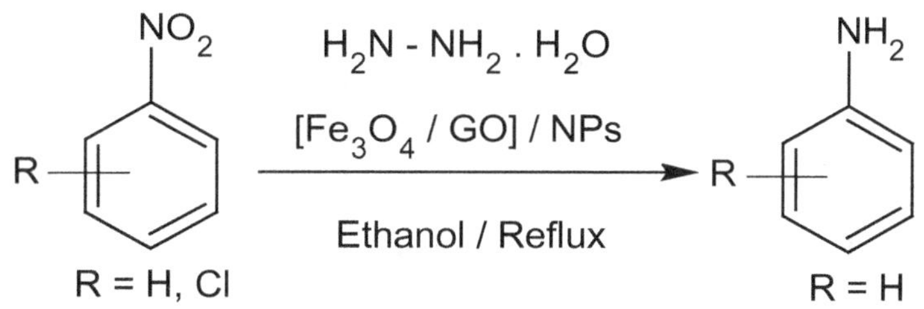

NO$_2$ H$_2$N - NH$_2$. H$_2$O   NH$_2$

[Fe$_3$O$_4$ / GO] / NPs

R ——   R ——

Ethanol / Reflux

R = H, Cl   R = H

[Magnetic Separation / 8 Recycles]   --Scheme-11

The as-prepared Fe$_3$O$_4$@GO composite (12nm) was tested as a catalyst to reduce a series of NAs with HH for the first time, which exhibited a great activity with a turnover frequency (TOF) of 3.63 min^{-1}, which is 45 times that of the commercial Fe$_3$O$_4$ NPs (Scheme-11). A superparamagnetic graphene-Fe$_3$O$_4$ nanocomposite (G-Fe$_3$O$_4$) was synthesized and used as an efficient catalyst for the reduction of NAs with HH for the first time (Scheme-12). The catalyst can be recovered and recycled many times without loss of activity (22).

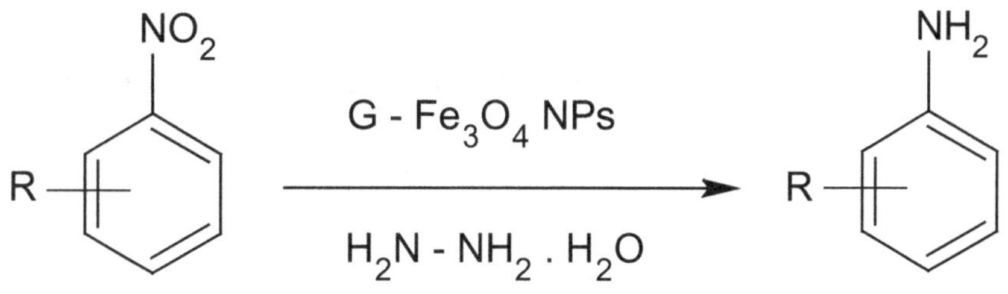

NO$_2$ G - Fe$_3$O$_4$ NPs   NH$_2$

R ——   R ——

H$_2$N - NH$_2$. H$_2$O

G = Supermagnetic Graphene - Fe$_3$O$_4$ Nanocomposites   --Scheme-12

Commercially available Fe_3O_4 NPs were utilized for efficient NA reductions, and could be recycled up to 10 times using magnetic separation while retaining the activity (99% AN yield in each case without any side-products) (Scheme-13).

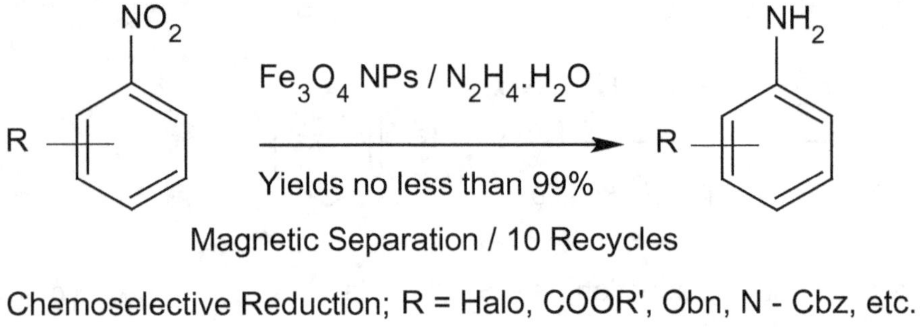

NO$_2$ → Fe$_3$O$_4$ NPs / N$_2$H$_4$.H$_2$O → NH$_2$

Yields no less than 99%

Magnetic Separation / 10 Recycles

Chemoselective Reduction; R = Halo, COOR', Obn, N - Cbz, etc. --Scheme-13

Excellent chemoselectivity for reduction of the nitro versus other functional groups, such as halogen, ester, O-benzyl, and N-Cbz groups, was investigated (23). Simple one-pot synthesis of Rh–Fe$_3$O$_4$ heterodimer nanocrystals (24) were achieved by controlled one-pot thermolysis and their use for efficient and selective reduction of NAs is presented (Scheme-14).

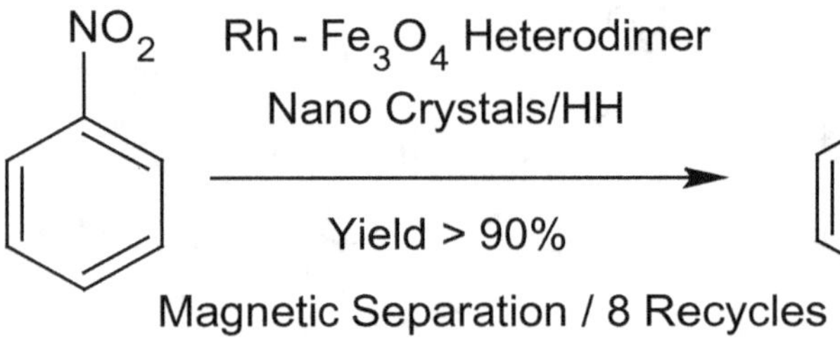

NO$_2$ → Rh - Fe$_3$O$_4$ Heterodimer Nano Crystals/HH → NH$_2$

Yield > 90%

Magnetic Separation / 8 Recycles   --Scheme-14

The recyclable nanocrystals exhibited excellent activities for the selective reduction of NAs to the corresponding ANs in high yields. Cantillo et al. have earlier reported that Iron oxide (Fe$_3$O$_4$) nanocrystals generated *in-situ* are used for the catalytic reductions of NAs to ANs with unparalleled efficiency. The procedure is chemoselective, avoids the use of precious metals, and can be applied under mild reflux conditions (65 or 80°C) or using sealed vessel microwaves to obtain quantitative AN yields (25). Reduction of NACs with HH in the presence of an iron oxide hydroxide (ferrihydrite) catalyst-II is reported (26).

The Rh-Fe$_3$O$_4$ nanocrystals were prepared by treatment of Rh(acac)$_3$ with Fe(acac)$_3$ in a mixture of oleylamine and oleic acid followed by the thermal decomposition.

Rh(acac)$_3$ + Fe(acac)$_3$ = Rh - Fe$_3$O$_4$ Nanocrystals

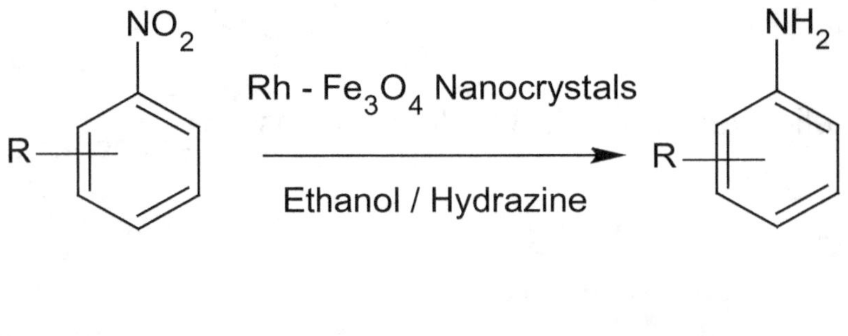

NO$_2$ → Rh - Fe$_3$O$_4$ Nanocrystals / Ethanol / Hydrazine → NH$_2$

--Scheme-15

The nanocrystals catalyzed the reduction of NAs with hydrazine in ethanol to give the corresponding ANs in 94-99% yield (9 examples) (Scheme-15) (27).

5M.01.10g: Hydrazine in Continuos Flow/Batch Reactor:

Continuous flow reactors have drawn attention for their applications in industries and research laboratories. Professor Kappe's group has contributed to this filed in recent times, especially using iron complexes as catalysts. In-situ generated iron oxide nanocrystals as efficient and selective catalysts are used for the reduction of NAs using a continuous flow method (28, 29). Kappe's group has published an important article on "Continuous-Flow Technology"—A tool for the safe manufacturing of active pharmaceutical ingredients. One can visit the following site for more publications on this subject. https://scholar. google.com/citations?user=MvCBzKQAAAAJ&hl=ja.

Highly efficient, reactive, stable and recyclable colloidal Fe_3O_4 nanocrystals are generated in-situ that remain in solution long enough to allow the efficient and selective reduction of NAs to ANs in a continuous-flow reactor (Scheme-16) (30).

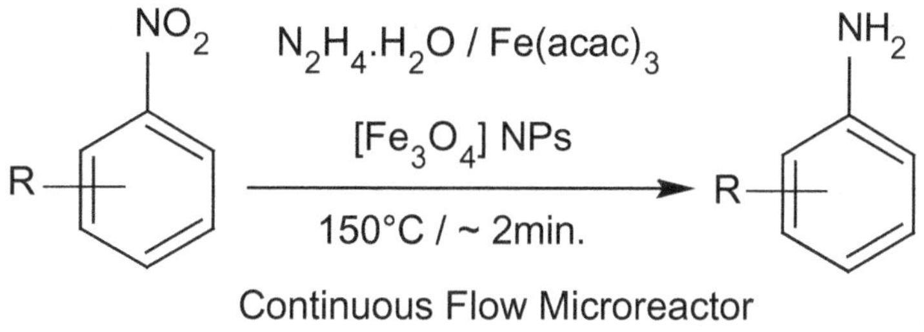

--Scheme-16

HH is used as hydrogen donor and the benefits of homogeneous and heterogeneous nanocatalysis are combined together under a continuous flow system. In situ generated $Fe(acac)_3$ NPs catalyzed the reduction of NAs with HH under MW conditions to give the corresponding ANs in 95–99% yield (20 examples) (Scheme-17a). In the reduction of NB to AN using the batch system, the catalyst was magnetically separated from the reaction mixture and reused seven times.

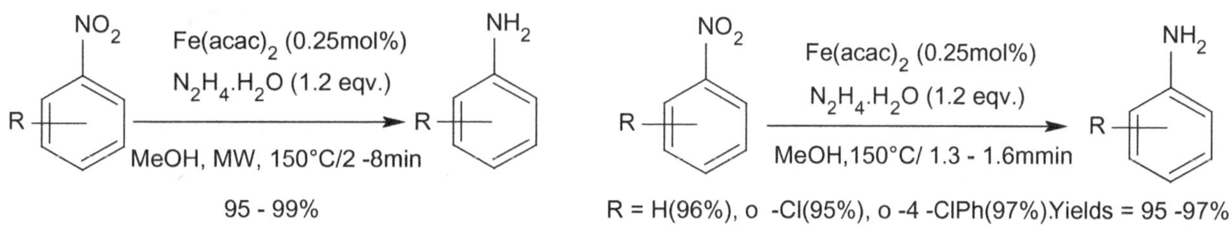

Schemes-17a, 17b

The reductions of various NAs were also performed using a continuous-flow system to afford the ANs in 95–97% yield (Scheme-17b). A simple method for the generation of alumina-supported Fe_3O_4-NPs, and the use of this material in hydrazine-mediated heterogeneously catalyzed reductions of NAs to ANs, under batch and continuous-flow conditions is presented. The bench-stable, reusable nano-$Fe_3O_4@Al_2O_3$ catalyst can selectively reduce functionalized NAs at 1mol% catalyst loading by using a 20 mol% excess of HH (30).

5M.01.10h: Hydrazine on Inorganic Soils:

Inorganic soils as montmorillonite K-10 clay and faujasite are used as supports for metal catalysts. NAs are reduced efficiently by $N_2H_4.H_2O$ inside the cages of trivalent metal ion-exchanged faujasite zeolites (Scheme-18).

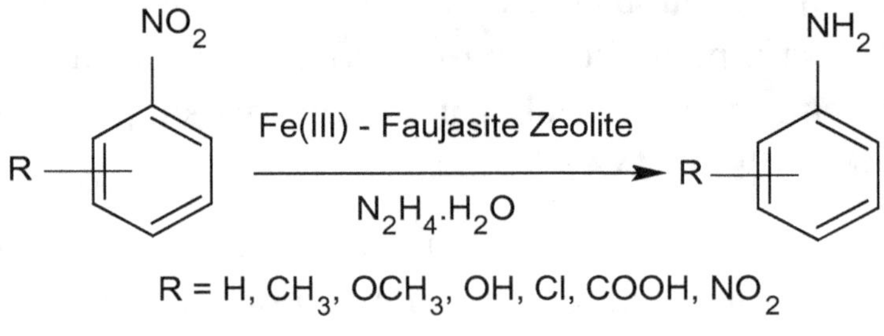

R = H, CH$_3$, OCH$_3$, OH, Cl, COOH, NO$_2$ --Scheme-18

Other advantages of employing zeolites as media during the title reduction such as higher yield, mildness of reaction conditions, reusability and chemoselectivity are also highlighted (31).

5MS.02.00: NA Reduction with Metals and Silanes/Sulfides:

As stated above, amongst the several available hydrogen sources, the following examples highlight the applications of silanes as reducing agents in case of reductions of NAs to ANs.

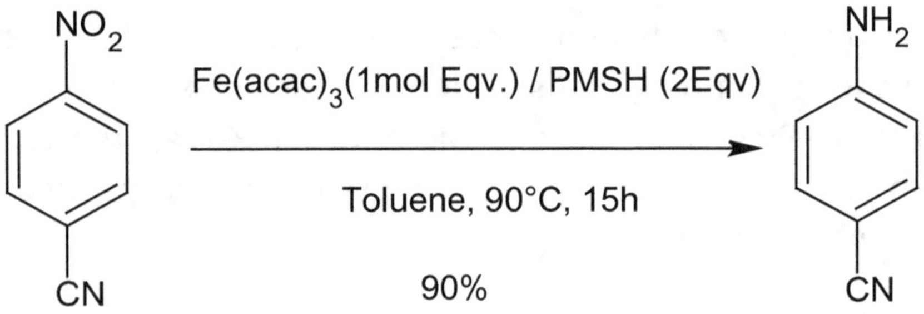

Pehlivan et al have recently developed iron-catalyzed selective and efficient method for reduction of NACs to ANs with a catalytic amount of Fe(acac)$_3$ and TMDS(Tetramethyl disiloxane) in THF at 60°C (32). TMDS was better than PMHS in nitro conversions and AN yields (Scheme-19a, b).

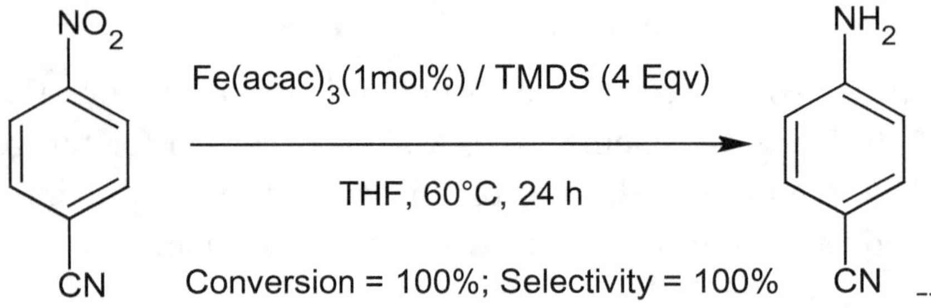

AN, HANs (4-Cl, 3-Br), 3-nitro and 3-amino-ANs were obtained in > 98- 99% yields. For other NAs with different substituents in 4-position and 2, 4-, 2,5-positions the yields ranged between 20% and 92%. The procedure is mild and highly chemoselective towards a wide range of functional groups. THF could be replaced by MeTHF, a friendlier solvent than THF itself. The products are isolated in their HCl salt form and pure enough to take further without purifications (33). The iron-catalyzed reduction of NACs to ANs applying organosilanes is reported. (Scheme-20)

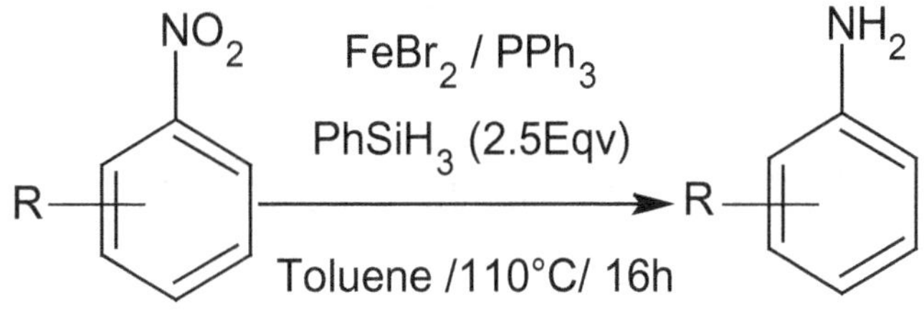

--Scheme-20

In the presence of FeX_2-R_3P as an iron catalyst a series of NAs are selectively reduced to corresponding ANs tolerating wide range of functional groups (34-37).

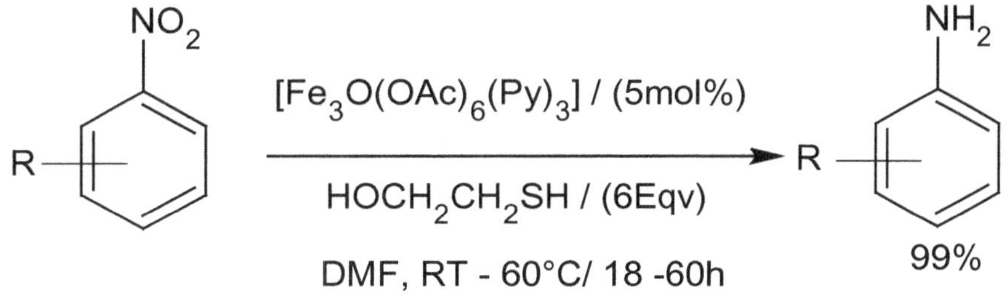

R = 4-Cl, 4-COMe, 4-Me, 4-CHO, 2-CHO   --Scheme-21

Earlier, we have seen that the addition of dimethylsulfide to the reduction system has enhanced the catalyst's activity and yields of ANs. Best results have been achieved

with $[Fe_3O$-$(OAc)_6(py)_3]$ (5mol %) and 2-mercaptoethanol as reductant (Scheme-21). A broad range of NACs have been reduced to ANs with this system (38). Chemoselective reduction of NAs by sodium dithionite ($Na_2S_2O_6$) using octylviologen as an electron transfer catalyst is reported (Scheme-22) (39).

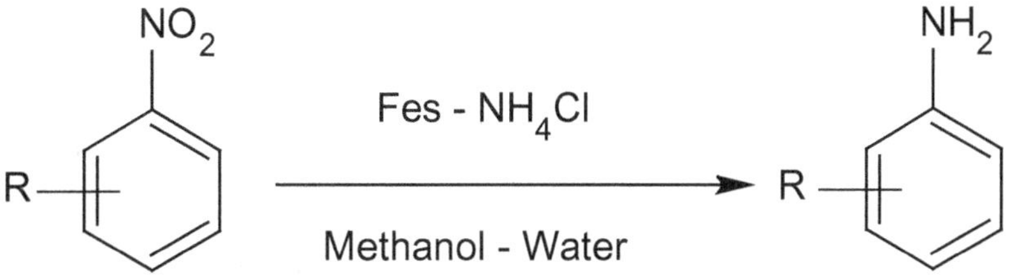

--Scheme-22

NAs are reduced to corresponding ANs with FeS-NH_4Cl-CH_3OH-H_2O system in good yield (40a). There are some other groups who have published their results about the chemoselective reductions of NAs to corresponding ANs using iron/iron salts/iron-NPs and different silane molecules as reducing agents or hydrogen sources, for which the details can be found in the given references (40b-40f). Tong et al have reported the reduction of NB in groundwater by Fe-NPs immobilized in PEG/nylon membrane. The highly reactive Fe-NPs immobilized in nylon membrane were synthesized and characterized, and the reduction of NB in groundwater was investigated. Analytical studies revealed that the Fe-NPs immobilized in membrane were mainly composed of Fe-oxides rather than zero-valent iron. Ground water remediation reduced NB concentration by about 70% in 20 min (41). NB and NA reductions using i) SiliaCatPlatinum-Hydrogel in waste water treatment (42), and ii) ultrasonic irradiation at 35 kHz are reported in good to high yields. These results are discussed under the sections "Solid Supports and Ultrasonication" (43).

5W.03.00: Metal Catalyzed Nitroarene Reductions in Water:

5MW.03.10: Water and Iron/Iron Salts:

The replacement of something unwanted is always at the forefront of new discoveries. In case of catalyst world the transition metals are imperative in the production of old products and development of new methods. The replacement of noble metal catalysts by non-precious metal-based complexes (Metals = Fe, Cu, Ni, Zn, Sn, Co, etc.) has attracted a great interest in research community in synthetic organic chemistry, especially for the remediation of wastewater (44a). Particularly, iron is of significant interest due to its abundance, non-toxicity and low price and some time the reductions are carried out in water or aqueous organic solvents. The following examples illustrate the unambiguous significance of water as a solvent or co-solvent in organic reactions, especially in the reduction of NAs to ANs. In a quest to complete the reduction of NB to AN various methods have been tried and constantly new methods are being developed.

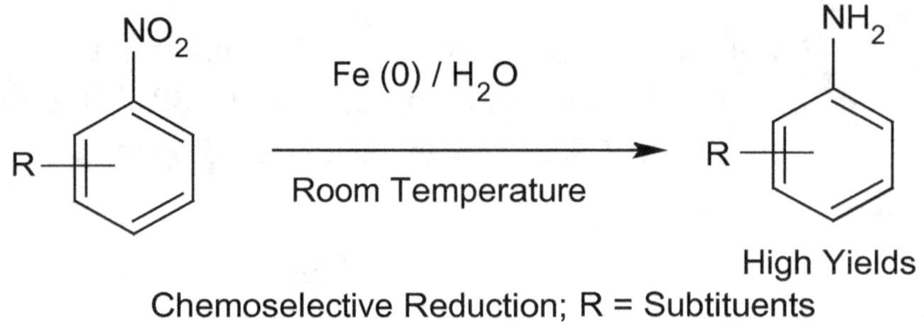

Chemoselective Reduction; R = Subtituents   -- Scheme-23

Highly selective reduction of NAs has been achieved using Fe(0)-NPs in water at room temperature (Scheme-23). A wide spectrum of reducible functionalities remained inert under the reaction conditions. During the reaction a change in shape of Fe-NPs was observed (44b). In a chemoselective reduction of NAs the solvent plays a crucial role (Scheme-24a) (45). A highly selective reduction of o-nitrotoluene to o-toluidine is accomplished by Fe/NH$_4$Cl as a reducing system (Scheme-24b) (46).

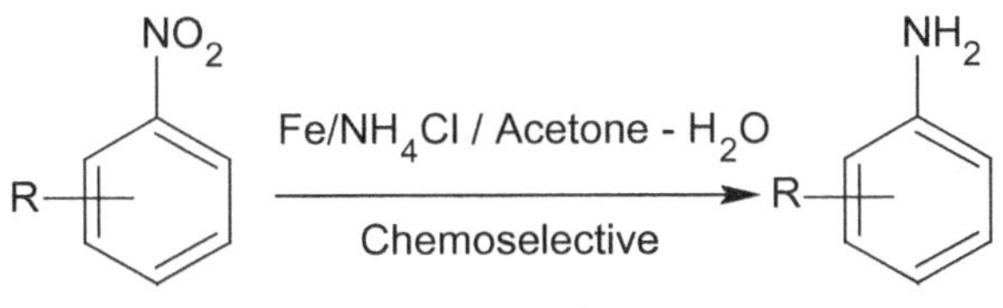

Scheme-24a   Scheme-24b

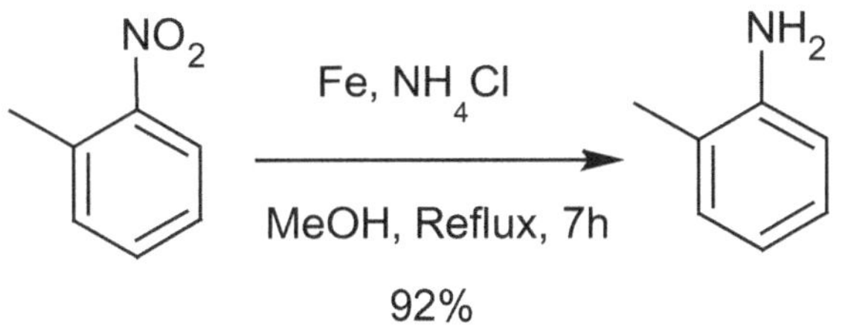

Catalytic reduction of 2-nitroaniline is reviewed in detail (47). A new, efficient, practical and highly chemoselective and green reduction of NAs to ANs with $FeSO_4$, $NaBH_4$, $H_3PW_{12}O_{40}$ in water at room temperature is reported in quantitative yields (Scheme-25).

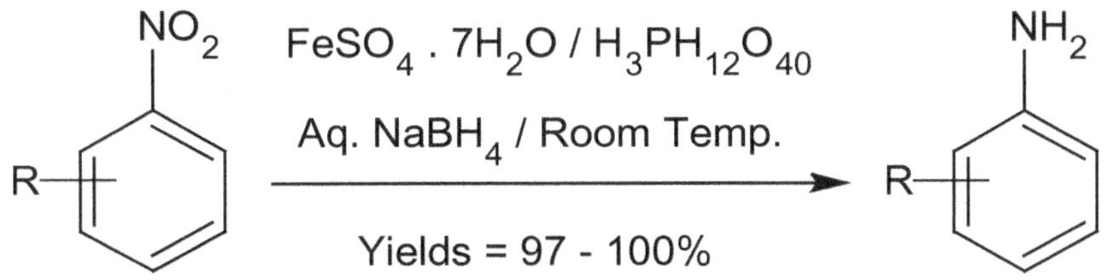

R = H(100%), m -Br(97%), m -Cl(99%), 0 -Me(99%) etc. --Scheme-25

The method is simple, inexpensive, no organic solvents, easily scaled-up and applicable for the large scale selective preparation of different substituted ANs (48).

The first well-defined iron-based catalyst system for the reduction of NAs to ANs has been developed applying iron and formic acid as reducing agent. A broad range of substrates including other reducible functional groups were converted to the corresponding ANs in good to excellent yields at mild conditions (Scheme-26).

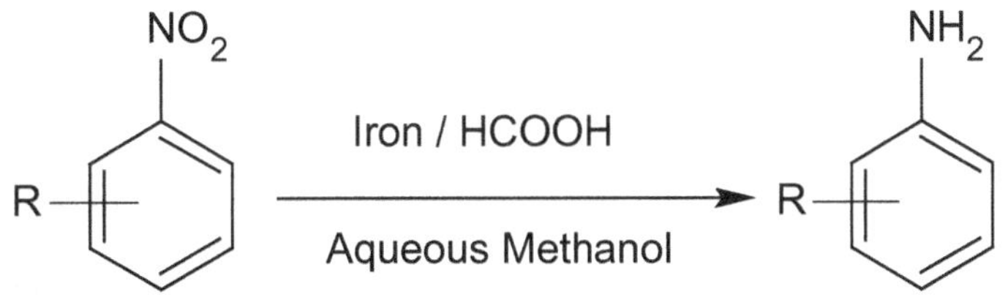

Chemoselective Reduction; R = Different Substituents. --Scheme-26

The process constitutes a rare example of base-free transfer hydrogenations (49). CO and water combinations are used for the chemoselective reductions of NAs (Scheme-27). In one case, to substitute catalytic hydrogenation with other methods, the iron catalyzed water gas shift reaction condition enables complete reduction of NB to AN in quantitative yield (50).

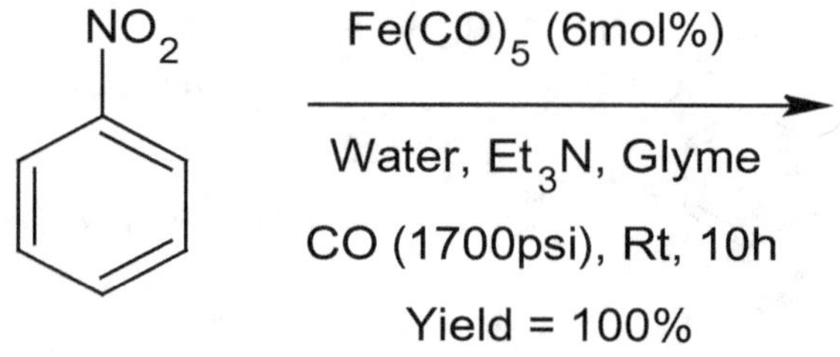

--Scheme-27

NB is also reduced using Fe/Cu bimetallic catalyst, which was effective in the pH range of 3 to 11. This is covered under another heading (51). An efficient Fe/CaCl$_2$ system enables the chemoselective reduction of NAs by CTH in the presence of sensitive functional groups with excellent yields (Scheme-28).

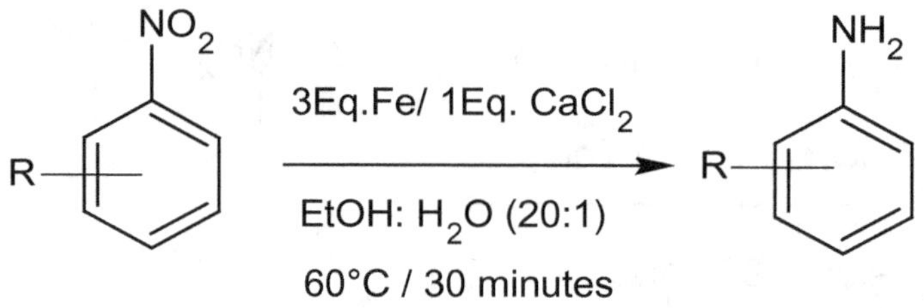

--Scheme-28

The simple procedure and easy purifications makes the protocol advantageous (52). A practical method for the reduction of NAs with nanosized (10nm) activated Fe powder in water at 210°C (near-critical water) has been developed. The reduction generates the aromatic amines in excellent yields (Scheme-29) (53).

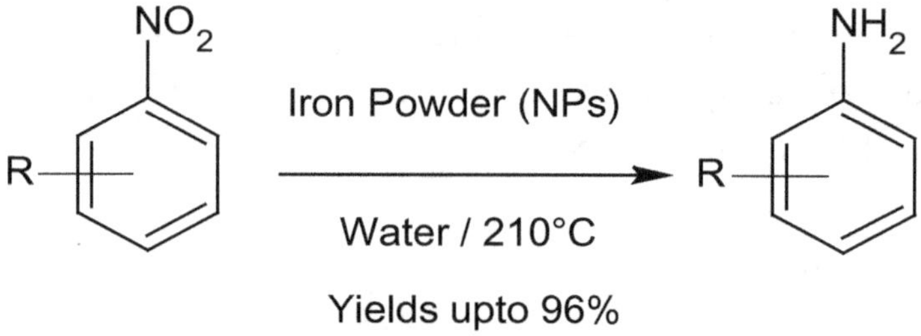

--Scheme-29

The reduction process was performed in a batch reactor system under nitrogen atmosphere in the presence of nanosized activated metallic iron powder (3mmol) in water at 210°C for 2 h (54-56). In another case, a highly selective reductions of NAs has been achieved using Fe(0)-NPs in water at room temperature (Scheme-30). Many other reducible functional groups remained unaffected under given reaction conditions (57). In some cases water as a solvent has performed better than organic solvents. For example, the reduction of 3-acetyl NB was tested in different solvents and was found that water is the best solvent for quantitative conversion of NAs into the corresponding ANs.

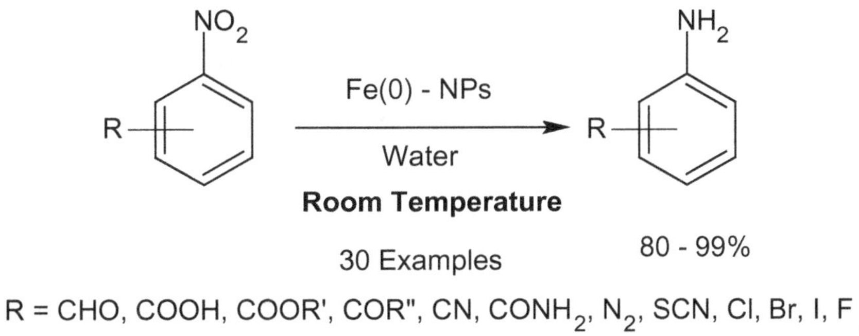

R = CHO, COOH, COOR', COR", CN, $CONH_2$, N_2, SCN, Cl, Br, I, F --Scheme-30

This is an example of the selective reduction of meta- (and para) nitroacetophenone to corresponding ANs in excellent yields (Scheme-31).

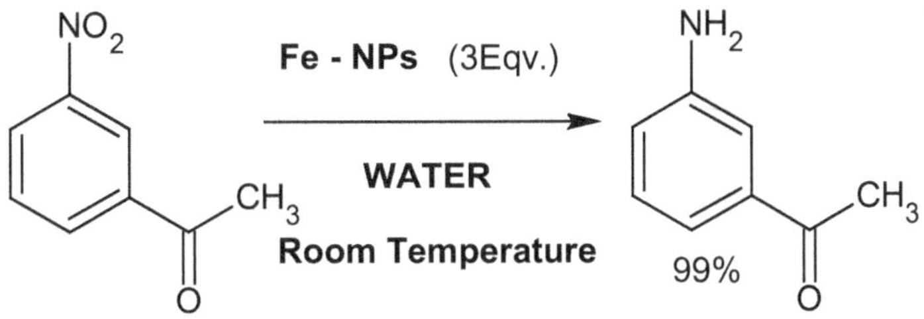

--Scheme-31

It is obvious that the mildest possible conditions, cheapest solvent at room temperature and excellene yields is dream come true. During the reaction a change in shape of Fe-NPs was observed (58).

In another case, the Fe catalyst was prepared by SBH reduction of ferrous sulphate hepta-hydrate [FeSO4.7H2O] and citric acid in water. The NAs and these Fe-NPs were stirred at room temperature under argon.

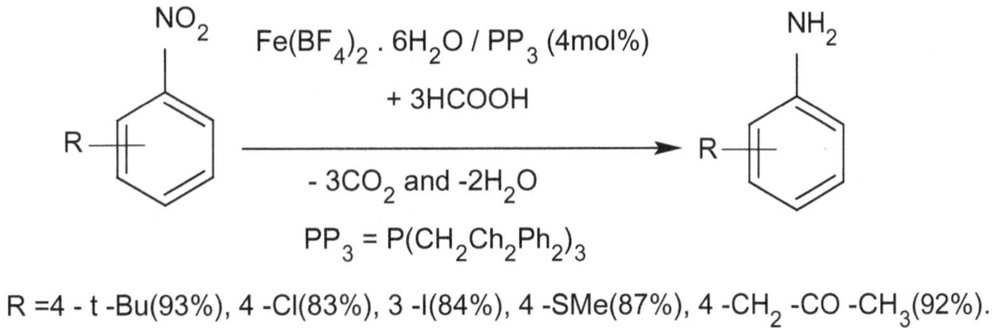

R = 4 - t -Bu(93%), 4 -Cl(83%), 3 -I(84%), 4 -SMe(87%), 4 -CH_2 -CO -CH_3(92%). --Scheme-32

The products were isolated by standard aqueous-organic work up. An iron-based catalyst enables the reduction of NAs to ANs with FA as the reducing agent under base-free conditions (59) (Scheme-32). An amino derivative which is an intermediate in the formal synthesis of Formoterol (Scheme-33) is accomplished using Fe and acetic acid as a reducing agent (60).

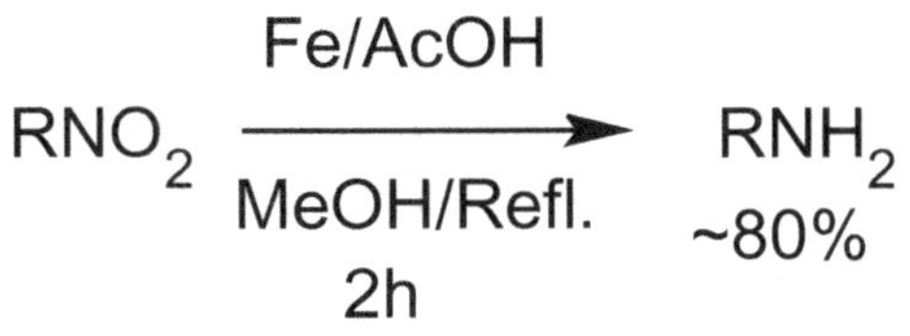

--Scheme-33

The R group in the equation above is the 3-nitroaryl moiety in the precursor/intermediate of Formoterol. The reaction required: i) the reduction of the nitro moiety in the presence of a phenyl benzyl ether, a secondary benzylic hydroxyl group, ii) a primary bromide, and iii) with no racemization at the stereogenic carbinol carbon atom. Based on previous studies, dimethyl sulfide was the poison of choice possessing all of the required characteristics for providing a highly chemoselective and high yielding reaction. The practicality and robustness of the process was demonstrated by preparing the final formamide product with high chemoselectivity, chemical yield, and the product purity on a multi-kilogram scale.

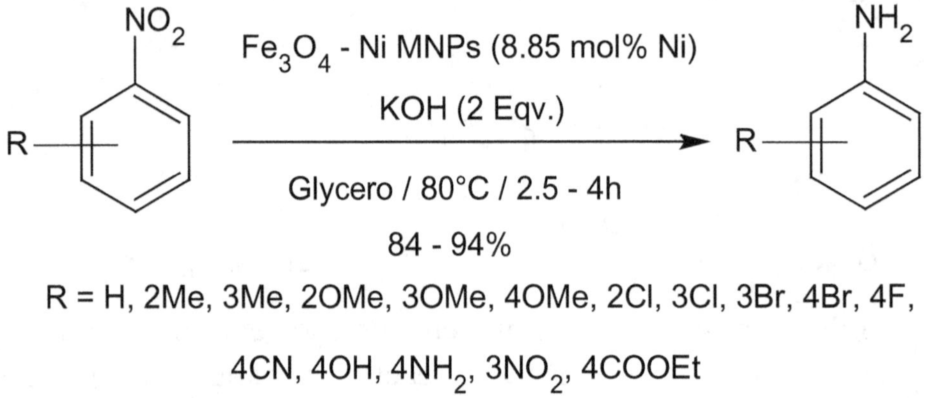

$$R = H, 2Me, 3Me, 2OMe, 3OMe, 4OMe, 2Cl, 3Cl, 3Br, 4Br, 4F,$$
$$4CN, 4OH, 4NH_2, 3NO_2, 4COOEt$$

--Scheme-34

Regio- and chemoselective reduction of NAs over recyclable magnetic ferrite-nickel NPs $(Fe_3O_4$-Ni$)$ by using glycerol as a hydrogen source in presence of aq. KOH is accomplished (Scheme-34). This is a facile, simple and environmentally friendly hydrogen-transfer reaction that allows aromatic amines to be synthesized from the precursor NAs (61-62).

5MW.04.00: Non-Noble Metal Catalysts Other Than Iron in Water:

5MW.04.10: Cobalt (Co): There are several heterogeneous catalytic protocols with H_2, CO/H_2O, trialkylammonium formate, 1, 4-dihydropyridine, and organo-silanes as hydrogen sources. Amongst them, the FA and its salts as HCOOM (M= K, Na), HCOONH_4, HCOOEt_3 etc. are abundantly available, cheap, and environmental friendly hydrogen sources.

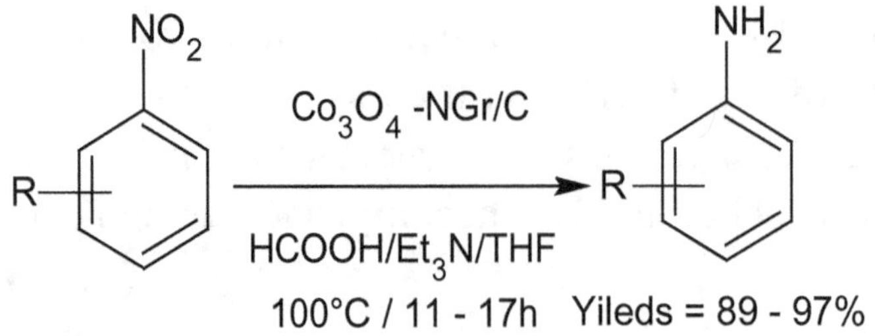

--Scheme-35

Here, the modified Co is used with formate salts as a reducing system (Scheme-35). For example, Co_3O_4-NGr@C-catalyzed hydrogenation of NAs using FA or HCOOEt_3 is a very good reaction with high yields of corresponding ANs (63).

5MW.04.11: Indium (In): Banik et al. have demonstrated a general procedure with the selective reduction of ethyl-4-nitrobenzoate to ethyl-4-aminobenzoate, as an example, in good to high yields. The reductions are carried out under reflux conditions in aqueous ethanol in the presence of indium metal and NH_4Cl, and the catalytic system is found chemoslective for nitro group so the other sensitive functional groups are left unaffected (Scheme-36).

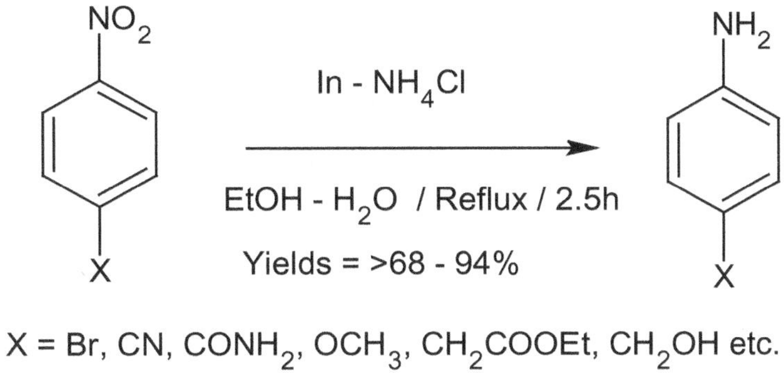

NO_2 → NH_2

In - NH_4Cl

EtOH - H_2O / Reflux / 2.5h

Yields = >68 - 94%

X = Br, CN, $CONH_2$, OCH_3, CH_2COOEt, CH_2OH etc. --Scheme-36

This method was extended for the preparation of large quantities of 6-aminochrysene in excellent yield (64).

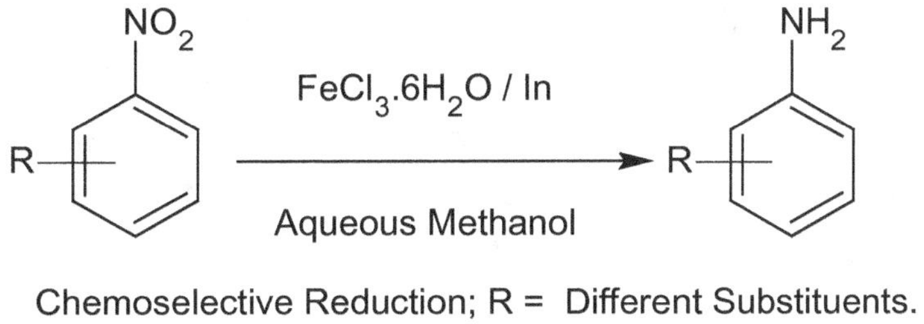

NO_2 → NH_2

$FeCl_3.6H_2O$ / In

Aqueous Methanol

Chemoselective Reduction; R = Different Substituents. --Scheme-37

A new reduction system consisting of NAs, $FeCl_3.6H_2O$/indium in aqueous methanol is performed (Scheme-37) (65).

5MW.04.12: Lanthanum (La): Preparation of specific tailor-made mixed oxides is one of the major topics of research in the field of heterogeneous catalysis. Liquid phase CTH of ANCs on perovskites prepared by microwave irradiation is successfully accomplished. Perovskites of the type $LaMO_3$ (M=Mn, Fe, Co, Cr, Al) are a class of mixed oxides, which have interesting catalytic and physicochemical properties (Scheme-38).

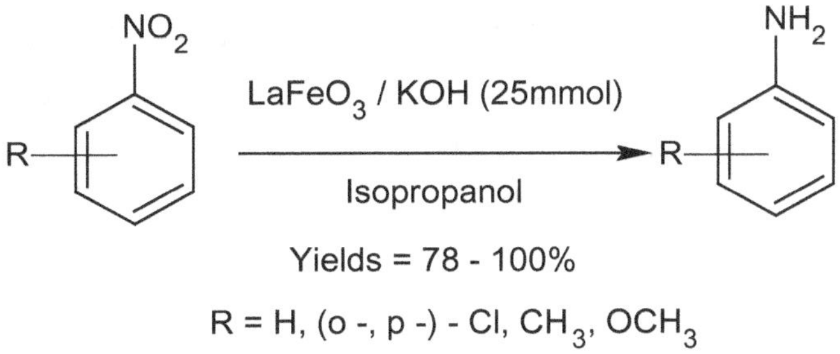

NO_2 → NH_2

$LaFeO_3$ / KOH (25mmol)

Isopropanol

Yields = 78 - 100%

R = H, (o -, p -) - Cl, CH_3, OCH_3 --Scheme-38

The LaMO$_3$ perovskites were used as catalysts for reduction of ANCs with 2-propanol as hydrogen donor and KOH as promoter. Kinetics of the NB reduction is also studied (66).

5MW.04.13: Lead (Pb): Efficient reduction of NAs and azoarenes to arylamines by Pb/NH$_4$Br under neutral conditions is investigated (Scheme-39) (67).

R = H, OH, Cl, Br, NH$_2$, COOH, COOMe, COCH$_3$, NHCOCH$_3$, OCH$_3$, CH$_3$ --Scheme-39

The authors have also reported the efficient reduction of diazoarenes with the same reagent but under sonication conditions. NAs and azo arenes are chemoselectively reduced to the corresponding amines using zinc and aqueous ammonium salts (Zn/NH$_4$Cl/H$_2$O) at room temperature (68).

5MW.04.14: Magnesium (Mg): Employing hydrazinium monoformate in the presence of magnesium powder, it was observed that, hydrazinium monoformate is more effective than hydrazine or formic acid or ammonium formate. The reduction of nitro group occurs without hydrogenolysis in the presence of low-cost magnesium compared to expensive metals like palladium, platinum, ruthenium, *etc* (69).

5MW.04.15: Nickel (Ni): P. Selvam et al. have reported the chemo- and regioselective reduction of NAs over NiHMA (nickel-incorporated hexagonal mesoporous aluminophosphate molecular sieves), FeHMA, CoHMA, Ni-MCM-41 using the propan-2-ol as the hydrogen donor (70). Ni-NPs were found highly active in the selective hydrogenation NB in aqueous phase. Among all the previously proposed transfer hydrogenation systems for NB reuction, NiHMA–IPA was the most efficient one, which gave 97 % yield of AN in 1.5 h at 356°C. Hu et al have reported that the Ni(0) nanocatalysts dispersed in aqueous phase are stable enough to be reused at least five times without significant loss of catalytic activity and selectivity during the catalytic recycles (71). Regio- and chemoselective reduction of NAs over recyclable magnetic ferrite-nickel nanoparticles (Fe$_3$O$_4$-Ni-NPs) by using glycerol as a hydrogen source are investigated. Glycerol is found as a versatile, green solvent as well as an efficient hydrogen donor (Scheme-40).

R = F, Cl, OH, CH_3, OCH_3 etc.   84 - 94%   --Scheme-40

The catalyst is prepared efficiently and glycerol works both as a solvent and as a hydrogen donor (72).

5MW.04.16: Raney-Ni (Ra-Ni): Aniline has been successfully produced by CTH of NB in glycerol, which was employed both as a green solvent and a hydrogen donor. As a solvent, glycerol also facilitated easy separation of the product and recycling of the catalyst (73). The specially prepared skeletal Ni-NPs stabilized by filamentous C are also found highly selective in NB reduction to AN with ~ 99% yield.

5MW.04.17: Tin (Sn): Pasha et al. have reported selective reduction of NAs, nitrosoarenes and arylhydroxyl amines to corresponding ANs by Sn/NH_4Br.

--Scheme-41

The reduction of NAs with $Sn-NH_4Br$ in methanol at 65°C is highly selective (Scheme-41) (74, 75). $SnCl_2 \cdot 2H_2O$ in a 1:1 mixture of EtOAc/MeOH is capable of mediating the tandem reduction–heterocyclization of a variety of 2-nitroacylbenzenes to their corresponding 2,1-benzisoxazoles in good to excellent yields under essentially neutral conditions (76). In another case, the NACs are readily reduced by $SnCl_2$, $2H_2O$ in alcohol or ethyl acetate or by anhydrous $SnCl_2$ in alcohol where other reducible or acid sensitive groups remain unaffected (77).

5MW.04.18: Titanium (Ti): An improved procedure which avoids prolonged reaction at high temperature and handling under reduced pressure was found suitable for the reduction of hetero-NACs and NACs with aqueous titanium (III) chloride (78).

5MW.04.19: Vanadium (V): During NB reductions the hydroxylamine intermediate gets accumulated in the reaction mass and might lead to exothermic decomposition. This problem of hydroxylamine accumulation in the catalytic hydrogenation of NAs is reduced to a large extent by using vanadium promoters during catalytic hydrogenations using Pt, Pd catalysts (79).

5MW.04.20: Zinc (Zn): Desai et al. have done studies on the reduction of NAs to corresponding ANs in good yields using ferric chloride - Zinc - dimethyl formamide - water system (Scheme-42) (80).

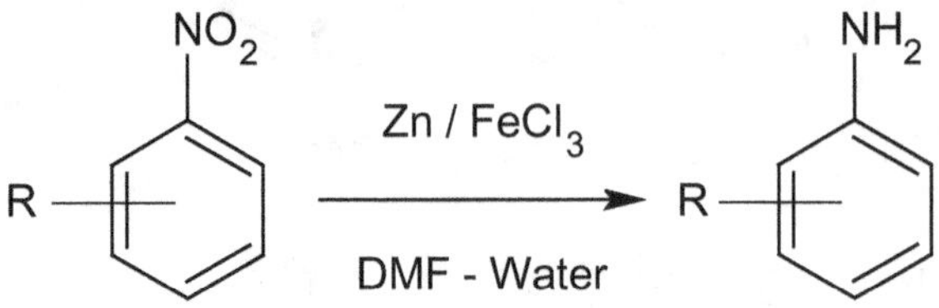

--Scheme-42

Khan et al. have reported chemoselective reduction of NAs and azo compounds in ionic liquids using zinc and ammonium salts (81). NAs can be reduced with a highly novel and chemoselective method in high yields to the corresponding ANs. Using abundantly available zinc metal and NH_4Cl in water (82) are used without any organic solvent at 80°C with a simple procedure at low cost (Scheme-43).

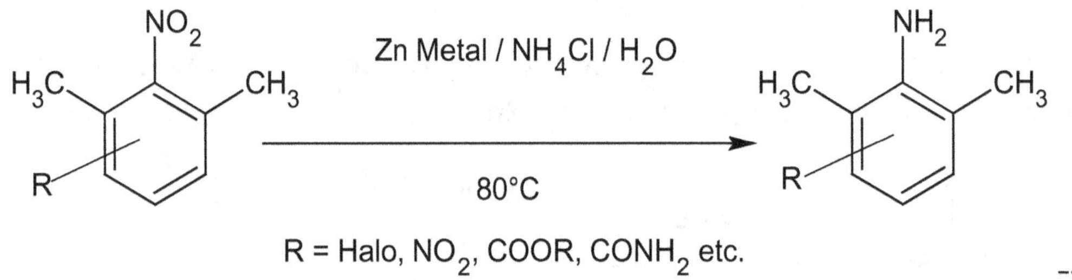

R = Halo, NO_2, COOR, $CONH_2$ etc.

--Scheme-43

The procedure is powerful enough to reduce sterically hindered 2,6-dimethyl-NB and is chemoselective for nitro groups, while the ester, amide and halide substituents on aromatic rings are unaffected (83). 2,5-dimethoxy-4-chloro-AN was prepared by reducing 2,5- dimethoxy -4-chloro -NB with zinc powder and NH_4Cl in mixed solution of ethanol and water (84a) (Scheme-44).

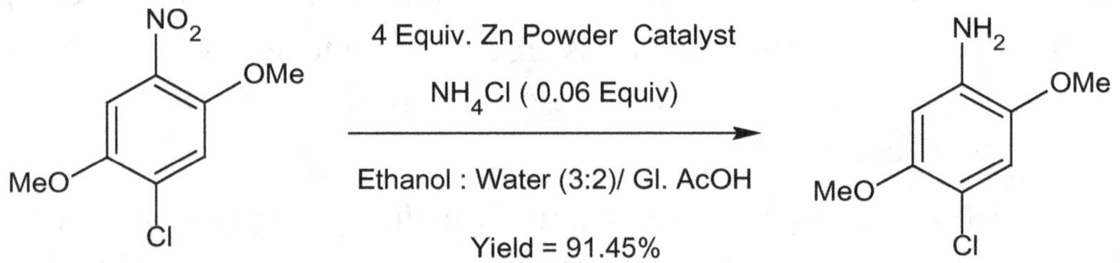

4 Equiv. Zn Powder Catalyst

NH_4Cl (0.06 Equiv)

Ethanol : Water (3:2)/ Gl. AcOH

Yield = 91.45%

--Scheme-44

Under optimum conditions, the yield of reduction product was 91. 45%. D. Wang, Y. Wang, et al. (Dept. of Chem.Univ. Sci. Technol, Suzhou 215009, China). Zinc ammonium formate is used for chemoselective reduction of NAs with yields up to 95% (84b). $(Ph_3P)_3$-$RuCl_2$ enables chemoselective reduction of nitro groups in the presence of Zn/ water as a stoichiometric reductant (85). Hydrazinium monoformate in the presence of commercial zinc dust is used for chemoselective reductions of many NAs (86). A mild, environmentally friendly method for the reduction of NAs to ANs is reported using zinc powder in aqueous solutions of chelating ethers (87).

5MW.04.21: Phenylhydroxylamine (PHA): PHA is prepared by *reducing NB* with zinc dust and aq. NH_4Cl (88) in yield: 62-68%, or by the electrolytic reduction of NB in aq. solution of acetic acid containing sodium acetate (89) (Scheme-45).

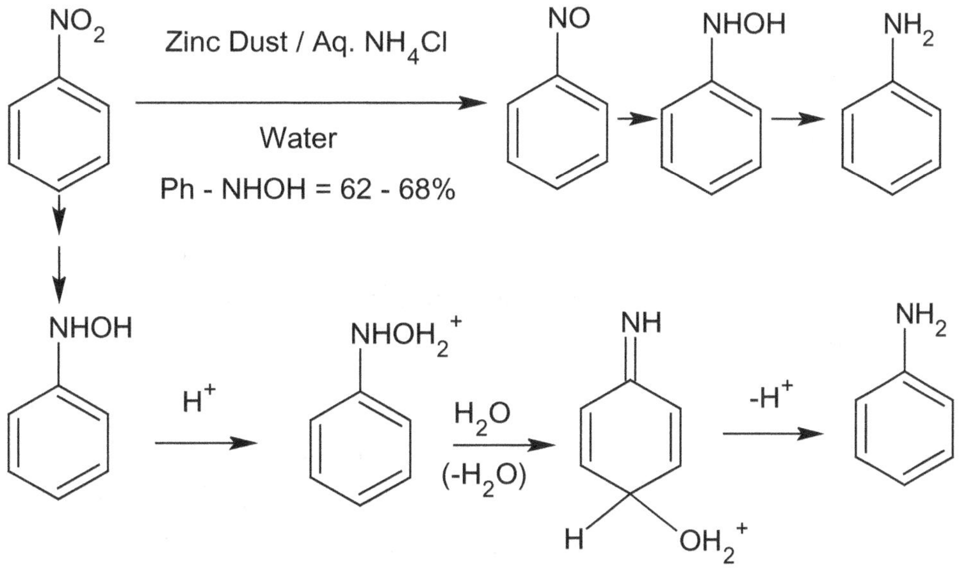

Nitrobenzene Reduction Process (Hughes - Ingold, 1951)   --Scheme-45

Production of N-phenylhydroxylamine (PHA, 95%) via zinc reduction of NB is performed electrolytically using a Zn packed bed and promoters as, ammonium salts under continuous-flow conditions at 60°C. The PHA batch yield owas 68%. The formation of NB and ammonium/zinc-complexes appears to be one of the key promoters of the reaction (90).

5MW.05.00: Nitroarene Reductions over Noble Metals in Water:

5MW.05.10: Iridium (Ir): We have seen under alcohols that some of the alcohols as methanol, ethanol, isopropanol, glycerol are used as solvents as well as hydrogen donors (91-93). Especially, glyecerol as hydrogen donor is used in the reduction of NAs with Ir-NHC based catalysts (94). Mata et al. have developed a dual catalysis with an Ir(III)-Au(I) heterodimetallic complex and applied for reductions of NAs by CTH using primary alcohols as hydrogen donors (95).

5MW.05.11: Palladium (Pd): Solid supported, recyclable (SS-Pd (0) showed high compatibility with various reducing agents ($NaBH_4$, Et_3SiH, and $NH_2NH_2 \cdot H_2O$) and insensitivity towards other functional groups (96). The 2-amino,5(4-fluoro)benzylamino nitro intermediate, in the total synthesis of Retigabine, is reduced to the aryltriamine derivative using Pd/C, which is further derivatized in one pot to ethyl carbamate drug molecule (97). Transfer hydrogenation of NACs using recyclable polymer-supported formate as hydrogen donor and Pd-C as a catalyst produces corresponding amines in high yields (90–98%) (98). Reduction of NAs to ANs with hypophosphites associated with

Pd/C hydrogenation in water is performed with good yield and selectivity within short reaction times (Scheme-1).

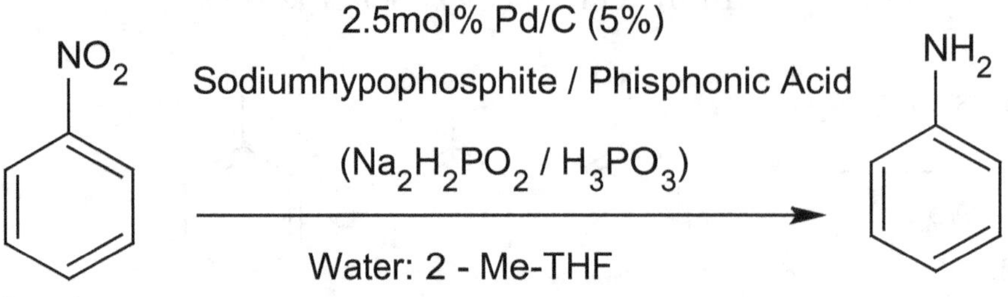

--Scheme-1

The reaction is facilitated with ultrasonication to achieve good results (99a). The in-house prepared catalyst ($Pd@SiO_2$ core–shell nanocatalysts) was found very stable at high temperatures and active and stable even after a long time use (99b).

5MW.05.12: Platinum (Pt): Just like alcohols, formic acid and its salts as formates (sodium and potassium formates, ammonium formate, triethylammonium formate, hydrazinium monoformate, etc.) are very important chemicals/reagents, especially for their ability to act as hydrogen donors. NACs were reduced to respective amines in high yields by using 5%Pt/C with FA or $HCOONH_4$ as hydrogen donor. The reduction of nitro groups occurs without dehalogenation of HANs and also the other reducible substituents remain unchanged under the reaction conditions (100).

5MW.05.13: Ruthenium (Ru): ANs are obtained in high yield and selectivity via Pt and Ru-catalyzed reduction of NAs using $HCOONH_4$.in methanol and aqueous HCOOMe in the presence of a tertiary amine. Especially, NB is reduced to AN in good yields with Pt/C/$HCOONH_4$ (101).

5MW.05.14: Silver (Ag): Anisotropic Ag-NPs act as efficient catalysts for 4-NP to 4-AP reduction with glycerol, a renewable biomass byproduct, as a green solvent and hydrogen source (102).

5MSi.06.00: Metals with "SILANES" as Hydrogen Sources:

Here both the noble and non-noble metal catalysts, except Iron, are used with silanes as hydrogen source. Silane is an inorganic compound with chemical formula SiH_4, which is a colorless, flammable gas with a sharp, repulsive smell, somewhat similar to that of acetic acid. PMHS is more air and moisture stable than other silanes and can be stored for long periods of time without loss of activity. Silanes are used in the fabrication of glass fibres, carbon fibres, polymers. In chemistry, they find application as hydrogen source. For example, SiH_4 reacts with oxygen to give SiO_2 and hydrogen molecule. $SiH_4 + O_2 = SiO_2 + 2H_2$. Hydrogen is pyrophoric gas but can be easily handled and used for the reduction of organic functional groups.

5MSi.06.10: Silanes with Non-Noble Metals:

5MSi.06.11: Chromium (Cr): A wide range of NACs can be reduced to ANs using Cr(II)/Mn(0) redox couple in the presence of trimethylsilyl chloride [TMSCl].

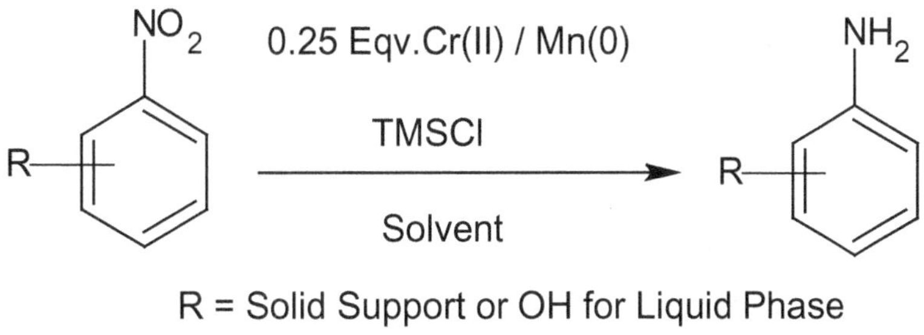

R = Solid Support or OH for Liquid Phase   --Scheme-2

Only 0.25 equivalents of chromium are required to reduce solid-supported NAs (R=solid support in the Scheme-2); and this represents a rare case of transfer catalysis between two solid phases. The reaction is also amenable to solution-phase synthesis (R=OH) (103).

5MSi.06.12: Indium (In): Controlling the type of indium salt and hydrosilane helps display high selectivity towards three intermediates (azoxybenzenes, azobenzenes and diphenylhydrazines) and ANs (Scheme-3) (104).

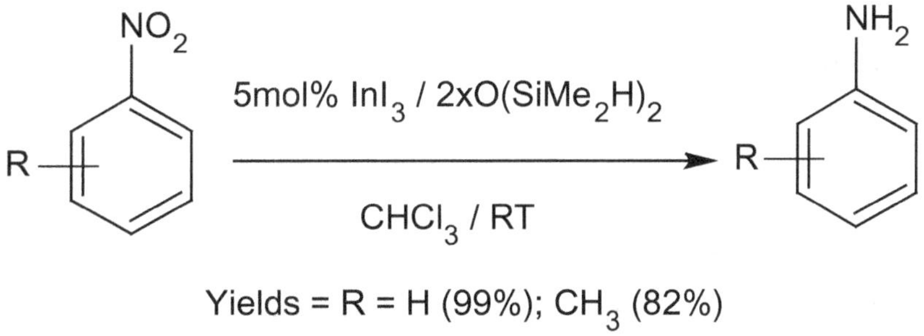

Yields = R = H (99%); CH$_3$ (82%)   --Scheme-3

The intermediate aniline derivative is used to prepare indoprofen, a nonsteroidal anti-inflammatory drug (105a). Controlling the type of indium salt and hydrosilane enables a highly selective reductionof NACs into three coupling compounds, azoxybenzenes, azobenzenes and diphenylhydrazines, and one reductive compound, ANs (105b).

5MSi.06.13: Iron (Fe): See details as discussed earlier in this chapter under "Iron".

5MSi.06.14: Lanthanum (La): Perovskite-type LaFeO$_3$ nanoparticles were readily synthesized via thermal decomposition of the La[Fe(CN)$_6$]·5H$_2$O complex (106a) and used as a reusable heterogeneous catalyst for highly efficient and selective reduction of various NACs into their corresponding amines(106b) and degradation of tobacco-specific nitrosamines (106c).

5MSi.06.15: Molybdenum (Mo): Another metal salt used in the selective reduction of the NACs is Molybdenum oxy chloride (MoO$_2$Cl$_2$) (107). An atom efficient and general protocol for the chemoselective hydrogenation of NAs to ANs catalyzed by well-defined diimino and diamino cubane-type Mo$_3$S$_4$clusters is reported. The as-prepared catalyst had high selectivity in the hydrogenation of functionalized NAs. Over thirty substituted ANs bearing synthetically different functional groups have been synthesized in 70 to 99% yield (108, 109).

5MSi.06.16: Nickel (Ni): A Nickel (II) dichloro complex bearing an abnormal N-heterocyclic carbene (*a*NHC) was synthesized. The NiCl$_2$-(*a*NHC)$_2$ complex is a stable, recyclable and an efficient catalyst for the chemoselective reduction of NAs with hydrosilanes to give ANs in good to excellent yields (110, 111).

5MSi.07.00: Noble Metals & Silanes:

5MSi.07.10: Gold (Au): Magnetically separable gold-nanoparticle catalyst was prepared, and it exhibited excellent activity for chemoselective reduction of NAs with hydrosilanes (Scheme-4)(112).

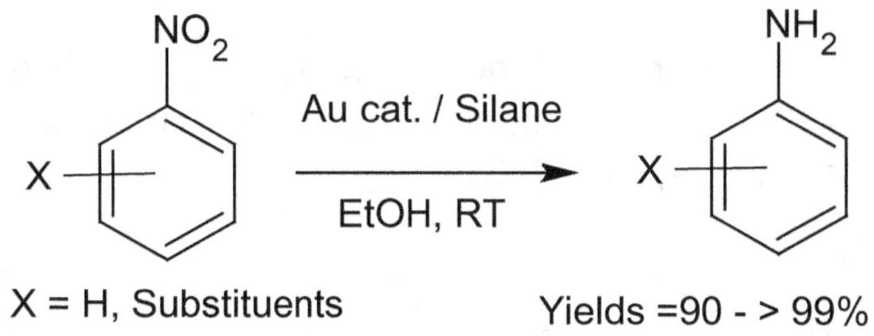

X = H, Substituents   Yields =90 - > 99% --Scheme-4

Au-Silane: Substituents and Yields in the above Scheme-100 are as given below = All at 4-position: H, F, OH, COCH$_3$, CN, OCH$_3$, OAc, OBn, CONMe$_2$ = > 99%; and 4-Cl (94%), 4-Br (95%), NHCBz (93%), 3-CH=CH$_2$ (90%). The NAs with other functional groups are also converted into the corresponding ANs in excellent yields (90->99%) (113). The Au/Au@polythiophene core/shell nanospheres were found to be very active catalysts for the hydrogenation of various NACs into corresponding ANs in the presence of NaBH$_4$. Both hydrophilic and hydrophobic NAs were efficiently hydrogenated under mild conditions (114).

5MSi.07.11: Iridium (Ir): Highly selective conversion of NAs to ANs using a simple reducing system combined with a trivalent indium salt (InI$_3$, InCl$_3$, InBr$_3$)) and a hydrosilane is reported. Controlling the type of indium salt and hydrosilane enables a highly selective reduction of NACs into three coupling compounds, azoxybenzenes, azobenzenes and diphenylhydrazines, and one reductive compound, the anilines (115).

5MSi.07.12: Palladium (Pd): Solid supported Pd(0) (SS-Pd) is found as an efficient recyclable *heterogeneous* catalyst for chemoselective reduction *of* NAs with triethylsilane as a reducing agent *(Scheme-5)*(116).

Yields = 50 - 99%

R = H, CH$_3$, OCH$_3$, CHO, CN, COOR, OBn etc.

NaBH$_4$ and Hydrazine were also used as H$_2$ donors --Scheme-5

Rahaim et al have reported a chemoselective method using palladium-catalyzed reduction of aromatic nitro groups to amines in high yield, at room temperature in the presence of aqueous KF and PMHS. Metal fluorides act as promoter in the catalytic reductions of NAs with silanes (Scheme-6).

--Scheme-6

Rahaim et al have found under optimized conditions, 2, 3, 4- methylcarboxylated-NBs and 2,6-dimethyl-NB, p-CNB, o-nitrotoluene gave corresponding ANs in 100% yields using Pd and Rh based catalysts and PMHS(117). Under identical conditions with KF as a promoter 3-nitrostyrene is reduced to 3-ethylAN in 100% yield (117c). Highly stable and active Pd-NPs were prepared by the reduction of NAP-Mg-PdCl$_4$ with PMHS, which serves as a chemoselective reducing agent for dinitrobenzenes to the corresponding nitroanilines and also for many other substituted ANs (118). Lipowitz et al have reported the reduction of NB to AN in high yields (89%) using Pd/C as a catalyst and PMHS as reducing agent under mild conditions within 1hour in ethanol at 40-60°C (119).

5MSi.07.13: Platimum (Pt): 3-Nitrostyrene was chemoselectively reduced to 3-aminostyrene using a mixture of dimethylphenyl-, diphenylmethyl-, and triphenylsilanes in the presence of H$_2$PtCl$_6$. The triphenylsilane was responsible for the hydride transfer. The other two silanes reduce the C=C in styrene (120). Novel polysiloxane gels, to which platinum species are encapsulated, are prepared by treatment of polymethylhydrosiloxane with alkenes in the presence of Karstedt's catalyst. These gels act as reusable catalysts in the reduction of functionalized NAs with H$_2$ to the corresponding ANs without leaking the catalyst species (121).

5MSi.07.14: Rhenium (Re): Highly chemo- and regioselective reduction of NACs using the silane/oxo-rhenium complexes to corresponding amines is reported. The method is quite tolerant towards a wide range of functional groups. This methodology also allowed the regioselective reduction of dinitrobenzenes to the corresponding nitroanilines and the reduction of an aromatic nitro group in the presence of an aliphatic nitro group (122).

5MSi.07.15: Rhodium (Rh): NAs are reduced to corresponding ANs using triethylasilane and Wilkinson's catalyst $[RhCl(PPh_3)_3]$. (123).

5MSi.07.16: Ruthenium (Ru): It is not with silanes but Ru-NPs-catalyzed reduction of NAs using ethanol, IPA and Aq. KOH as hydrogen sources is reported. (124,125).

5MSi.07.17: Thiols: In an interesting case, aniline is synthesized with 100% selectivity by Zn-Al double hydroxide with thiophenol as a reducing agent (Scheme-7). The heterogeneous reduction of NB by thiophenol catalyzed by the dianionic bis(2.sulfanyl-2,2-diphenylethanoxycarbonyl) dioxomolybdate(VI) complex, $[MOVIO_2(O_2CC(S)(C_6H_5)2h]_2$-, intercalated into a Zn(II)-Al(lll) layered double hydroxide host $[Zn_3$-Alx(OH)$_6]$x+, has been investigated under anaerobic conditions.

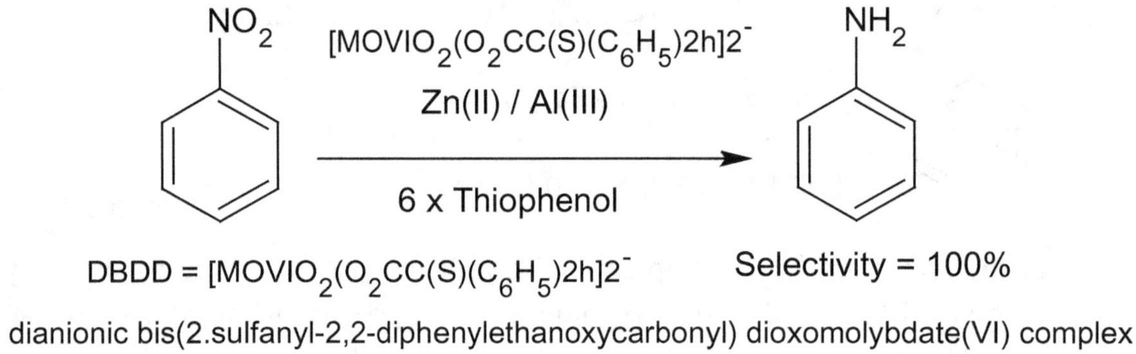

DBDD = $[MOVIO_2(O_2CC(S)(C_6H_5)2h]2^-$ Selectivity = 100%

dianionic bis(2.sulfanyl-2,2-diphenylethanoxycarbonyl) dioxomolybdate(VI) complex

Scheme-7

Aniline was found to be the only product formed through a reaction consuming six moles of thiophenol for each mole of aniline produced (126).

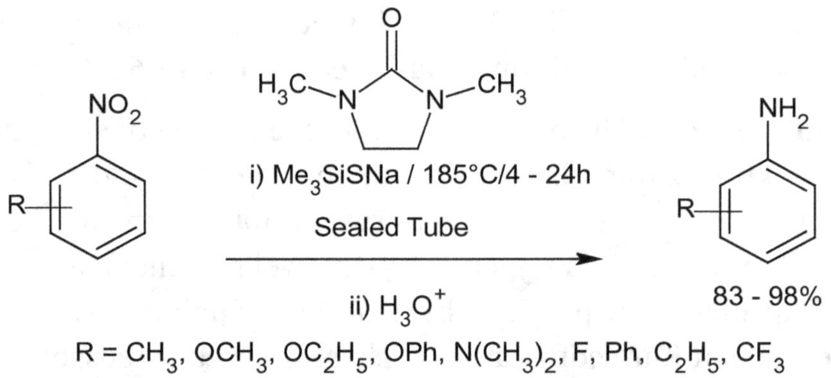

R = CH_3, OCH_3, OC_2H_5, OPh, $N(CH_3)_2$, F, Ph, C_2H_5, CF_3 --Scheme-8

Wong et al. have reported a new method with improved conditions and higher yields of ANs by reduction of NACs by sodium trimethylsilanethiolate (Scheme-8).

The Me_2SiSNa was freshly generated in-situ and reacted with the NAs at 185°C. The *o*-nitroanisole gave a mixture of *o*-aminoanisole and *o*-aminophenol under the experimental conditions but on optimizations of the conditions it gave 96% *o*-aminoanisole with just 4 equivalent of the silane reagent. Aminopyridines were also obtained in very high yields (68-78%) (127).

5MF.07.18: Metal-Free: The combination $HSiCl_3$ and a tertiary amine enable a metal-free reduction of NAs to ANs (Scheme-9).

$$ArNO_2 \xrightarrow[\text{ACN / 0°C - RT /18h}]{\text{HSiCl}_3 \text{ / DIPEA}} ArNH_2 \qquad \text{---Scheme-9}$$

The reaction is of general applicability and tolerates many functional groups (128).

References:

1). A. Bechamp. *Annales de Chimie et de Physique*, 42:186–196 (1854).

2a). C. V. Rode, et al. *Organic Process & Res. Development*, 03(06):465–470(1999).

2b). K. A. Kumar, D. C. Gowda, et al: *E- J. Chemistry*, 05(04):914-917(2008).

3). M. J. S. Mohamed, et al. *Industrial Engineering & Chemical Research*, 55(27):7267-7272(2016).

4). Y. Liu, et al. *Advanced Synthesis & Catalysis*, 347(02-03):217-219(2005).

5). G. G. Wu, et al. *Organic Lett*ers, 13(19):5220-5223(2011).

6). C. S. Keenan, et al. *Synthetic Communication*, 47(11):1085-1089(2017).

7a). H. Alper, K. E. Hashem, *J. American Chemical Society.*, 103:6514-6515(1981), 7b) J. F. Knifton, et al. *J. Organic Chemistry*, 41:1200-1206)1976).

8a). F. Ragaini, et al. *Organometallics:* 14:387-400(1995). 8b) F. Ragaini, *Organometallics:* 15:3572-3578(1996). 8c). J. Jiang, et al. *Green Chemistry*, 10(04):439-441(2008).

9a). G. B. Boldrini, et al. *J. Organometallic Chemistry,* 171(01):85–88(1979). 9b). G. Gao, Y. Tao, et al. *Green Chemistry*, 10:439-441(2008).

10a). P. S Kumbhar, et al. *Tetrahedron Letters,* 39(17):2573-2574(1998).10b). S. Natarajan, et al. *Synthetic Communications*, 22(22):3189-3195(1992).

11). Desai et al. *Synthetic Communications*, 31(08):1249-1251(2001).

12a). S. R. Boothroyd, M. A. Kerr, *Tetrahedron Letters*, 36(14):3 2411–2414(1995), 12b) S. R. Boothroyd, M. A. Kerr, *Cheminform* , vol. 26(31: pp. no-no, 2010.

13). A. Kamal, B. S. N. Reddy, *Chemistry Letters*, 07: 593-594(1998).

14). J. Hine, et al. *J. Organic Chemistry*, 50(25):5092-5096 (1985).

15a). C. Feng, et al. *Chinese Chemical Letters*: 24:539–541(2013).

15b). S. R. Boothroyd & M. A. Kerr, *Tetrahedron Letters:* 36:2411-2414(1985).

16). U. Sharma, et al. *Chemistry An European. J.* 17:59035907(2011).

17). M. Beller, et al. *Chemical Communications,* 47:10972-10974(2011).

18). X. Wu, H. Wang, et al. (College of Chemical Engineering,Qingdao University of Science and Technology, Qingdao 266042). http://en.cnki.com.cn/Article_en/CJFDTOTAL-JXSI200701008.htm

19). Q. Shi, et al. *Green Chemistry,* 08: 868-870(2006).

20). M. Lauwiner, et al. *Applied Catalysis A: General,* 172(01):141–148(1998).

21). G. He, W. Liu, et al. *Materials Research Bulletin,* 48(05):1885-1890(2013).

22). H. Zhang, et al. *Letters in Organic Chemistry,* 10(01):17-21(2013)

23). S. Kim, et al. *Chemistry An Asian J.,* 06(08):1921-1925(2011).

24). Y. Jang, S. Kim, et al. *Chemical Communications,* 47:3601-3603(2011).

25). D. Cantillo, et al. *J. Organic Chemistry,* 78(09):4530–4542(2013).

26). M. Benz, et al. *Applied Catalysis A: General,* 172(01):149-157(1998).

27a). Y. Uozumi, B. M. Kim, T. et al. *Synfacts,* (06): 0690-0690(2011).

28). D. Cantillo, C. O. Kappe, *Angewandte Chemie,* 124(40):10337–10340(2012).

29a). D. Cantillo, et al. *Angewandte Chemie Intl. Edition,* 51(40):10190–10193

(2012): 29b). Ibid, *Angewandte Chemie Intl Edition,* 54(23): 6688-6728(2015).

30). C. O. Kappe, et al. *ChemSusChem,* 07(11):3122–3131(2014).

31). M. Kumarraja, et al. *Applied Catalysis A: General,* 265(02):135–139(2004).

32). L. Pehlivan, et al. *Tetrahedron Letters,* 51:1939-1941(2010).

33). Y. Liu, et al. *Advanced Synthesis & Catalysis,* 347(02-03):217-219(2005).

34). L. Pehlivan, et al. *Tetrahedron Letters,* 51(15):1941(2010).

35). K. Junge, B. Wendt, et al. *Chemical Communications,* 46:1769-1771(2010).

36). K. Junge, et al. *Chemical Communications* (Camb), 46(10):1769-1771(2010).

37). X. F. Wu, et al. *Chemical Communications,* 47:12462-12463(2011).

38). S. Murata, et al. *J. Chemical Society, Perkin Trans.2,* 617-621(1989).

39). K. K. Park, *J. Organic Chemistry,* 60(19): 6202-6204(1995).

40a). D. G. Desai, et al. *Synthetic Communications,* 31(8):1249-1251(2001). 40b).G. Wienhofer, et al. *J. American Chemical Society,* 133:12875-12879(2011). 40c). L. Pehlivan, et al. *Tetrahedron,* 67(10):1971-1976(2011). 40d).Y. Sunada, *Angewandte*

Chemie Intl. Edition, 48(50):9511-9514(2009). 40e). P. K., Bala, et al. *Catalysis Letters,* 144(7):1258–1267(2014). 41f).J. Pesti & G. L. Larson, *Organic Process & Research Development,* 20(07):1164–1181(2016).

41). M. Tong, et al. *J. Contaminant Hydrology,* 122(01-04):16-25(2011).

42). V. Pandarus, et al. *Advanced Synthesis & Catalysis,* 353(08):1306–1316(2011).

43). A. B. Gamble, et al. *Synthetic Communications,* 37(16):2777-2786(2007).

44a). M. Anjum, et al. *Arabian J. Chemistry:* 2016. https://doi.org/10.1016/j.arabjc.2016.10.004

44b). D. R, Mukherjee, et al. *Chemical Communications,*48(64):7982-7984(2012).

45). J. Wu, et al. *Synthetic Communications,* 40(05):661-665(2010).

46). R. Krishnamurthy, et al. *Synthetic Communications,* 22(22):3189–3195(1992).

47). K. Naseem, et al. *Environmental Science & Pollution Research Intl.* 24(07):6446-6460(2017).

48). H. Aliyan, et al. *J. NanoStructures,* 01:21-26(2012).

49). G. Wienhofer, et al. *J. American Chemical Society,* 133(32):12875-12879(2011).

50). K. Cann, T. Cole, et al. *J. American Chemical Soci*ety, 100:3969-3971(1978).

51). W. Xu, et al. *J. Environmental Sciences* (China), 20(08):915-921(2008).

52). S. Chandrappa, T. Vinaya, et al. *Synlett,* 2010, 3019-3022(2010).

53). L. Wang, P. Li, et al. *Synthesis,* 13:2001-2004(2003).

54). C. Boix, M. Poliakoff, *J. Chemical Society, Perkin Trans*-1, 1487(1999).

55). H. Ohde, et al. *J. American Chemical Society,* 124:4540(2002).

56a). N. Pradhan, et al. *Langmuir,* 17:1800(2001), 56b).N. Pradhan, et al. *Colloids & Surfaces A:* 196:247(2002).

57). B. C Ranu, et al. *Chemical Communications* (Camb), 48(64):7982-7984(2012). 58a). Alan G. Jones, *J. Chemical Education,* 1975, 52 (10):668-669(1975). 58b).A. G. Nene, et al. *World J. Nano Science and Engineering,* 06:20-28(2016).

59). M. Beller, et al. *J. American Chemical Society, 133,* 12875-12879(2011).

60). B. Mereyala & K. Sambaru, *Indian J. Chemistry,* 44B:167-169(2005).

61). M. B. Gawande, et al. *Chemistry An European J,* 18:12628–12632(2012).

62). A. E. Diaz-Alvarez, *Applied Sciences,* 03:55-69(2013).

63). M. Beller, et al. *Green Chemistry,* 17:898(2015).

64). B. K. Banik, *Synthetic Communications,* 30(20):3745-3754(2000).

65). B. Y. Yoo, et al. *Synthetic Communications.* 33(17):2985-2988(2003).

66). A. S. Kulkarni, et al. *Applied Catalysis A: General,* 252(02):225–230(2003).

67). M. A. Pasha, et al. *Synthesis & Reactivity in Inorganic, Metal-Organic and Nano-Metal Chemistry,* 35:477–482(2005). 68a). Shodhganga Chapter-5: http://shodhganga. inflibnet.ac.in/bitstream/10603/91985/12/12_chapter%205.pdf. 68b). F. A. Khan, et al. *Tetrahedron Letters:* 44:7783–7787(2003).

69). K. Abiraj, et al. *Synthesis & Reactivity in Inorganic & Metal-Organic Chemistry,* 32(8):1409-1417(2002).

70). P. Selvam, et al. *Tetrahedron Letters,* 45:2003-2007(2004).

71). Y. Hu, Z. Hou, et al. *Science China Chemistry,* 53(07):1541-1548(2010).

72). M. B. Gawande, et al. *Chem. A European J.,* 18(40):12628–12632(2012).

73a). D. Tavor, et al. *Synthetic Communications,* 41(22):3490-3416(2011). 73b). A. Wolfson, et al. *Energy & Environmental Engineering:* 01:17-23(2013). 73c). D. A. Diaz-Alvarez, et al. *Applied Sciences,* 03(01):55-69(2013), and references cited therein. 73d). N. Mahata, et al. *Applied Catalysis A: General,* 351:204–209(2008).

74a). M A Pasha, et al., *Indian J. Chemistry,* 43B: 24-24(2004). 74b). Nagaraja D & Pasha M A, *Indian J Chemistry,* 43B: 593-594(2004).

75). G. Rai, et al. *Tetrahedron Letters,* 46(23):3987–3990(2005).

76). J. Chauhan, et al. *Tetrahedron Letters,* 53(37):4951-4954(2012).

77). F. D. Bellamy, et al. *Tetrahedron Letters,* 25(08):839-842(1984).

78). M. Somei, et al. *Chemical & Pharmaceutical Bulletin,* 28(08):2515-2518(1980).

79). P. Baumeister, et al. *Catalysis Letters,* 49(03-04):219-222(1997).

80). D. G. Desai, et al. *Synthetic Communications.* 29(06):1033-1036(1999).

81). B. K. Banik, et al. *Synthetic Communications.* 30:3745-3754(2000).

82a). H. K. Kadam, et al. *RSC Advances,* 05(101):83391-83407(2015). 82b). B. K. Banik, et al. *Organic Syntheses: 81,188-194* (2005).

83). T. Tsukinoki, H. Tsuzuki, *Green Chemistry,* 03:37-38(2001).

84a). T. Hirashima, O. Manabe, *Chemistry Letters,* 259-260(1975): 84b). D. C. Gowda, et al *Indian J. Chemistry,* 40B:71-75(2001).

85). T. Schabel, C. Belger, et al. *Organic Letters,* 15:2858-2861(2013).

86). S. Gowda, *Synthetic Communications,* 33(02):281-289(2003).

87). P. S. Kumar, P.M. L. Rai, *Chemical Papers,* 66(08):772–778(2012).

88). I. L. Finar, Organic chemistry, Vol1/6th edition, pp649-650(1973).

89a). http://shodhganga.inflibnet.ac.in/bitstream/10603/29496/7/07_chapter%20 %202.pdf. 89b). H. C. Bertsch, et al. US 3413349-A(1968).

90). L. Li, et al. *Industrial Engg & Chemical Research*, 46(21):6840–6846(2007).

91). E. Farnetti, et al. *Green Chemistry*, 11(05):704-709(2009).

92). A. Wolfson, et al. *Tetrahedron Letters*, 50:5951–5953(2009).

93). A. Azua, et al. *Organometallics*, 30:5532–5536(2011).

94). Azua, A., J. A. Mata, et al. *Organometallics*, 31:3911-3919(2012).

95). S. Sabater, J. A. Mata, et al. *Chemistry An-European J.* 18:6380-6385(2012).

96). A. K. Shil, et al. *Tetrahedron Letters* 53 (2012) 4858–4861(2012).

97). G. Blackburn-Munro, et al. *CNS Drug Reviews*, 11(01):01-20(2005).

98). K. Abiraj, *Synthetic Communications*, 35(02):223-230(2005).

99a). M. Baron, et al. *Green Chemistry*: 15:1006-1015(2013). 99b). Y. Hu, K. Tao, *J. Physical Chemistry, C,* 117 (17): 8974–8982(2013).

100). D. C. Gowda, et al. *Synthetic Communications*, 30(20): 3639-3644(2000).

101a). K. Van Aken, et al. *Beilstein J. Organic Chemistry:* (2006). 101b). A. B. Taleb & G. Jenner, *J. Molecular Catalysis*, 91(02):L149–L153 (1994). 101c). X. Liu, et al. *Chinese J. Catalysis:* 36:1461-1475(2015).

102). A. D. Verma, et al. *RSC Advances*, 06(105):103471-103477(2016).

103). A. Hari, B. L. Miller, *Angewandte Chemie Intl Edition*, 38(18):2777–2779(199).

104). N. Sakai, et al. *Chemical Communications:* 46:3173-3175(2010).

105a). Y. Ogiwara, Y. Sakurai, & N. Sakai, *Chemistry Letters*, 46(02):240-242(2017). 105b). N. Sakai, et al. *Chemical Communications*, 46(18): 3173-3175(2010).

106a). H. Young, et al. *J. Physical Chemistry C,* 111(17):6275–6280(2007). 106b). S. Farhadi, et al. *J. Molecular Catalysis A: Chemical*, 339(01-02):108-116(2011). 106c) K. Wang, et al. *Materials*, 09:326-337(2016).

107). A. C. Fernandes, *J. Molecular Catalysis-A-Chemical*, 272(01-02):60-63(2007).

108). E. Pedrajas, I. Sorribes, et al. *ChemCatChem*, 09(06): 1128–1134(2017).

109). L. Huang, Y. Wang, et al. Chin J Chem, 31(08):987-991(2013).

110). G. Vijaykumar & S. K. Mandal, *Dalton Trans.*, 45(17): 7421-7426(2016).

111). R. Lopes, et al. *ChemCatChem,* In Press June, 2017.

112). S. Park, et al. *Organic & Biomolecular Chemistry*: 11:395–399(2013).

113). S. Fountoulaki, et al. *ACS Catalysis*, 04(10):3504–3511(2014).

114). H. S. Shin, S. Huh, *ACS Appl. Mater. Interfaces*, 04(11):6324–6331(2012).

115). N. Sakai, et al. *Chemical Communications*, 46(18):3173-3175(2010).

116). A. K. Shil, et al. *Tetrahedron Letters*, 53(36):4858-4861(2012).

117a). R. J. Rahaim, R. E. Maleczka (Jr.), *Organic Letters*, 07:5087-5090(2005), 117b). R. J. Rahaim, et al. *Synthesis*, 19:3316-3340(2006). 117c). http://www2.chemistry.msu.edu/faculty/Maleczka/pdfs/47.pdf

118). D. Damodara, et al. *RSC Advances*, 04(43): 22567-22574 (2014).

119). J. Lipowitz, et al. *J. Organic Chemistry*, 38 (01):162-165(1973).

120). K. A. Andrianov, et al. *Bulletin Academy of Sciences of the USSR, Division of chemical science*, 20(05):1037-1039(1971).

121). Y. Motoyama, et al. *Organic Letters*, 11(06):1345–1348(2009).

122). R. G de Noronha, et al. *J. Organic Chemistry*, 74(18):6960-6964(2009).

123). H. R. Brinkman, et al. *Synthetic Communications*, 26(05):973-980(1996).

124). R. V. Jagadeesh, et al. *Chemistry - A European J.*, 17(51):14375–14379(2011).

125). J. H. Kim, et al. *Advanced Synthesis & Catalysis*, 354(13): 2412–2418(2012).

126). A. Cervilla, et al. *Intl. J. Chemical Kinetics*, 33(03):212-224(2001).

127). F.F. Wong, et al. *J. Organic Chemistry*, 57(19):5254-5255(1992).

128). M. Orlandi, F. Tosi, et al. *Organic Letters*: 17:3941-3943(2015).

Chapter-5S: Solid Supports
5SS.00.00. CTH of Nitroarenes over Solid Supports (SS):

Introduction: While working on this topic, I remembered common instances in our life such as a child is very active, playful but he cannot cross an overflowing river and needs to ride on someone's back. Similarly, although an old man can walk well but cannot cross the road without someone's helping hand. Exactly the same way, there are lots of metals but only some of them are being used as catalysts from the beginning of the innovative chemistry days. However, their performances remained limited to some extent for more than one reason and that is where they needed support from another co-participant (metal or non-metal, base, acid, etc.) as an additive or promoter or support in the form of a ligand, resin or polymeric material. It attracted attention from research groups all over the world and slowly progress was being made where the handicapped metal catalytic activities were enhanced by using them in different forms: i) initially in powder forms basically to increase the surface area and hence the activity, ii) later as nanoparticles, iii) as a partner with other metal to increase the activity through co-participation, iv) use of continuous flow technology and, v) by taking support from solid materials. In this last case, the metal catalysts are either dispersed in the matrix (resins, foams, polymers or inorganic solids as metal oxides) or deposited on the surface of a solid matter. By doing so not only the finely distributed metal particles have a greater active surface, but also the solid supports participate in enhancing the catalytic activity in the system. The best examples are the metal nanoparticles (M-NPs) on solid supports as silica (SiO_2), alumina (Al_2O_3), titania (TiO_2), iron oxide (ferrite or ferrihydrite) etc. There is a plethora of publications in the literature highlighting the use and applications of solid supports in catalysis. In chemistry, a catalyst support is the material usually a solid with a high surface area to which a catalyst is affixed. The activity of heterogeneous catalysts and nanomaterial-based catalysts occurs at the surface of the atoms. Consequently, great efforts are made to maximize the surface area of a catalyst by distributing it over the support. The support may be inert or participate in the catalytic reactions. Typical supports include various kinds of carbon, alumina, zirconia, titania and silica. In this context, the following references corroborate our views about the enhancement of the metal catalytic activity by another component. Catalytic hydrogenation with Raney-Ni, Raney-Co and the promoter effects of additives like platinum chloride on hydrogenation NB and nitrobenzoic acid is studied since a long time ago (1). https://en.wikipedia.org/wiki/Catalyst_support.

Here are few more recent examples which highlight the impact of gold (Au) using with and without solid supports. In this particular case, Valden et al. have reported the reduction and oxidations using Au metal supported on TiO_2 which is one of the most commonly used solid supports in catalysis. Gold clusters (1-6 nm) have been prepared on single crystalline surfaces of titania. The structure sensitivity of this reaction on gold clusters supported on titania is related to a quantum size effect with respect to the thickness of the gold islands. The islands with two layers of gold are most effective for catalyzing the oxidation of carbon monoxide. These results suggest that the supported clusters, in general, may have unusual catalytic properties as one dimension of the cluster becomes smaller than three atomic spacing (2).

5SS.01.00: Nitroarenes Reductions over different solid supports

5SS.01-10: Polymeric Solid Supports (SS): Usually it is difficult to collect and cite all the work that has been done on any subject, but here some of the significant contributions by different groups in the world related to metal catalysts on SS and their performances in the CTH of NAs to corresponding ANs are presented.

5SS.01-11: Polymers: As the name itself is suggestive that poly means many. Here the many units of the same molecule, known as a monomer, undergo a chain reaction of linking the monomers together to form a polymeric chain or material. It could be either single monomer repeating in the chain or if needed it could be a mixture of two monomer units. By definition, polymer is a mixture of compounds composed of the same repeating structural unit (monomer). Cotton, starch, cellulose, latex or natural rubber are examples of some of the natural polymers, while plastics, nylon, polyester, teflon, PVC (poly vinyl chloride), polyvinyl pyrrolidone (PVP), polyurethanes etc are synthetic polymers. Polymers have been widely used as catalysts or catalyst supports and the various applications can be categorized conveniently into four groups (a) catalysis by soluble linear polymers, (b) catalysis by ion exchange resins, (c) polymer-supported 'homogeneous' metal complex catalysts and (d) polymer-supported phase transfer catalysts. A brief review of each of these is given, with citation of some of the more recent and important examples which illustrate some of the basic principles involved (3). Some of the natural polymers used as metal catalyst support are cellulose, chitosan, ligands, etc., while the synthetic polymeric materials used as solid supports for catalysts are polyacrylonitrile, polystyrene-divinylbenzene(PS-DVB), polyvinyl pyrrolidone (PVP), phosphenated polystyrene, Amberlyst-A27, etc. Michalska & Webster have published a review on different polymeric materials used as supports for metal catalysts (4). Generally, whether it is an inorganic material as silica or alumina or organic based polymers, the impregnation of metals is done by following certain procedures in order to get finely distributed M-NPs on the support material. The size, shape and surface of the catalyst and the fine nature and surface area of the support are responsible factors for the net activity and selectivity of the catalysts.

5SS.01-12: Effect of Polymer/Solid Support with or without a Co-Catalyst:

Earlier, we have seen that the metal-free reduction of NAs is performed using different hydrogen sources and without any metal catalyst. For example, NACs are reduced to corresponding primary amines with HH under moderate N_2 pressure.

$$\text{R} \overset{\text{NO}_2}{\underset{}{\bigcirc}} \xrightarrow[\substack{150°\text{C} \\ \text{Yield} = \sim 90\%}]{\text{N}_2\text{H}_4\cdot\text{H}_2\text{O} / \text{N}_2 \ (2.0 \text{ Mpa})} \text{R} \overset{\text{NH}_2}{\underset{}{\bigcirc}}$$

---Scheme-1

Under identical reaction conditions reductions of several other aromatic nitro compounds to their amines have been achieved in moderate to high yields (Scheme-1) (5). However, in this section we are focusing on the use of metal catalysts supported on either polymeric supports or inorganic solid materials.

5SS.01-13: PVP-Metal(s): PVP=Polyvinylpyrrolidine: Most of the chemicals are commercially manufactured by using high pressures and high temperatures. Especially, the catalytic hydrogenation of NAs to corresponding amines is still the preferred choice of manufactures. However, with new developments in processes and technologies, it has become possible to achieve high yields and excellent selectivities of the desired amines. For example, metal ions from Pd, Au, Ag, Ru, Rh, Pt, Co, Ni, Cu, Fe, etc. alone or as bi-metallic colloids (Ag-AU, Pt-Pd, etc.) on solid supports have enabled to achieve the set goals under mild conditions. This has become possible due to the applications of the metals and metal-co-catalysts, chelators, crown ethers, PTCs, ligands as ionic liquids, solid supports as carbon, carbon nanofibres, silica, etc. and polymers as PVP have found successful applications as co-catalysts or additives or modifiers/promoters in CTH. With metals Fe, Pd, Ru etc. in their best possible nano form with PVP as a support are used in the selective hydrogenations of HNBs (o-CNB, p-CNB, etc) with an emphasis on the components of the system which would not cause hydrodehalogenation of in-situ formed HANs. Here is an interesting case. The liquid-phase hydrogenation of p-CNB with Pd/C in methanol at room temperature and atmospheric pressure produces p-CAN and HCl due to hydrodehalogenation of p-CAN. If you change the solvent from methanol to ethyl acetate, in the presence of 0.2 % Pd/C the chemoselectivity to p-CAN exceeds 90%. When Raney-Ni is used as a catalyst, instead of Pd/C, under the same conditions in methanol the process consumes one mole less of hydrogen and the sole isolated product (100%) is p-CAN, without any dehalogenation. It means Raney-Ni catalyst is chemoselective in the hydrogenation of halogen-containing NACs. Basically, both the type of catalyst and the solvent play important roles in the outcome of the hydrogenation product of HNBs (6). The reductions of o-CNB or p-CNB are carried out under high temperature and modest pressures in water, alcohol or water-alcohol mixtures in the presence of a catalyst, composed of Cu(en) or Zn(en) and PVP-Ru, which gave almost 100% selectivity to o-CAN or p-CAN (7).

In short, based on the above results it is possible to achieve 100% yields with highest NA conversions and AN selectivities by adding a catalytic amount of non-noble metals like Cu and Zn to PVP supported noble metal catalyst. The following examples illustrate these observations.

5SS.01-14: PVP-Pt/Ru: The PVP-Pt colloids were prepared and used as catalyst in the hydrogenation of *o*-CNB to *o*-CAN. The effects of PVP-Pt colloid particle size and Ru^{3+} ion on reaction results were examined (Scheme-2).

R = O-Cl   PVP = Poly(vinylpyrrolidone)

With Ru^{3+} addition Sel for *o*-CAN increased from 82% to 92%. --Scheme-2

The results showed that the selectivity to *o*-CAN changed little and the catalyst activity increased with decreasing of colloid particle size. With increasing of Ru^{3+} dosage, the selectivity to *o*-CAN increased gradually while the catalyst activity decreased. The selectivity of *o*-CAN increased from 82% to 92% after Ru^{3+} ion was added in the reaction system (8, 9).

5SS.01-15: PVP-Pt/Ni: Ni-colloids (10a) and Nickel acetate salt (10b) with Pt on PVP support have been tested for the catalytic hydrogenation of NAs. It was found that Ni_2^+ ions enhanced the yields and selectivity of *o*-CAN.

R = O-Cl Ni_2^+ increased *o*-CAN from 45 % to 66% & 83%. --Scheme-3a

R = O-Cl   Ni(en)Cl$_2$ gave o-CAN Sel = 94.00% --Scheme-3b

Modification of metal complex for hydrogenation of *o*-CNB over PVP-Pt colloidal clusters gave *o*-CAN with 83-94% selectivity (3a, 3b).

R = O-Cl   M$^+$ = Co$_2^+$; Cr$_3^+$; Sel = upto 91.5%   --Scheme-4

5SS.01-16: PVP-Pd/Pt: Under identical conditions, the hydrogenation of o-CNB over polymer-stabilized PVP-palladium–platinum (PV-Pd/Pt) bimetallic colloidal clusters (11) led to the formation of o-CAN with 91.5% selectivity (Scheme-4).

5SS.01-17: PVP-Ru/Zn: It is surprising to see that some of the metals or metal salts play a crucial and different role than others in enhancing the catalytic activity and also increasing the yield and selectivity of the end product. For example, under similar conditions as used in the above experiments (Scheme-5), PVP-Pt/Zn gives 100% o-CAN selectivity. Effect of metal complex on hydrogenation of o-CNB over PVP stabilized Ru colloids (PVP-Ru) has been studied at 320K and 4.0MPa.

R = O-Cl   Zn - DA = Zinc Complex with Diamine   Sel = ~ 100% --Scheme-5

The addition of Zn(II) complexes of diamines to PVP-Ru catalyst system leads to significant increase in the activity of PVP-Ru catalst, while the selectivity to o-CAN was maintained at ~100% (Scheme-5). Especially, the rate enhancement of more than 30 times has been achieved in the presence of Zn(trien)Cl$_2$ compared to neat PVP-Ru (12).

5SS.01-18: PVP-Pt/Fe or Ni: Just as in case of PVP-Pt-Zn (Zn salt), the catalytic hydrogenation of m- and p-CNBs with PVP-Pt/Ni and PVP-Pt/Fe gave quantitative selectivity to corresponding CANs. The metal complex effect on the hydrogenation of m-CNB and p-CNB over PVP-Pt colloidal catalyst has been studied under 303K and at atmospheric pressure (Scheme-6a, 6b).

R = m - Cl   Selectivity upto > 99.9% --Scheme-6a

X = m, p - Cl

p-CAN Sel = 99.9%   --Scheme-6b

The introduction of metal complexes to the catalytic system has great effect on the activity and the selectivity of the catalyst. Addition of complexes as $Ni(en)_3^{2+}$ or $Co(acac)_3^-$ to the hydrogenation of *m*-CNB increased the selectivities to *m*-CAN from 82.2% to 95.5% and then to > 99.9%. With iron (III) 8-hydroxyquinoline complex (Fe-8-HQ) the selectivity to *p*-CAN was >99.9% by the reduction of *p*-CNB (13).

5SS.01-19: PVP-Pt/La, Sm, etc: Effect of rare earths (La, Sm, Ce, etc.) on the hydrogenation properties of *p*-CNB over polymer-anchored platinum catalysts is investigated in ethanol at 303K and normal pressure (Scheme-7). Under optimized conditions, PVP–Sm–Pt catalyst exhibits the highest selectivity (95.8%) for *p*-CAN, when the molar ratio of NaOH to substrate is 0.12.

Rare Earth Ions = La_3^+, Sm_3^+, Pr_3^+, Ce_3^+, Nd_3^+

PVP-Sm-Pt ans PVP-Ce-Pt Showed highest slectivty of 95.8% and 98.2% resply. --Scheme-7

Under the same reaction conditions, PVP–Ce–Pt catalyst using PVP of large molecular weight (90,000) as support exhibits the highest activity and selectivity (14) for *p*-CAN (98.2%). Mo and La have been independently reported as good promoters of Ni-B nanoclusters in the hydrogenation of *p*-CNB. However, a bi-promoter effect of Mo and La in Ni-B catalysts has not been previously investigated. In this work, a series of La-Mo-doped, nanosized Ni-B catalysts with different La contents were prepared and used in the hydrogenation of *p*-CNB. The doping of La had a positive effect, particularly at La:Ni ratio at 0.2, on the NiMoB catalyst properties and activity in the hydrogenation of *p*-CNB (15). Hydrogenation of CNBs over Pd, Pt catalysts supported on cationic resins delivers about 85% selctivities for *o*- and *m*-CANs and 95% for *p*-CAN (16).

5SS.01-20: PVP-Pt/Pr: The PVP-Pt catalyst improved the *m*-CNB hydrogenation activity and significantly enhanced the reactive selectivity compared with the ethanol reflux method. The addition of a small amount of Pr^{3+}-modified PVP-Pt/Pr (Pt: Pr molar ratio of 1:0.14) further improved the activity of the catalyst (17).

5SS.01-21: Gold and Other Noble Metal catalysts: Gold is one of the metals most commonly used as a catalyst in catalytic transformations of organic functional groups. Here, the impacts of organic stabilizers on catalysis of Au-NPs from colloidal preparation and on hydrogenation of *p*-CNB are investigated (18). Efficient and chemoselective reduction of NAs under transfer hydrogenation using recyclable polymer-supported formate (PSF) and HH with ironoxidehydroxide (FeO(OH) is reported. Also, the aspects of immobilization of catalysts on polymeric supports are discussed which make positive contributions in these CTH processes (19).

Dehydrodehalogenation in the CTH of CNBs to CANs is still a big challenge to chemists with majority of catalysts. Recently, Wang et al. have overcome this problem using noble metal nanoclusters and inorganic semiconductor NPs, when the hydrodechlorination of *o*-CAN was completely suppressed even at 100% conversion of *o*-CNB. Further, they have reviewed this subject with a stress on supported metal catalysts and polymer-protected metal nanoclusters or colloid catalysts (20). In another study, it is revealed that the electron transfer from Pt-NPs to oxygen vacancies in the activated Pt/γ-Fe$_2$O$_3$ catalysts may play an important role in completely suppressing the hydrodehalogenation of HANs in the hydrogenation reactions (21). Recently, Li et al have reported the metal complex effect on the hydrogenation of chlorobenzene over a PVP-Pt nanocatalyst under 298 K and atmospheric pressure. The introduction of metal complexes and ligands exhibited great effect on the activity and the selectivity of the nanocatalyst (22). The PEI-stabilized Ag-NPs exhibited a higher stability (23) than those of PEG- and PVP-stabilized Ag-NPs in the diffusion-controlled catalytic reduction of 4-NP to 4-AP by SBH. A facile one-step redox polymerization method for the preparation of highly dispersed Pd polypyrrole (PPy) nanocapsules (Pd/PPy-NCs) has been demonstrated. The catalyst (2-4nm) exhibited a good catalytic activity in the reduction of 4-NP to 4-AP by aq. SBH with ten times recyclability of the catalyst (24).

5SS.01-22: Catalysts on Surfactants and Ethers: NACs reduction via hydride transfer using mesoporous mixed oxide catalysts is of interest for organic chemists. The catalysts NiCo$_2$O$_4$ were prepared by co-precipitation and surfactant approach. The catalytic activity of co-precipitated NiO, CoO, NiCo$_2$O$_4$ and surfactant-assisted NiCo$_2$O$_4$ was tested for the reduction of 4-CNB using a variety of hydrogen donor and basic promoters (25). A mild, environmentally friendly method for reduction of ANCs to arylamines is reported using zinc powder in aqueous solutions of chelating ethers, which acts as a ligand, a co-solvent and an activator of zinc. Water is the proton source. This procedure is also a new method for the activation of zinc for electron transfer reduction of NACs (Scheme-8).

NO$_2$ → NH$_2$

Zn Powder / Chelating Ether

Water : Ether Co-solvent

Ambient Temperature / Pressure

--Scheme-8

The reduction is accomplished in a neutral medium and the other reducing groups remained unaffected (26). Selective reduction of monosubstituted nitrobenzenes to anilines (27) by dihydrolipoamide-iron (II) is reported.

References:

1a). S. S. Scholnik , et al. *J. American Chemical Society*, 63(05):1192-1193(1941). 1b). S. H. Tucker, et al. *J. Chemical Education*, 27(09):489(1950). 1c). Ra-Co Hydrogenation: Applications. *J. Applied Chemistry*, 08(08):492-495(1958).

2a). M. Valden, et al. *Science*, 281(5383):1647-1650(1998). This article is updated till 2009, in science magazine and the site is given in the original article. http://science.sciencemag.org/content/281/5383/1647.full. 2b). M. Date, et al. *Angewandte Chemie Intl Edition*, 43(16):2129–2132(2004). 2c). Z. Xu, et al.*Nature*, 372:346-348(1994). 2d). O. Deutschmann, et al. 2009: Wiley-VCH Verlag GmbH & Co. KGaA, Weinheim, 110 Pages. 2e). The Indian Institutes, IITs and IISc have published a series of lectures on solid catalysts which is available at NPTEL – Chemical Engineering – Catalyst Science and Technology. http://nptel.ac.in/courses/103103026/pdf/mod2.pdf

3). D. C. Sherrington, *Polymer International*, 12(02):70-74(1980).

4). Z. M. Michalska, et al. *Platinum Metals Reviews*, 18(02):65-73(1974).

5). H. S. Kakati, et al. *Indian J. Chemical Technology*, 10(01):60-62(2003).

6). A. Tungler, T. Tarnai, et al. *Platinum Metals Reviews*, 42(03):108-115(1998). 7a). F. Li, et al. *Reaction Kinetics, Mechanisms and Catalysis*, 116(02):479–489(2015), 7b). M. Liu, et al. *J. Molecular Catalysis A: Chemical*, 138(02-03):295-303(1999). 7c). Y. Tang, et al. *J. Molecular Catalysis A: Chemical*, 159:115–120(2000).

8). PVP-Pt/Ru Colloids: W. Han, J. Zhang, M. Liu, *J. Qingdao University of Science and Technology*, (Natural Science Edition), 2011-03.

9a). M. Liu, et al. *J. Colloid and Interface Science*, 214(02):231-237(1999), 9b) *J. Catalysis*, 278:01–07(2011).

10a). X. Yang, H. Liu, *Applied Catalysis A: General*, 164(01-02):197–203(1997), 10b). X. Yang, et al. *J. Molecular Catalysis A: Chemical*, 144(01):123–127(1999).

11). X. Yang, et al. *J. Molecular Catalysis A: Chemical,* 147(01-02): 55–62(1999).

12). X. Yana, et al. *J. Molecular Catalysis A: Chemical,* 170:203–208(2001).

13). W. Tu, H. Liu, *J. Molecular Catalysis A: Chemical,* 159(01):115–120(2000).

14). X. Han, H. Jiang, et al. *J. Molecular Catalysis A: Chemical,* 193:103–108(2003).

15). D. Lee, Y. Chen, *Chinese J. Chemistry,* 34(11):2018-2028(2013).

16). M. Kralik, et al. *Chemical Papers,* 68(12):1690–1700 (2014).

17). F. Li, et al. *Reaction Kinetics, Mechs & Catalysis,* 116(02):479–489(2015).

19a). B. Basu, et al. *Molecular Diversity,* 09(04):259-262(2005), 19b). B. M. L. Dioos, et al. *Advanced Synthesis & Catalysis,* 348(12-13):1413–1446(2006), 19c). Q. Shi, et al. *Green Chemistry,* 08:868-870(2006).

18). R. Zhong, et al. *ACS Catalysis,* 04(11):3982–3993(2014).

20). X. Wang, et al. *Current Organic Chemistry,* 11(03):299-314(2007).

21). X. Chao, et al. *Current Organic Chemistry,* 16(02):280-296(2012).

22a). D. Li, et al. *Synthesis and Reactivity in Inorganic, Metal-Organic, and Nano-Metal Chemistry,* 16(06): 902-907(2016). 22b). A Review: W. Zang, et al. *Catalysis Science & Technology,* 05:2532-2553(2015).

23). A. Shahzad, et al. *Chemistry An European J.,* 10(11):2512–2517(2015).

24). Y. Xue, et al. *J. Colloid Interface Science,* 379(01):89-93(2012).

25). N. Chaubal et al. *J. Molecular Catalysis A: Chemical,* 261(02):232-241(2007).

26). P. S. Kumar, K. M. L. Rai, *Chemical Papers,* 66(08):772-778(2012).

27a). M. Kijima, et al. *Synthetic Communications,* 35(08):1121-1127(2005). 27b). M. Kijima, et al. *J. Org. Chem.,* 48(14):2407-2409(1983).

5SS.02-00: Catalysts on Inorganic Solid Supports (SS):

The catalysts could be either monometallic or bimetallic from transition metals group and dispersed in or mounted on the solid supports such as Zirconia, Titania, Molecular Sieves, Carbon, Graphene, etc.

5SS.02-10: Catalysts with an Additive: The addition of a second metal, non-metal, resins or even the catalysts prepared by using a particular method of immobilization, type of support, etc display higher catalytic activity and selectivity towards the desired end products. The following examples, illustrate these observations.

Calcined –Layered Double Hydroxides' (CLDHs) use as a solid catalyst is used quite often in the recent years, as they have been found superior to non-calcined versions, for chemoselective reduction of functional groups in NACs. Isopropanol is used as a hydrogen source (Scheme-9).

R = Reduction Sensitive Functional Groups as CN, CHO, COOR etc. --Scheme-9

The Ni^{II}-Al^{III} (mol ratio = 3) catalyst is more active in bringing about chemoselective reductions compared to other calcined LDHs (1).

Nickel-stabilized zirconia ($Zr_{0.8}Ni_{0.2}O_2$), a reusable solid catalyst efficiently catalyzes the chemoselective reduction of NAs, aldehydes and ketones using propan-2-ol and KOH in the liquid phase (2).

Conversio = 100%   Sel = 100% --Scheme-10

Various NACs are successfully and selectively reduced to amines with 100% conversion and selectivity in methanol at low temperature (~5°C), by using versatile system of 5% Ni–SiO_2 catalyst and $NaBH_4$ and in situ generation of Ni-boride (Ni-B/SiO_2) (Scheme-10). The catalytic efficiency of Ni loading (5%, 10% and 15%) with silica or titania as support materials is investigated for reduction of NB (3). Supported catalysts and their applications in synthetic organic chemistry are reviewed (4, 5).

Rh-NPs/SBA-NH_2 catalyst with <3 nm sized Rh-NPs has been synthesized using amine functionalized SBA-15 (SBA-NH_2) as support, rhodium acetyl acetonate as Rh precursor and sodium borohydride (SBH, $NaBH_4$) as a reducing agent. Rh-NPs/SBA-NH_2 effectively reduced the NAs into ANs with nearly 100% conversion and selectivity using in-situ produced H_2 from HH at room temperature in water medium (6). In another case, a novel, efficient and eco-friendly Pd/MCM-41 catalyst is used for reduction of NAs to ANs in excellent yields by CTH method (7). Nanostructured Pt catalysts those enable the selective reduction of NAs to ANs in high yields have been developed. The materials made of organosilica are physically doped with nanostructured Pt(0) to form a new class of SiliaCat Platinum-Hydrogel catalyst (8a). A new sol–gel entrapped Pd(0) catalyst is used for the selective hydrogenation of NAs under mild condition, except HNBs (8b).

A supported rhodium complex obtained by copolymerization of Rh(cod)(aaema) with acrylamides. The catalyst was used for the hydrogenation of HNBs under mild conditions (9) (H$_2$(20bar)/RT) in ethanol to afford HANs with high (85%, BAN) to excellent (98%, CAN).

Novel Cu-NPs were synthesized and tested in a series of NACs with SBH to afford the aryl amines in high yields (10a, 10b). Reduction of nitro compounds (NB conversion up to 100%) to the corresponding amines has been carried out efficiently with SBH in the presence of bis-thiourea complexes of bivalent cobalt, nickel, copper and zinc chlorides, [MII(tu)$_2$Cl$_2$]. The reactions were carried out under solvent-free conditions at room temperature to afford amines in high to excellent yields (80-96%). Comparison of the results showed that the reducing capability of SBH was influenced by bis-thiourea complexes as: Co>Ni>Cu>Zn complexes (11).

5SS.02-11: Reductions in Phases and Reactors: Some hydrogenations are carried out using different phases as gas-phase or liquid-phase and also in the reactors as fluidized bed reactor or a continuous flow reactor, where the results are better than in the normal cases Gaseous hydrogenation of NB over a Cu/SiO$_2$ catalyst in a two-stage and in a single-stage fluidized bed reactor, at 513–553 K and atmospheric pressure is studied. The placement of the second perforated plate in the fluidized bed reactor inhibits the back-mixing of gases and solids and consequently increases the local molar ratio of hydrogen to NB in the second stage. Thus, NB conversion, AN selectivity and the catalyst life increase in the two-stage fluidized bed reactor, as compared with those in the single-stage (12).

The hydrogenation of NB over Pt catalysts supported on alumina, activated carbon, and aluminum borate was investigated in a liquid phase batch reactor at 50°C and 1atm H$_2$. Under the conditions employed in this study, aluminum borate-supported Pt catalyst (Pt/Al$_2$-B$_2$O$_7$) was found to possess the highest activity (13).

Earlier we have seen that the gas phase (1atm, 453 K) hydrogenation of p-CNB over Pd(1–10% wt) supported on AC and non-reducible (SiO$_2$ and Al$_2$O$_3$) and reducible (ZnO) oxides has been examined. Pd/ZnO is found the best catalyst amongst these tested catalysts producing 100% p-CAN from p-CNB, while othrs agve lower yileds. (14). A highly dispersed Pd-NPs o TiO$_2$ is found an efficient catalsyt (15). NAs with other sensitive functional groups have been hydrogenated at near-complete conversion of the substrate on supported Au-NPs (Au/TiO$_2$ and Au/Fe$_2$O$_3$), using a batch reactor under H$_2$ pressure. Unlike other noble metals, gold shows high chemoselectivity towards reduction of the nitro group (16).

5SS.02-12: Carbon As SS: Simultaneous generation of filamentous carbon and Ni-NPs was achieved and was tested as a catalyst for the hydrogenation of NB.

CTH = Catalytic Transfer Hydrogenation   --Scheme-11

The catalysts exhibited excellent performance, producing clean AN (~99% yield) (Scheme-11). (17). N-Phenylhydroxylamine (PHA) and N-Acetylphenylhydroxylamine synthesis by CTH of NB using HH and 5% Rh/C as a catalyst in THF at 30°C is reported (18). Other noble metals as Pd, Pt, Ir are also used on carbon for chemoselective hydrogenations of NAs (19, 20). Activated carbons are attractive as support materials for precious metal catalysts in certain applications (21). The date pit active carbons are prepared by acid treatment/oxidation methods and used as Pd supports for 5% Pd by incipient wetness method. The catalyst exhibits the best activity for the liquid phase hydrogenation of NB to AN (22). A highly efficient method for removal of 4-NP through its conversion to 4-AP from water is studied using a novel catalyst–adsorbent composite of Au-NPs supported on functionalized mesoporous carbon (Au@CMK-3-O) (23). The novel nanocatalyst comprised of Co-NPs supported on a N-doped mesoporous carbon (Co/mCN-900) is prepared by simple one-pot pyrolysis of a homogeneous mixture of melamine, polyacrylonitrile, and $Co(NO_3)_2 \cdot 6H_2O$ under a N_2 atmosphere at 900°C. The catalyst was effective in hydrogenating NAs at milder conditions (i.e., 1 MPa H_2 and 120°C) as compared to previously reported Co- and Ni-based catalysts (24).

The C_{60}-stabilized Ni catalyst, denoted as Ni/C_{60}, was prepared by regular impregnation following the chemical reduction and employed to catalyze hydrogenation of NB, for which both the conversion and the AN selectivity reached above 99.9% within 40 min under 90°C and 2 MPa of H_2. C_{60} improved the AN selectivity and enhanced the stability of Ni/C_{60} ascribed to its hydrophobicity (25). Silver–Cerium Dioxide Core–Shell nanocomposite catalyst is prepared. As prepared core–shell nanocomposite comprising Ag-NPs core and a CeO_2-NPs shell catalyzes the chemoselective reduction of nitrostyrenes (26). Carbon nanotubes (CNTs) supported Pd-NPs (NP; 3-6nm) recyclable catalysts (Pd/CNTs) were prepared by a green and facile synthesis method based on hydrogen-bonding self-assembly, which exhibited relatively higher activity and selectivity (27). Multi-walled carbon nanotubes as a metal-free catalyst were functionalized with small organic molecules containing specific ketonic carbonyl groups through non-covalent van der Waals and π–π interactions. The carbonyl groups are active sites in the reduction of NB, and the catalysts functionalized with phenanthraquinone exhibit relatively high activity and selectivity (28). An improved hydrogenation of NAs using nano-structured iron- and cobalt-(Co-Co_3O_4/NGR@C or Fe_2O_3/NGR@C) based catalysts (b) and hydrogen gas in EtOH+H$_2$O or MeOH+H$_2$O as solvent(a) for the reduction of 4-NP to 4-AP(b) are reported. Solvent and solid supports are important for yield and selectivity of ANs (29).

5SS.02-13: Pt-Cu/CNT or AC: The promotional effect of Cu on the performance of supported Pt catalysts in the selective hydrogenation of o- and p-CNB was studied using carbon nanotubes (CNTs) and activated carbon (AC) as supports. Strong interactions between Pt and Cu result in partial coverage of the Pt surface by Cu. It is suggested that the superficial Cu species activate the N-O bond of nitro group in CNBs and also inhibit the adsorption of the desired product CAN. Thus, the bimetallic catalysts exhibited excellent performance in producing selectively CANs (>99%), in comparison to monometallic catalysts (91–98%). Moreover, the stability was much more improved in comparison to the monometallic catalysts (30). Pt-NPs appear as promising catalysts due to the fact that they combine a good activity in nitro reduction while keeping a certain degree of chemoselectivity. Herein, the most recent advances related to the preparation and application of Pt-NPs as catalysts in the hydrogenation of NACs are reviewed (31).

An Fe-modified Co–B/carbon nanotubes (CNTs) amorphous alloy catalyst (Fe–Co–B/CNTs) with the Co–B particles loaded inside of the CNTs was prepared by chemical reduction and used in the hydrogenation of m-CNB to m-CAN (~97%) over Co–B/CNTs and Fe–Co–B/CNTs. The addition of Fe in the Fe–Co–B/CNTs can improve the thermal stability of the amorphous alloy, promote the formation of active centers and enhance the electron interactions (32).

5SS.02-14: Carbon: Reviews: Considering the wide applications of metal catalytic hydrogenations using different forms of carbon as a solid support, the use of carbon nanotubes (CNT) and nanofibres and rGO as catalysts and catalysts supports is reviewed, covering i) applications, ii) the different preparation methods for supporting metallic catalysts on these supports, and iii) details of the catalytic results obtained with nanotubes or nanofibres based catalysts are reviewed (33, 34, 35). The mechanism of catalytic reduction has been described. Factors affecting the rate of reduction of NAs in the presence of M-NPs stabilized in polyelectrolyte brushes, polyionic liquids, micelles, dendrimers, and microgels have been discussed with a potential for further development in this area. Physical chemistry of catalytic reduction of NAs using various nanocatalytic systems: past, present, and future are summarized (36). Understanding nano effects in catalysis based on carbon nanotubes and grapheme 2D materials is summarized (37).

5SS.02-15: Biomass/Resins as SS: Nanopalladium on amino-functionalized siliceous mesocellular foam as an efficient heterogeneous catalyst for the transfer hydrogenation of NAs to ANs is reported. In all cases, ANs were obtained in high yields (e.g. 98% p-anisidine) using the naturally-occurring and renewable γ-terpinene as hydrogen donor. The process involving a recyclable catalyst is efficient, eco-friendly, economic, and is completed in short time (38). Liu et al. have reported one-pot synthesis of Ni–NiFe$_2$O$_4$/ carbon nanofibre composites from biomass for selective hydrogenation of NACs with high yield and selectivity (39). Development of catalytically active materials from bio-waste represents an important aspect of sustainable chemical research. Sustainable biomass-derived catalysts from inexpensive chitosan and abundant $Co(OAc)_2$ are developed and tested for selective catalytic hydrogenation of diversely functionalized NAs to ANs in high yield and selectivity. Three heterogeneous materials were synthesized and this green protocol has also been implemented for the synthesis of biologically important TRPC3 inhibitor and other functionalized NAs (40). Metal–organic films onto the solid surface have been successfully assembled through Layer-by-Layer (LbL) deposition. An improved method for reducing aromatic nitro compounds on solid-phase supports using sodium hydrosulfite is presented (41). The LbL films with 10 layers (denoted as (Pd-ligand)$_{10}$) were applied as catalysts for the hydrogenation of NAs (42).

5SS.03.00: Noble-Metals on Different Solid Supports & Different Hydrogen Sources:

Some other relevant information is covered under individual sub-headings as hydrogenations using molecular hydrogen, and other hydrogen sources as alcohols, hydrazine hydrate, formic acid etc. This is because of the fact that the metal catalysts are most frquenetly used on "Solid Supports". Also, as pointed out earlier, it is possible that some examples may be found at more than one place as they are relvant at both the places.

5SS.03.10: Gold (Au): A novel synthesis of Au-NPs supported on hybrid polymer/metal oxide as catalysts and exploring the roles of cyclodextrins (CDs) in catalyst preparation and its application for p-CNB hydrogenation are explored (43). The hydrogenation of NB at 298K and pressure of 40 bar of H_2 over 1wt% of Au/ZrO_2 catalysts prepared with CTMB has the activation energy 67.2 KJ/mol and gave moderate results (44). The Au-NPs size and support effects in hydrogenation over Al_2O_3 and TiO_2 are studied for the hydrogenation of NB. Small Au particle size and the metal-support interactions play important roles in high selectivity to AN (45, 46). NB reduction with > 95% selectivity to AN is reported by using a modern gold catalyzed synthesis method (47). Ferric hydroxide supported gold subnano clusters or quantum dots for enhanced catalytic performance for chemoselective hydrogenation of NAs is investigated. Under the same reaction conditions as 100°C and 1 MPa H_2 in the hydrogenation of NACs a 96-99% conversion (except for 4-nitrobenzonitrile) with 99% AN selectivity was obtained over the ferric hydroxide supported Au catalyst, and the TOF values were 2-6 times higher than that of the corresponding ferric oxide supported catalyst with 3-5 nm size Au particles (48). There are so many publications on gold catalyzed reductions of NAs. Therefore, without going into details of each article, some other reports about gold supported on different solid surfaces and catalytic reductions of NAs to corresponding ANs are as given below, only in the form of the titles of the articles and relevant references (49) for reader's ready reference.

49a). One-pot synthesis of indoles from NAs and aldehydes, catalyzed by Au-NPs son Fe$_2$O$_3$: Y. Yamane, X. Liu, *Organic Letters*, 11(22):5162–5165(2009).

49b). Titania-supported Au-NPs(Au/MTA) and NaBH$_4$ and TMDS for the reduction of NAs to ANs. S. Fountoulaki, et al. *ACS Catalysis*, 04(10):3504–3511(2014).

49c). Au/TiO$_2$/NH$_3$-BH$_3$ at 25°C in water or ethanol yielded 100%p-toluedine with 100% conversion of *p*-nitrotoluene. E. Vasilikogiannaki, et al. *Advanced Synthesis & Catalysis*, 2013, 355:907-911(2013).

49d). Au-SiO$_2$ for 4-NTP reduction to the corresponding aniline derivative (4-ATP): W. Xie, et al. *J. American Chemical Society*, 135(05):1657–1660(2013).

49e). Noble metal catalysts on titania, and N- doped Fe$_2$O$_3$, etc for the reduction of NAs to ANs. P. Serna, A. Corma, *ACS Catalysis*, 05(12):7114–7121(2015).

49f). Shape-controlled Au nanospheres, nanorods, nanoprisms for reduction of different NACs: S. Kundu, *J. Physical Chemistry C*, 113(13):5150–5156(2009).

49g). In Au/TiO$_2$ catalytic hydrogenation, with a bimetallic effect: L. Santosh, A. Corma, et al. *Chemistry A European J.*, 15(33):8196–8203(2009).

49h). (AuNPs-sPSB) catalyzed reduction of NAs into ANs (>99%). The porous δ and ε crystalline have highest activities. A. Noschese et al. *J. Catalysis*, 340:30-40(2016).

49i). Au/*meso*-CeO$_2$/IPA system for reduction of NAs under mild conditions.. X. Liu, S. Ye, et al. *Catalysis Science & Technology*, 03(12):3200-3206(2013).

49j). Au-NPs stabilized by PEG-tagged substrate for preparation of ANs in water at RT. W. Guo, et al. *Chemistry An Asian J.*, 10(11):2437-2443(2015).

49k). A highly efficient, chemoselective Au-Cu alloy for clean anilines without any side products: *X. Qi, S. Sarina, et al. ACS Catalysis*, 06(03):1744–1753(2016).

49l). Au-NPs supported on titania (Au-Nps/TiO$_2$) and hydrazine is used for the reduction of *p*-PHA to *p*-AP. H. Yazid et al. *Indian J. Chemistry*, 52A, 184-191(2013).

49m). A Review: metal organic framework (MOF)-encapsulated Au-NPs: A. Dhakshinamoorthy, A. M. Asiri, et al. *ACS Catalysis*, 07(04):2896–2919(2017).

49n). Au@Fe$_3$O$_4$ yolk–shell (2.5-10nm) efficient for NAs with electron withdrawing substituents: F. Lin, R. Doong, *J. Physical Chemistry C*, 121(14):7844–7853(2017).

49o). It is a matter of the support (Au/TiO$_2$ or Au/CeO$_2$) and the adsorption on the surface is discussed. M. Makosch, *ChemCatChem*, 04(01):59-63(2012).

49p). A review on Au-N-Pores as a better choice over Au-NPs. B.S. Takale, et al. *Organic & Organic & Biomolecular Chemistry*, **12**:2005-2027(2014).

5SS.03.11: Iridium (Ir): Selective hydrogenation of substituted NAs with an effect of meta-position and catalytic performance using Ir/ZrO$_2$ catalyst is reported (50).

5SS.03.12: Palladium (Pd): A facile synthesis of monodispersed Ni-Pd alloy NPs and their assembly on graphene (G) (G-Ni-Pd; 3.4nm) is found efficient in catalyzing the tandem dehydrogenation of ammonia borane (AB) and hydrogenation of NAs to arylamines in aqueous methanol at room temperature. Amongst the catalyst compositions, G-Ni$_{30}$Pd$_{70}$ was found to be the most active with the conversion yields reaching up to 100% (51). Photoinduced reduction of NAs using a transition-metal-loaded silicon semiconductor (Pd/Si) under visible light irradiation is reported (52).

5SS.03.13: Platinum (Pt): While going through the wealth of information on catalytic hydrogenation of NAs, for producing desired results, the catalysts must have extremely fine particle size (few nms), very high active surface area of the support, and efficient interactions with the nitro groups of the adsorbed molecules. Pt-NPs were successfully fabricated on the surfactant-free surface of Fe$_3$O$_4$. Characterizations revealed that Pt-NPs (3.1 nm) dispersed evenly on Fe$_3$O$_4$. This catalyst is active and selective for hydrogenation of CNBs, convenient, and suitable for cyclic utilization. The study showed that the activity of Pt is dependent mainly on its particle size and the Fe$_3$O$_4$ support is most favorable for this reaction (53).

Besides Bechamp process, the synthesis of 4-AP through the acid catalysed Bamberger rearaangement of the intermediate PHA is known. The PHA is formed in-situ during the reduction of NB to AN. Here, 4-AP is synthesized by using a carbon based solid acid which is responsible for the Bamberger rearrangement of in-situ formed PHA over Pt/C catalyst in the Liquid phase hydrogenation of NB to 4-AP in water. Under the optimal reaction conditions the NB conversion and PAP selectivity reached 61.0% and 77.8% respectively (54).

Thus, highly active and chemoselective hydrogenation catalysts based on Pt, Ir, Rh and Pd can be prepared by decorating the exposed metal faces with partially reduced support species by means of a simple catalyst activation procedure. The as-prepared catalysts will have a potential for applications in hydrogenation NAs (55). Enhanced chemoselective hydrogenation through tuning the interaction between ultra-small Pt-NPs supported on carbon nano tubes (CNTs), especially N-doped and with oxygen functionalities exhibit almost quantitative (100%) selectivity towards HANs produced by hydrogenation of HNBs (56). Clean and ultra-small sizes Pt nanoclusters with a diameter of 1.0–2.4 nm, supported on reduced graphene oxide (rGO) nanosheets were successfully synthesized by simple *in situ* thermolysis of a Pt-carbonyl complex. The as-prepared Pt-1nm/rGO shows high catalytic activity for the 100% selective hydrogenation of NB, with a higher TOF than observed earlier (57). Porous manganese oxide (OMS-2) and platinum supported on OMS-2 catalysts have been shown to facilitate the hydrogenation of the nitro group in CNBs to give CANs with no dehalogenation with the selectivity to CANs at 99.0% (58). A highly selective water-soluble bimetallic catalyst for hydrogenation of CNBs to CANs is developed.

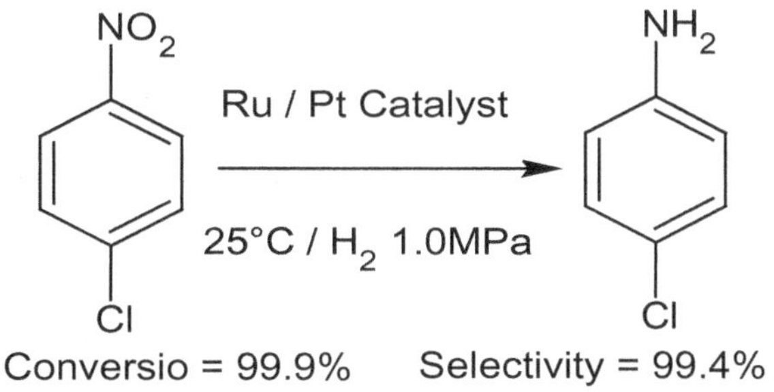

Conversio = 99.9%   Selectivity = 99.4% --Scheme-12

Under optimized and the mild conditions of 25°C, 1.0 MPa hydrogen pressure and in the presence of Ru/Pt catalyst the conversion of *p*-CNB reached 99.9%, with the selectivity to *p*-CAN of 99.4% (Scheme-12) (59).

5SS.03.14: Ruthenium (Ru): Electron-deficient Ru-NPs supported on Ru fullerite nanospheres (Ru/C_{60}) allow the successive and chemoselective hydrogenation of NB to AN, and then to cyclohexylamine. The hydrogenation is quicker in methanol than other alcohols and the chemoselectivity is mainly governed by the presence of surface hydrides on the electron-deficient Ru-NPs (60). The catalytic hydrogenation of *p*-CNB to *p*-CAN has been investigated over a series of ruthenium catalysts of widely varying dispersions and on supported bimetallics Ru-M (M=Sn, Pb, Ge). The alloying helps increase the catalyst activity (61).

5SS.04-00: Non-Noble Metals on Different Solid Supports & Different Hydrogen Sources:

5SS.04.10: Cobalt (Co): The challenging reduction of *m*-nitrostyrene containing an easily reducible vinyl-group was obtained under mild conditions with a near-quantitative yield of *m*-vinylaniline (100%) using hydrazine hydrate as the reducing agent over CoO_x/ACF as a structured catalyst (62). (ACF=Activated carbon fibres). Co-NPs supported on N-doped mesoporous carbon as highly efficient catalysts for the synthesis of aromatic amines are reported (63). A robust Co-NPs–nitrogen/carbon (Co–Nx/C-800-AT) catalyst showed extremely high activity, chemoselectivity, and stability toward the reduction of NACs with H_2, affording full conversion(100%) and >97% selectivity in water after 1.5 hours at 110°C and under a H_2 pressure of 3.5 bar for all cases (64). Metallic Co-NPs, as a non-precious metal catalyst imbedded into ordered mesoporous carbon (OMC) shows excellent hydrogenation performance in the reduction of 4-NP to 4-AP and NB to AN. The excellent catalytic performance of the CoNPs@OMC composite can be ascribed to synergistic effect between the high specific surface area, mesoporous structure and well-imbedded Co-NPs in the carbon matrix (65).

5SS.04.11: Copper (Cu): Synthesis of CuO and Cu_2O nano/microparticles from a single precursor and the effect of temperature on CuO/Cu_2O formation and morphology dependent NA reductions are investigated (66). Kadam, et al have presented the copper(II) bromide as a procatalyst for the *in-situ* preparation of active Cu -NPs for the efficient and chemoselective reduction of NAs using $NaBH_4$ in ethanol at room temperature (67).

5SS.04.12: Iron (Fe): Amongst iron oxides (γ-Fe_2O_3, α-Fe_2O_3, and FeO) γ-Fe_2O_3 has better activity than the other two, and Fe_3O_4 is more active than all of them and gives 100% selectivity to ANs in the chemoselective hydrogenation of NAs. Also, the surface area on Fe_3O_4 carries oxygen vacancies which play an important role in hydrogenation and the catalyst be magnetically separated and with excellent recycling stability (68). The specially prepared γ-Fe_2O_3/MC catalyst from Fe metal organic gel (Fe-MOG) had high catalytic activity and quantitative yield and selectivity (100%) to HANs from hydrogenation of Cl^-, Br^- and I^- functionalized NBs without any obvious dehalogenation (Scheme-13). MC = Mesoporous Carbon.

Sel. & Yileds = 100%   --Scheme-13

The hydrogenation reactions had a product yield and selectivity of 100% for the corresponding HANs using HH as a reducing agent and two harmless by-products as H_2O and N_2 (69). A highly chemoselective catalytic transfer hydrogenation (CTH) of NAs to corresponding amino derivatives is achieved with Fe–Ni bimetallic nanoparticles (Fe–Ni NPs) as the catalyst and SBH at room temperature. Their catalytic efficiency is ascribed to the presence of Ni sites on the bimetallic surface that not only hinder the surface corrosion of the iron sites but also facilitate efficient electron flow from the catalyst surface to the adsorbed nitro compounds (70).

5SS.04.13: Nickel (Ni): Nanosized silica gel supported Ni catalysts (3.7nm) were prepared and tested for the reduction of NB (100% conversion) to AN (99% Selectivity) in 5.5h at 90°C, 1.0MPa H_2 pressure, and NB:Ni = 305 (mole ratio). The Ni catalyst with > 11nm and Ra-Ni were inferior to Ni with 3.7nm and the mechanism is also discussed (71).

5SS.04.14: Ni-Rh/C: This bi-metallic catalyst directly reduced NB (100 mol% conversion) to cyclohexylamine (~92% sel.) at 140°C and 3.5MPa H_2 pressure (72). An economical catalyst with excellent selectivity (>99%) over Ni-Fe/CeO_2 nanocatalyst is used for hydrogen generation by dehydrogenation of HH in selective CTH of NAs (73). NiO-NPs with an average size of 12 nm and a high specific surface area of 88.5m^2/g were easily prepared with catalytic activity of nanosized NiO tested for chemoselective reduction of NACs into their corresponding amines using ethanol as a hydrogen donor and KOH as a promoter under microwave irradiation. This method is suitable for the large scale preparation of differently substituted ANs (74). Oxygen surface groups of activated carbon

(Ni/AC$_{OX}$) play a very good role in selective high yielding ANs (75). Ni(acac)$_2$ and PMHS are excellent for the chemoselective CTH of NACs to primary amines (Scheme-14).

R = CN, COOH, COCH$_3$, COOR, Cl, Br, CONH$_2$, CH$_2$=CH - etc. --Scheme-14

The catalyst has a good tolerance to other sensitive groups while producing the ANs in good to excellent yields with no by-product (76).

Over Ni/TiO$_2$, 99.9% conversion of o-CNB and 99.5% selectivity of o-CAN were achieved (77). Various NACs are successfully reduced to amines with 100% conversion and selectivity in methanol at low temperature (~5°C), by using versatile 5% Ni–B-SiO$_2$ catalyst (78). The IR studies indicate that in the absence of free -OH group onsilica surface, the nickel boride is anchored to the silica to facilitate the catalytic process and is better catalyst than Ni/TiO$_2$. Hydrogenation of NB to AN catalyzed by C$_{60}$-stabilized Ni catalyst, where the NB conversion and the AN selectivity both reached > 99.9% over hydrophobic Ni/C$_{60}$, within 40 min under 90°C and 2 MPa of H$_2$ (79). The continuous selective hydrogenation of the iodo-nitroaromatic refametinib active pharmaceutical ingredient (API) intermediate to the corresponding iodo-aniline was investigated using a conventional Raney-Co catalyst in a co-current trickle-bed reactor (80). Said et al have have demonstrated that the process is highly chemoselective and there was no dehydrohalogneation (F, I) occured at all. Highly selective hydrogenation of o-, m-, p-HNBs is performed using NI-B nanocatalyst under optimized pressure and temperature and found that p-CNB conversion reached 100% after 60min and the p-CAN selectivity was > 99% (81).

5SS.04.15: Ag-NPs/Fe$_2$O$_3$: Ag-NPs (3.93 wt %) on the surface of α-Fe$_2$O$_3$ are prepared and the Ag-NPs/Fe$_2$O$_3$ nanocatalysts were found good catalytic activity toward chemoselective hydrogenation of NAs in water (Scheme-15).

--Scheme-15

Water: Here, irrespective of the metal catalyst, the emphasis is on "Water" as a solvent and hydrogen donor. The catalytic system in water has a very good tolerance towards

H, Br, I, OH, OCH$_3$, COOH and CONH$_2$functional groups (82). Small and well dispersed aluminum oxyhydroxide matrix-entrapped Pt nanoparticles (Pt/AlO(OH)) were synthesized via a one-pot procedure, by the reduction of Pt4$^+$ followed by the hydrolysis of Al(O-sec-Bu)$_3$. The as-prepared catalyst was used for the hydrogenation of NB to AN- at 30°C and atmospheric H$_2$ pressure in a methanol-water which was found better than other solvents (Scheme-16).

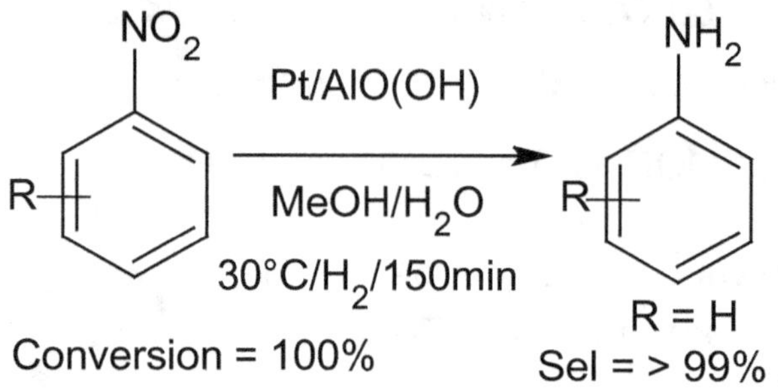

Scheme-16

A complete conversion of NB with a selectivity of 99.0% to AN was obtained (83). The promoting role of minor amount of water in solvent-free hydrogenation of HNBs is observed (84). A tris(triazolyl)-polyethylene glycol (tris-trz-PEG) amphiphilic ligand, with Fe, Co, Ni, Cu, Ru, Pd, Ag, were found efficient 4-NP reduction (85). There is a huge amount of data available in the literature, thanks to the untiring efforts of the scientists all over the world for their contributions in CTH with metal catalysts on solid supports. The recent developments in the synthesis of supported catalysts and their applications in CTH have been reviewed (86).

5SS.04.16: Zinc (Zn): Zn-Al-Hydrotalcite-supported Au$_{25}$ nanoclusters (2nm) as precatalysts for chemoselective hydrogenation of 3-nitrostyrene gave > 98% selectivity for 3-vinylaniline.This result is unprecedented for gold catalysts (87).

References:

1). T. M. Jyothi, et.al. *Bulletin Chemical Society*, Japan, 73(06):1425-1427 (2000).

2). T. T. Upadhya, et al. *Chemical Communications*, 1119-1120(1997).

3). A. Rahman, S. B. Jonnalagadda, *Catalysis Letters*, 123(03-04):264-268(2008).

4a). Y. R. deMiguel, *J. Chemical Society, Perkin Transactions-1*, 4213-4221(2000). 4b). Ibid, 00(23):3085-3094(2001).

5). P. Selvam, et al. *Tetrahedron Letters*, 45:2003-2007(2004).

6). S. Ganji, et al. *Catalysis Science & Technology*, 04:1813-1819(2014).

7). P. Selvam et al. *Applied Catalysis B-Environmental*, 49(04):251-255(2004).

8a). V. Pandarus, et al. *Adv. Synthesis & Catalysis*, 33(08): 1306–1316(2011).8b). V. Pandarus, et al. *Catalysis Science & Technology*, 01:1616-1623(2011).

9). M. M. Dell'Anna, et al. *Molecules,* 15:3311-3318(2010).

10). Z. Duan, et al. *Bulletin Korean Chemical Society*, 33(12):4003-4005(2012).

11). B. Zeynizadeh, et al. *J. Chemical Society Pakistan*, 38(04):679-684(2016).

12). S. Diao, et al. *Applied Catalysis A: General,* 286(01):30–35(2005).

13). C. Li, et al. *Applied Catalysis A: General,* 119(02):185-194(1994).

14). F. Cardenas-Lizana, et al. *ACS Catalysis,* 03(06):1386–139692013).

15). P. Chen, A. Khetan, *ACS Catalysis,* 07(02): 1197–1206(2017).

16). A. Corma, P. Serna, *Nature Protocols*, 01(06):2590-2595(2006).

17). N. Mahata, et al. *Applied Catalysis A: General,* 351(02):204–209(2008).

18). P. W. Oxley et al., *Organic Syntheses,* 67:187-189(1989).

19). N. R. Ayyangar et al., *Synthesis,* 938-941(1984); I. D. Entwistle et al., *Tetrahedron,* 34(02):213-215(1978). 20). A. Korte, et al. US 8816096 B2 (2014).

21). D. S. Cameron, et al. *Catalysis Today,* 07(02):113-137(1990).

22). N. Bouchenafa-Saïb, et al. Appl Catalysis A: General, 286(02):167–174(2005).

23). P. Guo, L. Tang, *J. Colloid and Interface Science,* 469 78–85(2016).

24). X. Cui, K. Liang, *J. Colloid and Interface Science,* 501:231-240(2017).

25). Y. Qu, et al. *Catalysis Communications,* 97:83-87(2017).

26). T. Mitsudome, *Angewandte Chemie Intl Edition,* 51(01):136-139(2012).

27). Z. Wang, H.Liu, et al. *J. Material Research.* 28(10):1326-1333(2013).

28). X. Gu, et al. *Catalysis Science & Technology,* 04(06):1730-1733(2014).

29a). D. Formenti, M. Beller, et al. *Catal Science & Technol,* 06(12):4473 4477(2016), 29b). J. Song, et al. *ACS Applied Materials & Interfaces,* 09(02):1692–1701(2017).

30). N. Mahata, et al. *Applied Catalysis A: General,* 464-465:28-34(2013).

31). P. Lara, K. Philippot, *Catalysis Science & Technology*, 04(08):2445-2465(2014).

32). F. Li, et al. Reaction Kinetics, Mechs & Catalysis, 120(02): 651–662(2017).

33). A Review: P. Serp, et al. *Applied Catalysis A: General,* 253(02):337-358(2003).

34). F. Rodríguez-Reinoso, *Carbon,* 36(03):159-175(1998).

35). E. Auer, et al. *Applied Catalysis A: General,* 173(02):259-271(1998).

36). R. Begum, R. Rehan, et al. *J. Nanoparticle Research,* 18:2312016).

37). F. Yang, et al. *National Science Review,* (2015) 2 (2): 183-201(2015).

38). O. Verho, et al. *ChemCatChem*, 06(01):205-211(2014).

39). W. Liu, et al. *Green Chemistry*, 17: 821-826(2015).

40). B. Sahoo, M. Beller, et al. *ChemSusChem*, 10(15):3035–3039(2016).

41). R. A. Scheuerman & D. Tumelty, *Tetrahedron Letters*, 41:6531-6535(2000).

42). X. Yang, et al. *Industrial Engg & Chemical Research*, 56(12):3429–3435(2017).

43a). C. H. Campos, B. F. Urbano, *J. Chemistry*, Volume 2017 (2017), Article ID 7941853, 9 pages. 43b).S. Menuel, et al. *Green Chemistry, 18:5500–5509(2016)*.

44). S. Gomez, et al. *J. Chilean Chemical Society,* 57(02):1194-1198(2012).

45). U. Hartfelder, *Catalysis Science & Technology,* 03(02):454-461(2013).

46). F. Moura de Oliveira, *Catalysts,* 06:215-225(2016).

47). A. Stephen, et al. John Wiley & Sons, 2012 - Science - 402 Pages, PP48-54.

48). L. Liu, B. Qiao, et al. *Dalton Transactions,* 19:2542-2548(2008).

49). As listed above in the text.

50). C. Campos, et al. *Catalysis Today,* 213:93–100(2013).

51). H. Goksu, S. F. Ho, *ACS Catalysis,* 04(06):1777-1782(2014).

52). K. Tsutsumi, et al. *ACS Catalysis,* 06(07):4394–4398(2016).

53). W. Du, *Industrial Engineering & Chemical Research,* 53(12):4589–4594(2014).

54). Y. Liu, *Chemical Engineering J.,* 229:105-110(2013).

55). S. Li, et al. *Chinese J. Chemistry,* 35(05):591–595(2017).

56). W. Shi, B. Zhang, *ACS Catalysis,* 06(11):7844–7854(2016).

57a). G. Wei, X. Zhao, et al. *Science China Materials,* 60(02):131-140(2017). 57b). Q. Yao, Y. Shi, et al. *Angewandte Chemie Intl Edition,* 51(01):136-139(2012).

58). I. J. McManus, H. Daly, *Faraday Discussions,* 188:451-466(2016).

59). Y. Zhou, et al. *China Petrol Processin Petrochem Technol,* 17(02):26-31(2015).

60). F. Leng, P. Serp, et al. *ACS Catalysis,* 06(09):6018–6024(2016).

61). F. Figueras, et al. *Appllied Catalysis,* 76(02):255-266(1991).

62). A. Parastaev, *Catalysis Today,* 279, Part 1:29–35(2017).

63a). X. Cui, et al. *J. Colloid and Interface Science,* 501: 231–240(2017). 63b).F. Zhang, C. Zhao, Journal of Catalysis, 348:212–222(2017). 63c).Xi. Wang, et al. *J. Molecular Catalysis A: Chemical,* 420:56-65(2016).

64a). P. Zhou, L. Jiang, et al. *Science Advances, 2017;3:* e1601945. 64b).F. A. Westerhaus, R.V. Jagadeesh, *Nature Chemistry, 05:537–543(2013)*.

65). J. Liu, et al. *J. Colloid and Interface Science,* 505:789-795(2017).

66). V. K. Vadivel, et al. *RSC Advances,* 6(88):85083(2016).

67). H.K. Kadam, S. G. Tilve, *RSC Advances,* 02(12):6057-6060(2012).

68). H. Niu, J. Lu, *Industrial Engig & Chemical Research,* 55(31):8527–8533(2016).

69). M. Tian, et al. *Green Chemistry,* 19(6):1548-1554(2017).

70). D. R. Petkar, et al. *RSC Advances,* 04(16):8004-8010(2014).

71). J. Wang, et al. *Industrial Engg & Chemical Research,* 49(10):4664-4669(2010).

72a). X. Lu, et al. *RSC Advances,* 06(19):15354-15361(2016).

73). D. Wu, et al. *ACS Applied Material Interfaces,* 09(19):16103–16108(2017).

74). S. Farhadi, M. Kazem, *Polyhedron,* 30(04): 606–613(2011).

75). Y. Ren, H. Wei, *Chemical Communications,* 53(12):1969-1972 (2017).

76). S. Sun, et al. *RSC Advances,* 05:84574-84577(2015).

77). J. Zhang, et al. *Catalysis Communications,* 08(03):345-350(2007).

78). A. Rahman, et al. *Catalysis Letters,* 123:264–268(2008).

79). Y. Qu, et al. *Catalysis Communications,* 97:83–87(2017).

80). M. B. Said, et al. *Organic Process & Res. Development,* 21(05):705–714(2017).

81). Y. Liu, et al. US 7381844 B2 (2008).

82). A. K. Patra, et al. *Applied Catalysis A: General,* 538:148-156(2017).

83). G. Fan, et al. *RSC Advances,* 04(21):10997-11002(2014).

84). J. Lyu, J. Wang, et al. *Chinese Chemical Letters,* 25(02):205-208(2014).

85). C. Wang, et al. *Angewandte Chemie Intl Edition,* 55(09):3091-5(2016).

86). P. Munnik, et al. *Chemical Reviews,* 115:6687-6718(2015).

87). Y. Tan, et al. *Angewandte Chemie Intl Edition,* 56(10): 2709–2713(2017).

5SS.04.17: Additional Recent References for CTH on SS:

1). PVP-Pd-NPs as efficient catalyst for nitroarene reduction under mild conditions in aqueous media. P. M. Uberman, *Green Chemistry,* 19(03):739-748(2017).

2). Highly chemoselective reduction of nitroarenes over non-noble metal nickel-molybdenum oxide catalysts. H. Huang, *Green Chemistry,* 19(03):809-815(2017).

3). Effect of N-Doping-Induced Metal–Support Interactions in Pd/TiO$_2$ Catalysts for NB Hydrogenation. P. Chen, *ACS Catalysis*, 07(02):1197-1206(2017).

4). Ultrasmall Pt Stabilized on Triphenylphosphine-Modified Silica for Hydrogenation. S. Jayakumar, et al. *Chemistry - A European Journal*, 23(32):7791-7797(2017).

5). Mono, bimetallic non-noble M-NPs into highly active and chemoselective hydrogenation catalysts. L. Liu, et al. *Journal of Catalysis*, 350:218-225(2017).

6). In situ spectroscopy of ligand exchange reactions at the surface of colloidal Au, Ag-NPs. A Review: R. Dinkel, et al. *J. Physics: Condensed Matter*, 29(13): (2017).

7). Alkali effects on the hydrogenation of functionalized NAs over high-loading Pt/FeOx catalysts. H. Wei, et al. *Chemical Science*, 08(07):5126-5131(2017).

8). Active sites on graphene-based materials as metal-free catalysts. S. Navalon, *Chemical Society Reviews*, 46(15):4501-4529(2017).

9). Reduction of NACs prepared by the carbonization of ordered mesoporous carbon as a heterogeneous catalyst. H. Fu, *J. Catalysis*, 344:313-324(2016).

10). Chemoselective transfer hydrogenation of nitroarenes by highly dispersed Ni-Co BMNPs. J.Zhang, et al. *Catalysis Communications*, 84:25-29(2016).

11). Ni-NCs Catalyst for the Selective Hydrogenation of NAs in the Presence of Sensitive Functional Groups. G. Hahn, *ChemCatChem*, 08(15):2461-2465(2016).

12). Bio-Based Aniline: A New Route to an "Old" Intermediate. *Focus on Catalysts*, 2017(07): 01-02(2017).

13). Microemulsion-Controlled Synthesis of Ir-Nanowires and Their Catalytic Activity in Hydrogenation of *o*-CNB. T. Lu, et al. *Langmuir*, 31(01):90–95(2015).

14). Single-Atom Catalysts: A New Frontier in Heterogeneous Catalysis. X. Yang, et al. *Accounts of Chemical Research*, 46(08):1740–1748(2013).

15). Catalytic Nanoreactors of Au@Fe$_3$O$_4$ Yolk–Shell Nanostructures for Efficient NA Reduction. F. Lin & R. Doong, *J. Physical Chemistry C*, 121(14):7844–7853(2017).

5S.05.00: Nitroarene Reductions with SULPHUR:

5S.05.10: NA Rediction with S or S Compounds: Sulfur has a symbol of "S" and atomic number 16. In nature sulfur is present in epsom and gypsum salts, galena, iron pyrites and other ores and minerals. Sulfuric acid is one of the most abundantly utilized inorganic acids in the industries and it is manufactured from sulfur. Sulfur is used in the creation of steel, rubber, production of inorganic chemicals, matches, fumigants and glass, dyes, fungicides and the production of agrichemicals. Medically, it is in lotions and skin cream ingredients. Sulfur has foul smell but it is very important part of human body or life. One of the most important applications or uses of sulfur, which are not listed above, are its application in the reduction of NAs to ANs, which is evident from the following examples.

The selective reduction of NAs to the corresponding amines is an important transformation since many aromatic amines exhibit biological activities and find a multitude of industrial applications. The reduction of NAs under mild conditions and non-destructive method is done using Zinin reaction. The reduction reaction of NACs by negative divalent sulfur in the form of hydrosulfide, sulfide, disulfide and polysulfide is called Zinin reduction (1). Unlike iron reduction, it is successfully used in special cases with more sensitive functional groups. Zinin reduction is of considerable practical value due to some inherent advantages of the method over other conventional processes as catalytic and iron reduction methods.

$$4 \times Ar\text{-}NO_2 + 6 \times S_2^- + 7H_2O \longrightarrow 4 \times Ar\text{-}NH_2 + 3\,S_2O_3^- + 6\,OH^-$$

For example, the continuous reduction 1-nitronaphtahlene with aqueous sodium sulfide is known (2). A highly chemoselective reduction of multi-nitro aromatic compounds to nitro aromatic amines is accomplished using a stoichiometric amount of sulfide, hydrogen sulfide and sodium dioxide. Mondal have done an extensive study on the chemoselective reduction of NAs using sulfur and sulfur compounds during his Ph.D. thesis which is available at the site: http://ethesis.nitrkl.ac.in/8490/1/2017-PhD-UMondal-512CH1008.pdf

Various NACs were reduced conveniently to the corresponding aniline derivatives with sodium dithionite using dioctyl viologen as an electron-transfer catalyst in CH_2Cl_2-water two-phase system (3, 4). Yadav et al (5) have developed a highly efficient method for the reduction of a variety of NAs in three immiscible liquid phases using aqueous sodium sulfide, and TBAB as the PTC. Compared to L-L PTC, the L-L-L PTC offers much higher rates of reaction, better selectivities (100%) and repeated use of catalyst (Scheme-1).

--Scheme-1

The reduction of CNBs by aqueous ammonium sulfide (6) to the corresponding CANs was carried out in toluene, under liquid-liquid mode with TBAB as a PTC. The selectivity of CANs was found to be 100%. The reaction rate of *m*-CNB was found to be highest among the three CNBs followed by *o*- and *p*-CNBs. A scalable and chemoselective reduction of a nitro functional group in the presence of an aryl imine using $(NH_4)_2S$/EtOH to afford the corresponding amino-imines in moderate to excellent yields is demonstrated (Scheme-2).

R = Imine, Vinyl, Ether, Halides

--Scheme-2

The other reducible groups such as aryl halides, styryl olefins, and ether linkages remained unaffected (7). Tin (II) complexes prepared by treatment of $SnCl_2$ or $Sn(SR)_2$ with appropriate amounts of RSH and Et_3N appear to be the best reducing agents for azides (to amines) reported so far. In general, azides react more rapidly than nitro substituents, whereas carbonyl groups, sulphones, nitriles, and esters are practically unreactive under the same conditions. Some mechanistic details of the reaction of Sn $(SPh)_3$- with aryl azides and NACs have also been elucidated (8). Beller et al. have demonstrated that the Cubane-Type Mo_3S_4 Cluster Catalysts, especially the cubane-type $[Mo_3S_4X_3(dmpe)_3]^+$clusters (dmpe=1,2-(bis)dimethylphosphinoethane), in combination with an azeotropic 5:2 mixture of HCOOH and NEt_3 as the reducing agent, act as selective cluster catalysts (X=H) or precatalysts (X=Cl) for the CTH of functionalized NAs, without the formation of hazardous hydroxylamines (9).

A simple and convenient, chemoselective method is developed for the reduction of a variety of NAs on Al_2O_3 support in presence of sodium hydrogen sulphide under microwave conditions (10). NACs can be reduced with CO in water-methanol and a sulphur compound at 120–150°C and 10–15 MPa pressure, with 100% NB conversion and AN is obtained with selectivity over 97%. The ratio of catalytic effectiveness of sulfur compounds is as follows: S: CS_2:H_2S: COS = 1:1.3:10:10 (Scheme-3).

R = H; NO_2 Base = NaOH or MeONa

Conversion = 100%   Selectivity = 97%; when R =H --Scheme-3

As an additive, Vanadium helps improve the catalytic activity and aniline selectivity. Aromatic dinitroderivatives undergo this reaction and selectivity to one of two main products (phenylenediamine or nitroaniline), which can be switched by the choice of reaction conditions (11). Macho et al. have continued their efforts in this matter and found that the sulfur based catalytic system was found to be very effective. Its low price and non-sensitivity to the common catalytic poisons makes it more advantageous than the other catalytic systems.

R = H, OMe,o or p - CH$_3$, ; 4 - Cl,   R' = CH$_3$, Ph; R" = H ; R', R" = Ph

Conversions upto 99.8%   Selectivity upto 90%   --Scheme-4

The results from one stage condensation reaction of NAs and carbonyl compounds with CO+H$_2$O are presented. Schiff's bases are formed by this reaction (Scheme-4). High conversion of nitro- and nitroso-arenes and selectivity to Schiff's bases up to 90% was observed with the sulfur based catalytic system (COS, or H$_2$S, + Et$_3$N+NH$_4$VO$_3$) at pressure 12 MPa of CO (measured at 25°C) and the temperature range from 90 to 165°C. A comparable reaction rate and conversion of nitro- and nitroso-arenes up to 99.8% was also reached (after 4 h, at 160°C) with benzaldehyde using Pd/(PdCl$_2$+FeCl$_3$+Et$_3$N) system as a catalyst (12). Haung et al have achieved very good results using commercial MoS$_2$ as a highly selective catalyst for the reduction of NBs and HNBs (F, Cl, Br and I) to corresponding ANs in excellent yields without dehalogenation products. The reduction of p-CNB was studied over MoS$_2$ and Pd/C respectively with HH. The yield of p-CAN was much higher with MoS$_2$ than that with Pd/C at full conversion of p-CNB (13).

5S.05.11: Reduction with S or S Compounds on Sold Supports: An improved method for reducing NAs on solid-phase supports using sodium hydrosulfite (Na$_2$S$_2$O$_4$) is presented. Conditions have been optimized to enable the use of this reagent for reductions on both polyethyleneglycol-polystyrene (PEG) resins and traditional polystyrene (PS) resins (14, 15). Maity et al. have reported the reduction of 4-NT by aqueous ammonium sulfide and S8 with anion exchange resin as triphasic catalyst is reported (16).

Alumina supported NaOH catalyses this transformation (Scheme-5) (17). A chemoselective method was developed to reduce ANCs to the corresponding ANs in high yields using sulfur and a base, without using hydrogen and transition metal catalysts (18) (Scheme-6).

--Scheme-6

Chemoselective reduction of nitro groups in the presence of activated heteroaryl halides was achieved via catalytic hydrogenation with a commercially available sulfided Pt catalyst (Scheme-7).

Dechlorination not observed. Sel = > 99% --Scheme-7

The optimized conditions employ low temperature-pressure, and catalyst loading (<0.1 mol % Pt) to afford heteroaromatic amines with > 99% selectivity and minimal or no hydrodehalogenation by-products (19).

A novel thioacetate mediated one-step reductive acetamidation of NAs is developed and applied to an efficient synthesis of acetaminophen, which is known as "Paracetamol" an antipyretic, analgesic drug and currently being produced at multi-thousand tons per year.

R = H or Substituents --Scheme-8

The reaction also proceeds well (21) without a solvent in the presence of a catalytic amount of surfactant (Scheme-8). The segment of the bulk analgesics market was between 75,000 and 80,000 metric tons *per year and* with a global market value of Paracetamol was over $350 million in 2002. We know that, according to Zion Research report, the global acetaminophen market was valued at around USD 801.3 million in 2014 and is expected to reach USD 999.4 million in 2020, growing at a CAGR of around 3.8% between 2015 and 2020. In terms of volume, the global acetaminophen market stood at above 149.3 kilo tons in 2014. http://www.marketresearchstore.com

References:

1). W. G. Dauben, Organic Reactions, John Wiley & Sons, Inc, New York, Vol. 20, p-455(1973).

2). G. M. Tomokkin, B. I. Kissin, *Khim Prom*, 3:79(1960).

3). K. K. Park, et al. *Tetrahedron Letters*, 34(46):7445-7446(1993).

4). F. Liu, et al. *European Polymer Journal*, 33(03):311–315(1997).

5). G. D. Yadav, *Advanced Synthesis & Catalysis*, 347(09):1235–1241(2005).

6a). N. C. Pradhan, et al. *Industrial Engineering & Chemical Research*,29:1103-1108(1990). 6b). S. K. Maity, et al. CHEMCON: Ankleshwar, Gujarat, India, 2006. 6c). S. K. Maity, et al. *Applied Catalysis A: General*, 301:251–258(2006).

7). W. P. Gallagher, M. Marlatt, et al. *Organic Process Research & Development*, 16(10):1665-1668(2012).

8). M. Bartra, P. Romea, et al. *Tetrahedron*, 46(02):587-594(1990).

9). M. Beller, *Angewandte Chemie International Edition*, 51(31):7794-7798(2012).

10). S. Ravikant, et al. *Synthetic Communications*, 32(18):2849-2853(2002).

11). V. Macho, et al. *J. Molecular Catalysis*, 88(02):177–184(1994).

12). V. Macho, et al. *J. Molecular Catalysis A: Chemical*, 209(01-02):69-73(2004).

13). L. Huang, et al. *Chinese J. Chemistry*, 31(08):987-991(2013).

14). A. Hari, B. L. Miller, *Tetrahedron Letters*, 40(02):245-248(1999).

15). R. A. Scheuerman & D. Tumelty, *Tetrahedron Letters*, 41:6531-6535(2000).

16). S. K. Maity, et al. *Chemical Engineering J.* 141 (10): 187-193(2008).

17). K. Niknam, et al. *Phosphorus, Sulfur, and Silicon and the Related Elements*, 178(06):1385-1389 (2003).

18). M. A. McLaughlin, D. M. Barnes, *Tetrahedron Letters*, 47(51):9095-9097(2006).

19). A. J. Kasparian, C. Savarin *J. Organic Chemistry*, 76(23):9841–9844(2011).

20). A. Bhattacharya et.al. *Tetrahedron Letters*, 47(11):1861–1864(2006).

5W: CTH of Nitroarenes with Water:

5W.00.00: Introduction:

Water is one of the five elements of nature and one of the most important parts of human life. As such about 70-75% of earth is occupied or her surface is covered by water. Not only that, in general water is extremely important for all living beings' survival, may be next to oxygen and in some organisms up to 90% of their body weight is from water. Human body weight is made up of about 60-70% of water. Generally, infants have higher percentage of water than adults, ranging between 70-78%. Oceans, rivers, streams, lakes (natural or man-made), ponds, deep valleys constitute the major reservoirs of water storage on earth's surface. Because of the growing population and waste disposal have caused concerns about the water pollution. Especially, some of the chemicals are highly toxic in nature and need remediation for maintaining the purity of water in the water reservoirs. Amongst many chemicals, NAs and NA based commercial products as dyes & paints, agriculture products are highly toxic and are mostly present in water sources on the ground. Therefore, water remediation is of great importance both for the nature and human life. Remediation is a process in which the polluted ground water is treated with special chemical processes so that the pollutants are converted into non-toxic or harmless products. Already there are quite a few water remediation processes in place. However, the growing needs require more and more advance technologies to treat water. Chemical treatments of NAs are being constantly evolved. In the following sections we will look at the new developments in the reduction of NAs, some of which may be suitable for water treatments.

5W.01.10: Waste Water Remediation:

AN and its many analogs are manufactured at multi tons scale every year and the waste, including the toxic NB and NAs go into waste water and become a cause of concern for water pollution and environmental issues. Although there are quite a few pollutants from which the water gets polluted, but here we are focusing on waste water remediation by polluted by organic compounds, especially NAs and ANs. Powerful remedies to convert these harmful pollutants into harmless chemicals are important. There are many ways to deal with this problem such as chemical treatments for degradation of pollutants in wastewater degraded by ultrasonication or photocatalysis, which are reviewed (5). Some of them are discussed below.

5W.01.11: Degradation of Nitroarenes in Wastewater:

5W.01.11a: By Zero Valent Iron: Fe (0): The reductive degradation of NB by zero-valent iron was investigated and found that the degradation was influenced by pH and NB concentration. The optimum pH value was found to be 3.0 for the reductive degradation of NB. Analytical study revealed that AN formation goes through the azo, azoxy intermediates and the reduction mechanisms are also suggested (6). The degradation of NB by advanced catalytic oxidation reagent, generated in the system Fe(0) - EDTA-O_2-H_2O, degraded NB to CO_2, H_2O and other compounds like low molecular acids (7). Under optimized conditions the NB conversion at 200°C to CO_2, H_2O and others substances achieved was 94%. Reduction of nitrate to ammonia by zero-valent iron can be achieved in the presence of buffers at acidic pH at about 5 and also in dilute HCl. The reduction of nitrate to NH_3 occurs with nearly complete conversion at room temperature and pressure under aerobic conditions in the presence of iron and either HCl or pH buffer. There was no nitrate reduction in the absence of surfactants (8). NB is a major environmental pollutant, and its degradation is difficult to achieve. A solution to the industrial scale problem of destroying NB through its commercial scale reduction to AN is accomplished using zero valent iron Fe(0) and $CoCl_2.2H_2O$. The process has been successfully employed to reduce NB to AN in synthetic wastewater in both batch and continuous flow reactors. The concentration of NB studied was that which would be present in industrial wastewater (123ppm). The AN thus formed in the waste systems is further destroyed by bio-organic strains/enzymes (9).

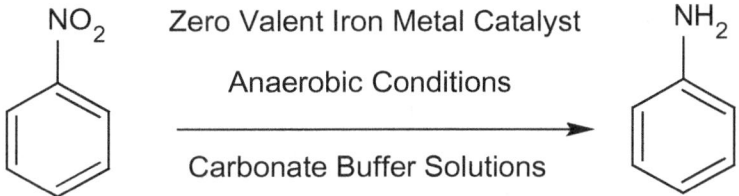

Aniline Formation Through Nitrosobenzene Only   --Scheme-1

The properties of iron metal that make it useful in remediation of chlorinated solvents and the reduction of other groundwater contaminants such as nitro NACs are known. Hence, the reduction of NACs by zero-valent Iron metal is investigated in aqueous medium. NB is reduced by iron under anaerobic conditions to AN with nitrosobenzene (NOB) as an intermediate product (Scheme-1). Coupling products such as azobenzene and azoxybenzene were not detected (10). Reduction of eleven NA pesticides was studied with zero-valent iron powder. Average half-lives ranged from 2.8 to 6.3h and the parent compounds were completely reduced after 48–96h. The 2, 6-dinitro groups in some of the herbicides were rapidly reduced to the corresponding diamines, with a negligible amount of partially reduced monoamino or nitroso products (11).

5W.01.11b: NACs Degradation with Natural Products: Korean red pine wood carbon/charcoal is used in treating NB in waste water under nitrogen atmosphere and in the presence of a buffer (12). The use of cordierite or Cu-cordierite for heterogeneous catalytic ozonation enhances significantly the degradation efficiency and the TOC removal of NB in aqueous solution relative to ozonation alone (13). This is attributed to the synergistic effects between ozone and the catalysts, and the modification process with Cu can increase the catalytic activity of cordierite for the ozonation of NB. Vasicine, an abundantly available quinazoline alkaloid from the leaves of *Adhatoda vasica*, has been successfully employed for metal- and base-free reduction of NAs to the corresponding ANs in water. The method is chemoselective and tolerates a wide range of reducible functional groups. The dinitroarenes are selectively reduced to the corresponding nitroanilines under the present reaction conditions (14).

5W.01.11c: Biodegradation: NB was completely degraded by biodegradation using mixed cultures and a sequential anaerobic-aerobic treatment process. Under anaerobic conditions in a fixed-bed column AN was formed from NB through gratuitous reduction by cells of sewage sludge in the presence of glucose as an accelerator and hydrogen source. In another case alcohols were used as hydrogen donors (15). Method of controlled reduction of NACs by enzymatic reaction with oxygen sensitive nitroreductase enzymes is presented (16). Methylviologen-pendant iron porphyrins as models of a reduction enzyme with six-electron reduction of NB to AN- is reported (17). In the reduction of NACs by NAD (P) H: quinone oxidoreductase (NQO1), the role of electron-accepting potency and structural parameters in the substrate specificity is evaluated. The multi-parameter regression analysis shows that the reactivity of NACs increases with an increase in their single-electron reduction potential and the torsion angle between nitro group(s) and the aromatic ring. Further, the factor enhancing the reactivity of NACs is their ability to bind at the dicumarol/quinone binding site in the active center of NQO1 (18). Peres and Ju have independently reviewed the biodegradation of nitroaromatic pollutants and covered from their biodegradation pathways to remediation. The bacteria and enzymes in the bioremediation of NAs are reviewed (19).

5W.01.11d: Photocatalytic Degradation:

Degradation of pollutants like NB, 2-NP, 4-NP and 2, 4-diNP in aqueous medium by means of direct photolysis, and the photolysis in the presence of H_2O_2 were investigated. The rates of the oxidations of these compounds depend on their electronic structures in the ground and excited states. Comparative study of this method with Fenton reductions revealed that the Fenton method is preferable, as it is very difficult to decompose NB by other means (22). The water pollution and its impact on environment and living being are felt for a long time. Some remedial steps were taken as per time and resources available during contemporary times. There are many ways to deal with this problem such as chemical treatments through chemical or enzymatic degradation of pollutants, which are reviewed in detail (23).

M-ETS-4 (M = Fe, Co, Ni, Cu and Ag) have been synthesized from Ti-O-Ti by photocatalytic process, characterized and used for the degradation and mineralization of aqueous solution of NB as a model pollutant compound in aqueous system. The photocatalytic activity of ETS-4 is attributed to the active −Ti−O−Ti− wires in its framework. Ag ions are found to show pronounced improvement in the photocatalytic activity compared to other transition metal ions. COD study reveals ~59% mineralization in 240 min of irradiation with Ag-ETS-4 is achieved (24). $Zn_{1-x}Cd_xS$ nanocrystals with tunable band structure are synthesized and their visible-light-assisted water splitting into H_2 and reduction of NACs in water is demonstrated. Interestingly, for the first time, the water splitting activity of the $Zn_{1-x}Cd_xS$ NCs has been applied for efficient reduction of NA pollutants in water by utilizing water as a source of hydrogen under visible-light-assisted photocatalytic conditions (25).

5W.01.11e: Ultrasound/Sonication:

The reduction by elemental iron Fe (0) and a combination of the two processes were used to facilitate the degradation of NB and AN in water facilitated by Ultrasound, which enhances the NB reduction rates (20). The integrated high gravity-ultrasonic/ozonation/electrolysis technology was applied in the pre-treatment of waste water containing NB. At pH 11 for 180 min the degradation of NB and COD reached 99% and 80% respectively (21).

5W.02.00: Green Chemistry in Water:

5W.02.10: Water as a Solvent & Hydrogen Source:

5W.02.10a (i): Water as a Solvent: Water is probably the most abundantly available component on earth, readily available anywhere and highly economical at any scale. Here, the water is used as a solvent and in some cases as a hydrogen donor or hydrogen source in the reductions of NAs. Water as a solvent is the best choice of chemists for performing reactions in aqueous medium for the following reasons:

i). It is non- flammable, non-toxic, and non-carcinogenic.

ii). Water is the least expensive and most easily accessible solvent,

iii). With water the organic substrates could be easily isolated by phase separation.

iv). Water helps in enabling the more facile control of an exothermic reaction.

v). The network of H- bonds in the system can influence the reactivity of substrates.

vi). Above features and the high qualify water as the choice of solvent for a sustainable chemistry.

vii). It is also found that the organic reactions exhibited improved reaction rates and product selectivity in water compared to the reactions in organic solvents. For example, a) the hydrogenation of t-butyl benzene with Rh-NPs encapsulated in a porous carbon shell were better in water compared to octane, acetone, and ethanol as solvents, b) the transfer hydrogenations of styrene and NB over Pd-based catalyst in methanol and the selective hydrogenation of p-CNB in ethanol over silica supported metal catalysts and of *p*-NP were higher with the addition of water in stead of only the respective organic solvents.

viii). The presence of water in the reaction medium has always enhanced the reaction efficiencies and yields of the end products. For example, the hydrogenations of water-insoluble NB and CNBs could be performed selectively and efficiently in H_2O–CO_2 system over the supported Ni catalysts at 35–50°C, without using any harmful organic solvents

ix). Water is green solvent in which several types of reactions have been performed as pericyclic reactions, multicomponent reactions, free radical reactions, carbocation reactions, biochemical reactions (in all living cells in about 90% water), acid-base reactions, precipitation reactions, oxidation-reduction reactions, Barbier-Grignard type carbonyl alkylation reactions etc. Water is a preferable solvent because it is abundantly available, safe, and for its lowest cost and environmental benefits, etc.

5W.02.10b: Developments in Organic Synthesis in Water as a Hydrogen Source or Donor:

Here, we will be focusing on the developments in chemical processes using different catalytic systems, mostly based on metal catalyzed reductions of NACs in a biphasic aqueous medium or purely in water. In both the cases water acts as a solvent or hydrogen donor to nitro groups or in some cases it exhibits both the functions. The processes described here can be applied to both for wastewater remediation- wherever applicable, and in the chemistry research laboratories and for manufacturing some of the ANs after optimizing the conditions suitable for industrial productions. Commercially, AN is manufactured by using different methods. For details see chapter-1. A new procedure for the preparation of anilines in good yield (up to 97%) by reduction of the corresponding NAs using Zn in H_2O at 250–300°C is described. The procedure is powerful enough to reduce sterically hindered 2-nitro-*m*-xylene and is chemoselective for the NO_2 groups in *o*-CNB and *m*-nitrostyrene.

R = o,o'-dialkyl, o-Cl, m-vinyl   --Scheme-2

The process involves the reduction of H_2O by Zn to generate H_2 followed by hydrogenation of the NO_2 group catalysed by residual traces of Zn (Sheme-2). Highly chemo-selective reductions of NAs to ANs are accomplished using metallic zinc in near-critical water, without affecting other functional groups (26, 27). NACs are reduced in high yields using a user-friendly combination of Raney-Ni alloy and NH_4Cl in water (28) at 80–90°C. The results of continued efforts in using water as a reaction solvent with its superior qualities over other organic solvents has been reported by different groups. More recently, Lina et al. have published the deactivation of Ni/TiO_2 catalyst in the hydrogenation of NB in water and improvement in its stability by coating a layer of hydrophobic carbon (29). The extent of Ni contents helps enhance the conversion rates to about 98%, and the selectivities up to 99.6% and the reaction rates up to 100%. A simple, practical and eco-friendly reduction of NAs with Zinc in the presence of PEG immobilized on SiO_2 as a new solid–liquid PTC in water is reported with aniline selectivity up to 92% (Scheme-3).

Chemoselectivity = 60 - 92% --Scheme-3

The reduction reactions proceeded efficiently with excellent chemoselectivity (68-92%) without affecting other sensitive groups. Other metal complexes, especially Sn, Zn, Ti, Ni and Samarium etc. have been used for this purpose (30). Natural organic matters as anthrapogenic surfactants and model quinones are used as an effective remedy for the removal such pollutants in the presence of zero valent iron (31).

A chemoselective, efficient, non-hazardous and mild protocol is developed for the reduction of NAs to aryl amines (Scheme-4) using "Iron Activated Water".

R = 4 -CH$_3$, >99.9% Conversion   Sel = > 99%   --Scheme-4

Water functions as a terminal hydrogen source without any external catalyst, acid, salts or base and in the absence of a solvent. In the course of the reaction, the zero valent iron was oxidized to magnetite (32). The GC purity of *p*-toluidine was >99.9%. A variety of differently substituted NAs have been synthesized (34) in good to excellent yields (81-97%) (33). An unprecedented palladium-catalyzed chemoselective reduction and reductive amination of NAs with water as a hydrogen source mediated by diboronic acid have been discovered (Scheme-5).

$$(\text{Het})\text{ArNHCHR} \xleftarrow[\text{ACN, H}_2\text{O}]{\substack{\text{Pd/C / B}_2(\text{OH})_4 \\ + \text{RCHO}}} (\text{Het})\text{ArNO}_2 \xrightarrow[\text{ACN, H}_2\text{O}]{\text{Pd/C / B}_2(\text{OH})_4} (\text{Het})\text{ArNH}_2$$

--Scheme-5

By using this process, a series of aryl amines containing various reducible functional groups were obtained in good to excellent yields (35). With p-NCB as a substrate, 100% dechlorination efficiency was achieved for particles with 2.0% Ni, with Ni/Fe bimetallic NPs catalyzed reduction/dehalogenation in water (36). Recently, p-CNB reduction is performed in water using Fe–Pd bimetallic NPs stabilized with sodium carboxymethyl cellulose. Fe–Pd bimetallic NPs were able to remove 100% of p-NCB within 40 min, small amounts of p-CAN as an intermediate were detected in the final stage of degradation, with no other products except AN and inorganic chloride were detected (37). There are many examples of the selective reductions of NAs using homogeneous transition metal (Ru, Rh) metal catalysts and $CO+H_2O$, CO_2+H_2O and $scCO_2+H_2O$, which are covered under this sub-heading in other section. The accumulation of hydroxylamines during the catalytic hydrogenation of several ANCs could be reduced from >40% to <1% by the addition of catalytic amounts of vanadium promoters, resulting in a faster reaction and purer products (38, 39, 40).

A chemoselective, new efficient, green and practical method for the room-temperature reduction of ANCs employing $FeSO_4.7H_2O$, $NaBH_4$, $H_3PW_{12}O_{40}$ system in H_2O under mild conditions is reported. The method is simple, inexpensive, easily scalable and applicable for the large scale preparation of different substituted anilines (41). Binaphthyl stabilized Pt-B-NPs have been synthesized, characterized and utilized as heterogeneous catalyst for an efficient chemoselective reduction of NAs at room temperature in water, with tolerance for other groups. About 25 examples are treated with this catalyst and the ANs are obtained in high yields as 84-96%. The Pt-B-NPs catalyst is stable and recyclable (42). Highly robust magnetically recoverable Ag/Fe_2O_3 nanocatalyst is prepared and used for chemoselective hydrogenation of NAs in water. LC–MS study suggested that the catalytic reaction pathway is through NO_2, NHOH, NH_2 and certainly skips the nitrosoarene intermediate step (43). Quinolines are synthesized from cheap and readily available 2-nitrobenzyl alcohol without a transition-metal catalyst in water. The reaction features an intramolecular redox process, which generates the key intermediate leading to product formation quinolones in water (44).

Under certain conditions, water itself is an excellent source of hydrogen or works as hydrogen donor. It also enhances or improves the hydrogen transfer from alcohols to organic substrates as styrene and NB. Pd catalysed reduction of styrene and NB under CTH in a fixed-bed reactor increased conversions up to 100% and selectivity up to ~95%, under proper methanol water ratios. Additionally, the H_2 atom utilization of the methanol donor in the presence of water is higher than the other donors (45).

R = H, m,p - Me, MeO, Cl,Br, I,   Yields = 90 - 96% --Scheme-6

L. Wang et.al have reported the reduction of NAs by using nanosized activated metallic iron powder in water at 210°C (near-critical water) to the corresponding aromatic amines in high yields (Scheme-6) (46, 47). The discovery of NB reduction by Russian chemist Nikolai Zinin in 1842 was pivotal to the development of the AN dye industry with all the advances in synthetic organic chemistry that ensued. As an example, 2, 4-dinitrotoluene was reduced 2, 4-diaminotoluene with Fe/HCl/H$_2$O/ethanol under reflux conditions.

--Scheme-7

For example, here, it has been demonstrated that aromatic nitro compounds could be efficiently reduced to the corresponding amines (Scheme-7) very efficiently using reagent system consisting of Al-NiCl$_2$.6H$_2$O-THF (48).

ANs and intermediates are obtained by the reduction of NAs in high yields using activated metals as Cu, Li, Mg, Al, Zn, Mo, Ti, Nb etc. Organic Reaction in Water, Part 5: Novel reduction methods of ANs are discussed (49, 50, 51, 52). Water as hydride source in the reduction of NB to AN catalyzed by cis-[Rh(CO)$_2$(2-picoline)$_2$](PF$_6$) in aqueous 2-picoline under CO atmosphere as a water gas shift reaction conditions is investigated with kinetic studies (53). The effect of water on the hydrogenation of o-CNB in different solvents gives 100% o-CNB conversion and 99.5% o-CAN selectivity (54). Many examples of NA reductions in water with different catalysts are reported (55). Pd/ Fe-powder, iodine, and pyridine was used in the reduction of NB to AN with CO+H$_2$O at 180°C and 2.5-4 MPa pressure for 2h. The system gives NB conversion of 98-100% and 100% selectivity with respect to AN (56). NAs were readily and chemoselectively transformed into ANs in excellent yields under mild conditions with Pt/CO+H2O system. Triethylamine, SnCl$_4$ and PPh$_3$ are essential for the high catalytic activity (57). Rapid and inexpensive, simple, practical and eco-friendly methods for reduction of NAs to ANs using FeCl$_3$-Zinc-DMF-Water and Zinc in Water are developed (58, 59, 60, 61). A practical and chemoslective method is developed for NA reduction using Fe/HCl and FeZn/FeSO$_4$ (63). NB is reduced to AN in high yield (93%) using (Tin(II) and HCl

(Sn + 2HCl = SnCl$_2$ + H$_2$). Actually, SnCl$_2$ itself is a very good reducing agent (64), as it generates hydrogen when reacts with HCl (SnCl$_2$ + 2HCl = SnCl$_4$ + H$_2$).

Samarium has a special place in synthetic chemistry for its applications in different forms, in different reactions including reactions in water (Scheme-8) (65). The reactivities of various nitrogen functional groups towards SmI$_2$ were also examined. Aq. phase reductions of NAs using Fe, Sn, Zn and Ni(OAc) and acids or ammonium salts have been reported in high yields of end products (66, 67, 68, 69, 70, 71).

SmI$_2$ / THF - H$_2$O / IPA

RT

IPA = Isopropylamine   Yield = 99% --Scheme-8

3-Nitro-4-methoxy acetanilide (NMA) was selectively reduced to 3-amino-4-methoxy acetanilides (AMA) with metallic Cu sub-microparticles (0.36 μm) in NaBH$_4$ solution in water. The catalyst gave AMA selectivity of 96.1% at the NMA conversion of 100% after reacting at 303 K for 40 min (72). Fe/ppm and Pd-NPs are used in the reductions of nitro groups in water at room temperature in a homogenous system (73, 74). The evolution of nanomicelles using new "designer surfactants" makes it possible to conduct numerous organic transformations in aqueous medium instead of organic solvents. This leads to a strong decrease of waste production and allows replacing toxic dipolar aprotic solvents. The technology lends itself easily to scale-up in process chemistry (75). The recent developments in this context are reviewed in an article, "Switching From Organic Solvents to Water at an Industrial Scale". Here, an alternative medium to replace toxic polar aprotic *solvent* is presented with the use of nonionic designer surfactant (e.g. TPGS-750-M) in water instead of traditional organic solvents. Applications to several commonly used transformations in active pharmaceutical ingredient (API) synthesis, Suzuki–Miyaura cross-couplings, nitro group reductions, and aromatic nucleophilic substitutions are covered including case studies of successful uses to attain improved green processes. Future directions and most recent results using tailor-made reagents suitable for micellar environments are given (76). Facile synthesis of Zn$_{1-x}$Cd$_x$S nanocrystals and their visible-light-assisted water splitting into H$_2$ and reduction of NACs in water is demonstrated (77).

References:

1). H. C. Hailes, *Organic Process & Research Development*, 11(01):114–120(2007).

2). A. Chanda, V. V. Fokin, *Chemical Reviews* 109 (02):725-748(2009).

3). P. E. Savage, *Chemical Reviews*, 99(02):603–622(1999).

4). Comprehensive Organic Reactions in Aqueous Media, IInd ed, John Wiley & Sons, Inc., Hoboken, New Jersey, 2007, PP01-15.

5). 0. Legrini, et al. *Chemical Reviews*, 93:671-698(1993).

6). Y. Mu, H. Yu, et al. *Chemosphere*, 54:789–794(2004).

7). *Anal Universiţii din Bucuresti-Chimie*, Anul XVIII, Vol.-I: 27-33 (2009).

8). I. F. Cheng, *Chemosphere*, 35(11):2689–2695(1997).

9a). R. Mantha, et al. *Environmental Science & Technology*, 35(15):3231-3236(2001). 9b). J. Klaussen, et al. *Chemosphere*: 44:544-547(2001).

10). A. Agrawal, et al. *Environmental Science & Technology*, 30(01):153–160(1995).

11). Y. Keum, Q. X. Li, *Chemosphere*, 54:255–263(2004).

12a). X. Yu, et al. *J. Hazardous Materials*: Vol 198(2011). CN102531141 A(2012).

13). L. Zhao, et al. *Environmental Science & Technology*, 43:2047–2053(2009).

14). S. Sharma, *J. Organic Chemistry*, 79(19):9433–9439(20114).

15). D. Olaf, et al. *Biodegradation*, 04(03):187-194(1993).

16a). M. M. Shah, US Patent 5777190(1998), 16b). M. M. Shah, US-6130083(2000).

17). H. Koga, et al. *Dalton Transactions* , 03(06) 1153-1160(2003).

18). N. Cenas, et al. *Microbiology & Mole Biology Reviews:* (2006). ncenas@bchi.lt.

19a). C. M. Peres, et al. *Biotechnology Annual Reviews:* 06:197-220(2000). 19b). K. Ju, et al. *Microbiology & Molecular Biology Reviews*, 74(02):250–272(2010).

20a). H. M. Hung, et al. *Environmental Science & Technology*, 34(09):1758–1763(2000). 20b). L. Zhao, et al. *Ibid*: 43:5094–5099(2009).

21). J. Weizhou, et al. *China Petroleum Processing and Petrochemical Technology, Environment Protection*, 14(03):96-101(2012).

22a). E. Lipczynska-Kochany, *Environmental Technology:* 12:87-92, (1991), 22b). Ibid, *Chemosphere*, 24(09):1369-1380 (1992).

23). 0. Legrini, et al. *Chemical Reviews*, 93:671-698(1993).

24a). P.K.Surolia&R.V.Jasra,*Desalination&WaterTreatment*,57(34):15989-15998(2016). 24b). Ibid, *Desalination & Water Treatment*, 57(46):22081-22098(2016).

25). M. Kaur, et al. *ACS Sustainable Chemical Engineerig*, 05(05):4293–4303(2017).

26). C. Boix, M. Poliakoff, *J. Chemical Society, Perkin Trans. 1*, 1487-1490(1999).

27). T. Tsukinoki & H. Tsuzuk, *Green Chemistry*, 03:37–38(2001).

28). K. Bhaumik, et al. *Canadian Journal of Chemistry*, 81(03):197-198(2003).

29). W. Lina, et al. *Journal of Catalysis*, 291:149–154(2012).

30). K. A. Reza, et al. *Iranian J. Chemistry & Chemical Engineering*. 30(2): (2011).

31). P. G. Tratnyek, et al. *Water Research*, 35(18):4435–4443(2001).

32). R. D. Patil, et al. *Organic Chemistry: Current Research*, 04:154-158(2015).

33). M. Strotmann, et al. *Synthetic Communications*, 30(22):4173-4176(2000).

34). T. Tsukinoki, H. Tsuzuki, *Green Chemistry*, 03:37-38(2001).

35). Y. Zhou, et al. *Tetrahedron*, 73(27-28): 3898–3904(2017).

36). X. Xu, et al. *Desalination*, 242(01-03):346-354(2009).

37). T. Dong, et al. *Desalination*, 271(01-03):11-19(2011).

38). K. Nomura, *J. Molecular Catalysis A: Chemical*, 130(01-02): 01–28(1998).

39). P. Baumeister, et al. *Catalysis Letters*, 49(03-04):219-222(1997).

40). J. Hu, Y. Ding, *RSC Advances*, 06(04):3235-3242(2016).

41). R. Fazaelei, et al. *J. Nanostructutres*, 01(01):21-26(2011).

42). S. S. Kotha, *Tetrahedron Letters*, 57(13):1410-1413(2016).

43). A. K. Patra, *Applied Catalysis A: General*, 538:148–156(2017).

44). M. Zhu, et al. *Tetrahedron Letters*, 56(48):6758-6761(2015).

45). Y. Xiang, L. Ma et al. *Applied Catalysis A: General*, 375(2):289-294(2010).

46). L. Wang, *Synthesis*, 13:2001-2004(2003), and references cited therein.

47). C. Bolm, et al. *Chemical Reviews*, 104:6217-6254(2004).

48). P. Sarmah, N.C. Barua, *Tetrahedron Letters*, 31(28):4065–4066(1990).

49). S.H. Pyo, et al. *Bulletin Korean Chemical Society*, 1995, 16(2):181-183(1995).

50). T. Tsukinoki, H. Tsuzuki, *Green Chemistry*, 03:37-38(2001).

51). A. Saha, B. Ranu, *J. Organic Chemistry*, 73 (17): 6867-6870(2008).

52). R. R. Dey, *Chemical Communications* (London), 48(64):7982-7984 (2012).

53). C. Longo, *Polyhedron*, 04(19): 497-493 (2000).

54). H. Chenga, F. Zhao, *Applied Catalysis A: General*, 455:08–15(2013).

55). T. Aditya, et al. *Chemical Communications*, 51:9410-9431(2015).

56). J. Skupinska, et al. *Reaction Kinetics and Catalysis Letters*, 72(01):21-27(2001).

57). Y. Watanabe, *Tetrahedron Letters*, 24(38):4121–4122(1983).

58). C. Boix, et al. *New J. Chemistry*, 23: 641-643(1999).

59). D. G. Desai, et al. *Synthetic Communications,* 29(06):1033-1036(1999).

60). A. R. Kiasat, et al. *Indian J. Chemistry & Chemical Engg,* 30(02):37-41(2011).

61). Y. Liu, B. Liu, et al. *Molecules,* 28:16(5):3563-3568(2011).

62). S. Ahammed et al. *J. Organic Chemistry,* 76(17):7235-7239(2011).

63). Y. Liu, et al. Advanced Synthesis & Catalysis, 347(02-03):217-21992005).

64a). Comprehensive Practical Organic Chemistry: Preparations And Quantitative Analysis. V. K. Ahluwalia, et al. Universities Press, 2000, 332 pages, PP 195-196. 64b). A. B. Gamble, et al. *Synthetic Communications:* 37:2777-2786(2007).

65a). E. D. Brady, et al. *J. American Chemical Society,* 2002, 124 (24): 7007–7015(2002). 65b). Z. Hou, et al. *J. Organic Chemistry,* 53(13):3118–3120(1988). 65c). J. Souppe, et al. *J. Organometallic Chemistry,* 250(01):227-236(1983).

66a). T. Tsukinoki, H. Tsuzuki, *Green Chemistry*: 03:37(2001), 66b) P. S. Kumar, et al. *Chemical Papers,* 66:772(2012), 66c) S. M. Kelly, et al. *Organic Letters,* 16:98(2014). *66d)*. H. K. Kadam, et al. *RSC Advances:* 05:83391–83407(2015).

67). D. Setamdideh et al. *Oriental J. Chemistry,* 27(03):991-996 (2011).

68a). Y. Ogata, *J. Organic Chemistry,* 47(18):3577-3581(1982). 68b). A. B. Gamble, et al. *Synthetic Communications:* 37: 2777-2786(2007).

69). S. Yamabe, S. Yamazaki, *J. Physical Organic Chemistry,* 29(07):361-367(2016).

70). R. M. Deshpande, et al. *J. Organic Chemistry,* 69(14):4835–4838(2004).

71). M. Tafesh, M. Beller et al. *Tetrahedron Letters,* 36(51):9305-9308(1995).

72). Y. Feng, et al. *The Canadian J Chemical Engineering,* 95(08):1562–1568(2017).

73). D. Dandu, A. Racha, *RSC Advances,* 04(43):22567-22574(2014).

74). C. M. Gabriel, et al. *Organic Process Research & Devel,* 21(02):247–252(2017).

75). N. Krause, *Current Opinion Green and Sustainable Chemistry,* 07:18-22(2017).

76). M Parmentier, *Current Opinion Green & Sustainable Chemistry,* 07:13-17(2017).

77). M. Kaur, et al. *ACS Sustainable Chemical Engineering,* 05(5):4293-4303(2017).

5ANPs.01.00: Bi-Metal and Bi-Metallic Alloy-NPS (ANPs) Catalyzed Dehydrogenation of Other Hydrogen Sources:

Bimetallic catalysts and bimetallic nanoalloys as $Ni_{70}Pd_{30}$ (1), Ni-Fe/CeO_2(2) , Ni–Pt alloy (3, 4), Au-Cu (5), Cr(II)/Mn(6) Pt-Ni NPs, (7), Fe–Ni –NPs and Ni–B (8), Ni-Pd alloy NPs(9), Ni-Pd (10), Cu-MgO(11), Pd-Rh and Pd_{13}-Pb_9 (12), have been used in the dehydrogenation of hydrogen sources as hydrazine hydrate, boranes, borohydrides, formic acid etc. and used the hydrogen generated in the chemoslective reductions of NAs to corresponding ANs in good, high to excellent NACs conversions and the yields and selectivity of the corresponding ANs. Syntheses, properties, and applications of Bimetallic Nanocrystals have been discussed in detail (13).

References:

1). D. Bhattacharjee, et al. *RSC Advances*, 06(69):64364-64373(2016).

2). D. Wu, et al. *ACS Applied Materials & Interfaces*, 09(19):16103–16108(2017).

3). Y. Moon, H. Mai, *ChemNanoMat*, 03(03):196–203(2017).

4). S. K. Singh & Q. Xu, *Inorg. Chem.*, 49(13):6148–6152(2010).

5). M. Hajfathalian, et al. *J. Physical Chemistry C*, 119(30):17308-17315(2015).

6). A. Hari, & B. L. Miller, *Angewandte Chemie Intl Edition*, 38(18):2777-2779(1999).

7). S. K. Ghosh, et al. *Applied Catalysis A: General*, 268(01-02):61- 66(2004).

8). D. R. Petkar, et al. RSC Advances, 04(16):8004-8010(2014).

9). H. Wen, et al. *Catalysis Communications*, 10(08):1207-1211(2009).

10). US20160279619-A1(2016).

11). K. H. P. Reddy, et al. *Catalysis Communications*, 95:21-25(2017).

12). S. Furukawa, Y. Yoshida, et al. *ACS Catalysis*, 04(05):1441-1450(2014).

13). K. D. Gilroy, et al. *Chemical Reviews*, 116 (18):10414-10472(2016).

CHAPTER-6

CTH OF NITROARENES USING DIFFERENT TECHNIQUES

6.00.00: Biocatalysis:

Introduction: Catalysis is a chemical reaction in which the reaction rate is governed, especially increased or enhanced by another chemical or substance which is other than the substrates and is known as a catalyst. Biocatalysis is a process in which the chemical or biochemical reaction rate is enhanced by a natural substance and not by a synthetic chemical. For example, some of the proteins in our body act as the enzymes. The enzymes present in our body/gut do speed up the process of breaking down the food we eat into small fragments and micronutrients and make them available for digestion. Here, the enzymes are the best examples of natural catalysts or biocatalysts. Another example of biocatalyst is the fermentation process used in wine brewing, which is an age old (about 4000-7000 years old) biochemical process. Fermentation is a process in which a chemical breakdown of a substance (as sugars, molasses etc.) is done by bacteria, yeasts, or other microorganisms. Usually sugar, jiggery, molasses, starch from corn, wheat, grains and cellulose from natural materials are used as feedstocks for brewing wine or even in the production of ethanol. Synthetic ethanol can also be produced from non-renewable sources like coal and gas. In recent times, the biochemical technologies are gaining importance as they can be used as a substitute for hazardous, expensive, polluting chemical processes. Biochemical processes are used for making fine chemicals, which in turn are used as starting materials for specialty chemicals, particularly pharmaceuticals, biopharmaceuticals and agrochemicals. The worldwide chemical business is valued at US$2500 billion, while the fine chemical business is pegged at US$85 billion and will keep growing at about 10-15% per year in the future. http://www.reuters.com/article/us-wine-oldest-idUSTRE70A0XS2011011. https://en.wikipedia.org/wiki/History_of_wine. https://en.wikipedia.org/wiki/Fine_chemical.

In recent years, the development of co-immobilized multi-enzymatic systems is increasingly driven by economic and environmental constraints that provide an impetus to develop alternatives to conventional multistep synthetic methods. Co-immobilization provides benefits that span numerous biotechnological applications, from biosensing of molecules to cofactor recycling and to combination of multiple biocatalysts for the synthesis of valuable products. Particularly such system (co-immobilized coupled enzyme system in biotechnology) is used in selectively reducing nitrobenzene (NB) to phenyl hydroxylamine (PHA) (1).

The bio-catalysis will impact the chemical business worldwide through speedy production processes, greener synthetic routes, cost-effective processes, lesser or no pollutions, sustainable technologies and business models. These industrial processes/technologies are needed daily for the production of agrochemicals for increasing food productivity for growing populations, pharmaceuticals/drugs to treat multiple diseases, in the production of flavors, fragrances, polymers, paint, dyes, plastics and materials in the electronics industry etc. The following chemical reactions used in the synthesis/production of aniline by the reduction of NB have been accomplished by biocatalysts. Let us have a look at these processes.

6.01.00: Selective Reductions of NAs Using Biocatalysis:

Highly selective and efficient reduction of NB to AN-with a biocatalyzed cathode is achieved. NB is a toxic compound that is often found as a pollutant in the environment. The present substrate removal strategies suffer from high cost or slow conversion rates. Wang et al. have developed a bioelectrochemical system with microbially catalyzed cathode, which produces AN as the sole product. The changes in the additives as $NaHCO_3$ in place of glucose reduced the NB conversion or degradation but retained the AN selectivity (Scheme-1).

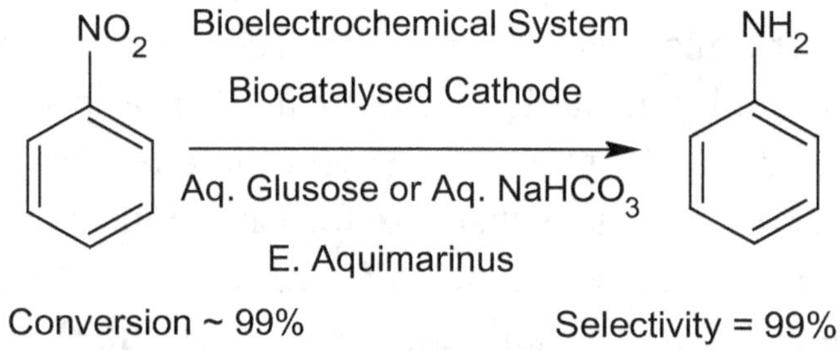

Scheme-1

16S-rRNA based analysis of the biofilm on the cathode indicated that the cathode was dominated by an Enterococcus species closely related to Enterococcus aquimarinus. With abiotic cathode reduction NB intermediates as NOB and PHA were detected (2).

The reduction of NB using natural resources is known for a long time. For example, the reduction of NB with dextrose in alkaline medium is reported (3) in 1935.

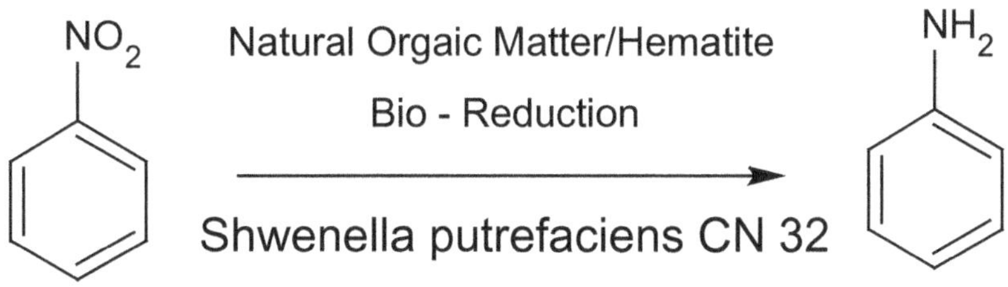

--Scheme-2

Similarly, the bio-reduction of NB, natural organic matter (NOM), and hematite by Shewanella putrefaciens CN32 species is studied (Scheme-2) and found that NOM-mediated reduction of NB was more important than Fe(II)-mediated reduction (4-6).

Nitroreductases such as xanthine hydrogenase and quinone reductase are known to reduce NAs to the corresponding ANs. Cervilla et al. have reported that molybdo-reductase is a novel and highly efficient NB reducing agent (Scheme-3).

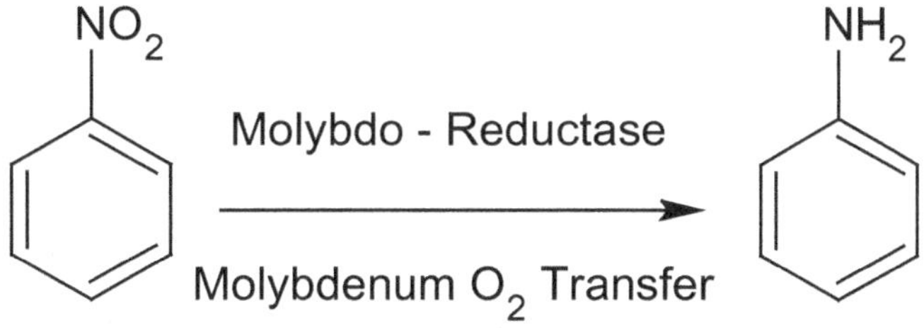

--Scheme-3

It is found that an enzyme catalyzes the reduction of NB to AN by a molybdenum-mediated oxygen atom transfer reaction (7). Experimental proof of concept was obtained that NB can be reduced to AN by a mixed reductive microbial culture using H_2 as the sole electron donor source.

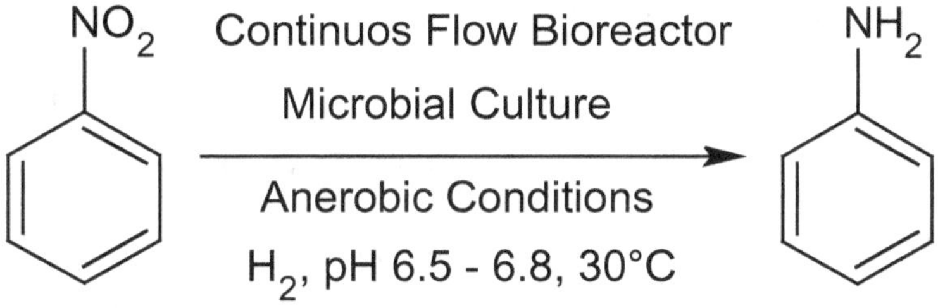

--Scheme-4

In a continuous-flow anaerobic bioreactor, both pH and the temperature affected NB reduction with a pH at 6.5–6.8 and 30°C (Scheme-4). The efficiency of NB degradation increased with H_2 up to 10% (v/v). An increase in sulfate concentration decreased the removal rate of NB (8). Effects of water-miscible ionic liquids on cell growth and NB reduction using Clostridium sporogenes have been tested (9). 2-Hydroxy ethyl trimethyl-ammonium dimethyl phosphate ([EtOHNMe$_3$][Me$_2$PO$_4$]) and N,N-dimethylethanolammonium acetate (DMEAA) increased the growth rate of C. sporogenes and they were found sufficiently non-toxic to allow efficient reduction of NB using harvested cells, providing AN yields up to 79%.

Yield = 79% --Scheme-5

The high product yield with reactions in [Emim][EtSO$_4$] represented a significant improvement over conventional solvents, and the ionic liquid appeared to suppress unproductive substrate consumption by an unknown mechanism (Scheme-5). NB occurs as a pollutant in wastewaters originating from numerous industrial and agricultural activities. In this study, Mu et al. have investigated the use of a bioelectrochemical system (BES) to remove nitrobenzene at a cathode coupled to microbial oxidation of acetate at an anode (10).

References:

1). L. Betancor, *Biotechnology & Genetic Engineering Review*, 27(01):95-114(2010).

2). A. J. Wang, et al. *Environ Science & Technology*, 45(23):10186-10193(2011).

3). N. Opolonick, *Industrial Engineering.& Chemistry*, 27 (09):1045–1046(1935).

4). F. Williams, et al. *Environ. Science & Technology*, 44:184–190(2010).

5). F. Luan, Y. Liu, et al. *Environ. Science & Technology*, 49 (03):1418–1426 (2015).

6). F. Luan et al. *Environ. Science & Technology*, 49 (06):3557–3565(2015).

7). A. Cervilla, A. Corma, et al. *J. Am. Chem. Soc.*, 117(25): 6781–6782(1995).

8). H. Cao, Yu. Li, et al. *Biotechnology Letters*, 26(04):307-310(2014).

9). O. Dipeolu, E. Green, & G. Stephens, *Green Chemistry*, 11:397-401 (2009).

10). Y. Mu, et al. *Environmental. Science & Technology*, 43:8690–8695(2009).

6.01.10: Aniline Analogs through Bioctalysis:

Transformation of *p*-CNB by Escherichia coli is reported (Scheme-6). Washed cells of E. coli, the resting culture and the homogenate of disintegrated cells transform *p*-CNB into *p*-CAN. The growing culture of E. coli (Eh = -210 mV) reduces the nitro group in *p*-CNB, however, there is no reduction in the absence of E. coli cells.

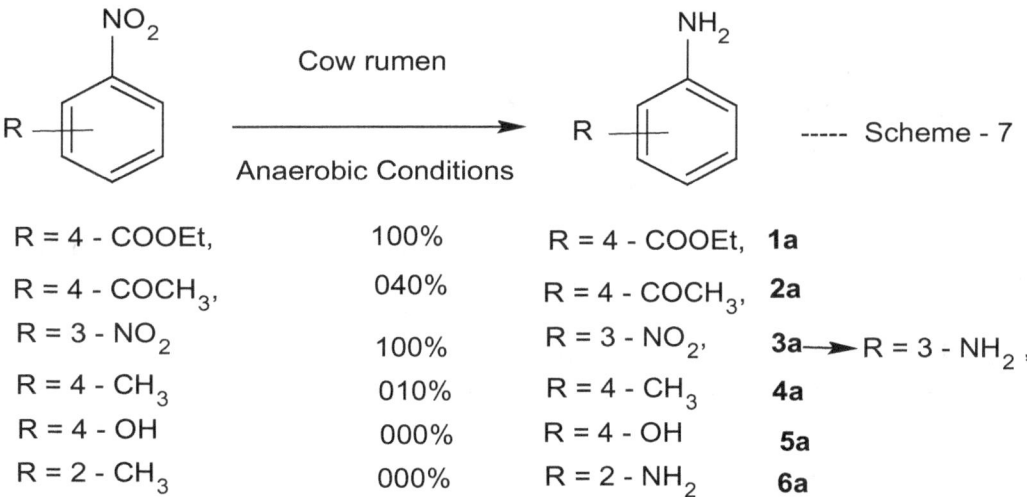

--Scheme-6

The rate at which E. coli reduces the nitro group of *p*-CNB depends on the redox potential of the medium (Scheme-6). It is likely that any microorganism is capable of reducing *p*-CNB at a low value of the redox potential (11). Tuan et al have used a bioelectrochemical process in the reduction of *p*-CNB to *p*-CAN, where the microbial-mediated electron transfer at the negative cathode potential enhances the *p*-CNB conversions. Under optimized conditions of voltage and time, the *p*-CNB reduction and *p*-CAN formation efficiency reached 99% and 94.1%, respectively (12). Enzymatically, NAs are directly reduced to the corresponding ANs. Some of the NAs are transformed into imidazoles with baker's yeast (13). Anaerobic biotransformation by microorganims of the bovine rumen fluid is used in the reduction of NAs to respective amines. This enzymatic process using ruminal contents has been reported in association with the bioreduction of nitro groups (Scheme-7).

R = 4 - COOEt,	100%	R = 4 - COOEt,	**1a**
R = 4 - COCH$_3$,	040%	R = 4 - COCH$_3$,	**2a**
R = 3 - NO$_2$	100%	R = 3 - NO$_2$,	**3a**➝R = 3 - NH$_2$,
R = 4 - CH$_3$	010%	R = 4 - CH$_3$	**4a**
R = 4 - OH	000%	R = 4 - OH	**5a**
R = 2 - CH$_3$	000%	R = 2 - NH$_2$	**6a**

The numbers in % are the biotransformation of the nitro compound into aminlines in 24h

The biotransformation reactions catalyzed by this system were dependent of both the electronic characteristics and the area/volume of the nitro-substrates confirming the processes are enzymatic. The semi-preparative scale biotransformation went by in good yield showing the rumen fluid may be employed in the synthesis of amines under very mild conditions, and it may also have applications in the bioremediation of NACs (13). 2-AP and its antimicrobial derivative 2-aminophenoxazin-3-one (2-APO) have been synthesized using zinc metal and silica-immobilized enzymes in microfluidic devices (15). With continuing efforts in this field of solid supported enzymatic biocatalysis, selective synthesis of PHA with > 90% selectivity by the enzymatic hydroxylation of

NB with cofactor recycling via co-immobilization of enzymes in silica nanospheres is reported(16) A variety of NAs have been regioselectively reduced using baker's yeast (*Saccharamyces cerevisiae*), in some cases with very high selectivity (Scheme-8).

--Scheme-8

The origin of the selectivity together with a possible mechanism for the reduction is proposed (17). The role of electronic and structural parameters of NACs in their two-electron reduction by NAD-(P)H-quinone is studied. The multi-parameter regression analysis shows that the reactivity of NACs (n=38) increases with an increase in their single-electron reduction potential and the torsion angle between nitro group(s) and the aromatic ring (18). Experimentally, it is observed that the reduction of NACs by different *Pseudomonas* species CBS3 under aerobic conditions or under argon atmosphere yields selective formation of aniline and its derivatives. For example, mononitro-compounds were reduced to ANs, while 1-chloro-2, 4-DNB was reduced via the two possible chloronitroanilines to 1-chloro-2, 4-diaminobenzene. In case of 2, 4, 6-trinitrotoluene, two monoaminodinitrotoluenes and one diaminomononitrotoluene were obtained. Cells of *Pseudomonas* sp. CBS3 cultivated on complex medium showed higher nitro-reducing activity than those cultivated on mineral salts medium with 4-chlorobenzoate as a substrate, which is normally used as medium for this strain (19). Reduction of NACs by anaerobic bacteria isolated from the human gastrointestinal tract, especially Clostridium leptum, Clostridium paraputrificum, Clostridium clostridiiforme, and a Eubacterium sp. produced aromatic amines (20).

A new bioelectrochemical system (BES), a membrane-free, continuous feeding up-flow biocatalyzed electrolysis reactor (UBER) was developed to reduce NAs to the corresponding ANs. Granular graphite and carbon brush were used as cathode and anode, respectively. NB was efficiently reduced to AN in high yields (21). As enzyme-catalyzed reactions exhibit higher enantioselectivity, regioselectivity, substrate specificity, and stability and they require mild conditions to react while prompting higher reaction efficiency and the product yields. Naturally occurring enzymes are remarkable biocatalysts with numerous potential applications in industry and medicines. However, many of their catalyst properties often need to be further tailored to meet the specific requirements of a given application. Within this context, directed evolution has emerged over the past decade as a powerful tool for engineering enzymes with new or improved functions. Zhao et al have reviewed these developments (22).

Nitroreductases as enzymes with environmental, biotechnological and clinical importance are of great value in the biocatalyzed reductions. These enzymes are capable of catalyzing the reduction of NAs using flavin mononucleotide (FMN) or flavin adenine dinucleotide (FAD) as prosthetic groups and nicotinamide adenine dinucleotide (NADH) or nicotinamide adenine dinucleotide phosphate (NADPH) as reducing agents (23). Besides aldehydes and ketones, some NAs were reduced using whole plant cells from *Lens culinaris* seeds. The NACs showed low (2%) to high (> 99%) conversion depending upon the nature and position of the aromatic ring substituents. Ester hydrolysis by the *Lens culinaris* was quite effective with the ester *p*-nitrophenyl acetate (> 99% conversion).

Lens culinaris: A new biocatalyst for reducing carbonyl and nitro groups is reported (24). The combined action of silica-immobilized PHA mutase and zinc in a flow-through system catalyzes the conversion of NACs to the corresponding *o*-APs, including a novel analog of chloramphenicol and an antibacterial agent. The combinatorial synthesis of 2-aminophenoxazin-3-one (APO) in a microfluidic device is reported. Individual microfluidic chips containing metallic zinc, silica-immobilized PHA mutase and silica-immobilized soybean peroxidase are connected in series to create a chemo-enzymatic system for synthesis. Zinc catalyzes the initial reduction of NB to PHA which undergoes a biocatalytic conversion to 2-AP, followed by enzymatic polymerization to APO. Silica-immobilization of enzymes allows the rapid stabilization and integration of the biocatalyst within a microfluidic device with minimal preparation. The system proved suitable for synthesis of a complex natural product (APO) from a simple substrate (NB) under continuous flow conditions. More information about APs (25) can be found in this book under the section/chapter "Amino Phenols".

References:

11). P. I. Gvozdiak, et al. *Mikrobiologiia:* 52(01):22-26.(1983).

12). Y. Tuan, et al. *Environmental Science & Technology, 36(14):1847-1854 (2015).*

13). A. Navarro-Ocana, *J. Chemical Society, Perkin Trans.,* 01:2754-2756(2001).

14). A. Rodríguez, et al., *Green & Sustainable Chemistry,* 01:47-53 (2011).

15). H. R. Luckarift, et al. *Biotechnology & Bioengineering,* 98(03):701-705(2007).

16). L. Betancor, et al. *Biotechnoly &Genetic Engg Reviews,* 27(01):95-114(2010).

17). C. L. Davey, L. W. Powell, et al. *Tetrahedron Letters,* 35(42):7867–7870(1994).

18). L. Miseviciene, Z. Anusevicius, *Acta Biochimica Pol.* 53(03):569-576 (2006).

19). A. Schackmann, *Applied Microbiology & Biotechnology.,* 34(06):809-813(1991).

20). F Rafil, et al. *Applied Environmental Microbiology,* 57(04):962–968 (1991).

21). A. Wang, et.al. *J. Hazardous Materials,* Vol: 199–200: 401–409 (2012).

22). S. B. Rubin-Pitel & H. Zhao, *Combinatorial Chemistry & High Throughput Screening*, 09:247-257(2006).

23). L. M. Oliveira, et al. Current Research: Techn & Education Topics in Appl. *Microbiology and Microbial Biotechnology*, A. Mendez-Vilas (Ed), Formatex, PP1008-1019(2010). http://www.formatex.info/microbiology2/1008-1019.pdf.

24). D. A. Ferreira, et al. *Biotechnology & Bioprocess Engg*, 17(02):407-412 (2012).

25a). H. R. Lukarift, et al. *Chemical Communications* (Camb): 03:383-384(2005). 25b). H. R. Luckarift, B. S. Ku, et al. *Biotechnology and Bioengineering*, 98 (03):701–705, (2007). 25c). Biosilica-immobilized enzymes for biocatalysis: Lorena Betancor, Submitted for publication as a chapter in book review *"Recent Advances in Biocatalysis and Biotransformation"*, J. M. Palomo (Ed.), published by Research Signpost (2007). 25d). Immobilized Biocatalysis: R. F. Lafuente, *Molecules*, 22:601, 5 Pages, (2017). doi:10.3390/molecules22040601 25e). *Biocatalysis in the Pharmaceutical & Biotechnology Industries*,. edited by R. N. Patel, CRC Press, 12-Dec-2010 - Science - 893 pages.

6.01.11: Application of Bio-Mass in Chemistry/Chemical Industry:

As we grow, the factor of growth is always there with us. We have gone through the evolution stages and today reached a very advanced position in human life history in many fields such as telecommunication, medicines, IT, transport, space etc. Since the discovery of DNA structure during fifties of last century lots of developments have occurred in the pharmaceutical industry. Similarly, knowing the abilities of microbes in fermentations for a long time and also with recent knowledge of some of microbe genomes their use in the bio-transformations is seen frequently. Biotechnology holds some salient features as cost-effective/economical processes, less hazardous and mostly pollution free, which can help replace the old chemical processes in manufacturing chemicals, pharmaceuticals and agriculture products. Erickson et al. have highlighted the perspectives on opportunities in industrial biotechnology in renewable chemicals (26a). Enzyme immobilization on ordered mesoporous supports enhances their operational stability and enables the use of enzymes as reusable and robust biocatalysts even for continuous processes. Current status and future scope is outlined. Hartman et al., have reported biocatalysis with enzymes immobilized on mesoporous hosts with the status quo and future trends (26b). According to Erickson et al., the industrial biotechnology encompasses the application of biotechnology-based tools to traditional industrial processes ("bioprocessing") and the manufacturing of bio-based products (such as fuels, chemicals and plastics) from renewable feedstocks (27-32).

We have seen that metal catalysts supported on solid supports (TiO_2, SiO_2 etc.) and also on natural polymeric materials give better performance than single metal catalysts.

Similarly, NB is hydrogenated over biocatalyzed cathodes, natural polyphenols, or M-NPs even over core-shell have been used for the reduction of NB to AN. Wang et al. have successfully found an efficient reduction method for NB to AN with a biocatalyzed cathode with AN selectivity up to ~99%. 16S rRNA based analysis of the biofilm on the cathode indicated that the cathode was dominated by an Enterococcus species closely related to Enterococcus aquimarinus (33).

A novel heterogeneous Pd-NPs catalyst stabilized by collagen fibres (CF) was synthesized. Epigallocatechin-3-gallate (EGCG), a typical natural polyphenol, was grafted onto the CF surface to improve the stabilization/immobilization of Pd(0)-NPs. These well-dispersed Pd(0)-NPs were found to be active, selective and recyclable catalysts for the hydrogenation of NAs under mild reaction conditions (34). A concept of bio-inspired catalytic hydrogenation of NAs by mimicking the catalytic behavior of enzymes with NiS_{2+x} NPs and polymeric melon (g-C_3N_4) is investigated. The g-C_3N_4-supported NiS_{2+x} NPs functioned as ligand-free and noble metal-free catalysts and offered high efficiency comparable to noble metal-based catalysts, but at a much better selectivity. Thus bio-inspired noble metal-free reduction of NAs using NiS_{2+x}/g-C_3N_4 is experimentally realized (35).

With growing needs, Clostridium sporogenes is an anaerobic bacterium with unique properties among living organisms known since 1930s and used for reduction of NAs and other functional groups under mild conditions compared to chemical catalysts. https://www.research.manchester.ac.uk/portal/files/54593560/FULL_TEXT.PDF Biocatalysis has many advantages over chemical catalysis as the enzymes are active at low concentrations, efficient and the biggest advantage of using enzymes comes from their selectivities, as chemoselectivity, regioselectivity and enantioselectivity (36a,b). There are some certain limitations as well, as high temperature and pH ranges and mostly efficient in water and require their natural cofactors such as NAD(P)H or ATP which are usually unstable and expensive. Biotransformation can be carried out by whole cells which eliminate a need to recycle cofactors as inexpensive equivalents such as carbohydrates can be used to drive the reaction. On the other hand, the isolated enzymes have higher activities and tolerance to organic solvents (37-40).

Microbes are commonly used in chemical transformations. For example, one of the most used species is Clostrodia on nitroaryl reductase activities in several Clostridia (41). Nitroreductases (NRs) are enzymes that catalyze the reduction of NAs to their corresponding nitroso, hydroxylamine, and in limited cases to amines involving six electrons and six protons in overall reduction process.

$$PhNO_2 + 2H+ +2e- = PhNO +H_2O; \quad PhNO +2e^- +2H+ = PhNHOH;$$

$$PhNHOH+2e=+2H+ =PhNH_2+H_2O.$$

One such example is the reduction of trinitrotoluene to triaminotoluene. The reduction pathway (42) is as reported earlier and is as shown in the equation here: NB +NAD (P)H = NO-NHOH-AN + NADP+.

Later, an enzymatic reduction of 2, 4, 6-trinitrotoluene (TNT) and related NAs (4-methylnitrobenzene, 4-chloronitrotoluene, 4-acetylnitrotoluene, 1,3-dinitrobenzene, and 1,4-dinitrobenzene) is successfully reported (43). A nitro reductase system found in animals is known to reduce chloramphenicol, *p*-nitrobenzoic acid and a variety of other NACs to the corresponding amines (44). The biocatalytic activity of nitroreductase from *Salmonella typhimurium* (NRSal) was investigated for the reduction of carbonyl compounds, alkenes, and NAs with substrate conversion efficiency of > 95%. NRSal also demonstrated the first single isolated enzyme-catalyzed reduction of nitrobenzene to aniline through the formation of NOBs and PHAs as intermediates. However, chemical condensation of the two intermediates to produce azoxybenzene currently limits the yield of AN (45).

Nadeau et al have studied the biocatalytic process for the production of *o*-APs from chloramphenicol and analogs. The biocatalyst is an enzyme system that makes use of a nitroreductase enzyme that initially reduces the NAs to the PHAs and a mutase enzyme that converts the PHA to an *o*-AP. The biocatalyst can also consist of a coupled, two-step metal and enzyme reaction in which the metal, such as zinc, catalyzes the transformation of the NAs to the PHAs and the mutase then catalyzes the transformation of PHA to the corresponding *o*-AP. Thus the limitations observed in earlier cases (as listed above) can be overcome with suitable choices of the enzyme, nitroreductase and favorable conditions (46). A highly selective and controllable synthesis of PHA by the reduction of NAs with an electron-withdrawing group using a new bacterial nitroreductase enzyme BaNTR1 is reported. The corresponding PHA are obtained under mild reaction conditions with excellent selectivity (>99%). This method therefore represents a green and efficient method for the synthesis of PHAs, which can be further transformed into the corresponding ANs using other options including normal one step hydrogenation (47). Nitroreductases have great potential for the highly efficient reduction of NAs to PHAs. The challenging regioselective reduction of polyNAs with nitroreductase by structure-based engineering is successfully applied in the synthesis of PHAs. The structure-based engineering of Escherichia coli nitroreductase to alter its regioselectivity, in order to achieve reduction of a target nitro group is studied. When 2, 4-dinitrotoluene was used as the substrate, the wild-type enzyme regioselectively reduced the 4-NO_2 group, but the T41L/N71S/F124W mutant primarily reduced the 2-NO_2 group without loss of activity. The preferential regioselectivity is attributed to the nature (hydrophobicity) and conformational changes in some residues (48). NAs are toxic chemicals to human life and hence naturally reductive processes are needed for their chemical modifications and one such way is the reduction of NAs with nitroreductase enzymes for toxic remediation.

References:

26a). B. Erickson, et al. *Biotechnology J.* 07(02):176–185 (2012).

26b). M. Hartmann & D. Jung, *J. Materials Chemistry*, 20(05):844-857(2010).

27). J. Becker, et al. *Current Opinions in Biotechnology*, 23(04):631-40(2012).

28). J. Nielsen, et al. *Current Opinions in Biotechnology*, 24(03):398-404(2013).

29). B. Thompson, et al. *Current Opinions in Biotechnology*, 00:17-23(2014).

30). J. Zhou, et al. *Current Opinions in Biotechnology*, 25:17-23(2014).

31). J. Y. Lee, et al.. *J. Microbiology & Biotechnology*, 26(05):807-822(2016).

32). Catalytic Conversion of Biomass: Focus on Chemistry: CATCHBIO, L. Joppen, Article No. 18175, April 04, 2017.

33). A. Wang, et al. *Environmental Science & Technology*, 45:10186–10193(2011).

34). H Wu, et al. *Applied Catalysis A: General*, 366(01): 44-56(2009).

35). Y. Zhang, X. Li, et al. *RSC Advances*, **04**(105): 60873-60877(2014).

36a). Ramesh N. Patel, *Biocatalysis in the Pharmaceutical and Biotechnology* Industries: CRC Press, 2006, Science - 893 pages. 36b). Ramesh N. Patel, Green Biocatalysis: John Wiley & Sons, 2016, *Technology & Engineering* - 792 pages.

37). K. Faber, *Biotransformations in Organic Chemistry*: A Textbook, 6th edition, pp.03-25. Berlin: Springer-Verlag ((2011).

38). Y. Asano, et al. *J. Biotechnology.* 14:65-72(2002).

39). K. Buchholz, et al. *Engineering & Life. Sciences*, 05:309-323(2005).

40). N. J. Turner, et al. *Trends in Biotechnology*, 21:474-478(2003).

41a). L. Angermaier & H. Simon, *Z. Physiology & Chemistry*: 364:961-975(1983). 41b). L. Angermaier & H. Simon, *Z. Physiology & Chemistry*: 364:1653-1663(1983).

42). R. L Koder, et al. *Biochimica Biophysica Acta,* 1387(01-02):395-405(1998).

43). R. G. Riefler, et al. *Environmental. Science & Technology*, 34:3900-3906(2000).

44). J. R. Fouts, et al. *J. Pharma & Experimental Therapeutics,* 119:197-207(1957).

45). Y. Yanto & M. Hall, *Organic & Biomolecular Chemistry*, 08(08):1826-1832(2010).

46). L.J. Nadeau, et al. US 8071340 B1, 2011.

47). J. H. Xu, et al. *Chemical Communications* (Camb): 50(22):2861-2864(2014).

48). J. Bai et al. *ChemBioChem*, 16(08):1219-1225(2015).

6.01.12: New Development in Biocatalytic Reduction of NAs.

Recent efforts in scientific community are on biocatalytic transformations of useful products, some of the recent developments are presented here.

The reduction of NACs is an important transformation that being used in the synthesis of amines which are interesting intermediates of pharmaceuticals and other derivatives. Linseed *(Linum usitatissimum L.)* is a natural product and acts as a biocatalyst in many processes and hence can be used in the reduction of NACs. Linseed, *Linum usitatissimum*, were used to catalyze bio-transformations of NAs (as NB, *o*-nitroacetophenone, *m*-nitroacetophenone and *p*-nitroacetophenone, and the *o*-aminoacetophenone, and *p*-aminoacetophenone) were obtained with excellent chemoselectivity (92.1 to ≥99%) (49). http://www.orientjchem.org/vol30no2/linseed-linum-usitatissimum-l-as-a-biocatalyst-for-reduction-of-nitroaromatic-compounds. Pseudomonas pseudoalcaligenes JS45 grows on NB as a sole source of carbon, nitrogen, and energy. The catabolic pathway involves reduction to PHA followed by rearrangement to *o*-aminophenol and ring fission (Nishino et al.). NOB was not detected as an intermediate of NB reduction, but NOB is a substrate for the enzyme and the specific activity for NOB is higher than that for NB. These results suggest that NOB is formed but is immediately reduced to PHA, which was the only product detected after incubation of the purified enzyme with NB and NADPH. PHA does not serve as a substrate for further reduction by this enzyme. The products and intermediates are consistent with two two-electron reductions of the parent compound. Furthermore, the inducible control of enzyme synthesis suggests that NB is the physiological substrate for this enzyme (50a, b).

Nanobiocatalysis is a technology in which enzymes are incorporated into nanostructured materials and it has emerged as a rapidly growing area. Nanostructures, including nanoporous media, nanofibres, carbon nanotubes and NPs, have manifested great efficiency in the manipulation of the nanoscale environment of the enzyme and thus promise exciting advances in many areas of enzyme technology. Nanobiocatalysis and its potential applications in various fields, such as trypsin digestion in proteomic analysis, antifouling, and biofuel cells are reviewed (51). Lignin/Zeolites (a) and Chitosan-Cu-NPs (b) have been used as natural components in the reduction of NAs to ANs. Aromatic amines can be directly produced from lignin by *ex situ* catalytic fast pyrolysis with ammonia over zeolite catalysts. Cu-NPs were supported into a chitosan/poly(vinyl alcohol) matrix. This system enables the catalytic reduction of NAs to ANs, which are intermediates for many pharmaceutical products. The energy of activation of this reaction was lower when compared to other catalysts and the catalytic efficiency was kept even after 6 consecutive reuse cycles. This novel catalytic system shows several advantages over other metal based catalysts (52). Quinazoline alkaloid (from the leaves of *Adhatoda vasica*) (a) and Glutothione (b) have been used in the metal-free hydrogenation of NAs to ANs in water. The alkaloid based method is chemoselective and tolerates a wide range of reducible functional groups, such as ketones, nitriles, esters, halogens, and heterocyclic rings. Dinitroarenes are also reduced region-selectively to the corresponding mononitroanilines under the present reaction conditions. The water soluble glutathione (a naturally occurring antioxidant) capped M-NPs (M-GS, where M=Pd, Pt, Au and Ag; GS=glutathione) with size 2.4±0.2 nm were synthesized by borohydride reduction of metal ions in the presence of glutathione as capping ligand and used as catalyst for the hydrogenation of nitroaniline in aqueous phase (53).

Porous activated carbon derived from biomass feedstock (beet-root) shows excellent performance as an efficient support for magnetite-NPs (Fe_3O_4@BRAC) catalyst for the reduction of NAs. The process is highly chemoselective with nitro groups in the presence of other functional groups (RNO_2; R = H, OH, NH_2, CH_3, and COOH). The reaction is carried out under microwave irradiation (85°C/5-10min) in the presence of KOH as a base and IPA acting as a hydrogen donor as well as a solvent and also tested with other solvents. The reaction system not only exhibits excellent activity with high AN yields but also represents a green and durable catalytic process, which facilitates facile operation, easy separation, and catalyst recycle (54).

Also, a metal-free reduction of NAs to ANs is chemoselectively achieved in excellent yields with B_2pin_2 in IPA and t-KOBu-t (55). Co-operative transformation of NAs and biomass-based alcohols catalyzed by a simple Cu-Ni-AlO$_x$ catalyst is envisioned. It is interesting to see that glycerol acts both as a solvent and hydrogen donor and directly gets transformed into 1, 3-dihydroxypropan-2-one (56). Heterogenous bio-mass derived catalysts for selective hydrogenation of NAs is of great interest in the recent times as it is an important aspect of sustainable chemical research. Sahoo et al have developed such a process utilizing a newly prepared catalyst and molecular hydrogen. A variety of diversely functionalized NAs including few pharmaceutically active compounds were selectively converted to aromatic amines in high yield and selectivity with excellent functional group tolerance. As an example, this green protocol has also been implemented for the synthesis of biologically important TRPC3 inhibitor (57). CuO-NPs are synthesized using the peel of *Musa balbisiana* and applied in the reduction of NAs in water as a green solvent with a high yield of conversion (74–96%). The catalyst was found to be active for several runs as confirmed by several pieces of experimental evidence (58). Baker's Yeast (*Saccharomyces cerevisiae*) is used in the reduction of NAs and other N-O containing functional groups (59). Nitro- and nitrosoarenes can be reduced using baker's yeast (*Saccharomyces cerevisiae*) under two distinct sets of conditions. Generally, the reactions are performed in water under pH6, however, if the reduction is carried out in aqueous methanol at pH 12, the N-nitroso derivative gives 95% AN. Similarly other NB derivatives are also reduced using base and glucose as the hydrogen source (60).

References:

49). L. C. Tavares et al. *Oriental J. Chemistry*, 30(02):2014).

50a). J. C. Spain et al., *J. Bacteriology*, 177(13):3837–3842(1995). 50b).S. F. Nishino and J. C. Spain, *Applied Environmental Microbiology*, 59:2520-2525(1993).

51). J. Kim, et al. Trends in Biotechnology, 26(11):639-646(2008).

52a). L. Xu, *ACS Sustainable Chemical Engineering*, 05(04):2960–2969(2017).

52b). A. R. Fajardo et al. *Carbohydrate Polymers*: 161:187-196(2017).

53a). S. Sharma, et al. *J. Colloid Interface Science*, 441:25-29(2015),

53b). S. Sharma et al. *Chemistry*, 79 (19):9433–9439(2014).

54). P. V. Kumar, et al. *ACS Sustainable Chemical Engg*, 04 (12):6772–6782(2016).

55). H. Lu, et al. *Organic Letters*, 18(11):2774–2776(2016).

56). X. Dai, et al. *RSC Advances*, 05(11):7970-7975(2015).

57). B. Sahoo, et al. *ChemSusChem*, Accepted manuscript online: 26 June 2017.

58). C. Tamuly, et al. *RSC Advances*, 04(95): 53229-53236 (2014).

59). J. A. Blackie, et al. *Tetrahedron Letters*, 38(17):3043–3046(1997).

60). J. C. Spain, Biodegradation of Nitroaromatic Compounds. *Annual Review of Microbiology*, Vol.49:523-555(1995).

6.02.00: Electrochemical Reduction of Nitroarenes:

Electrochemical reductions of NB have been studied since a long time ago (1). NB reduction is studied using Ag electrodes in aqueous alkaline solution, and it has been confirmed that AN formed through PHA as an intermediate (2). The efficient electrochemical reduction of NB and azoxybenzene to AN in neutral and basic aqueous methanolic solutions at devarda copper and Raney-Ni electrodes through electrocatalytic hydrogenolysis of N=O and N=N bonds is investigated. Aniline was obtained in high chemical (85–100%) and current (80–100%) yields, thus showing that hydrogenolysis of the N=O bond of PHA and of the N=N bond of hydrazobenzene is an efficient electrochemical process at these electrodes (3). Electrochemical reduction of NAs as NB and 4-NP is performed at room temperature in ionic liquids on a gold microelectrode. NB was reduced reversibly by one electron and further by two electrons in a chemically irreversible step. The more complicated reduction of 4-nitrophenol revealed three reductive peaks (two irreversible and one reversible) which were successfully simulated using the digital simulation program, DigiSim® using a mechanism of rapid self-protonation. Room-temperature ionic liquids (RTILs) have been shown to have a significant effect on the redox potentials of compounds such as 1,4-dinitrobenzene (DNB), which can be reduced in two one-electron steps(4).

A series of nitropyrenes and other NAs were reduced electrochemically with a dropping mercury electrode and the relationship between polarographic reduction potential and mutagenicity of NAs is established (5). An electrochemical investigation is made using ion-pairing effects on the reduction of NAs in IPA solutions (6). The electrochemical reduction of p-CNB in IPA also is studied by means of cyclic voltammetry at a glassy-carbon electrode with LiSCN, NaSCN, KSCN, and Bun4NSCN as support electrolytes. Using the above salts the shifts in anodic potentials are studied. Based on the results, it is suggested that the electrochemical reduction of p-CNB in IPA proceeds via formation of the radical anion, followed by a second electron transfer to give the dianion. The relevance of these observations to the mechanism of reduction promoted by alkoxide ions is briefly discussed. Newly synthesized dinitrobenzo[a]pyrenes and 6-aza analogues of 1- and 3-nitrobenzo[a]pyrenes showed strong mutagenic activity in Salmonella assays (TA98 and TA98NR). Nitroreduction is essential for metabolic activation of NAs. The structure activity relationships of mono-, di- and trinitrophenanthrene were studied. Electrochemical ease of nitroreduction and dihedral angles of nitrosubstituents to aromatic rings is found to be important factor to determine their mutagenic potency.

The potential relevance of toxicity caused by redox cycling could be solved by using this method (7). The effect of the hypophosphite ion on the electrochemical reduction of NB on Ni was evaluated from a cyclic voltametric study and from constant potential electrolysis in an aqueous-ethanol alkaline medium. It is proposed that the Ni-modified surface which is formed upon hypophosphite oxidation is responsible for the non-reducibility of nitrosobenzene (NOB). Under experimental conditions it leads to AN intermediates as NOB, (yield 33% and selectivity = 82%), which is known to undergo hydrolysis under alkaline condition to give AN. The formation of NOB leads to an electrode poisoning effect in the electrolysis process (8).

The electrochemical behaviors of NB at a pyrolytic graphite electrode modified with CNTs and some with metal/metal oxide modified electrodes were studied using cyclic voltammetry and constant-potential electrolysis technique. The results showed that CNTs exhibited high activity for NB reduction to AN, and the electrochemical reduction of NB at CNT-modified electrode followed the pathway of NB → PHA → AN. These methods have high NB conversions and AN formation (9).

Since NACs are a substantial hazard to the environment and to the supply of clean drinking water, Huang et al have developed a method for reduction of NACs by use of iron oxide coated electrodes, and demonstrate that single sheet iron oxides formed from layered iron(II)-iron(III) hydroxides have unusual electrocatalytic reactivity. Electrodes were produced by coating of single sheet iron oxides on indium, tin oxide electrodes. Under optimized conditions, the fast mass transfer favors the initial reduction of the NACs which is well explained by a diffusion layer model. Reduction was found to comprise two consecutive reactions: i) a fast four-electron first-order reduction of the nitro-group to the hydroxylamine-intermediate, followed by ii) a slower two-electron zero-order reduction resulting in the final amino product. The zero-order of the latter reduction was attributed to saturation of the electrode surface with hydroxylamine-intermediates which have a more negative half-wave potential than the parent compound. For reduction of NACs, the SSI electrode is found superior to metal electrodes due to low cost and high stability, and superior to carbon-based electrodes in terms of high coulombic efficiency and low potential (10).

6.02.10: Bio-Electrocatalytic Reduction:

NB is a toxic compound that is often found as a pollutant in the environment. The present removal strategies suffer from high cost or slow conversion rate. An efficient method for the conversion of NB to AN, a less toxic end product that can easily be mineralized, using a fed-batch bioelectrochemical system with microbially catalyzed cathode is developed. When a voltage of 0.5 V was applied in the presence of glucose, 88.2 (0.60% of the supplied NB (0.5 mM) was transformed to AN within 24 h, which was 10.25 and 2.90 times higher than an abiotic cathode and open circuit controlled experiment, respectively. Aniline was the only product detected during bioelectrochemical reduction of NB (efficiency 98.70 (+/- 0.87%), whereas under abiotic conditions NOB was observed as an intermediate (decreased efficiency to 73.75 (+/- 3.2%). When glucose was replaced by $NaHCO_3$, the rate of NB degradation decreased by 10%, although the selective transformation of NB to AN was still achieved (98.93%) (+/- 0.77%). 16S rRNA based analysis of the biofilm on the cathode indicated that the cathode was dominated by an Enterococcus species closely related to Enterococcus aquimarinus (11).

References:

1). W. H. Harwood, et al. *Industrial & Engineering Chemistry, Process Design & Development.,* 02(01):72–77(1963).

2). C. Nishihara, & H. Shindo, *J. Electroanalytical Chemistry & Interfacial Electrochemistry,* 202 (01–02): 231–239(1986).

3). A. Cyr, et al. *Electrochimica Acta,* 35(01):147-152(1990).

4a). D. S. Silvester, et al. *J. Electroanalytical Chemistry,* 596:131-140(2006): 4b). Y. Zhang, et al. *Chemical Research in Chinese Universities,* 30(02):293-296(2014). 4c). A. Atifi, et al. *Analytical Chemistry, 86* (13):6617–6625(2014).

5). G. Klopman, et al. *Mutatant Research,* 126(02):139-144(1984).

6). M. Maggini, et al. *J. Chemical Society, Perkin Trans. 2,* 267-269(1986).

7). K. Fukuhara, et al. *Bulletin National Institute of Health Science*, (115):72-85(1997). 8). M. Cristina, & F. Oliveira, *Electrochimica Acta*, 48:1829-1835(2003).

9a). Y. Li, et al. *J. Hazardous Materials*, 148(01-02):158-163(2007). 9b). Y. sang, et al. *Sensors and Actuators B: Chemical*, 23446-52(2016). 9c).Q. Zhang, et al. *J. Hazardous Materials*, 285:185-190(2014). 9d). X. Duan, et al. *J. Taiwan Institute of Chemical Engineers*, 45(06):2975-2985(2014). 9e). Y. Chen, et al. *Frontiers of Environmental Science & Engineering*, 09(05):897-904(2015).

10). L. Z. Huang, et al. *J Hazardous Materials*, 306:175-183(2016).

11). A. Wang, et al. *Environmental Science & Technology*, 45:10186–10193(2011).

6.03.00: Microwave Irradiations (MWI):

The name itself is self explanatory that the MWs are some kind of waves. Infact, they are electromagnetic waves that transmit energy with wavelengths at micro level. Radiation is the emission or transmission of energy in the form of waves or particles through space or through a material medium. For example, light waves, heat waves, radio waves, etc. are examples of waves. In case of MWs, the waves are a form of electromagnetic radiation in nature with wavelengths ranging from one meter to one millimeter and with frequencies between 300 MHz (100 cm) and 300 GHz (0.1 cm). https://en.wikipedia.org/wiki/Microwave. We generally, receive heat energy from microwave irradiations, so the objects, items subjected to these waves receive heat energy and get warm or become really hot at 100-200°C. Many chemical reactions are accelerated by using microwave irradiations, so that the reactions are completed in a few minutes compared to normal heating/refluxing conditions for long hours. Usually, microwave heating of pre-prepared reaction mixture gives high yields. Here are few examples, where microwave is used for reduction of NAs.

6.03.10: Reduction of Nitroarenes (NAs) Under MWI:

MWIs are used either for the preparation of a catalyst which is subsequently used in the reduction of NAs or they are used directly for performing the NA reduction in the presence of the substrate and the catalyst. Generally, the hydrogen sources vary from molecular hydrogen, hydrazine, alcohols, hydrocarbons as cyclohexene or cyclohexadiene, sulfides, formates to sodium borohydride. Sometimes alcohols work as a solvent and metals as salts or NPs are used as catalysts. An ethanolic mixture of molybdenum hexacarbonyl [$Mo(Co)_6$] and DBU mediates the reduction of NAs to the corresponding ANs in excellent yields in 15-30 minutes under MWI(1). MW-activated Ni/C catalysts for highly selective hydrogenation of NB to cyclohexylamine are reported. Biocarbon supported Ni catalysts (10%Ni/CSC-II) have been prepared by facile impregnation of Ni species by MW-heating and used for selective hydrogenation of NB to cyclohexylamine. The catalyst exhibits the best catalytic activity to achieve 100mol% conversion of NB and 96.7% selectivity of cyclohexylamine under reaction conditions of 2.0 MPa H_2 and 200°C, ascribed to high dispersion of Ni species and formation of nanosized Ni particles on the support aided by microwave-heating. Thus as-prepared Ni/CSC catalyst is highly activated under the experimental conditions. Therefore, it doesn't require the addition of precious metals like Ru or Rh as a promoter (2). NiO-NPs with an average size of 12 nm and a high specific surface area of 88.5 m^2/g were easily prepared via the thermal decomposition of the complex Ni(dmgH)$_2$ and were characterized by TGA, XRD, FT-IR, TEM and BET surface area measurement. This nanosized transition metal oxide was used as a new heterogeneous catalyst for the reduction of NAs under microwave irradiation. The efficient and selective reduction of ANCs into their corresponding amines was observed by using ethanol as a hydrogen donor and KOH as a promoter under MWIs. This highly regio- and chemoselective method is fast, simple, inexpensive, high yielding, clean and compatible with several sensitive functionalities. This method is suitable for the large scale preparation of different substituted ANs as well as other aryl amines. In addition, the catalytic activity of nanosized NiO is higher than that of the bulk sample (3).

Highly efficient iron oxide (Fe$_3$O$_4$) nanocrystals generated *in-situ* from an inexpensive and readily available iron source catalyze the reduction of NAs to ANs with unparalleled efficiency. The procedure is chemoselective, avoids the use of precious metals, and can be applied under mild reflux conditions (65°C or 80°C) or using sealed vessel MW heating in an elevated temperature regime (150°C) with hydrazine as hydrogen donor. Utilizing MW conditions, a variety of functionalized ANs have been prepared in nearly quantitative yields within 2–8 min at 150°C. Selected examples of ANs of industrial importance have been prepared in a continuous regime using this protocol (4). Preparation of specific tailor-made mixed oxides is one of the major topics of research in the field of heterogeneous catalysis. Perovskites are a class of mixed oxides, which have interesting catalytic and physicochemical properties. A method is developed using MWIs to synthesize perovskites of the type LaMO$_3$ (M=Mn, Fe, Co, Cr, Al) in just 15 min. The LaMO$_3$ perovskites were used as catalysts for reduction of ANCs with propan-2-ol or IPA as hydrogen donor and KOH as promoter. Kinetics of the NB reduction has also been studied (5).

A fast, easy and rapid environmentally friendly method for the reduction of NACs to their corresponding amines in good yield and short time is developed by using microwave irradiation, in the presence of zinc dust, ammonium chloride and solvent free condition (6). The ANs with substituents such as Cl, Br, CH$_3$, OCH$_3$, etc are prepared in high yields (84-95%). Also, a variety of NAs are reduced on Al$_2$O$_3$ support in presence of sodium hydrogen sulfide under MW condition (7) to give the corresponding aromatic amines in high yields. A generally applicable method for the introduction of gaseous hydrogen into a sealed reaction system under MWI allows the hydrogenation of various substrates in short reaction times under moderate temperatures between 80°C and 100°C and 50 psi of hydrogen (8). A MW-assisted, Pd-catalyzed (9) catalytic transfer hydrogenation of different homo- or heteronuclear organic compounds using formate salts as a hydrogen source was performed in ([bmim][PF$_6$].

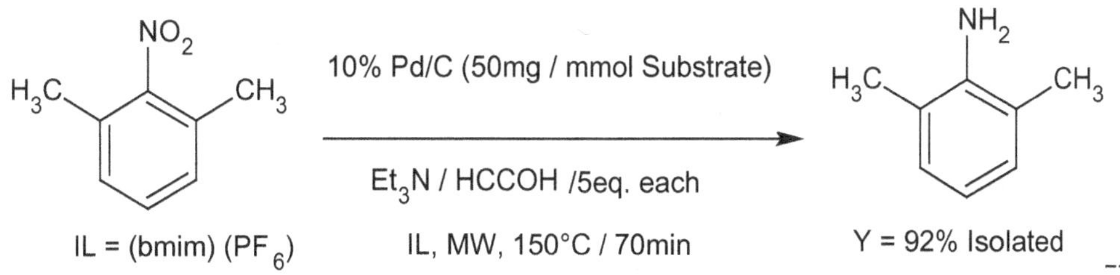

----Scheme-9

Essentially pure products could be isolated in moderate to excellent yields by simple liquid-liquid extraction (Scheme-9).

MW mediated reduction of nitro and azido arenes to *N*-arylformamides using Zn–HCOONH$_4$ is described. In the absence of MW conditions, this methodology affords amines. This protocol has been extended to the synthesis of pyrrolo-[2,1-*c*][1,4] benzodiazepines and 4(3*H*)-quinazolinones (10). Many heterocyclcic substituted NAs were reduced to aryl amines in 79-91% yields using HCOONH$_4$ in ethanol under MW at 80-120°C for 2-40min with Pd/C as a catalyst (11).

A method for the rapid and safe reduction of heteroaromatic and aromatic nitro groups to amines is described using catalytic transfer hydrogenation under MW heating conditions and hydrocarbons as hydrogen donors. Commonly available Pd/C or Pt/C catalyst are extremely effective with 1,4-cyclohexadiene as the hydrogen transfer source in methanol as a solvent, where the nitro substrate conversions are > 99% with yields of corresponding ANs between 89-99%. In the case of substrates containing potentially labile aromatic halogens, Pt/C is effective and results in little or no dehalogenation (Scheme-10).

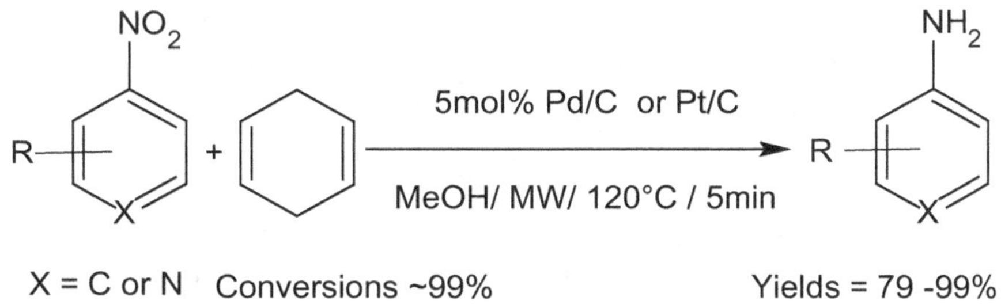

--Scheme-10

In general, the reactions are complete within 5 min at 120°C (12). Similar results are obtained for NB and alkyl-NBs or HNBs with >99% conversions yields up to 99%.

References:

1). J. Spencer, et al. *Synlett*, (16):2557-2558(2007).

2). X. Lu, *et al. Scientific Reports*-7, Article number: 2676(2017). doi:10.1038/s41598-017-02519.

3). S. Farhadi, et al. *Polyhedron,* 30(04):606-613(2011).

4). D. Cantillo, et al. *J. Organic Chemistry,* 78(09):4530–4542(2013).

5). S. Kulkarni, R. V. Jayaram, *Applied Catalysis A: General,* 252(02):225-230(2003).

6). M. Abdullah, et al. *Iraqi National J. Chemistry,* 43:418-423(2011).

7). S. R. Kanth, et al. *Synthetic Communications,* 32(18):2849-2853(2002).

8). G. S. Vanier, *Synlett,* 131-135(2007).

9). H. Berthold, T. Schotten, et al. *Synthesis,* 1607-1610(2002).

10). A. Kamal, et al. *Tetrahedron Letters,* 45(34):6517–6521(2004).

11). E. Cini, et al. *Catalysts,* 07:89-116(2017).

12). F. Quinn, et al. *Tetrahedron Letters,* 51:786-789(2010).

6.04.00: Photocatalysis:

6.04.10: Nitroarene Reduction under Photocatalysis:

Amongst techniques, microwave, ultrasonication and photocatalysis are found very useful with many applications in different fields. Especially, solar light is composed of light emitting diodes and it does wonderful things in the world. There are so many reactions which are catalyzed by the sunlight, and one of them is the photosynthesis process in the plant world. So far photolysis is concerned in chemistry, it is there for quite some time but it has really attracted attention since 1970's and continuing strongly till date with openings for more opportunities for its applications. The degradation of organic pollutants and the production of hydrogen and oxygen from water with semiconductor photocatalysts such as TiO_2, ZnO and CdS have been extensively studied. However, there are few studies on the photocatalytic reactions of organic synthesis under visible light with new catalysts such as Au-NPs. have been extensively used in the catalytic studies, and especially in the reduction of NAs to ANs, which we have covered under different sections in this book, especially CTH with hydrazine, alcohols, reductions on solid supports and in water. Gold can catalyze both oxidation, reduction reactions at elevated temperature, as Au-NPs are quite stable at high temperatures. Moreover, Au-NPs can strongly absorb visible light and can help run reactions at ambient temperatures. This implies that this photocatalytic process can produce compounds that would have been unstable intermediates in a thermal reaction at high temperatures due to the surface plasmon resonance (SPR) effect (1). Tanaka et al have reported their results on functionalization of a plasmonic Au/TiO_2 photocatalyst with an Ag co-catalyst for quantitative reduction of NB to AN in 2-propanol suspensions under irradiation of visible light. A functionalized plasmonic Au/TiO$_2$ photocatalyst with an Ag co-catalyst was successfully prepared by the combination of two types of photodeposition methods (Scheme-11).

--Scheme-11

The catalyst was found highly efficient and it quantitatively converted NB to AN, and 2-propanol to acetone under irradiation of visible light (2). Organic synthesis through semiconductor photocatalysis has become an important research area in photochemistry in the last two decades. The developments in this field are reviewed (3) with an emphasis on the use of irradiated semiconductors and different operation conditions in the reduction of organic compounds including the photocatalytic reduction of NAs.

Ag clusters (1.5nm) were photodeposited on TiO$_2$ particles in a highly dispersed state. The loading of a small amount of the Ag clusters (0.24 wt %) dramatically enhanced both the activity for the TiO$_2$ photocatalytic reduction of NB and the product selectivity to AN using methanol as hydrogen donor (4). The essential action mechanism of the Ag clusters is discussed. A simple hydrothermal method of preparing highly photocatalytic graphene-ZnO-Au nanocomposites (G-ZnO-Au-NCs) has been developed in two steps. Upon UV light activation, G-ZnO-Au-NCs in methanol generates electron–hole pairs. Methanol (hydroxyl group) assists in trapping holes, enabling photogenerated electrons to catalyze reduction of NB to AN, in 97.8% yield during a reaction course of 140 min. The as prepared catalyst is 3-4.5 times more efficient than commercial TiO$_2$ and ZnO-NSs. It is found that aniline is formed through the nitroso and N-hydroxylamine intermediates and the process has great potential in removal of pollutants like NB and in the production of AN (5).

Zand et al have presented their work at a conference about a highly efficient TiO$_2$-NPs and solar light. Except for another nitro group, the process is highly chemoselective to Cl, NH$_2$, OH, etc. substituted NAs (Scheme-12). TiO$_2$-NPs were synthesized directly under high intensity ultrasound irradiation without thermal treatment by the agglomeration of monodispersed TiO$_2$ on bicrystalline anatase and brookite and characterized by in-house tools (6).

--Scheme-12

Efficient hydrogen production and photocatalytic reduction of NB were achieved by using a Pt-deposited amino-functionalised Ti(IV) metal–organic framework (Pt/Ti-MOF-NH$_2$) under visible-light irradiation. The Pt with 1.5wt% enhances reaction rates through the electron transfer from the organic linker to deposited Pt as a co-catalyst by way of titanium-oxo clusters. In addition, the Pt/Ti-MOF-NH$_2$ photocatalyst was found to catalyze photocatalytic reduction of NB under visible-light irradiation (7). NB was effectively and selectively reduced to AN in an acidic aqueous suspension of titanium (IV) oxide photocatalyst in the presence of oxalic acid as a hole scavenger and the AN yield was improved in the presence of a small amount of dioxygen (8). Just like ultrasound applications in preparation of M-NPs for various applications, photocatalytically M-NPs are prepared. M-NPs have already revolutionized the world in many aspects as electronics, telecommunications, information technology, instrumentation, etc. Photochemically prepared M-NPs are used in the reduction of NAs in the presence of a hydrogen donor as hydrazine or an alcohol (9). Meta catalysts are activated using plasmonic energy (10). Functionalization of a plasmonic Au/TiO$_2$ photocatalyst with Ag co-catalyst for quantitative reduction of NB to AN in 2-propanol suspensions under irradiation of visible light is investigated (Scheme-13).

--Scheme-13

A functionalized plasmonic Au/TiO$_2$ photocatalyst with an Ag co-catalyst was successfully prepared by the combination of two types of photodeposition methods, and it quantitatively converted NB and 2-propanol to ANand acetone under irradiation of visible light (11). Surface modification of TiO$_2$-NPs for photochemical reduction of NB is investigated. Modification of the TiO$_2$ surface with arginine resulted in enhanced NB adsorption and photodecomposition in methanol as compared to unmodified TiO$_2$. These results indicate that surface modification of nanocrystalline TiO$_2$ with electron-donating chelating agents is an effective route to enhance (12) photodecomposition of NACs. The photoreduction of NB on TiO$_2$ NPs modified with asparagine (Asp), serine (Ser), phenylalanine (Phe) and tyrosine (Tyr), which were found to bind to TiO$_2$ via carboxyl group, have been investigated under high-pressure mercury irradiation. Modification of TiO$_2$ with Asp, Ser and Phe resulted in enhanced photocatalytic degradation of NB and high selective activity to AN compared to using bare TiO$_2$.

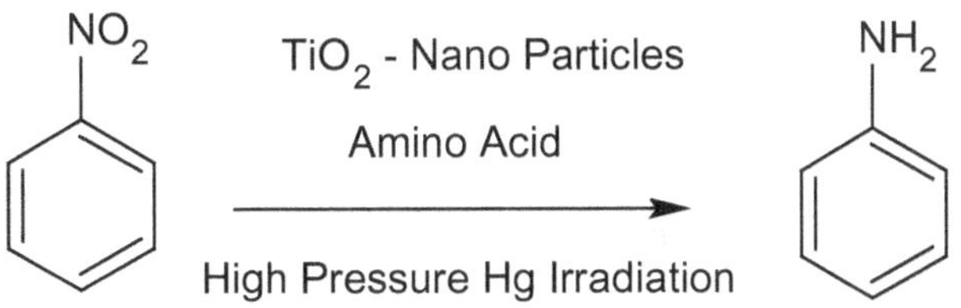

--Scheme-14

Furthermore, NB degradation followed a reductive approach over Asp, Ser, Phe-modified TiO_2 in the presence or absence of methanol (Scheme-14). The result indicates that modification of TiO_2 with electron-donating groups is an effective way to enhance (13) photoreduction of NACs. Photocatalytic reduction of NB over TiO_2 is studied with by-product identification and possible pathways. The photocatalytic reduction of NB (ethanol + TiO_2 + UV-light) in non-aqueous suspension yielded AN, and acetaldehyde as main products. However, various by-products as intermediates of NB photoreduction (i.e. nitrosobenzene and hydroxyaniline) and others (i.e. indoles and quinolines) were also detected. These last results were explained in terms of reactions between aniline and oxidation products of ethanol (14). ANC were chemoselectively reduced to the corresponding amines by using N-doped TiO_2 and potassium iodide as photocatalysts in the presence of methanol (Scheme-15).

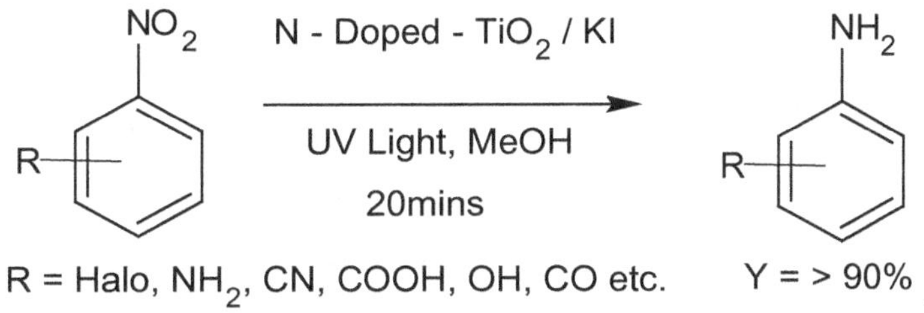

R = Halo, NH_2, CN, COOH, OH, CO etc.   Y = > 90%   --Scheme-15

The novel method is highly efficient with very short reaction time (<20 min), excellent yields (>90%) and wide range of functional group tolerance (15, 16).

The $Ag/AgBr/TiO_2/NZ$ and $Ag/AgBr/NZ$ Nanocomposites photocatalysts were prepared and characterized. The photoreaction performed in the presence of $NaBH_4$ which is a strong reducing agent in aq. medium and the effect of pH, illumination source, $NaBH_4$ concentration and temperature were also studied (Scheme-16).

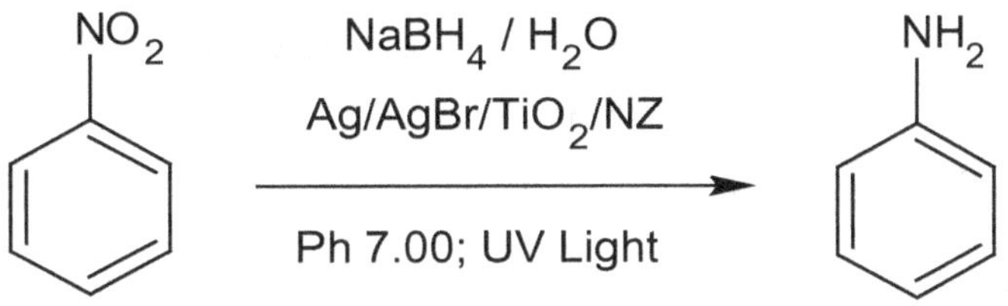

--Scheme-16

The results showed that the best efficiency obtains over the $Ag/AgBr/TiO_2/NZ$ in natural pH under UV light (17). Homogeneous photocatalysis of ethanol for reduction of NB by titanium complexes is studied. NB is reduced to AN using Ti(IV) catalyst in ethanol and irradiation with near-UV light (Scheme-17).

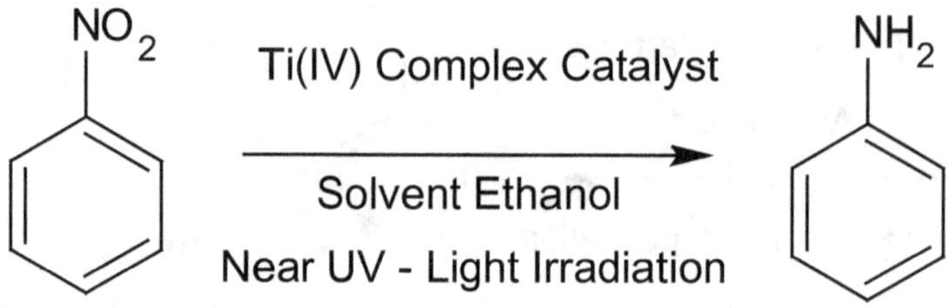

--Scheme-17

The quantum yield of AN formation is weakly dependent on the NB concentration and increases with increasing Ti(IV) content in solution (18). Irradiation of suspensions of CdS in water-ethanol mixtures containing NB with light at $\lambda > 320$ AN is formed in the presence of solvato complexes of V(III) (γPhNH$_2$) and V(IV) (γPhNH$_2$). Process for the production of aromatic primary amines by reduction of NACs using metal catalyst and carbon monoxide and water is reported (19). For example, AN is prepared from NB with $CO+H_2O$ as a reducing agent (Scheme-18).

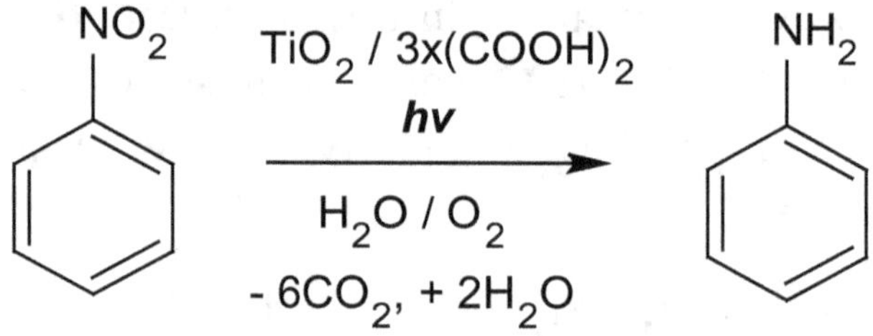

--Scheme-18

Ragaini et al have done a lot of work using $CO+H_2O$ as a reducing system (20).

Nickel-NPs-decorated phosphorous-doped graphitic carbon nitride (Ni@g-PC$_3$N$_4$) was synthesized and used as an efficient photoactive catalyst for the reduction of various NBs under visible light irradiation. Hydrazine monohydrate was used as the source of protons and electrons for the intended reaction. Using Ni@g-PC$_3$N$_4$ and hydrazine hydrate under visible light irradiation after 8h aniline was obtained in 96.5% yield. As for the reduction of NB to AN, six protons and six electrons are required, so 3/2 NH$_2$NH$_2$·H$_2$O mole were consumed per mole of the reactant (21). Visible-light-induced photocatalytic reduction of NACs to the corresponding ANs at room temperature using reduced graphene oxide (rGO) immobilized iron(II) bipyridine complex as photocatalyst is described. The rGO-immobilized iron catalyst exhibited superior catalytic activity than homogeneous iron(II) bipyridine complex and much higher than metal free rGO photocatalysts. The heterogeneous photocatalyst was found to be robust and could easily be recovered and reused for several runs without any significant loss in photocatalytic activity (22).

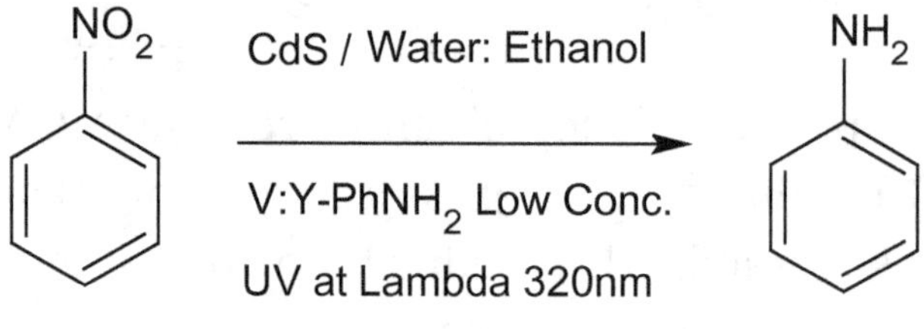

--Scheme-19

The photocatalytic reduction of NB is accompanied by oxidation of the alcohol and generation of hydrogen. Mechanistically, the electron transfers in the system from V surface to nitro groups are observed (23) (Scheme-19). Highly efficient photocatalytic reduction of 4-nitroaniline (4-NA) to *p*-phenylenediamine over microcrystalline $SrBi_2Nb_2O_9$ was observed upon purging with N_2 under UV-light irradiation. Its photocatalytic activity was higher than that of commercial TiO_2.

--Scheme-20

The catalysts were stable and ammonium oxalate was indispensable for the photocatalytic reduction (24) of 4-NA and its high efficiency might be ascribed to its relatively high conduction band (Sceme-20). The transition-metal-loaded Si-NPs for the photocatalytic reduction of NACs in the presence of formic acid under visible light irradiations are investigated. Formic acid acts both as a hydrogen source and a sacrificial reagent for the introduction of electrons into the generated holes of semiconductors. In particular, Pd-loaded silicon (Pd/Si) was the most suitable catalyst for the conversion of NB to AN, compared to Pt/Si, Ru/Si, and Pd/C (25).

Continuous selective hydrogenation of Refametinib-Iodo-nitroaniline, a key intermediate (DIM-NA), over Raney-Co catalyst at Kg/day scale in a continuous flow tricle-bed reactor with online UV–Visible conversion control is reported. The conditions were optimized such that high conversion, high selectivity and no hydrodehalogenation occurred (26). Selective photocatalytic reductions of NAs using $PbBiO_2X$ and the blue light irradiation of heterogeneous photocatalysts $PbBiO_2X$ (X = Cl, Br) in the presence of triethanolamine as an electron donor leads to hydrogen evolution (Scheme-21).

--Scheme-21

The process is selective, clean and complete for the reduction of NACs to their corresponding ANs (27). The liquid-phase reduction of NB is investigated under solar light in deaerated conditions at atmospheric pressure and room temperature in a batch reactor in acetonitrile as well as in aqueous medium. The effects of NB concentration, metal content, amount of photo catalyst, sustainability and pH role have been investigated. The product analyzed showed AN in 65% yield (28).

A new method for UV-irradiated degradation of NB by titania photocatalysts was proposed. The titania-NPs were coated on a quartz tube through the introduction of tetraethyl orthosilicate into the matrix. The dependence of NB photodegradation on pH, temperature, concentration, and air feeding was discussed, and the physical properties such as the activation energy, entropy, enthalpy, adsorption constant, and rate constant were acquired by conducting the reactions in a variety of experimental conditions. The optimum efficiency of the photodegradation with the NB residue as low as 8.8% was achieved according to the experimental conditions indicated. The photodegradation pathways were also investigated through HPLC, GC/MS, ion chromatography (IC), and chemical oxygen demand (COD) analyses (29). The photochemical reduction of the NB and monosubstituted NBs has been investigated in 2-propanol under a nitrogen atmosphere. The NBs with electron-withdrawing groups on *meta-* and *para-*positions were photoreduced to the corresponding ANs, and NB and the *p*-substituted NBs with electron-donating groups were photoreduced to the corresponding phenylhydroxylamines. 4-NA and 4-NP were unreactive toward the photoreduction under given conditions (30).

Selective reduction of NAs to intermediate azo compounds is accomplished using Au-NPs/ZrO$_2$ catalyst under the influence of visible or UV light and in isopropyl alcohol as a solvent. The process occurs with high selectivity and at ambient temperature (40°C) and pressure, and enables the selection of intermediates that are unstable in thermal reactions (31). As a part of AN-analogs, azoxy or azobenzenes are obtained in high yields (96%) with photocatalytic reduction of NAs (32).

6.04.11: Direct Amination of Arenes under Photoirradiations:

Reaction mechanism of amination of benzene and substituted benzenes by aq. ammonia over Pt/TiO$_2$ photocatalyst oxidizes an ammonia to form an amide radical (·NH$_2$) and a proton, and the amide radical attacks an aromatic ring to produce an intermediate, followed by the abstraction of the hydrogen atom from it on the platinum sites to produce AN and a hydrogen molecule (33) The reactions were performed on NB, benzonitrile and HNBs.

Alkyl, aryl amines are used in place of NH$_3$ for direct amination of aromatic compounds, especially of phenols under visible-ight-promoted and via oxidative cross-dehydrogenative C-N coupling reactions (34). Yasuda et al have reported the stereochemical studies on the amination of arenes with ammonia and alkylamines *via* photochemical electron transfer (35). Photoamination of 9-methoxyphenanthrene with ammonia in the presence of *m*-dicyanobenzene (DCNB) gives *cis-* and *trans*-9-amino-10-methoxy-9,10-dihydrophenanthrene in a ratio of 75: 25. The photoamination of phenanthrene and anthracene proceeds with *trans-* addition of ammonia with high selectivities by nucleophilic addition of amine to the cation radical of the arene, generated from photochemical electron transfer to DCNB. Further reduction steps produce the aryl amines. The developments in the C–N bond construction from abundantly available sources as alkanes and arenes to generate amines have been reviewed (36).

6.04.12: 4-Nitrophenol (4-NP) Reduction under Photocatalysis:

The reduction is performed using hydrazine, $NaBH_4$, alcohols as ethanol, isopropanol etc. as hydrogen sources under photocatalytic atmosphere. The Ag-NPs/TiO_2 dyads are found very efficient catalyst for the reduction of 4-NP with hydrazine under photoirradiations. The photoefficiency enhancement under UV light irradiation (37) was attributed to the electron transfer from the TiO_2 semiconductor surface to the adsorbed acceptor reactant (4-NP) through the deposited Ag-NPs. The Ag/TiO_2NPs were successfully used for 4-NP reduction in the presence of aqueous $NaBH_4$ under visible-light irradiation with 98% 4-NP conversion (38). In a particular case, Ag-NPs of various sizes were prepared at room temperature using silver nitrate as a precursor, various molar ratios of sodium citrate as a surfactant stabilizing material and sodium borohydride as a reducing agent. The prepared samples were applied in photo catalysis of 4-NP. As expected, due to their high surface energy, the smaller particle sizes were more active in the photo catalytic application (39). Photocatalytic reduction of 4-NP using CdS in the presence of Na_2SO_3 under blue light irradiation or visible light is performed. The CdS semiconductor prepared at higher *EN* contents (90 mol%) presented the highest photocatalytic activity for the reduction of 4-NP to 4-AP. Their photo-efficiency enhancement is attributed to the quantum confinement effect (40) caused by the small particle size (~15 nm).

6.04.13: Natural Products in CTH under Photocatalysis:

Generally, the CTH of 4-NP requires a metal catalyst for desired results. Gazi et al have reported the reduction of 4-NP using the visible light irradiation under a metal-free condition, but in the presence of Eosin Y loaded resin. The effectiveness of the catalytic process depends upon the adsorption and the electron transfer. From the control experiments, it is identified that the photocatalyst is primarily serving as an electron carrier in the reaction mechanism (41). Natural leaves-assisted synthesis of nitrogen-doped, carbon-rich nanodots-sensitized, Ag-loaded anatase TiO_2 square nanosheets with dominant {001} facets and their enhanced catalytic applications are reported. Extending the UV response of anatase TiO_2 photocatalysts into the visible light range can play a pivotal role in promoting the practical applications of these catalysts. Nitrogen and carbon co-doped, Ag(1.5 -15nm) loaded anatase TiO_2 (Ag@NC–TiO_2) single-crystal nanosheets dominated by {001} facets were prepared for the first time using leaves as the nitrogen and carbon source by a facile, low cost method (42). The catalyst was more active than NC–TiO_2, TiO_2 and P25 under visible light irradiation, which was ascribed to a synergistic effect between N, C and Ag. The Ag@NC–TiO_2 heteronanosheets (43) are dispersible in aqueous solution and efficiently reduce 4-NP.

--Scheme-22

Ag-NPs and Au-NPs (< 10nm) have been grown on calcium alginate gel beads using a green photochemical approach and evaluated, for the first time, on the well-known 4-NP reduction to 4-AP in the presence of excess borohydride (Scheme-22). The gel served as both a reductant and a stabilizer. The catalytic efficiency of alginate-based Ag catalyst was much more compared to that of the Au catalyst, but both have the potential for industrial applications (44). Some other groups have presented their work on photocatalyzed reduction of NB and NAs to intermediates and ANs (45).

6.04.14: Photochemical Degradation of NB

The effects of surface modification of nanocrystalline TiO_2 with specific chelating agents on photocatalytic degradation of NB was investigated in order to design a selective and effective catalyst for removal of NACs from contaminated waste streams. Arginine, lauryl sulfate, and salicylic acid were found to bind to TiO_2 via their oxygen-containing functional groups. Modification of the TiO_2 surface with arginine resulted in enhanced NB adsorption and photodecomposition, and compared to unmodified TiO_2 the NB degradation followed a reductive pathway over arginine-modified TiO_2 and was further enhanced upon addition of methanol. No degradation of arginine was detected under the experimental conditions. Arginine improved the coupling between NB and TiO_2 and facilitated the transfer of photogenerated electrons from the TiO_2 conduction band to the adsorbed NB. These results indicate that surface modification of nanocrystalline TiO_2 with electron-donating chelating agents is an effective route to enhance photodecomposition of NACs. The biodegradation of NB by aerobic granular was investigated in this study. Micrococcus luteus, Bacillus sp., Pseudomonas sp., Comamonas sp., Acinetobacter sp. and Rhodococcus sp. are the main NB degrading bacteria in the granular sludge. Sequencing Batch Reactor (SBR) with granular sludge was an efficient, reliable and stable process for NB treatment. These results may provide a guideline for NB wastewater treatment (46). NB and NAs are used in the productions of many dyes. There are quite a few reports on photoinduced degradation of dyes. One of them is zinc oxide (ZnO), as it has displayed excellent photocatalytic performance (47) for the photodegradation of dyes (superior to Degussa P25 TiO_2), which can be easily prepared in large quantity by direct calcination of zinc acetate ($Zn(Ac)_2 \cdot 2H_2O$).

References:

1). H. Zhu, H. Ke, *Angewandte Chemie Intl Edition*, 49:9657-9661 (2010).

2). A. Tanaka, Y. Nishino, *Chemical Communications*, 49:2551-2553(2013).

3). S. O. Flores, et al. *J. Advanced Oxidation Technologies*, 13(03):321-340(2010).

4). H. Tada, et al. *Langmuir*, 20(19):7898–7900(2004).

5). P. Roy, et al. *Environmental Science & Technology*, 47(12):6688–6695(2013).

6). Z. Zand, et al. *Proceedings of the 4th Intl Conference on Nanostructures*, (ICNS4), 12-14 March, 2012, Kish Island, I.R. Iran.

7a). T. Toyao, M. Saito, et al. *Catalysis Science & Technology*, 03(08):2092-2097(2013).
7b). S. Chen, et al. *Chinese Journal of Chemistry*, 28(01):21-26(2010).

8). H. Kominami, et al. *Chemistry Letters*, 38(05):410-411(2009).

9). M. Sakamoto, *Bulletin Chemical Society*, Japan, 83(10):1133–1154(2010).

10). Y. Kim, et al. *Nano Letters*, 16(o5):3399–3407(2016).

11). A. Tanaka, Y. Nishino, *Chemical Communications*, 49:2551-2553(2013).

12). T. Rajh, et al. *Environmental Science & Technology*, 34(22):4797–4803(2000).

13). H. Huang, J, Zhou, et al. *J Hazardous Materials,* 15:178(1-3):994-998(2010). 14). S. O. Flores, et al. *Topics in Catalysis*, 44(01):507-511(2007).

15). H. Wang, et al. *Catalysis Communications,* 10(06):989–994(2009).

16). H. Toshikazu, et al. *Nippon Kagakkai Koen Yokoshu*, Vol 81(2), 1094 (2000).

17). M. Padervand, et al. *Intl J. Materials and Chemistry*, 01(01):49-52(2011).

18). A. I. Kryukov, et al. *Theoretical, Experimental Chemistry*, 28(5-6):319-321(1992).

19). A. F. M. Iqbal, US3944615 (1976).

20). F. Ragaini, *Dalton Transactions*, 6251-6266(2009).

21). A. Kumar, P. Kumar, et al. *Nanomaterials* (Basel), 06(04): 59(2016).

22). A. Kumar, P. Kumar, et al. *Applied Surface Science,* 386:103-114(2016).

23). S. Kuchmii, et al. *Theoretical, Experimental Chemistry*, 25(05):509-514(1989).

24). W. Wu, et al. *Catalysis Communications:* 17:39-42(2012).

25). K. Tsutsumi, *ACS Catalysis*, 06(07):4394–4398(2016).

26). M. B. Said, *Org. Process Research & Devevelopment*, 21(05):705-714(2017).

27). S. Fuldner, et al. *Green Chemistry*, 13:640-643(2011).

28). B. Srinivas, et al. *Indian J. environmental protection*, 31(01): 01-15(2011).

29). T. Whang, T. Shi, et al. *Intl. J. Photoenergy*, Volume 2012 (2012), ArticleID 681941,8 pages. doi:10.1155/2012/681941.

30). S. Hashimoto, et al. *Bulletin Chemical Society, Jpn*, 45(02):549-553(1972).

31). H. Zhu et al. *Angewandte Chemie*, 49(50):9657-9661(2010).

32). X. Guo, et al. *Angewandte Chemie*, 126(7):2004–2008(2014).

33a). H. Yuzawa, et al. *J. Physical Chemistry C*, 117(21):11047–11058(2013).

33b). *M. Yasuda, et al. J. Organic Chemistry*, 52(05):753–759(1987).

34a). Y. Zhao, et al. *Organic Letters*, 18(14):3326–3329(2016).

34b). Y. Zhao, et al. *ACS Catalysis*, 07(04):2446–2451(2017).

35). M. Yasuda, et al. *J. Chemical Society, Perkin Transactions 2*, 305-310(1992).

36a). P. Stavropoulos, *Comments on Inorganic Chemistry*, 37(01):01-57(2017).

36b). H. Yi, et al. *Chemical Reviews*, 117(13):9016–9085(2017).

37). A. Hernández-Gordillo, *Journal of Hazardous Materials*, 268:84-91(2014).

38). M. M. Mohamed, et al. *Applied Catalysis B: Environ*, 142-143:432-441(2013).

39). F. A. Al-Marhaby, et al *World J. Nanoscience & Engg*, 06(01):29-37(2016).

40). A. H-Gordillo, *J. Photochem & Photobiology A: Chemical*, 257:44-49(2013).

41). S. Gazi, et al. *Applied Catalysis B: Environmental*, 105(03-04):317-325(2011).

42). Z. Jiang, et al. *J. Materials Chemistry A*, **01**(47):14963-14972(2013).

43). K. Kim, et al. *Nanoscale*, 05(14):6254-60(2013).

44). S. Saha, A. Pal, *Langmuir*, 26(04):2885–2893(2010).

45a). A. Maldotti, et al. *J. Photochemistry & Photobiology A: Chemistry*, 133(01):129-133(2000). 45b). S. Liu, et al. *ACS Applied Materials & Interfaces*, 05(10):4309–4319(2013). 45c).S. Liu, et al. *J. Phyical Cheistry C*, 117(16):8251–8261(2013). 45d).S. Hashimoto, et al. *Bulletin Chemical Society, Japan*,: 41:1249-1251(1968), 45e). Ibid, *Tetrahedron Letters*, 3509-3512 (1970).

46). *D. Zhao, C. Liu, et al. Desalination*, 281:17–22(2011).

47). C. Tian, Q. Zhang, et al. *Chemical Communications*, 48(23):2858-2860(2012).

6.05.00: Ultrasonication:

6.05.10: Nitroarene Reductions under Ultrasonication:

Ultrasonication is used for enhancing the rate of reactions in the synthesis of organic compounds, especially in the reduction of NAs to the corresponding ANs. The M-NPs have attracted a great deal of attention in the recent past, almost in every field of science, especially in chemistry, pharma industry etc. Ultrasonication is used in the preparation of nanosize metal or metal oxides by sonication process. The preparative methods for NPs are described in the respective journals (1). A comparative study and development of an improved method for the reduction of nitroarenes into aryl amines by Al/NH_4X (X = Cl, Br, I) in methanol under ultrasonic conditions is performed (Scheme-23). A plausible mechanism of the reaction is envisaged (2).

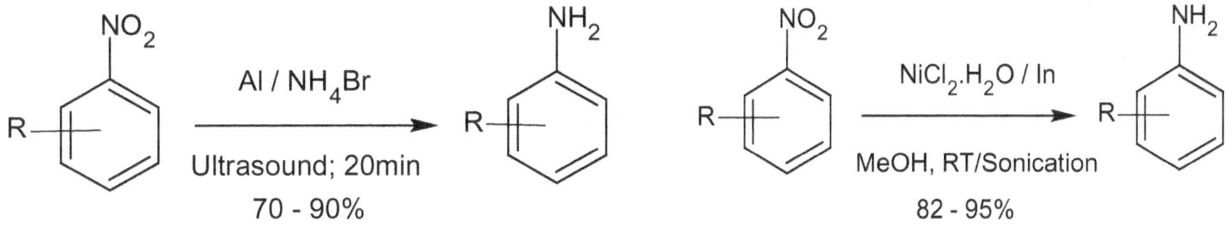

Scheme-23

An efficient and chemoselective reduction of various NACs to the corresponding amines by treatment with $NiCl_2·6H_2O/In$, with high yields and selectivity over other labile substituents, is accomplished under sonication. This $NiCl_2·6H_2O/In$ system provides an useful alternative (3) to other presently used procedures, since the reduction of NACs proceeds expeditiously and in high yields under mild conditions (Scheme-23). A convenient, environmentally benign, and highly efficient protocol for the preparation of N-aryl- hydroxylamines from the corresponding NAs in a $Zn/HCOONH_4/CH_3CN$ system under ultrasound is described. The advantages of the present method include high chemoselectivity, simple and practical work-up procedure and high yield (4). The selective reduction of NAs in the presence of sensitive functionalities under ultrasonic irradiation at 35 kHz is reported in yields of 39-98%. Iron powder proved superior to stannous chloride with high tolerance of sensitive functional groups and high yields of the desired aryl amines in relatively short reaction times (5).

NAs were rapidly reduced to the corresponding aromatic amines using stannous chloride in ionic liquid as a safe and recyclable reaction medium under sonication. The method is environmentally benign and the sensitive functional groups remain unaffected (6). The combined uses of Microwaves and Ultrasound or Ultrasonication with improved tools in process chemistry and organic synthesis have been reviewed. Microwave heating and ultrasonic waves are among the most simple, inexpensive, and valuable tools in applied chemistry. Besides saving energy, these green techniques promote faster and more selective transformations. The authors have pondered on if these two techniques could be combined together to harvest the benefits of both processes. Indeed, they have been successful in utilizing them in a combined form and further made a review on apparatus currently available for simultaneous or tandem irradiations and explain how it can be utilized in organic synthesis and analysis (7).

References:

1a). R. V. Kumar, Y. Mastai, et al. *J. Materials Chemistry:* 11:1209-1213(2001), 1b). T. R. Bastami, M. H. Entezari, *Ultrasonic Sonochemistry:* 19:560–569(2012).

2a). M. A. Pasha, V. P. Jayashankara, *Ultrasonic Sonochemistry:* 13:42–46(2006). 2b). D. Nagaraja, M. A. Pasha, *Tetrahedron Letters:* 40:7855-7856(1999).

3). J. H. Han, et al. *Bulletin Korean Chemical Society,* 27(08):1115-1116(2006).

4). Q. Shi, et al. *J. Chemical Research in Chinese Univs,* 25(02):183-188 (2009).

5). A. B. Gamble, et al. *Synthetic Communications,* 37:2777-2786(2007).

6). G. Rai, et al. *Tetrahedron Letters,* 46(23):3987–3990(2005).

7). G. Cravotto, P. Cintas, *Chemistry A European J.,* 13(07):1902–1909(2007).

ABBREVIATIONS

AN	=	Aniline
2 or 3- or 4-AP	=	2 or 3 or 4-Aminophenol
3-AS	=	3-Aminostyrene
ANCs	=	Aromatic Nitro Compounds
o-, *m-*, *p-*CAN	=	*o-*, *m-*, *p-*Chlororoaniline
CB	=	Chlorobenzene
*o-*CNB	=	*o-*Chloronitrobenzene
*m-*CNB	=	*m-*Chloronitrobenzene
*p-*CNB	=	*p-*Chloronitrobenzene
2,4 or 3,4-DCB	=	2,4- or 3,4-Dichlorobenzene
2,4 or 3,4-DCNB	=	2,4 or 3,4-Dichloronitrobenzene
DNB	=	Dinitrobenzene
DPCO	=	Dense Phase CO_2.
FA	=	Formic acid
HAN	=	Haloaniline
HAB	=	Haloaminobenzene
HATs	=	Haloaminotoluenes
HNA	=	Halonitroaniline or Halonitroarene
o-, *m-*, *p-* HNB	=	*o-*, *m-*, *p-* Halonitrobenzene
HNTs	=	Halonitrotoluenes
HH	=	Hydrazine hydrate
HAB	=	Hydroxyaminobenzene
IPA	=	Isopropanol or 2-Propanol
M-NPs	=	Metal-Nanoparticles
NA	=	Nitroarene or Nitroaromatic
NAs	=	Nitroarenes or Nitroaromatics

NACs	=	Nitroaromatic Compounds
NB	=	Nitrobenzene
NCs	=	Nanocomposites
NPs	=	Nanoparticles
NPs	=	Nitrophenols
4-NP	=	4-Nitrophenol or PNP = *p*-Nitrophenol
NOB	=	Nitrosobenzene
NS	=	Nitrostyrene
3-NS	=	3-Nitrostyrene
NTs	=	Nitrotoluenes
PAP	=	*p*-Aminophenol
PGMs	=	Platinum Group Metals
PHA	=	N-Phenylhydroxylamine or
SS	=	Solid Support
SNAs	=	Substituted Nitroaromatics or Substituted Nitro Arenes
SANs	=	Substituted Anilines
$scCO_2$	=	Supercritical Carbon Dioxide
p- TSA	=	*p*-Toluenesulfonic acid
PTC	=	Phase Transfer Catalyst
TOF	=	Turn Over Frequency